Asymmetric Synthesis of Nitrogen Heterocycles

Edited by
Jacques Royer

Further Reading

Cornils, B., Herrmann, W. A., Muhler, M., Wong, C.-H. (eds.)
Catalysis from A to Z
A Concise Encyclopedia

2007
Hardcover
ISBN: 978-3-527-31438-6

Knipe, C., Watts, W. E. (eds.)
Organic Reaction Mechanisms, 2005

Hardcover
ISBN: 978-0-470-03403-3
Online Buch Wiley Interscience
ISBN: 978-0-470-06661-4

Mikami, K., Lautens, M. (eds.)
New Frontiers in Asymmetric Catalysis

2007
Hardcover
ISBN: 978-0-471-68026-0

Sheldon, R. A., Arends, I., Hanefeld, U.
Green Chemistry and Catalysis

2007
Hardcover
ISBN: 978-3-527-30715-9

Hiersemann, M., Nubbemeyer, U. (eds.)
The Claisen Rearrangement
Methods and Applications

2007
Hardcover
ISBN: 978-3-527-30825-5

Dalko, P. I. (ed.)
Enantioselective Organocatalysis
Reactions and Experimental Procedures

2007
Hardcover
ISBN: 978-3-527-31522-2

Roberts, S. M.
Catalysts for Fine Chemical Synthesis
Volume 5: Regio- and Stereo-Controlled Oxidations and Reductions

2007
Hardcover
ISBN: 978-0-470-09022-0

Wyatt, P., Warren, S.
Organic Synthesis
Strategy and Control

2007
E-Book
ISBN: 978-0-470-06120-6

Asymmetric Synthesis of Nitrogen Heterocycles

Edited by
Jacques Royer

WILEY-VCH Verlag GmbH & Co. KGaA

The Editor

Prof. Dr. Jacques Royer
Université Paris Descartes
Faculté de Pharmacie
Laboratoire de Chimie
CNRS/UMR 8638
4, Avenue de l Observatoire
75270 Paris Cedex 06
France

All books published by **Wiley-VCH** are carefully produced. Nevertheless, authors, editors, and publisher do not warrant the information contained in these books, including this book, to be free of errors. Readers are advised to keep in mind that statements, data, illustrations, procedural details or other items may inadvertently be inaccurate.

Library of Congress Card No.:
applied for

British Library Cataloguing-in-Publication Data
A catalogue record for this book is available from the British Library.

Bibliographic information published by the Deutsche Nationalbibliothek
The Deutsche Nationalbibliothek lists this publication in the Deutsche Nationalbibliografie; detailed bibliographic data are available on the Internet at http://dnb.d-nb.de.

© 2009 WILEY-VCH Verlag GmbH & Co. KGaA, Weinheim

All rights reserved (including those of translation into other languages). No part of this book may be reproduced in any form – by photoprinting, microfilm, or any other means – nor transmitted or translated into a machine language without written permission from the publishers. Registered names, trademarks, etc. used in this book, even when not specifically marked as such, are not to be considered unprotected by law.

Printed in the Federal Republic of Germany
Printed on acid-free paper

Typesetting Laserwords Private Limited, Chennai, India
Printing betz-druck GmbH, Darmstadt
Binding Litges & Dopf Buchbinderei GmbH, Heppenheim
Cover Design Grafik-Design Schulz, Fußgönheim

ISBN: 978-3-527-32036-3

Foreword

It is a great pleasure to see a new volume of Asymmetric Synthesis of Nitrogen Heterocycles going to press.

Since the nineteenth century, the study of nitrogen-containing products, especially alkaloids, has often provided the impetus for great advances in organic chemistry. On one hand, the unceasing proliferation of alkaloid literature accounts for the interest in this class of natural products. On the other hand, heterocyclic chemistry, in general, represents a special topic in medicinal chemistry. Moreover, optically active amines constitute an important class of compounds that find a wide range of interest as chiral building blocks and as part of many pharmaceutical products. Indeed, asymmetric synthesis is more than an academic specialty.

It is the goal of this book to bring all these aspects up to date and to extend the scope of asymmetric synthesis of nitrogen heterocycles to a large variety of structures, that is, with respect to ring size and number of heteroatoms. The objective of the Editor, who is also contributor of a chapter, has been reached in receiving the collaboration of leaders in this field from Europe, China and Japan.

This volume is divided into two parts. Part One describes in four chapters asymmetric synthesis of nitrogen heterocycles containing only one heteroatom in 3-,4-,5-,6-,7- (and more) membered rings. Part Two consists of four chapters covering the asymmetric synthesis of nitrogen heterocycles with more than one heteroatom.

I am sure that this treatise, which highlights an important field of enantioselective chemistry, should be of value to both academic and pharmaceutical chemists as well as PhD students.

Paris, October 2008   Professor *H.-P. Husson*

Contents

Preface *XIII*

List of Contributors *XV*

Part One Asymmetric Synthesis of Nitrogen Heterocycles Containing Only One Heteroatom *1*

1 Asymmetric Synthesis of Three- and Four-Membered Ring Heterocycles *3*
Giuliana Cardillo, Luca Gentilucci and Alessandra Tolomelli
1.1 Substituted Aziridines *3*
1.1.1 Generalities *3*
1.1.2 Asymmetric Aziridination via Cyclization Methods *6*
1.1.2.1 Cyclization of a Nucleophilic N on an Electrophilic C (Pathway A) *7*
1.1.2.2 Cyclization of a Stabilized Anion on an Electrophilic N (Pathway B) *9*
1.1.3 Asymmetric Aziridination via Cycloaddition Methods *12*
1.1.3.1 Addition of Nitrenes to Alkenes *12*
1.1.3.2 Reaction between Carbenes and Imines *16*
1.1.4 Ring Transformation Methods *27*
1.1.4.1 Aziridines from Epoxides *27*
1.1.4.2 Aziridines from Other Heterocycles *30*
1.1.5 Racemate Resolution *31*
1.1.6 Asymmetric Synthesis of Azirines *33*
1.1.6.1 The Neber Reaction *33*
1.1.6.2 Thermal or Photochemical Treatment of Vinyl Azides *34*
1.1.6.3 Elimination from Aziridines *34*
1.1.6.4 Resolution of Racemic Azirines *35*
1.1.6.5 Oxidation of Aziridines *36*
1.2 Substituted Monocyclic Azetidines and Carbocyclic-Fused systems *36*
1.2.1 Generalities *36*
1.2.2 Cyclization Methods: Introduction *38*

Asymmetric Synthesis of Nitrogen Heterocycles. Edited by Jacques Royer
Copyright © 2009 WILEY-VCH Verlag GmbH & Co. KGaA, Weinheim
ISBN: 978-3-527-32036-3

1.2.2.1	Cyclization methods: Enantiopure Azetidines via Formation of C–N Bond	39
1.2.2.2	Cyclization Methods: Enantiopure Azetidines via Formation of C–C Bond	40
1.2.3	Azetidines by Resolution of Racemates	42
1.2.4	Azetidines by Ring Transformation	44
	References	45

2 Asymmetric Synthesis of Five-Membered Ring Heterocycles 51
Pei-Qiang Huang

2.1	Monocyclic Pyrrolidines and Pyrrolidinones	51
2.1.1	Generalities	51
2.1.2	Cyclization Methods	52
2.1.2.1	Cyclization via C_1/C_5–N Bond Formation	52
2.1.2.2	Cyclization via C_2–C_3 Bond Formation	61
2.1.2.3	Cyclization Involving C_3–C_4 Bond Formation	62
2.1.3	Cycloaddition Methods	64
2.1.3.1	Cycloaddition Approach	64
2.1.3.2	Annulation Approach	66
2.1.4	Ring Transformation Methods	67
2.1.4.1	Ring Expansion Methods	67
2.1.4.2	Ring Contraction Methods	68
2.1.5	Substitution of Already Formed Heterocycle	68
2.1.5.1	By Nucleophilic Reaction of Pyrrolidinium Ions	71
2.1.5.2	By Nucleophilic Reaction of Cyclic Imides	72
2.1.5.3	By Nucleophilic Addition/Cycloaddition of Pyrrolidine Nitrones	74
2.1.5.4	By Functionalization of 2-Pyrrolines	76
2.1.5.5	By Enantioselective Reactions	77
2.1.5.6	By Functionalization at C_3/C_4 Positions of Pyrrolidines	77
2.2	Pyrrolines	79
2.2.1	Synthesis of Pyrrolines by Cyclization and Annulation Reactions	79
2.2.2	Synthesis of Pyrrolines by Substitution of Already Formed Heterocycles	80
2.3	Fused Bicyclic Systems with Bridgehead Nitrogen	82
2.3.1	Pyrrolizidines	82
2.3.1.1	Through Extension of Methods for the Synthesis of Pyrrolidines	82
2.3.1.2	Other Methods for the Synthesis of Pyrrolizidines	83
2.3.1.3	Asymmetric Synthesis of Polyhydroxylated Pyrrolizidines	85
2.4	Acknowledgments	87
	References	87

3 Asymmetric Synthesis of Six-Membered Ring Heterocycles 95
Naoki Toyooka

3.1	Introduction	95
3.2	Dihydropyridines	95

3.3	Tetrahydropyridines	98
3.3.1	Ring-Closing Metathesis (RCM)	98
3.3.2	Reduction of Pyridine Derivatives	101
3.3.3	Deracemization Processes	101
3.3.4	Michael Addition Followed by Elimination	102
3.3.5	Enamine Reaction	102
3.3.6	Electrocyclization	105
3.4	Monocyclic Piperidines and Carbocyclic Fused Systems	107
3.4.1	Generalities	107
3.4.2	Cyclization Methods	107
3.4.2.1	Nitrogen as a Nucleophile	108
3.4.2.2	C–C Bond Formation	113
3.4.3	Cycloaddition Methods	117
3.4.3.1	[4 + 2] Azadiene Cycloaddition	117
3.4.3.2	[4 + 2] Acylnitroso Cycloaddition	117
3.4.3.3	[3 + 2] Cycloaddition	119
3.4.4	Ring Transformation Methods	119
3.4.4.1	Ring Enlargement of Pyrrolidines to Substituted Piperidines	120
3.4.4.2	Ring Transformation of Lactones to 2-Piperidones	121
3.4.4.3	Ring Enlargement of γ-Lactam to 2-Piperidones	122
3.4.5	Substitution of Already Formed Heterocycle	123
3.4.5.1	Phenylglycinol-Derived Oxazolidine	123
3.4.5.2	Asymmetric Michael Addition	125
3.4.5.3	Nitrone Cycloaddition	127
3.4.5.4	Iminium Strategies	128
3.4.5.5	Oxidative Methods	131
3.5	Fused Tri- or Bicyclic System with Bridgehead Nitrogen	132
	References	135

4	**Asymmetric Synthesis of Seven- and More-Membered Ring Heterocycles**	**139**
	Yves Troin and Marie-Eve Sinibaldi	
4.1	Substituted Azepines	139
4.1.1	Generalities	139
4.1.2	Cyclization Methods	142
4.1.2.1	Lactamization: C–N Bond Formation	142
4.1.2.2	Radical Cyclization	146
4.1.2.3	Intramolecular Cyclization	149
4.1.2.4	Oxidative Phenol Coupling Reaction	155
4.1.2.5	The Ring Closure Metathesis	155
4.1.3	Cycloaddition Methods	161
4.1.3.1	[5 + 2] Cycloaddition	162
4.1.3.2	[4 + 3] Cycloaddition	162
4.1.3.3	Nitrone Cycloaddition	162

4.1.3.4	Intramolecular Diels–Alder Reactions (IMDA) – [4 + 2] Cycloaddition *163*	
4.1.4	Ring Transformation Methods *163*	
4.1.4.1	Classical Methods *164*	
4.1.4.2	Ring Expansion *166*	
4.1.4.3	Substitution of Already Formed Heterocycles *169*	
4.2	Substituted Azocines *171*	
4.2.1	Azocines from Intramolecular Nucleophilic Substitution *172*	
4.2.2	Ring Transformations Methods *173*	
4.2.3	Cycloaddition Approaches to Azocines *174*	
4.2.4	Ring-Closing Metathesis *175*	
4.3	Substituted Large Nitrogen-Containing Rings *177*	
	References *181*	

Part Two Asymmetric Synthesis of Nitrogen Heterocycles with More Than One Heteroatom *187*

5 Asymmetric Synthesis of Three- and Four-Membered Ring Heterocycles with More Than One Heteroatom *189*
Steve Lanners and Gilles Hanquet

5.1	Introduction *189*	
5.2	Three-Membered N-Heterocycles with Two Heteroatoms *189*	
5.2.1	Diaziridines *190*	
5.2.1.1	Substrate-Controlled Diastereoselective Diaziridination Using Chiral Enantiomerically Pure Amines *191*	
5.2.1.2	Substrate-Controlled Diastereoselective Diaziridination Using Chiral Enantiomerically Pure Ketones *192*	
5.2.2	Diazirines *193*	
5.2.3	Oxaziridines *194*	
5.2.3.1	Chiral Peracidic Oxidation of Achiral Imines *195*	
5.2.3.2	Achiral Peracidic Oxidation of Chiral Nonracemic Imines *196*	
5.2.3.3	Photocyclization of Nitrones *207*	
5.3	Four-Membered N-Heterocycles with Two Heteroatoms *208*	
5.3.1	Diazetidines *208*	
5.3.2	Oxazetidines *210*	
5.3.2.1	Thiazetidines *212*	
5.4	Conclusions *217*	
	References *217*	

6 Asymmetric Synthesis of Five-Membered Ring Heterocycles with More Than One Heteroatom *223*
Catherine Kadouri-Puchot and Claude Agami

6.1	Five-Membered Heterocycles with N and O Atoms *223*	
6.1.1	Oxazolidines *223*	

6.1.1.1	*N*-Alkyloxazolidines	*224*
6.1.1.2	*N*-Tosyl and *N*-Boc Oxazolidines	*228*
6.1.2	Oxazolines (4,5-dihydrooxazoles)	*230*
6.1.3	Oxazolidinones	*235*
6.1.3.1	Oxazolidin-2-ones	*235*
6.1.3.2	Oxazolidin-4-ones and 5-ones	*242*
6.1.4	Isoxazolines and Isoxazolidines	*243*
6.1.4.1	Isoxazolidines	*244*
6.1.4.2	Isoxazolines	*246*
6.2	Five-Membered Heterocycles with Two N Atoms	*249*
6.2.1	Imidazolidines and Imidazolidinones	*249*
6.2.1.1	Imidazolidines	*249*
6.2.1.2	Imidazolidinones	*252*
6.2.2	Pyrazolidines and Pyrazolines	*255*
6.2.2.1	Pyrazolidines	*255*
6.2.2.2	Pyrazolines	*257*
6.2.3	Pyrazolidinones	*260*
6.3	Five-Membered Heterocycles with N and S Atoms	*263*
6.3.1	Thiazolidines	*263*
6.3.1.1	Iminothiazolidines	*267*
6.3.1.2	Thiazolidinethiones	*268*
6.3.1.3	Thiazolidinones	*269*
6.3.2	Thiazolines	*270*
6.3.2.1	2-Thiazolines	*270*
6.3.2.2	3-Thiazolines	*275*
6.3.3	Sultams	*276*
	References	*281*
7	**Asymmetric Synthesis of Six-Membered Ring Nitrogen Heterocycles with More Than One Heteroatom** *293*	
	Péter Mátyus and Pál Tapolcsányi	
7.1	Six-Membered Rings with Another Heteroatom in the Same Ring	*293*
7.1.1	Pyridazines	*293*
7.1.1.1	Ring Closure of Optically Active Precursors	*294*
7.1.1.2	Diels–Alder Reactions	*299*
7.1.2	Pyrimidines	*302*
7.1.2.1	Formation of the Pyrimidine Ring	*302*
7.1.2.2	Stereoselective Transformation by the Involvement of the Pyrimidine Ring	*307*
7.1.3	Piperazines	*311*
7.1.3.1	Formation of the Piperazine Ring	*311*
7.1.3.2	Stereoselective Transformation of the Piperazine Ring	*327*
7.1.4	Oxadiazines	*332*
7.1.4.1	1,2,5-Oxadiazines	*332*
7.1.4.2	1,3,4-Oxadiazines	*332*

7.1.5	Morpholines 335
7.1.5.1	Formation of the Morpholine Ring 335
7.1.5.2	Asymmetric Transformations with the Involvement of the Morpholine Ring 352
	References 359

8 Asymmetric Synthesis of Seven-Membered Rings with More Than One Heteroatom 367
Jacques Royer

8.1	Diazepines 367
8.1.1	1,2-Diazepines 367
8.1.2	1,3-Diazepines 368
8.1.3	1,4-Diazepines 371
8.1.3.1	1,4-Benzodiazepines 371
8.1.3.2	Other 1,4-Diazepines 376
8.2	Oxazepines 378
8.2.1	1,2-Oxazepine 378
8.2.1.1	Diels–Alder Cycloaddition 378
8.2.1.2	Intramolecular 3 + 2 Cycloaddition 379
8.2.1.3	Pd-Catalyzed 4 + 3 Cycloaddition 379
8.2.1.4	Rearrangements 379
8.2.2	1,3-Oxazepines 380
8.2.2.1	N,O-Acetals 380
8.2.3	1,4-Oxazepines 383
8.2.3.1	Amino Alcohol Double Condensation 383
8.2.3.2	Other Cyclization Methods 383
8.2.3.3	Pd-Catalyzed Allene Cyclization 384
8.2.3.4	Radical Cyclization 385
8.2.3.5	Ring Enlargement 385
8.2.3.6	Cycloaddition 386
8.3	Thiazepines 386
8.3.1	1,2-Thiazepines 387
8.3.2	1,3-Thiazepines 387
8.3.3	1,4-Thiazepines 388
8.3.3.1	From Mercaptopropionic Acid Derivatives 388
8.3.3.2	From Amino Thiols 390
8.3.3.3	Others 391
	References 392

Index 399

Preface

Nitrogen heterocycles are a huge family of compounds. Among them, numerous natural products as well as synthetic derivatives exhibiting some interesting biological activities can be found. Chirality is often encountered in such products since bioorganic substances such as receptors and enzymes are essentially asymmetric and thus their ligands are better to be asymmetric in order to fit with this asymmetric environment. On the other hand, natural nitrogen compounds are also mainly asymmetric.

Thus, the asymmetric synthesis of nitrogen heterocycles is a frequent preoccupation of organic chemists from both the academic and industrial areas. While the synthesis of nitrogen heterocycles uses several types of known classical reactions, it also has its own specificity shown through various strategies.

In this context, I had been keen on editing a book dealing with the *"asymmetric synthesis of nitrogen heterocycles"*. While several books dealing with nitrogen heterocycles already exist, no book or review article proposes an overview of the different methods used in their preparation in the asymmetric form according to their structure. It then appeared that there would be a need to have a review gathering such information.

I requested some colleagues to contribute the chapters in this book. They were all approached on the basis of their experience in and authoritative knowledge of specific nitrogen heterocycles.

In an attempt to cover most of the classical nitrogen heterocycles, the book has been divided into two parts. The first part deals with heterocycles bearing only one heteroatom in their ring; it is organized into chapters according to the size of the ring: aziridine, azetidine, pyrrolidine, piperidine, azepine and larger rings. The second part deals with heterocycles that contain at least two heteroatoms (one being nitrogen), here again, the chapters correspond to the size of the ring: from three to seven-membered rings. Each chapter is also carefully organized with the aim to provide easy access to information about the different heterocyclic structures. In summary, the main idea of this book is to furnish a comprehensive handbook giving firsthand information to researchers wanting to prepare chiral nitrogen heterocycles.

Asymmetric Synthesis of Nitrogen Heterocycles. Edited by Jacques Royer
Copyright © 2009 WILEY-VCH Verlag GmbH & Co. KGaA, Weinheim
ISBN: 978-3-527-32036-3

This book would be useful for organic chemists interested in the asymmetric synthesis of heterocyclic compounds including natural products and to those working in pharmaceutical companies or in academic institutions as well. It would also be helpful for graduate students.

I am deeply indebted to my colleagues who, knowing the time-consuming nature of this important piece of work, have contributed the articles for this book.

Paris, October, 2008   *Jacques Royer*

List of Contributors

Claude Agami
University Pierre et Marie Curie
Paris
France

Giuliana Cardillo
University of Bologna
Bologna
Italy

Luca Gentilucci
University of Bologna
Bologna
Italy

Gilles Hanquet
CNRS-University of Strasbourg
Strasbourg
France

Pei-Qiang Huang
Xiamen University
Xiamen
China

Catherine Kadouri-Puchot
University Pierre et Marie Curie
Paris
France

Steve Lanners
University of Namur
Namur
Belgium

Péter Mátyus
University of Budapest
Budapest
Hungary

Jacques Royer
CNRS-University Paris Descartes
Paris
France

Marie-Eve Sinibaldi
Clermont Université
UBP, CNRS (UMR 6504)
Laboratoire SEESIB
63177 Aubière Cedex
France

Pál Tapolcsányi
Semmelweis University
Budapest
Hungary

Alessandra Tolomelli
University of Bologna
Bologna
Italy

Naoki Toyooka
University of Toyama
Toyama
Japan

Yves Troin
Clermont Université
ENSCCF
Laboratoire de Chimie des
Hétérocycles et des Glucides
EA 987
63174 Aubière Cedex
France

Part One
Asymmetric Synthesis of Nitrogen Heterocycles Containing Only One Heteroatom

1
Asymmetric Synthesis of Three- and Four-Membered Ring Heterocycles

Giuliana Cardillo, Luca Gentilucci and Alessandra Tolomelli

1.1
Substituted Aziridines

1.1.1
Generalities

Small heterocyclic rings constitute systems of central importance in theoretical, synthetic organic, bioorganic, and medicinal chemistry, and in particular aziridines and azirines are very useful and interesting systems as they occur in a number of natural and biologically active substances and as they are useful building blocks and versatile synthetic intermediates. Therefore, the development of efficient and stereoselective methods for synthesis and elaboration of aziridines is an inviting ongoing challenge [1]. Very often, stereogenic centers within such strained heterocycles can be used to direct the stereochemical outcome of subsequent transformations.

Aziridines and their dehydro derivatives, 2H-azirines, can be regarded as representatives of the first and most simple heterocyclic systems [2].

Aziridine

C–C 1.48 Å
C–N 1.49 Å
Endocyclic angle 60°
Exocyclic angle 153–160°
Nitrogen inversion barrier 64.8 kJ mol^{-1}
Nitrogen deviation from planarity 69.7°

Azirine

C–C 1.46 Å
C=N 1.23–1.28 Å
C–N 1.59 Å
Endocyclic angle 48–60°
Exocyclic angle 142–160°

While numerous members of the aziridine ring systems are known and have been fully characterized, derivatives of the azirine ring system are mainly known as useful intermediates and only few examples of naturally occurring azirine derivatives have been reported.

Aziridines are present as structural motifs in a variety of strongly biologically active compounds such as azinomycins A and B [3], which are potent antitumors as well as antibiotic agents against both Gram-positive and Gram-negative bacteria and have been isolated from the fermentation broth of *Streptomyces griseofuscus* S42227.

Azinomycin A

Azinomycin B

The antineoplastic activity of mitomycins A, B, and C [4], produced by *Streptomyces caespitosus*, is associated with the high reactivity of the strained heterocycle. Furthermore, some synthetic aziridines show strong activity as enzyme inhibitors [5], or are versatile intermediates for bioactive compounds.

Mitomycin A

Mitomycin B

Mitomycin C

The first example of azirine-ring-containing natural compound was azirinomycin [6], an unstable antibacterial agent isolated from *Streptomyces aureus* fermentation broth in 1971. Only several years later, the cytotoxic compound dysidazine [7], the second structure showing the azirine motif, was extracted from *Dysidea fragilis*, a sponge collected in Fiji islands.

Since their discovery by Gabriel [8], aziridines have attracted attention as starting materials for further transformations in organic synthesis. The ring strain of aziridines, which amounts to 26–27 kcal mol^{-1}, renders these compounds susceptible to ring opening with excellent stereo- and regiocontrol and allows their use as precursors of a variety of nitrogen-containing compounds such as amino acids, aminoalcohols, and β-lactams.

The main transformations of three-membered ring [9] are reported below:
1. Hydrogenolysis: this allows chiral amine derivatives to be obtained via the regiospecific cleavage of a C–N bond. The process induces ring opening usually with inversion of the reacting stereocenter and without any modification of the stereochemistry at the unaffected carbon.

2. **Hetero- and carbon nucleophile ring opening**: this gives access to a variety of optically active ramified amines or amino alcohols, amino thiols, diamines, etc. While the reactivity of N-unsubstituted aziridines is relatively low, high reactivity is associated with aziridines incorporating an electron-withdrawing group on the nitrogen atom. For instance, the presence of an acyl group strongly activates the ring toward opening by a nucleophile. This reaction is generally favored by the presence of Lewis acids and proceeds with inversion of configuration at the stereogenic center of the aziridine. Unlike their acyclic amide counterparts, acylaziridines are highly pyramidalized at nitrogen, which makes the acyl-aziridine nitrogen more basic. The stereoselectivity is usually high and the regioselectivity depends upon the ring substituents and the nature of the nucleophile.

3. **Ring expansion**: this is another important reaction characteristic of N-acylaziridines, which represents the isomerization to the corresponding oxazolines, protected form of chiral amino alcohols. This reaction generally occurs in the presence of a Lewis acid and leads to the five-membered ring with retention of configuration.

4. **aza-Payne reaction**: this is the rearrangement of aziridine-2-methanols under basic conditions to the corresponding epoxide-2-amines. These last compounds can be further transformed by reaction with nucleophiles.

5. **1,3-dipolar cycloaddition of aziridine-2,2-dicarboxylates**: this is an interesting reaction that involves generation of 1,3-dipoles from three-membered rings and π systems, giving access to larger size heterocycles.

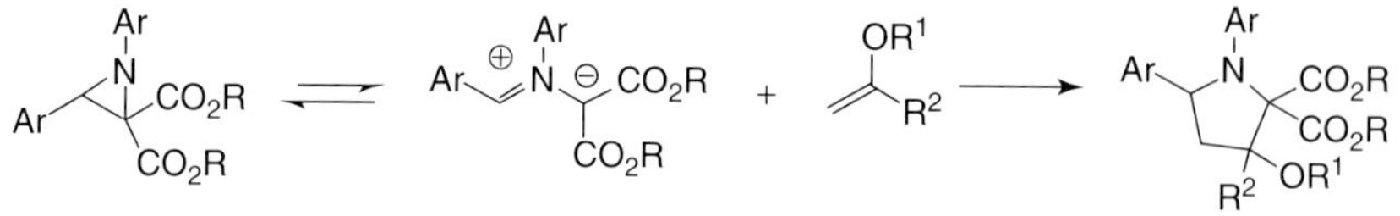

6. Carbonylation: this allows the formation of β-lactams via insertion of the carbonyl group and inversion of configuration at the reacting carbon terminal.

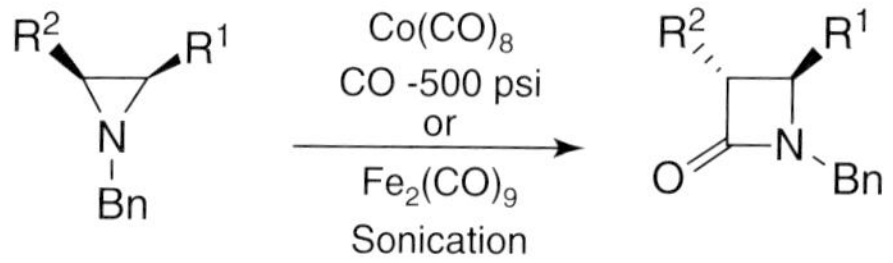

7. Formation and reactivity of aziridinyl anions: these are stable species at low temperature, which may react with a variety of electrophiles allowing introduction of functionalized chains on the heterocycle.

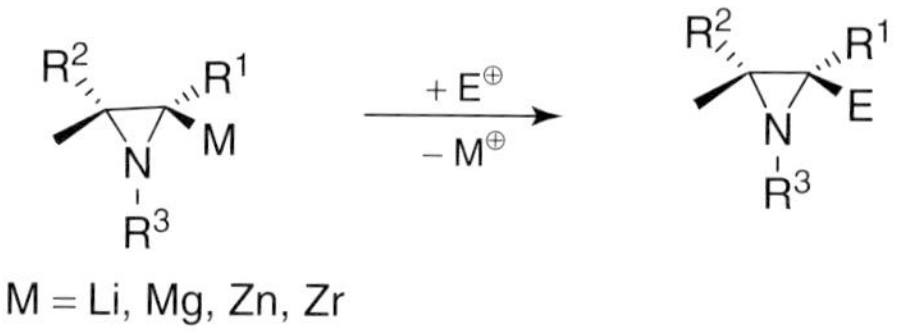

M = Li, Mg, Zn, Zr

1.1.2
Asymmetric Aziridination via Cyclization Methods

General approaches to the asymmetric synthesis of aziridines through cyclization methods can be divided into two main categories: (A) nitrogen nucleophilic cyclization on the adjacent position bearing a leaving group and (B) ring closure to three-membered ring via attack of a stabilized carbanion on the electrophilic nitrogen bearing a leaving group. The last approach is particularly suitable for the preparation of aziridine-2-carboxylates.

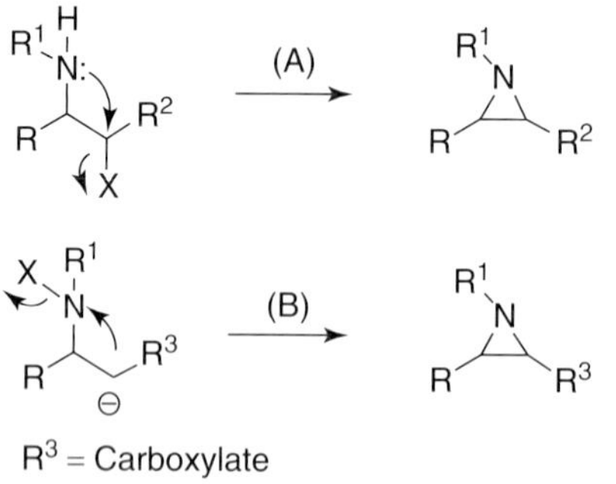

R^3 = Carboxylate

In both cases, the asymmetric induction may be exerted by the chirality of the substrate, by the introduction of a chiral auxiliary, by the presence of a chiral metal catalyst, or by the application of organocatalytic processes.

1.1.2.1 Cyclization of a Nucleophilic N on an Electrophilic C (Pathway A)

Cyclization of Amino Alcohols The most general and conceptually simple method for the synthesis of optically active aziridines is the cyclization of amino alcohols and amino halides. The availability of enantiopure amino alcohols directly from the chiral pool or by simple reduction of amino acids makes this approach extensively exploited. Protection of the amino function is sometimes required, and in these cases sulfonyl or phosphinyl groups are usually preferred to carbamates or acyl groups, to avoid competitive formation of larger heterocycles such as oxazolines or oxazolidinones.

Thus, conversion of the hydroxyl moiety into a good leaving group such as tosyl [10], mesyl, nosyl, tetrahydropyranyl (THP) [11], diphenylphosphinyl (Dpp) [12], ^{t}butyldimethylsilyloxy (TBS) [2] allowed the preparation of an activated intermediate to be easily converted to aziridine under basic conditions through an intramolecular S_N2 reaction.

When reduction of the carboxylic function is needed to produce the proper aminoalcohol, N-tosyl amino acids have been used as reagents of choice to avoid difficulties in the isolation of water-soluble amino alcohols [10].

Activation of the hydroxyl function may be preformed also under Mitsunobu conditions. Treatment of N-Boc-amino alcohols, easily obtained by reduction of the corresponding amino acids, with triphenylphosphine and Diethyl azodicarboxylate (DEAD) afforded optically active N-Boc-aziridines in good yields [13].

Aziridine-2-carboxylates have been successfully obtained starting from serine or threonine. Starting from D-threonine, Rapoport and coworkers [14] synthesized enantiopure N-benzyl-aziridine-2-esters by treatment with triphenylphosphine. On the other hand, ring closure of N-protected-serine esters with diethoxyphenylphosphorane (DTPP) gave aziridines-2-tert-butylate in satisfactory yield [15].

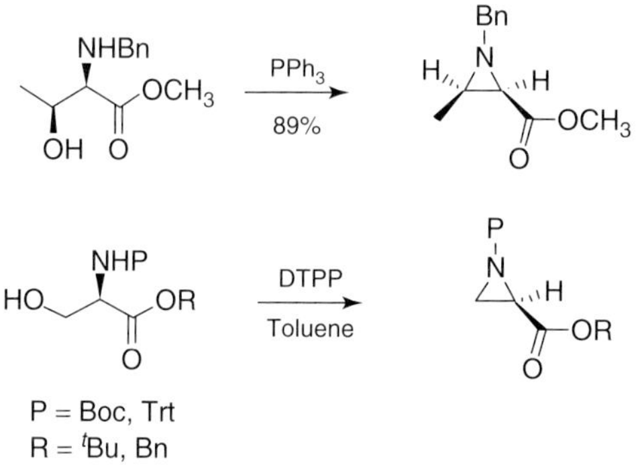

P = Boc, Trt
R = ^{t}Bu, Bn

Generally, enantiopure aziridine-2-carboxylates are synthesized as precursors of α- or β-amino acids in a multistep sequence directed to the preparation of bioactive peptides and peptidomimetics. To this purpose, Palomo and coworkers reported the easy formation of the aziridine ring by treatment of the dipeptide serine–glycine benzyl ester via nosylation of the oxygen atom under basic conditions. This procedure allowed the authors to reduce protection/deprotection steps, directly performing the reaction on glycine benzylester derivative [16].

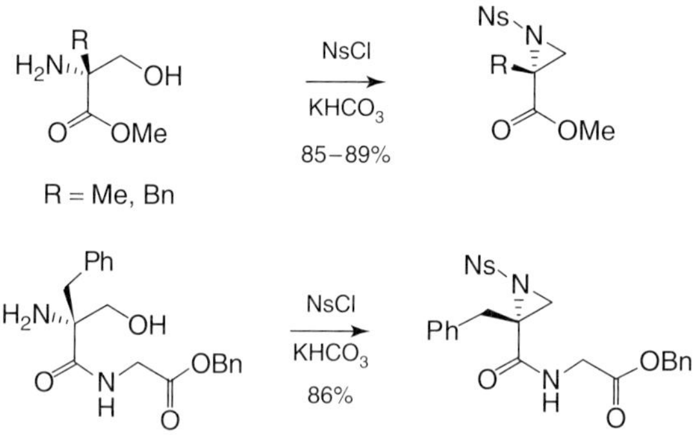

R = Me, Bn

Asymmetric Synthesis of Aziridines via Gabriel-Cromwell Reaction The Gabriel-Cromwell reaction is a general and convenient method for aziridination of α, β-unsaturated compounds. Since its introduction in 1952 [17]. many variations on the standard procedure have been explored to broaden the field of application. The reaction occurs via addition of bromine to the double bond followed by treatment with an amine. The mechanism proceeds with the formation of a dibromo derivative that converts to a α-bromo-alkene through elimination of bromhydric acid. The conjugate addition of a second molecule of amine, followed by nucleophilic displacement of the bromine, leads to aziridine-carboxylate formation.

1.1 Substituted Aziridines

The asymmetric version of this reaction has been described using chiral auxiliaries such as camphor sultam [18]. Stereodefined 3-unsubstituted-aziridine-2-carboxylic acid may be prepared starting from acryloyl camphor sultam derivatives and by removal of the chiral auxiliary in a nondestructive manner in the final step. Repetition of aziridination protocol with N-crotonyl-camphor sultam resulted in the formation of 1:1 mixtures of easily separable aziridines.

The asymmetric induction controlled by means of ephedrine-derived Helmchen's auxiliary gave better results [19]. Gabriel-Cromwell reaction on chiral imides in DMSO afforded via diastereoselective and high yielding procedure, optically active *trans*-aziridines, easily purified by flash chromatography. The nondestructive cleavage of the chiral auxiliary with lithium benzyloxide gave the corresponding enantiopure benzyl aziridine-2-carboxylates.

1.1.2.2 Cyclization of a Stabilized Anion on an Electrophilic N (Pathway B)

Starting from the unsaturated derivatives reported above, the same authors performed the diastereoselective synthesis of aziridine-2-imides via conjugate addition of O-benzylhydroxylamine followed by cyclization to three-membered rings [20]. This second step was induced through the formation of the stabilized aluminum enolate of the addition product that spontaneously attacked the electrophilic nitrogen bearing the O-benzyl leaving group. Removal and recovery of the chiral auxiliary were performed as reported above.

In a similar way, optically active 1H-α-keto aziridines were synthesized from conjugated enones via Sc[(R)-BNP]$_3$-catalyzed enantioselective Michael addition of O-methylhydroxylamine, followed by La(OiPr)$_3$-catalyzed ring closure [21].

The products obtained by addition of chiral hydroxylamine to acrylates have been transformed into 2- and 2,3-disubstituted-N-alkylaziridinecarboxylate through an efficient diastereoselective 3-exo-tet ring closure induced by O-acylation of the diastereomeric adduct followed by enolization. This two-step protocol afforded optically active aziridines with excellent diastereoselectivity (93:7). Attempts to perform a one-pot two-step reaction resulted in a lower stereoselectivity (67:33) [22].

A further development of the same synthetic approach is represented by the synthesis of 3′-unsubstituted-N-Boc-aziridines, which has been carried out in one step by conjugate addition of sodium or lithium anion of N-Boc-O-benzoylhydroxylamine to chiral acryloyl-imides [23].

1.1 Substituted Aziridines

The Michael-type addition of (+)-(R)-o-methoxyphenylsulfonylethylsulfimide to unsaturated carbonyl compounds afforded optically active *trans*-acylaziridines with modest stereoselectivity by displacement of the diphenylsulfide group from the intermediate enolates [24].

The addition of N, O-bis(trimethylsilyl)hydroxylamine to alkylidene malonates in the presence of a catalytic amount of $Cu(OTf)_2$ and chiral bisoxazoline as ligand, followed by base-induced cyclization, represents a useful route to enantiomerically enriched aziridine-2,2-dicarboxylates. This two-step protocol afforded the three-membered rings with good yield and enantiomeric excesses up to 80%, depending on the substituent on the alkylidene double bond [25].

In general, the conjugate addition of hydroxylamines to unsaturated carbonyl derivatives is one of the most convenient methods for the stereoselective synthesis of β-hydroxylamino-carbonyl building blocks, which are useful intermediates in the preparation of enantiopure aziridines. Asymmetric methods dealing with stoichiometric use of chiral auxiliaries in the presence of Lewis acids or with the use of catalytic amount of chiral metal complexes have been exhaustively reviewed [26].

1 Asymmetric Synthesis of Three- and Four-Membered Ring Heterocycles

The organocatalytic version of the aza-Michael addition was less explored. MacMillan and coworkers reported the first organocatalytic addition of N-silyloxy-carbamates to unsaturated aldehydes, giving access to aziridine precursors [27].

In a recent report, the conjugate addition of O-benzylhydroxylamine to chalcones was performed in the presence of thiourea-derived catalysts. Although a screening of solvents, O-substituted hydroxylamines, and chiral ligands was performed, moderate enantiomeric excesses up to 60% could be observed [28].

The organocatalysis may be applied to aziridination when the substrate to be transformed is an aldehyde. In fact, the highly chemo- and enantioselective organocatalytic aziridination of a variety of α, β-unsaturated aldehydes with acetyloxy carbamates was developed. The reaction was catalyzed by chiral pyrrolidine derivatives and gave Boc- or Cbz-2-formylaziridines in yields ranging from 60 to 70%, with dr 4:1–19:1 and 84–99% ee [29].

R	R'	Yield	dr	ee
Me	Boc	54%	6:1	94%
Et	Cbz	60%	5:1	97%
nPr	Cbz	62%	10:1	99%
nBu	Cbz	70%	5:1	96%
BzO~~~	Boc	60%	5:1	98%

1.1.3
Asymmetric Aziridination via Cycloaddition Methods

1.1.3.1 Addition of Nitrenes to Alkenes

One of the most important pathways to aziridines is the addition of a nitrene to an alkene; however, this reaction may not be well controlled stereochemically owing to the rapid interconversion of the singlet and triplet nitrene states. Anyway, several methods for nitrene generation have been successfully developed with the aim to obtain stereospecific aziridination. The most common involve photolysis, thermolysis, or chemical modification of nitrogen derivatives.

Many chiral metal catalysts have been used to induce nitrene formation from N-substituted iminoiodinanes, although this method produces stoichiometric amount of iodobenzene and yields N-protected aziridines. The use of azide precursors gives some advantage in terms of atom efficiency and environmental impact since molecular nitrogen is the only side product. Nosyloxycarbamates and N-aminophtalimides are also alternative sources of nitrene.

(a) $TsNH_2 + PhI(OAc)_2 \xrightarrow[\text{MeOH}]{\text{KOH}} PhI=NTs$

$PhI=NTs \xrightarrow{L^*Cu\ PF_6^{\ominus}} [^*LCu=NTs]^{\oplus}\ PF_6^{\ominus} + PhI$

(b) phthalimide-N-NH$_2$ $\xrightarrow{Pb(OAc)_4}$ phthalimide-N-N:

(c) $RO_2CN_3 \xrightarrow{h\nu\ \text{or}\ \Delta}$:N-CO$_2$R
    $NsONHCO_2R \xrightarrow{\text{Base}}$

N-(p-toluenesulfonyl)iminophenyliodinane (PhI = NTs) [30] (source a) proved to be superior to other imido group donor as precursor and yielded excellent stereoselectivity. In 1991, Evans and coworkers disclosed that low-valent copper complexes catalyze the aziridination of several different olefins by this reagent. Development of the enantioselective process consists in the use of chiral bisoxazoline catalysts. Some selected results are reported in Table 1.1.

$R^2R^1C=CR^3 + PhI=NTs \xrightarrow[\text{CH}_3\text{CN}]{\text{5–10\% Cu-ligand}}$ aziridine(Ts, R^1, R^2, R^3) + PhI

Aryl-substituted olefins have been found as good substrates, which can be efficiently transformed into N-tosyl-aziridines with enantioselectivities up to 97% ee (entries 1–4) [31]. The reactions have been carried out with 5% of chiral catalyst derived from copper(I) triflate and bisoxazoline as ligand. Unfortunately, tartrate-derived bisoxazoline gave only low enantiomeric excess (2–49% ee). A complete study on the effect of bisoxazoline substituents and reaction conditions on this reaction has been reported by Page and coworkers [32]. Some improvements of both enantioselectivity and chemical yields were obtained when [N-(4-nitrobenzenesulfonyl)imino]phenyliodinane was employed instead of the

Table 1.1 Enantioselective aziridination with N-aryl-iminophenyliodinane and chiral metal complexes

Entry	Chiral ligand	Olefin	Catalyst	Nitrene precursor	Yield	ee
1	a: R = Ph; b: R = CMe₃ (bis-oxazoline)	Ph-CH=CH-CO₂Ph	a	PhI=NTs	64	97
2		(α)Nap-CH=CH-CO₂Me	a	PhI=NTs	76	95
3		Ph-CH=CH-Me	b	PhI=NTs	62	70
4		Ph-CH=CH₂	b	PhI=NTs	89	63
5		Ph-CH=CH-Me	b	p-NO₂-PhI=Ts	83	80
6		Ph-CH=CH₂	b	p-NO₂-PhI=Ts	94	66
7		Ph-CH=CH₂		PhI=NTs	79	66
8	salen-type (Cl-substituted)	NC-chromene		PhI=NTs	75	>98
9		dihydronaphthalene		PhI=NTs	70	87
10	binaphthyl salen (Ph, OH)	Ph-CH=CH₂	Mn	PhI=NTs	76	94

commonly used p-tolyl analog (entries 5–6) [33]. In a similar way, Jacobsen and coworkers [34] obtained excellent results in the aziridination of benzylidene derivatives with chiral copper complexes deriving from bis(benzylidene)imine of 1,2-diaminocyclohexane (entries 7–9). The asymmetric aziridination of styrene derivatives has also been successfully performed with Salen-Mn(III) complexes (entry 10) [35].

Phtalimidonitrene (source b), generated from N-aminophatalimide by oxidation with lead tetraacetate, reacted with N-enoylbornane[10,2]sultams (Oppolzer auxiliary) to give the corresponding N-phtalimidoaziridine adducts in 12–94% yield and diastereofacial selectivity up to >95% [36]. In a similar way, excellent diastereoselection could be obtained by the addition of phtalimidonitrene to sugar-derived α,β-unsaturated esters [37].

Chen and coworkers applied this methodology to the diastereoselective aziridination of α, β-unsaturated amides linked to a camphor pyrazolidinone-derived chiral auxiliary [38]. The reactions carried out in 5 min afforded excellent yield (86–95%) of diastereomeric aziridines with high selectivity (up to >90% de). In pursuing this work, the same author reported the enantioselective version of this protocol, by performing the lead tetraacetate oxidative addition on N-enoyl oxazolidinones in the presence of camphor-derived chiral ligands [39].

Besides the use of a chiral auxiliary linked to the alkene reagent or of chiral complexes of lead tetraacetate, another possibility is the use of chiral aziridinating agents. Atkinson and coworkers [40] reported that enantiopure 3-acetoxyaminoquinazolinones react with β-trimethylsilylstyrene affording 11:1 ratio of diastereomeric aziridines.

Organic azides (source b) may be considered as ideal sources of nitrenes although they are not very reactive and harsh conditions, such as heating or UV irradiation, are generally needed for molecular nitrogen dissociation. Initial attempts to perform asymmetric aziridination using azides and chiral metal catalysts have been reported by Jacobsen with Cu-diimine complexes and by Müller and co-workers using chiral Rh-complexes. In both cases, the nitrene generation was induced by UV irradiation. On the other hand, excellent yields (up to 99%) and enantioselectivities (up to >99% ee) have been obtained by Katsuki and coworkers in the presence of ruthenium(CO)(salen) complexes and tosyl azide. Under these conditions, neither heating nor irradiation was required for nitrene formation. The method is of general application since it has been successfully applied to a number of different olefins, changing the substituent of the chiral ligand and generating the nitrene reagent starting from different organic azides.

R^1 = Ph, BrPh, $C_{10}H_7$, nC_6H_{13}, indene

R^2 = Ts, Ns, SES

Finally, nitrenes may be formed from N-protected nosyloxycarbamate by treatment with a base such as CaO. Pallacani and Tardella applied this method to the synthesis of a variety of substituted N-protected aziridines [41]. Recently, excellent diastereoselectivity was observed in the aziridination of 2-L-α-aminoacyl-(E)-acrylonitriles under parallel solution-phase conditions [42].

1.1.3.2 Reaction between Carbenes and Imines

Among several synthetic routes, the classic methods involving imines have been upgraded to their asymmetric version inducing enantiocontrol by the use of chiral imines, chiral nucleophiles, or chiral catalysts. All these approaches share as common feature the addition of a nucleophile on the electrophilic imine carbon, followed by cyclization to the three-membered heterocycle.

aza-Darzens-Type Reactions Involving α-Haloenolates The aza-Darzens reaction between a α-haloenolate and an imine can be considered a not fully investigated tool for the asymmetric synthesis of aziridines.

Davis and coworkers obtained excellent results in the synthesis of N-sulfinyl-aziridine-2-carboxylates by reacting enantiopure chiral sulfinimines with α-bromo-enolates. The method is general for aliphatic, aromatic as well as α,β-unsaturated imines, giving good yields and de up to 98%, N-sulfinyl-heterocycles, which are easily transformable into activated N-tosyl-aziridines [43].

Following the same procedure, enantiopure aziridine-2-phosphonates were obtained from sulfinimines and halomethylphosphonates. A further improvement could be obtained employing N(2,4,6-trimethylphenylsulfinyl)imines and lithium diethyliodophosphonate [44].

Despite the excellent results obtained with chiral sulfinimines, development of new chiral imines could overcome shortcomings as sensitivity to oxidative conditions and destructive chiral auxiliary removal. The reaction between chiral N-phosphonimines and α-bromo-enolates gave N-phosphonylaziridines with excellent yield and high stereocontrol. The electrophilicity of the imine can be controlled by introducing electron-donating or electron-withdrawing groups onto phosphonate chiral auxiliary [45].

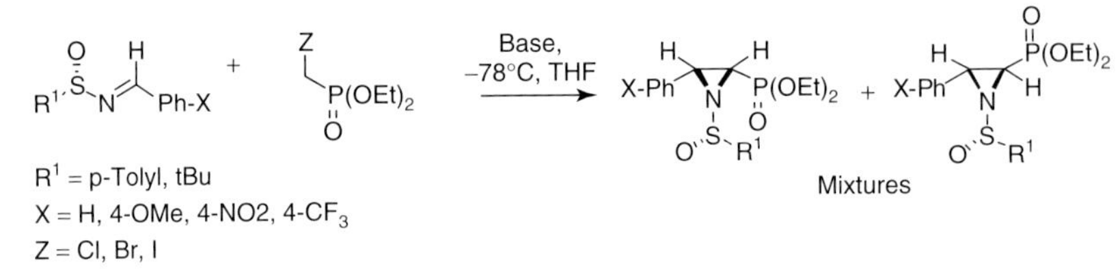

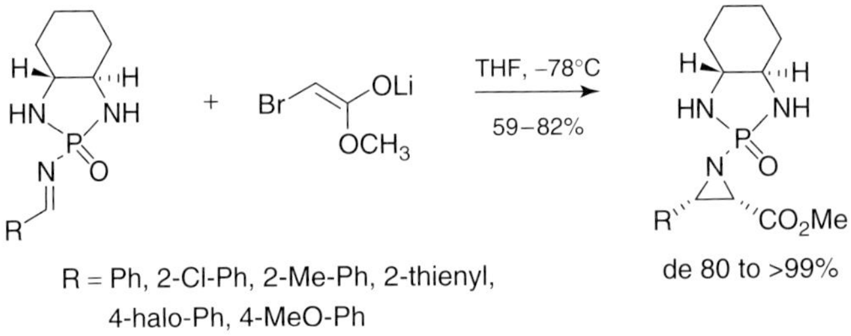

1 Asymmetric Synthesis of Three- and Four-Membered Ring Heterocycles

R^1 = p-Tolyl, tBu
X = H, 4-OMe, 4-NO2, 4-CF$_3$
Z = Cl, Br, I

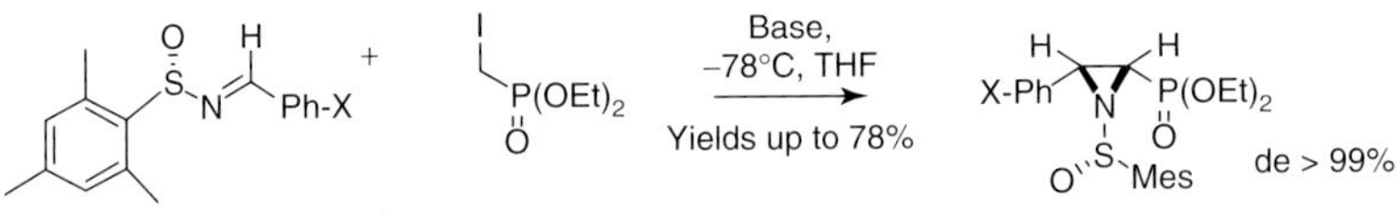

Chiral heterosubstituted aziridines have been obtained by the coupling of lithium enolates derived from (α-chloroalkyl)heterocycles with various enantiopure imines, which are obtained from nonracemic phenylethylamine. The reaction afforded chiral aziridines with complete stereocontrol. The steric hindrance and the coordination power of the alkyl group linked to the iminic nitrogen are responsible for the stereochemistry of the final product [46].

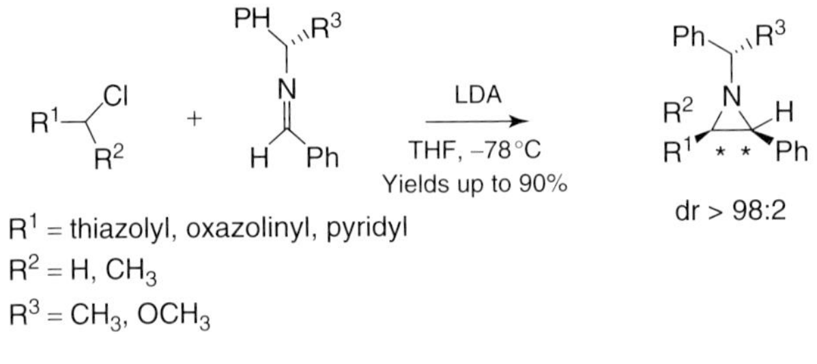

R^1 = thiazolyl, oxazolinyl, pyridyl
R^2 = H, CH$_3$
R^3 = CH$_3$, OCH$_3$

In a similar way, the addition of chloromethyllithium to the imine derived from 2-pyridinecarboxaldehyde and chiral aminoalcohol or aminoesters gave disubstituted aziridines with good yields and excellent diastereoselectivity. In the latter case, double addition of the organometallic reagent occurred, affording aziridines having a keto function in the side chain [47].

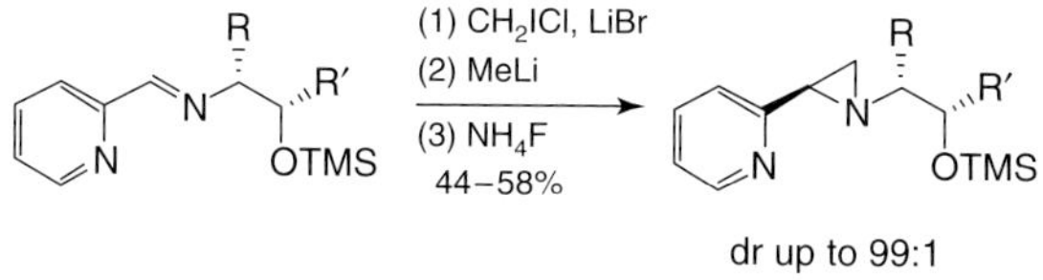

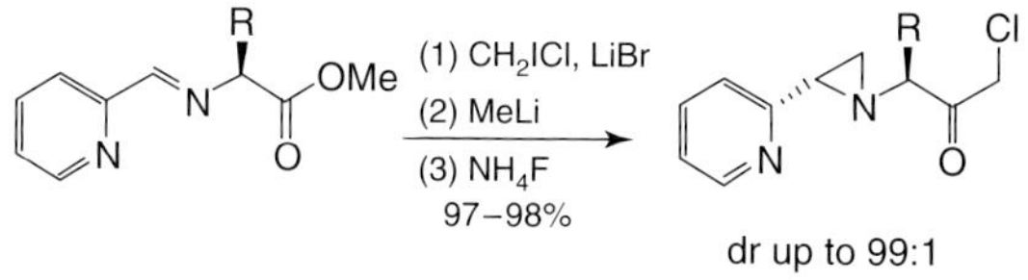

1.1 Substituted Aziridines

dr up to 99:1

dr up to 99:1

Enantiopure N-acylaziridines have been obtained starting from aldimine bearing the chiral auxiliary into the carbon side chain. Under these conditions, complete inversion of diastereocontrol was induced by changing the metal counterion of the bromoenolate from lithium to zinc [48].

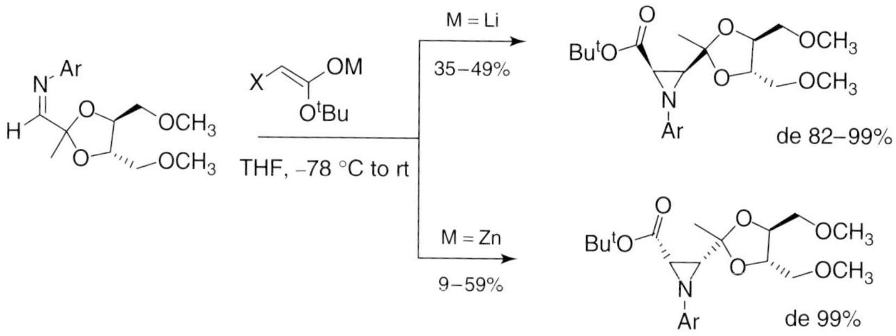

Similarly to aza-Darzens mechanism, the addition of organometallics as Grignards to chiral sulfinylimines bearing a α-halogen leaving group followed by treatment with a base represents a high yielding route to optically active N-sulfinylaziridines. In this reaction, spontaneous or base-induced cyclization of the non-isolated intermediate β-halo-N-sulfinamides affords the three-membered rings in high yield and excellent diastereomeric ratio [49].

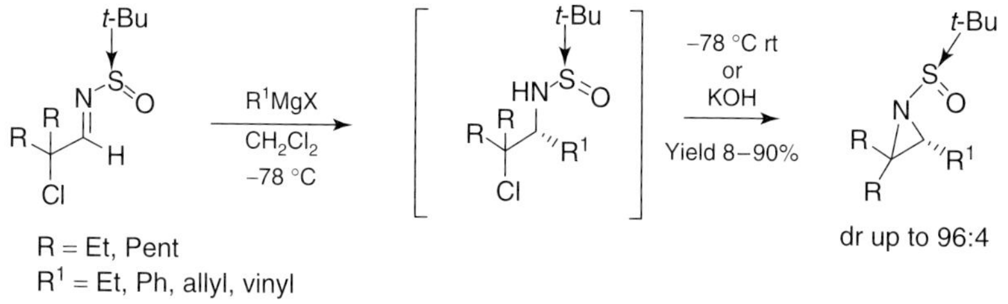

R = Et, Pent
R^1 = Et, Ph, allyl, vinyl

dr up to 96:4

An application of aza-Darzens-type reaction has been performed to obtain polyfunctionalized aziridines through the base-induced dimerization of oxiranylaldimines. This highly diastereoselective process is noteworthy since both nucleophile and electrophile originate from the same precursor. The nucleophilic moiety is the 1-aza-allylanion having the epoxide at the β carbon and the leaving group is represented by the oxirane oxygen atom [50].

Asymmetric induction in the condensation between imine and haloacetates may be also obtained introducing chiral auxiliaries into the α-haloacetate counterpart.

For instance, (+)-8-phenylmenthyl esters gave aziridine-2-carboxylates only in 40% yield and as diastereomeric mixtures of cis/trans heterocycles. Anyway, the diastereomeric excess of the trans isomer reached 85% [51].

On the other hand, camphorsultam-derived α-bromoenolates reacted with N-Dpp-imine to afford a single cis diastereoisomer in high yield. Removal of the chiral auxiliary was simply performed with LiOH at room temperature. The method was successfully applied to aromatic, para-substituted aromatic, unsaturated, and aliphatic enolates. The introduction of an ortho substituent into the aromatic ring determined the stereoselectivity inversion, often giving exclusively trans products [52].

Finally, enantiopure cis N-alkoxy-aziridine-carboxylates have been obtained via aza-Darzens like reaction between the anion of optically pure chloroallyl phosphonamide and oximes. The reaction occurred in good to excellent yields through the approach of the oxime on the less hindered left cleft of the phosphonamide anion, giving a single diastereoisomer [53].

aza-Darzens-Type Reactions Involving Diazo Compounds This methodology, involving the formation of a metal–carbene intermediate complex that adds to an imine, shows several advantages of other synthetic approaches since the reagents are synthetically accessible and highly reactive and the only by-product is represented by molecular nitrogen. The asymmetric version of this reaction has been successfully developed using enantiopure starting materials, stoichiometric amounts of chiral auxiliaries linked to the reagent backbone, or by catalyzing the reaction with chiral Lewis acids.

The protocol recently reported by Johnston and coworkers, based on the Bronsted acid-catalyzed annulation, afforded enantiopure aziridine-2-carboxylates by using glyceraldehyde as chiral auxiliary. The diazocompound reacted as enolate synthon without any decomposition, giving the product in 83% yield and complete diastereoselectivity [54].

The asymmetric metal-catalyzed transfer of diazocarbonyl-derived carbene to imines represents the most explored approach. To this purpose, Jacobsen and coworkers investigated copper(II) complexes catalyzed aziridination of N-benzylidene aniline and diazoacetate [55]. Although cis-aziridines were obtained only in modest yields (5–65%) and stereoselectivities (2–67% ee), a deep exploration of the reaction mechanism allowed to suggest the generation of a transient bis(dihydrooxazole)copper carbene complex that reacts with the imine to form a metal-complexed azomethine ylide. This intermediate may undergo intramolecular ring closure to optically active aziridine or it may dissociate to free azomethine ylide, precursor of the formation of racemic aziridine. This investigation outlined the influence of the chiral ligand on the metal-complexed azomethine ylide evolution, thus providing useful information for enantioselectivity enhancement.

On the other hand, Jørgensen and coworkers explored the reaction of imines with diazoacetate. In the course of this study, they proposed a second possible mechanism involving the coordination of the Lewis acid to the nitrogen atom of the imine, followed by the nucleophilic attack of diazoacetate on the C=N double bond and by the ring closure on the carbon bearing N_2 leaving group [56].

Extension of this approach to the reaction of α-imino esters in the presence of chiral ligands allowed the development of a catalytic diastereo- and enantioselective aziridination of imines derived from α-ethylglyoxylate. Bisoxazolines, phosphinoxazolines, and bis(phosphino)-binaphtyl ligands were tested in combination with $AgSbF_6$ or $CuClO_4$. High diastereoselectivity in cis-aziridine formation was observed using (R)-Tol-BINAP (cis/trans 19:1) with a good enantiomeric excess (72%), while the trans isomer was obtained as major product in the presence of (4R, 5S)-Ph-BOX [57].

One of the best generally applicable method for the catalytic asymmetric aziridination was presented by Wulff and Antilla, using a catalyst prepared from VAPOL and borane-tetrahydrofurane. Under these conditions, excellent yields and high asymmetric induction were obtained in the reaction of benzhydrylimines with ethyl diazoacetate using 1% mol of the chiral catalyst [58]. A further enhancement of cis/trans selectivities (up to >50:1), yields (up to 91%), and enantiomeric excesses (90–98%) could be obtained using VAPOL or VANOL ligands in the presence of triphenylborate [59].

Aziridination by Reaction of Imines with Ylides Besides α-halo derivatives and diazo compounds, also ylides were reacted with imines in the asymmetric ring-closure to aziridine. The initial attack of the ylide to the electrophilic imine carbon affords a betaine, which evolves to aziridine via intramolecular ring closure and elimination of ylide heteroatom.

This methodology has been mainly applied to the preparation of enantiopure terminal-, alkyl-, aryl-, propargyl-, and vinyl-substituted heterocycles, by using methylene sulfur ylides as reagents.

The asymmetric induction in this reaction has been obtained introducing a stereocenter on the nitrogen imine side chain or generating sulfur ylides from chiral sulfides.

Following the first approach, Garcia Ruano and coworkers performed the reaction between enantiopure N-tolylsulfinylimines and dimethyloxosulfonium methylides and dimethylsulfonium methylides [60]. Optically active imines were easily generated by applying the "*DAG methodology*", where diacetone-*d*-glucose was used as an inducer of chirality [61]. Under these conditions, the formation of terminal aziridines occurred with good to excellent diastereoselectivities (up to 95:5) and the enhancement of the stereocontrol was obtained by increasing the bulkiness of substituents. The opposite diastereoselectivity observed in the aziridination of dimethyloxosulfonium methylides and dimethylsulfonium methylides was explained by the authors, who suggested a thermodynamic control for the reaction of the first reagent and a kinetic control for the second one.

In a similar way, vinyl aziridines were obtained by Stockman and coworkers by treatment of chiral *tert*-butylsulfinylimines with the ylide generated by deprotonation of *S*-allyl tetrahydrothiophenium bromide. Using these methodologies, good yields (44–82%) and satisfactory *cis/trans* selectivities, always around 20/80, could be observed. On the other hand, aziridines were always obtained in excellent diastereoselectivity (up to >95%), thus demonstrating the efficiency of *tert*-butylsulfinyl group as activating and directing group [62].

1.1 Substituted Aziridines

[Scheme: N-sulfinyl imine + allyl sulfonium bromide, tBuOLi, 44–82%, giving vinyl aziridines; cis/trans 20/80; Up to >95% de]

Dai and coworkers [63] were able to perform the enantioselective synthesis of acetylenylaziridines owing to the introduction of chirality on ylides. Propargylic sulfonium ylides, generated in situ under phase transfer conditions, reacted with N-sulfonylimines to give acetylenylaziridines in excellent yields (80–98%). In most cases complete diastereoselectivity could be achieved to give exclusively *cis* heterocycles, although with enantiomeric excesses not higher than 85%.

[Scheme: R-CH=N-Ts + camphor-derived propargyl sulfonium bromide (SiMe$_3$), CsCO$_3$, CH$_2$Cl$_2$, giving cis acetylenyl N-Ts aziridine; Yields 80–98%; Up to 85% ee]

In a similar way, Corey–Chaykovsky reaction between N-sulfonylimine, aryl-methylbromide, chiral sulfide, and a base by solid–liquid phase transfer conditions allowed Saito and coworkers [64] to synthesize enantiopure aziridines. The reaction occurs via the formation of a sulfonium ylide from the coupling of the sulfide with the halide, followed by deprotonation with the inorganic base. Excellent yields were obtained with imines bearing electron-withdrawing substituents on nitrogen atom.

[Scheme: R^1-CH=N-R^2 + R^3-CH$_2$Br + chiral p-Tol oxathiane, K$_2$CO$_3$, giving N-R^2 aziridine with R^1, R^3 substituents; Yield 53 –> 99%; trans-cis up to 79:21; trans 85–98% ee]

On the basis of the excellent results reported for oxathianes as precursors of enantiopure ylides in asymmetric synthesis, Solladiè-Cavallo and coworkers [65] developed a two-step asymmetric process for the preparation of enantiopure disubstituted N-tosyl-aziridines using a phosphazene base to generate the ylide. Although stechiometric amounts of oxathiane are required in this reaction, complete recovery and recycle of this reagent are possible. Furthermore, no unstable or hazardous reagents are involved. Under these conditions, complete conversion of the starting material into *cis/trans* mixtures of aziridines was observed. Both diastereoisomers have exceptionally high enantiomeric purities (98.7–99.9%).

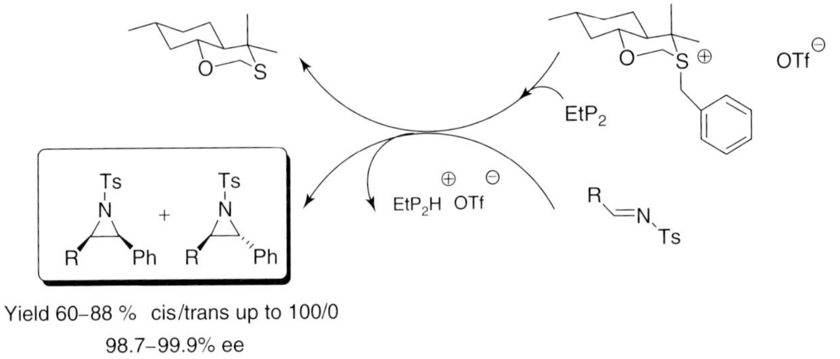

Yield 60–88 % cis/trans up to 100/0
98.7–99.9% ee

Generation of the two reagents for aziridination was also carried out by treatment of an aminosulfoxonium-substituted β, γ-unsaturated α-amino acid with a base. Fragmentation of the anion indeed affords a conjugated allyl aminosulfonium ylide and an *N-tert*-butylsulfonyl-imino ester. Recombination of these two molecules gave *cis*-vinyl aziridine in almost quantitative yield and excellent diastereoselectivity and enantioselectivity [66].

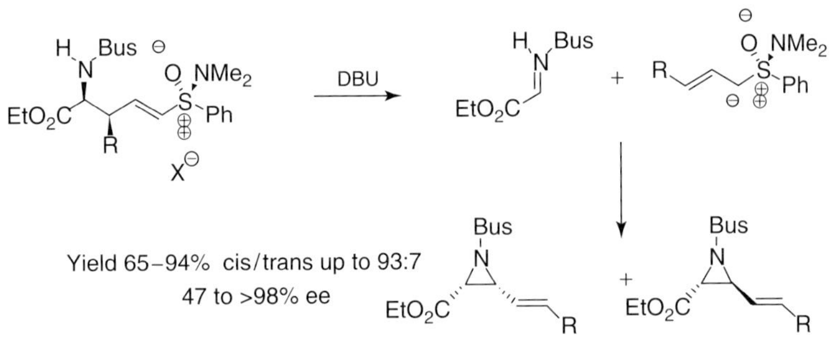

Yield 65–94% cis/trans up to 93:7
47 to >98% ee

The asymmetric aziridination involving addition of sulfur ylides to imines normally requires stoichiometric amounts of enantiopure reagents. The first catalytic asymmetric application of this reaction was reported by Aggarwal and coworkers, wherein imines bearing an electron-withdrawing group on the nitrogen atom reacted with diazocompounds in the presence of chiral sulfide (20 mol%) and rhodium or copper salts (1 mol) [67].

The reaction proceeds following the catalytic cycle as reported in the figure below, affording optically active aziridines in excellent yields.

A complete study on the relevant factors governing stereocontrol allowed to establish that the origin of diastereoselectivity lies for semistabilized ylides (benzylic) in the nature of transition states leading to betaines, while for stabilized ylides (ester/amide) in the nature of the transition states leading to ring closure [68]. On the other hand, the enantioselectivity is always very high and may be attributed to both steric and electronic factors in ylide preferred conformation (trimethylsi-lylethanesulfonyl (SES)). It is noteworthy that the parallel reaction of diazocompounds with imines is very limited and does not significatively affect yield and stereocontrol. Although higher yields could be obtained using stoichiometric amount of chiral sulfide, reduction to catalytic amount did not result in lower stereoselectivity. To overcome potential problems due to hazardousness of large-scale reactions, the diazo compound was generated in situ and a new class of sulfides, compatible with reaction conditions, was developed [69].

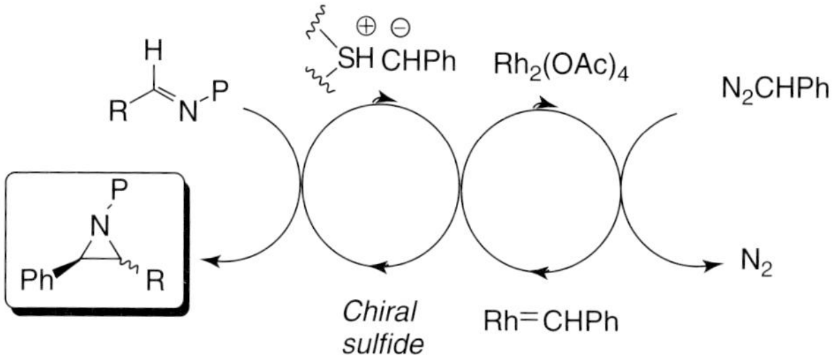

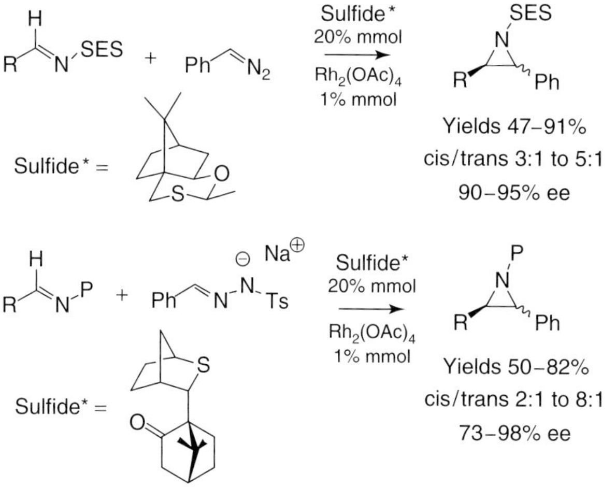

Yields 47–91%
cis/trans 3:1 to 5:1
90–95% ee

Yields 50–82%
cis/trans 2:1 to 8:1
73–98% ee

1.1.4
Ring Transformation Methods

1.1.4.1 Aziridines from Epoxides

The transformation of oxiranes to aziridines through a Staudinger-type reaction [70] represents a useful and well-known method for synthesis of these nitrogen-containing heterocycles.

The reaction occurs via the regioselective ring opening of the oxirane moiety by means of an azide followed by closure to aziridine by treatment with triphenylphosphine. Overall, both carbons of the initial epoxide are inverted.

1 Asymmetric Synthesis of Three- and Four-Membered Ring Heterocycles

The absolute stereocontrol of this methodology suggests that starting from enantiopure oxiranes, enantiopure aziridines can be obtained. In particular, epoxides deriving from allylic alcohols, which are readily available in optically active form employing Sharpless epoxidation technique, can be considered excellent starting materials. The initial introduction of chirality may be guided by simple variation of the starting allylic alcohol (Z or E) and tartrate (D or L) geometries. The excellent regio- and stereospecificity of the following Staudinger reaction allows to obtain only one enantiomer of the aziridine by choosing the proper precursors. This methodology has been successfully applied to the preparation of aziridine-2-carboxylates [71], which represent an interesting class of compounds for their potential application as mimetics and precursors of both α- and β-amino acids. In a similar way, the same authors applied this methodology to the preparation of enantiopure 2,3-dicarboxylic acid, the only example of naturally occurring aziridine-carboxylic acid, isolated as metabolite of Streptomyces MD398-A1 [72].

This methodology has been successfully applied to the preparation of building blocks for the synthesis of bioactive derivatives as carbapenems [73] or lipooxygenase pathway intermediates [74].

Enantiopure azides, useful starting material for the preparation of bicyclic analog of aziridine-2-carboxylates, have been obtained using readily available carbohydrates as a source of chirality. Thus, ring opening of 2,3-sulfite-furanoside by an azide group, followed by tosylation of the resulting hydroxyl moiety, gave a reactive substrate for a Staudinger-type reaction, leading to carbohydrate-fused aziridines [75].

In general, the presence of an azido moiety vicinal to a good oxygenated leaving group is a sufficient requirement for the synthesis of aziridine rings via Staudinger-type mechanism. Recently, the preparation of polyfunctionalized azetidin-2-ones, bearing an aziridine ring and an hydroxyl moiety on the side chain, has been reported via aza-Payne displacement induced by triethylphosphine [76].

The direct conversion of chiral epoxides to aziridines can be also performed using a cyclic guanidine derivative as nitrogen source. The reaction involves the formation of a spiro intermediate, which undergoes acid catalysis fragmentation to aziridine and urea. Application of this procedure to (R)-styrene oxide gave (S)-aziridine in 41% yield [77].

Following a very close mechanism, aziridine-2-esters have been obtained from enantiopure guanidine ylides and a variety of aryl aldehydes. This method afforded *trans*-aziridines as major diastereoisomers in excellent yields (up to 95%) and high enantiomeric excess (72–97%).

An efficient and practical route to enantiopure aminoalcohols starting from racemic terminal oxiranes via enantioselective ring opening with trimethylsilylazide in the presence of chromium-salen was presented by Jacobsen and coworkers. This kinetic resolution allowed the preparation of azido-alcohols with excellent enantiomeric excesses (80–98% ee) [78].

The same group explored the possibility to identify alternative nitrogen sources to overcome azide practical concerns and reported a general catalytic method for the preparation of enantiomerically enriched aziridines starting from racemic epoxides and N-Boc-2-nitrobenzenesulfonamide. The reaction, promoted by (S,S)-[(salen)Co-Ac], provided in few steps enantiopure N-nosylaziridines in yields ranging from 58 to 86% and enantiomeric excesses always higher than 99%. Moreover, the presence of the N-nosyl protecting group imparts to the heterocycle a particular reactivity toward nucleophilic addition [79].

1.1.4.2 Aziridines from Other Heterocycles

4-Isoxazolines are useful sinthons for the preparation of 2-acylaziridines through thermal rearrangement. This transformation was first reported by Baldwin and coworkers but its application to asymmetric synthetic purposes was scarcely developed. In order to accelerate this rearrangement, catalysts for N–O bond cleavage have been tested and $CO_2(CO)_8$ in anhydrous acetonitrile gave excellent results. The transformation proceeds with complete diastereoselectivity when a stereogenic center is present in the substituent on the N atom [80].

1.1.5
Racemate Resolution

Starting from racemic mixtures, mono- and disubstituted enantiomerically pure aziridines can be obtained by chemical or enzymatic resolution. Concerning chemical methods, an efficient resolution of N-alkyl-aziridine-2-carboxylates has been carried out by host–guest molecular association with optically active host compounds derived from tartaric acid [81].

Efficient methods for the kinetic resolution of aziridines have also been obtained with the use of biocatalysts. Racemic substituted aziridine-methanol, aziridine-carboxylate, and aziridine-carboxamide derivatives have been easily separated. Enzymatic hydrolysis catalyzed by *Candida Cylindracea Lipase* (CCL) [82] has been performed both on N-unsubstituted aziridine-carboxylates and on more reactive N-chloro, N-acyl, or N-sulfonyl derivatives.

In a similar way Lipase PS-C II, immobilized on porous ceramic particles, has been reported to catalyze the resolution of (2R*,3S*) and (2R*,3R*)-3-methyl-3-phenyl-2-aziridinemethanol [83]. The temperature control on alcohol acetylation by means of vinylacetate suggests that enantioselectivity of this lipase-catalyzed kinetic resolution is favored by low temperature [84].

Biotransformation of racemic 1,2-trans-N-substituted-aziridine-2-carboxamides were carried out with a standard cell concentration of *Rhodococcus rhodochrous* IFO15564, an amidase-containing commercially available bacterium. Owing to the concomitant presence of a nitrile hydratase in these bacterial strains, the biotransformation was also successfully performed on trans-N-substituted-aziridine-2-carbonitriles [85].

In a similar way, enantiopure (2R,3S)-3-aryl-aziridine-2-carboxamides were obtained from racemic 2,3-trans-aziridine-2-carbonitriles and amides under the catalysis of *Rhodococcus erythropolis* AJ270 whole cells. This highly efficient and enantioselective hydrolysis occurred under very mild conditions in aqueous phosphate buffer at pH 7.0 at 30 °C [86].

1.1.6
Asymmetric Synthesis of Azirines

1.1.6.1 The Neber Reaction

2H-Azirines have been first reported by Neber et al. in 1932 [87]. The Neber reaction possibly occurs either through an internal concerted nucleophilic displacement (route a) or via a electrocyclization of a vinylnitrene (route b), a reactive species formed by base-promoted loss of the leaving group on the nitrogen atom of oxime sulfonates and hydrazonium halides [88].

The first optically active 2H-azirine was synthesized by Neber reaction starting from the O-mesyl amidoxime derivative carrying a chiral phenylglycine (Phg) ester as a chiral auxiliary. Treatment of this derivative with base gave the 3-amino-2H-azirine in good yield and 96:4 stereoselectivity [89].

A remarkable asymmetric synthesis of azirine 2-carboxylates has been performed with a stoichiometric amount of dihydroquinidine or quinine as chiral tertiary base. The enantiomeric excess obtained ranged between 44 and 82%. Good results were also obtained when a catalytic amount (10 mol%) of quinidine was used. The hydroxy group of the base proved to be fundamental for a good stereoselectivity. Indeed, other chiral tertiary bases deprived of such hydroxy group, sparteine, brucine, and strychnine, did not provide any optically active azirine [90]. Later, this strategy has also been applied to the first synthesis of enantiomerically enriched 2-phosphinyl-2H-azirines [91].

Finally, optically active 2H-azirines substituted in the 3-position with a phosphine oxide group or a phosphonate in the 2-position have been obtained with moderate enantiomeric excess by Neber reaction, starting from easily available oximes, and using chiral polymer-bound amines [92].

1.1.6.2 Thermal or Photochemical Treatment of Vinyl Azides

The thermal and/or photochemical treatment of vinyl azides has become a general method for the synthesis of 2H-azirines. In a similar way as for the Neber reaction, 4π-electron vinylnitrenes are thought to be the intermediates, which would then undergo electrocyclization to 2H-azirines.

Optically active 3-amino 2H-azirines can be obtained starting from mono- or disubstituted thioamides with a chiral substituent at the amino group, by treatment with phosgene/triethylamine and sodium azide [93]. This reaction is based on a previously reported synthetic protocol that likely proceeds through α-chloro enamine and vinyl azide intermediates, which are not isolated [94].

The highly toxic phosgene can be substituted by diphenyl phosphorochloridate (DPPCl), (route a) [95]. Further, diphenyl phosphorazidate (DPPA) has been used as an alternative azide source allowing to obtain the azirines in a single step with very good yields (route b) [96].

1.1.6.3 Elimination from Aziridines

Aziridines carrying a leaving group at the nitrogen (N-chloro, N-sulfonyl, and N-acyl groups) are prone to elimination when treated with a base, giving 2H-azirines. This strategy has also been used for the asymmetric synthesis of azirine-carboxylates by the elimination of N-haloaziridines [97].

An alternative approach was based on the elimination of the SiMe₃ and the N-quinazolinone substituents from a chiral aziridine promoted by cesium fluoride. The resulting optically active azirines were not isolated, but directly treated with nucleophiles to yield aziridines in high enantiomeric excess.

On the other hand, the treatment of chiral N-sulfinylaziridines with TMSCl followed by LDA gave 2H-azirine-2-carboxylates under complete regioselectivity. This procedure has been applied to the first asymmetric synthesis of the marine cytotoxic antibiotic (R)-$(-)$-dysidazirine and its (S)-$(+)$ epimer [98].

Most methods developed for the preparation of azirines cannot be utilized for the asymmetric synthesis of 2H-azirine-3-carboxylates. On the contrary, the dehydrochlorination of methyl 2-chloroaziridine 2-carboxylates provided the first examples of enantiopure 2-substituted 2H-azirine 3-carboxylates [99].

1.1.6.4 Resolution of Racemic Azirines

Enantiomerically, pure 2H-azirines have been recently obtained by enzymatic methods. Thus, the kinetic resolution of the racemic 2H-azirinemethanol with Amano lipase at low temperature gave optically pure (S)-(1)-phenyl-2H-azirine-2-methanol and the (R)-(2)-acetate derivative [100].

1.1.6.5 Oxidation of Aziridines

One of the first asymmetric syntheses of 2H-azirine-2-carboxylates described in the literature is the Swern oxidation of 3-alkylaziridine-2-carboxylates to the corresponding 2H-azirines. The oxidation of either the (Z) or the (E) isomers with $COCl_2$/DMSO, followed by NEt_3, proceeded with complete regioselectivity, in good yields and with retention of configuration of the surviving stereogenic center [101].

The Swern oxidation has been later utilized for the efficient synthesis of (+)-2H-azirine 3-phosphonate. This compound represents a new kind of chiral iminodienophile that on reaction with dienes such as trans-piperylene affords bicyclic aziridine adducts [102].

1.2 Substituted Monocyclic Azetidines and Carbocyclic-Fused systems

1.2.1 Generalities

Azetidines are four-membered nitrogen-containing analogs of cyclobutane whose nonplanar cyclic structure has been elucidated by electron diffraction and X-ray christallographic studies.

Unsaturated derivatives of azetidine are also known as *1-azetines, 2-azetines, and azetes*. Considerable attention has been paid in particular to the well-known amide derivatives azetidin-2-ones (β-lactams) that constitute systems of central importance due to their antibacterial properties. They have been the subject of many exhaustive reviews and books of bioorganic and medicinal chemistry owing to the widespread interest shown by scientists [103]. For this reason, in this chapter, their asymmetric synthesis are not treated. On the other hand, some azetidinones have been reported as starting materials for the preparation of enantiopure azetidines.

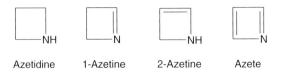

1.2 Substituted Monocyclic Azetidines and Carbocyclic-Fused systems | 37

In comparison with strained highly reactive three-membered aziridines, the four-membered rings are more stable, unreactive toward reduction and susceptible of ring cleavage only at high temperature.

Nevertheless, azetidines are unstable toward mineral acids and ring cleavage by nucleophiles may be performed on protonated rings or in the presence of Lewis acid activation.

Azetidines are typical cyclic amines, appreciably more basic than both smaller and larger rings, showing in aqueous solution a $pK_a = 11.29$.

Naturally occurring azetidine derivatives are rare, and only in 1991 biologically active sphingosine-like compounds from marine origins, penaresidin A and B, were isolated as mixture of isomers. Tested as the mixture, they induced activation of myofibrils from rabbit skeletal muscle elevating the ATPase activity. After few years, a related compound, penazetidine A, was isolated from Indopacific marine sponge *Penares sollasi*, which possesses potent protein kinase C inhibitory activity.

The isolation and characterization of the polyoxin group nucleosides have been reported by Isono and coworkers [104]. Nucleoside polyoxins A, F, H, and K represent a class of antifungal antibiotics. An unusual common feature is the presence of an unsaturated azetidine-containing amino acid peptidically linked to polyoxin C.

Finally, azetidine 2-carboxylic acids and 2-phosphonic acids have been recently proposed as conformationally constrained analogs of α–amino acids in peptide chemistry [105]. Moreover, chiral C_2-symmetric 2,4-disubstituted azetidine derivatives showed excellent catalytic ability in the asymmetric addition of organozinc to carbonyl compounds.

1.2.2
Cyclization Methods: Introduction

General synthetic methods for the preparation of azetidine ring are based on intramolecular displacement of a leaving group on carbon by a γ-amino function. Aminoalcohols and aminohalo-derivatives are among the most important classes of starting materials for heterocycle formation. Halo-alkyloxiranes have also been converted to 3-hydroxy-azetidines via epoxide ring opening by an amine followed by intramolecular nucleophilic displacement.

In a similar way, reaction of an amine or a sulfonamide with a 1,3-dihalogeno derivative results in the dialkylation of the nitrogen atom, providing a useful method for the preparation of NH, N-alkyl, or N-tosyl azetidines.

A general enantioselective synthesis of this class of four-membered rings is still lacking and the stereoselective methodologies presented so far suffer from indirect and lengthy procedures such as reduction of enatiopure β-lactams, bis-alkylation of 1,3-sulfonates with primary amines, or intramolecular N-alkylation involving 1,3-amino alcohols. Anyway, the transformation of these methodologies into asymmetric procedures has been performed mainly by cyclization of optically active precursors or by resolution of racemic azetidine mixtures. In the following sections, some selected examples are reported.

1.2.2.1 Cyclization methods: Enantiopure Azetidines via Formation of C–N Bond

The most important and easy synthetic pathway to azetidines involves the ring closure of aminoalcohols induced by transformation of the hydroxyl moiety into a good leaving group. Many optically active amino alcohols are commercially available; nevertheless, they can also be easily obtained by asymmetric reduction of β-aminoketones or β-amino acids [106]. This approach has also been applied to the synthesis of bioactive alkaloids core and fused-azetidine rings present in bridged nucleosides [107].

When the reacting hydroxy derivative was obtained by reduction of hydroxyaspartate [108], the presence of two possible leaving groups generated a competition between three-membered and four-membered rings. A strong effect of the starting aminoalcohol stereochemistry on the regioselectivity of the process was demonstrated.

On polyfunctionalized amino alcohols, precursors of sphingosine-derived alkaloids named penaresidins, crucial cyclization has been induced under mild Mitsunobu conditions to yield enantiopure azetidines [109]. Under the same conditions, enantiopure ethynylazetidines were obtained in high yields from 2-ethynyl-1,3-aminoalcohols [110].

Enantiopure 1,3-diols, obtained by hydrogenation of 1,3-diketones in the presence of chiral ligands, have been successfully used as 2,4-disubstituted azetidine precursors. Treatment with methanesulfonyl chloride followed by reaction with an excess of benzylamine afforded N-benzyl heterocycles in yields ranging form 60 to 85% [111].

Iodomethylation of enantiopure α-(dibenzyl)aminoaldehydes by means of samarium/diiodomethane under mild conditions gave optically active aminoiodohydrins, as precursors of azetidinium tetrafluroborate salts. These salts are versatile building blocks that can be transformed into aminoepoxides, 1,3-oxazolidines, or turned into N-benzyl-hydroxyazetidines by simple hydrogenolysis [112].

Ring closure of aminoallenes has attracted much attention in the development of stereoselective processes to five- or six-membered nitrogen-containing heterocycles. In a similar way, allenes bearing shorter carbon chain may lead to small size rings. Treatment of β-aminoallene and iodobenzene in the presence of palladium afforded exclusively 2,4-cis-azetidines in excellent yield [113].

1.2.2.2 Cyclization Methods: Enantiopure Azetidines via Formation of C–C Bond

Considerable attention has been paid to the synthesis of enantiopure azetidines bearing a nitrile or a phosphonic acid linked to the α position of the ring, owing to the potential application of these molecules as precursors or mimics of cyclic amino acids. To this purpose, Couty and coworkers [114] developed an easy three-step methodology starting from readily available β-amino alcohols.

N-cyanomethylation or N-phosphomethylation of the starting material was followed by substitution of the alcoholic moiety by thionyl chloride. Stereoselective 4-exo-tet ring closure through intramolecular alkylation of the methylene group gave enantiopure azetidines.

1.2 Substituted Monocyclic Azetidines and Carbocyclic-Fused systems

2-Cyano azetidines are versatile building blocks that can be easily transformed into functionalized heterocycles. In fact, treatment of cyano azetidines with phenyllithium cleanly afforded 2-acyl azetidines. Unfortunately, under these conditions, complete epimerization at C2 could not be avoided. On the other hand, hydrolysis of the cyano derivatives into carboxylic acids required harsh conditions, and prolonged heating in concentrated acid was necessary to completely hydrolyze the intermediate amide. Although drastic conditions were applied, neither ring opening nor epimerization was observed. These derivatives were successfully introduced into peptidic sequences as constrained mimetics of natural amino acids. Finally, reduction of the nitrile to alcohol followed by mesylation allowed the expansion of enantiopure azetidines to 3-mesyloxy pyrrolidines.

42 | *1 Asymmetric Synthesis of Three- and Four-Membered Ring Heterocycles*

Starting from the same N-cyanomethylated intermediates, the same authors reported the preparation of a new class of functionalized heterocycles via Wittig olefination of a transient amino-aldehyde followed by intramolecular Michael addition of the deprotonated methylene position. Unfortunately, low diastereoselectivity could be observed because of the base-catalyzed equilibration between stereoisomers [115].

1.2.3
Azetidines by Resolution of Racemates

Azetidines can be obtained in enantiomerically pure form through enzymatic or chemical resolution of racemic mixtures.

Starting from the Baldwin's adaptation of Cromwell general method of preparation of azetidine from 1,3-dihalogeno compounds, racemic azetidine-2-carboxylates were obtained in 96% by reaction with benzhydrylamine under microwave irradiation in CH_3CN. The resolution [116] was conveniently carried out by using L-tyrosine hydrazide as a resolving agent. Enantiomerically pure azetidines have been converted into (R)- or (S)-oxazaborolines, useful for the enantioselective reduction of prochiral ketones.

(S)-1-Phenylethylamine has been used as a chiral auxiliary as well as a nitrogen atom donor in the synthesis of an enantiomeric pair of azetidine-2,4-dicarboxylic acids from racemic dibromoderivatives; the absolute configuration of one of which has been assigned on the basis of the X-ray structure and the known absolute configuration of the (S)-1-phenylethylamine moiety [117]. These C_2-symmetric disubstituted heterocycles have been successfully exploited as rigid core for the preparation of chiral ligands in the asymmetric addition of diethylzinc to aldehydes.

1.2 Substituted Monocyclic Azetidines and Carbocyclic-Fused systems | 43

The application of hydroxy-azetidines as chiral ligands for zinc-catalyzed enantioselective additions was also previously reported by Martens and coworkers. Starting form (S)-azetidinecarboxylic acid, a constituent of the natural mugineic acid and one of the few commercially available azetidines, the chiral catalyst was prepared by enantioselective catalytic reduction of ketone moiety in the presence of oxazaborolidines [118].

Resolution of racemic *trans*-azetidine-2,4-dicarboxylic acids, synthesized following the same procedure, was achieved by transesterification of N-substituted dimethylesters with (-)-8-phenylmenthol and chromatographic separation of the resulting diastereoisomers [119].

Starting from dicarboxylic derivatives, *cis*- and *trans*-dihydroxy-substituted heterocycles were obtained by reduction with LiAlH$_4$, followed by treatment with benzylamine. The dihydroxy-meso compound was desymmetrized and transformed into monoacetate by the immobilized mammalian lipase from *Porcine pancreas* (S-PPL). The best results were obtained when diisopropylether was used as co-solvent in the presence of vinyl acetate. Optimized procedure allowed to obtain enantiomeric excess higher than 98% by stopping the reaction at conversion around 55%. Longer reaction times showed the formation of a significant amount of *meso*-diacetate derivative. Using the same enzyme, the trans isomer was resolved by a double kinetic resolution, stopping the reaction at moderate degree of conversion. In this case, the diacetate was isolated by the higher enantiomeric excess, while the starting dihydroxyderivative was isolated with 94.5% ee after recrystallization [120].

Enzymatic resolution of N-alkyl-azetidine-2-carboxylates could be accomplished via transacylation of ammonia catalyzed by *Candida Antarctica* lipase. Treatment of racemic azetidine-esters with an alcoholic saturated solution of ammonia afforded enantiopure unreacted (R)-esters and highly enriched (S) amides [121].

1.2.4
Azetidines by Ring Transformation

Transformation of heterocyclic precursors represents a further possibility for the synthesis of this class of compounds. One of the most important features is the ring expansion of activated aziridines.

Azetidine derivatives carrying a carbonyl group on the ring backbone occupy a special place in heterocyclic chemistry. Besides the well-known natural and synthetic azetidin-2-one derivatives, whose asymmetric synthesis has been extensively reviewed in the past, much less attention has been paid to azetidin-3-ones. Recently, a useful protocol for the synthesis of these heterocycles has been reported by De Kimpe and coworkers [122], starting from readily accessible N-alkylidene-tribromopropylamines. Anyway, both classes of compounds represent starting materials for the preparation of azetidine derivatives. Optimization of the conditions for chiral nonracemic azetidinones reduction with metal hydrides allowed to identify DIBAL-H or AlH$_2$Cl as the reagents of choice [123].

Natural-cis and unnatural-trans polyoximic acids have been synthesized starting from D-serine through L-3-azetidinone-2-hydroxymethyl chiron using a rhodium-mediated diazoketene insertion reaction. By choosing the proper reagents, the following Horner-Wadsworth-Emmons and Wittig reactions were suited to exclusively obtain the cis or the trans isomer [124].

References

1. For reviews on aziridine synthesis see: (a) Yudin, A. K. (ed) (2006) *Aziridines and Epoxides in Organic Synthesis*, Wiley-VCH Verlag GmbH, Weinheim; (b) Osborn, H. M. I. and Sweeney, J. (1997) *Tetrahedron: Asymmetry*, **8**, 1693–715; (c) Tanner, D. (1994) *Angew. Chem. Int. Ed. Engl.*, **33**, 599–619; (d) Atkinson, R. S. (1999) *Tetrahedron*, **55**, 1519–59; (e) Osborn, H. M. I. and Sweeney, J. (1997) *Tetrahedron: Asymmetry*, **8**, 1693–715; (f) Kemp, J. G. (1991) *Comprehensive Organic Synthesis* (eds B. M. Trost and I. Fleming), Pergamon, Oxford, Vol. **7**, Chapter 3.5; (g) Watson, I. D. G., Yu, L. and Yudin, A. K. (2006) *Acc. Chem. Res.*, **39**, 194–206; (h) Singh, G. S., D'hooghe, M. and De Kimpe, N. (2007) *Chem. Rev.*, **107**, 2080–135.

2. Katritzky, A., Ramsden, C., Scriven E. and Taylor R. (eds) (2008) *Comprehensive Organic Chemistry*, Vol. **1**, Elsevier.

3 Hodgkinson, T. J. and Shipman, M. (2001) *Tetrahedron*, **57**, 4467.
4 Kasai, M. and Kono, M. (1992) *Synlett*, 778.
5 Schirmeister, T. (1999) *Biopolymers*, **51**, 87.
6 Stapley, E. D., Hendlin, D., Jackson, M., Miller, A. K., Hernandez, S. and Mata, J. M. (1971) *J. Antibiot.*, **24**, 42–47.
7 Molinski, T. F. and Ireland, C. M. (1988) *J. Org. Chem.*, **53**, 2103–5.
8 Gabriel, S. (1888) *Chem. Ber.*, **21**, 1049.
9 For reviews on the reactivity of aziridines see: (a) Cardillo, G., Gentilucci, L. and Tolomelli, A. (2003) *Aldrichimica Acta*, **36**, 39–50; (b) Lee, W. K. and Ha, H. J. (2003) *Aldrichimica Acta*, **36**, 57–63; (c) McCoull, W. and Davis, F. A. (2000) *Synthesis*, **10**, 1347–65; (d) Kulkarni, Y. S. (1999) *Aldrichimica Acta*, **32**, 18–27; (e) Padwa, A. (1991) *Comprehensive Organic Synthesis*, (eds B. M. Trost and I. Fleming), Pergamon, Oxford, Vol. **4**, Chapter 4.9; (f) Righi, G. and Bonini, C. (2000) *Targets in Heter. Syst.*, **4**, 139–65.
10 Berry, M. B. and Craig, D. (1992) *Synlett*, 41–44.
11 Kim, B. M., Bae, S. J., So, S. M., Yoo, H. T., Chang, S. K., Lee, J. H. and Kang, J. (2001) *Org. Lett.*, **3**, 2349–51.
12 (a) Osborn, H. M. I., Cantrill, A. A., Sweeney, J. B. and Howson, W. (1994) *Tetrahedron*, **35**, 3159–62; (b) Osborn, H. M. I. and Sweeney, J. B. (1994) *Synlett*, 145–47.
13 Ho, M., Chung, J. K. K. and Tang, N. (1993) *Tetrahedron Lett.*, **34**, 6513–16.
14 Shaw, K. J., Luly, J. R. and Rapoport, H. (1985) *J. Org. Chem.*, **50**, 4515–23.
15 (a) Kuyl-Yeheskiely, E., Lodder, M., van der Marel, G. A. and van Boom, J. H. (1992) *Tetrahedron Lett.*, **33**, 3013–16; (b) Baldwin, J. E., Farthing, C. N., Russell, A. T., Schonfield, C. J. and Spivey, A. C. (1996) *Tetrahedron Lett.*, **37**, 3761–64.
16 Palomo, C., Aizpurua, J. M., Balentova, E., Jimenez, A., Oyarbide, J., Fratila, R. M. and Miranda, J. I. (2007) *Org. Lett.*, **9**, 101–4.
17 (a) Nagel, D. L., Woller, P. B. and Cromwell, N. H. (1971) *J. Org. Chem.*, **36**, 3911–17; (b) Tarburton, P., Woller, P. B., Badger, R. C., Doomes, E. and Cromwell, N. H. (1977) *J. Heterocycl. Chem.*, **14**, 459–64.
18 Garner, P., Dogan, O. and Pillai, S. (1994) *Tetrahedron Lett.*, **35**, 1653.
19 Cardillo, G., Gentilucci, L., Tomasini, C. and Visa Castejon-Bordas, M. P. (1996) *Tetrahedron: Asymmetry*, **3**, 755–62.
20 Cardillo, G., Casolari, S., Gentilucci, L. and Tomasini, C. (1996) *Angew. Chem. Int. Ed. Engl.*, **35**, 1848–49.
21 Sugihara, H., Daikai, K., Lin, X. L., Furuno, H. and Inanaga, J. (2002) *Tetrahedron Lett.*, **43**, 2735–39.
22 Bew, S. P., Hughes, D. L., Savic, V., Soapi, K. M. and Wilson, M. A. (2006) *Chem. Commun.*, 3513–15.
23 Cardillo, G., Gentilucci, L., Ratera Bastardas, I. and Tolomelli, A. (1998) *Tetrahedron*, **54**, 8217–22.
24 (a) Fukurawa, N., Yoshimura, T., Ohtsu, M., Akasaka, T. and Oae, S. (1980) *Tetrahedron*, **36**, 73–80; (b) Fukurawa, N. and Oae, S. (1975) *Synthesis*, 30–32.
25 Cardillo, G., Fabbroni, S., Gentilucci, L., Gianotti, M., Percacciante, R. and Tomelli, A. (2002) *Tetrahedron: Asymmetry*, **13**, 1407–10.
26 (a) Sibi, M. P. and Manyem, S. (2000) *Tetrahedron*, **56**, 8033–61; (b) Almasi, D., Alonso, D. A. and Najera, C. (2007) *Tetrahedron: Asymmetry*, **18**, 299–365.
27 Chen, Y. K., Yoshida, M. and MacMillan, D. W. C. (2006) *J. Am. Chem. Soc.*, **128**, 9328–29.
28 Pettersen, D., Piana, F., Bernardi, L., Fini, F., Fochi, M., Sgarzani, V. and Ricci, A. (2007) *Tetrahedron Lett.*, **48**, 7805–8.
29 Vesely, J., Ibrahem, I., Zhao, G. L., Rios, R. and Cordova, A. (2007) *Angew. Chem. Int. Ed. Engl.*, **46**, 778–81.

30. Evans, D. A., Faul, M. M. and Bilodeau, M. T. (**1991**) *J. Org. Chem.*, **56**, 6744–46.
31. (a) Evans, D. A., Faul, M. M., Bilodeau, M. T., Anderson, B. A. and Barnes, D. M. (**1993**) *J. Am. Chem. Soc.*, **115**, 5328–29; (b) Evans, D. A., Miller, S. J., Lectka, T. and von Matt, P. (**1999**) *J. Am. Chem. Soc.*, **121**, 7559–73.
32. Taylor, S., Gullick, J., McMorn, P., Bethell, D., Bulman Page, P. C., Hancock, F. E., King, F. and Hutching, G. J. (**2001**) *J. Chem. Soc., Perkin Trans. 2*, 1714–23.
33. Sodergren, M. J., Alonso, D. A. and Andersson, P. G. (**1997**) *Tetrahedron: Asymmetry*, **8**, 3563–65.
34. Li, Z., Conser, K. R. and Jacobsen, E. N. (**1993**) *J. Am. Chem. Soc.*, **115**, 5326–27.
35. Nishikori, H. and Katsuki, T. (**1996**) *Tetrahedron Lett.*, **37**, 9245–48.
36. Kapron, J. T., Santarsiero, B. D. and Vederas, J. C. (**1993**) *J. Chem. Soc. Chem. Commun.*, 1074–76.
37. Chilmonczyk, Z., Egli, M., Behringer, C. and Dreiding, A. S. (**1989**) *Helv. Chim. Acta*, **72**, 1095–106.
38. Yang, K. S. and Chen, K. (**2001**) *J. Org. Chem.*, **66**, 1676–79.
39. Yang, K. S. and Chen, K. (**2002**) *Org. Lett.*, **4**, 1107–9.
40. Atkinson, R. S., Cogan, M. P. and Lochrie, I. S. T. (**1996**) *Tetrahedron Lett.*, **37**, 5179–82.
41. Fazio, A., Loreto, M. A., Tardella, P. A. and Tofani, D. (**2000**) *Tetrahedron*, **56**, 4515–19.
42. Fioravanti, S., Massari, D., Morreale, A., Pellacani, L. and Tardella, P. A. (**2008**) *Tetrahedron*, **64**, 3204–11.
43. (a) Davis, F. A., Zhou, P., Liang, C. H. and Reddy, R. E. (**1995**) *Tetrahedron: Asymmetry*, **6**, 1511–14; (b) Davis, F. A., Liu, H., Zhou, P., Fang, T., Reddy, G. V. and Zhang, Y. (**1999**) *J. Org. Chem.*, **64**, 7559–67; (c) Davis, F. A., Deng, J., Zhang, Y. and Haltiwanger, R. C. (**2002**) *Tetrahedron*, **58**, 7135–43.
44. (a) Davis, F. A., Wu, Y., Yan, H., McCoull, W. and Prasad, K. R. (**2003**) *J. Org. Chem.*, **68**, 2410; (b) Davis, F. A., Ramachandar, T. and Wu, Y. (**2003**) *J. Org. Chem.*, **68**, 6894–98.
45. Kattuboina, A. and Li, G. (**2008**) *Tetrahedron Lett.*, **49**, 1573–77.
46. De Vitis, L., Florio, S., Granito, C., Ronzini, L., Troisi, L., Capriati, V., Luisi, R. and Pilati, T. (**2004**) *Tetrahedron*, **60**, 1175–82.
47. Savoia, D., Alvaro, G., Di Fabio, R., Gualandi, A. and Fiorelli, C. (**2006**) *J. Org. Chem.*, **71**, 9373–81.
48. Fujisawa, T., Hayakawa, R. and Shimizu, M. (**1992**) *Tetrahedron Lett.*, **51**, 7903–6.
49. Giubellina, N., Mangelinckx, S., Tornroos, K. W. and De Kimpe, N. (**2006**) *J. Org. Chem.*, **71**, 5881–87.
50. Alickmann, D., Frohlich, R. and Wurthwein, E. U. (**2001**) *Org. Lett.*, **3**, 1527–30.
51. Takagi, R., Kimura, J., Shinohara, Y., Ohba, Y., Takezono, K., Hiraga, Y., Kojima, S. and Ohkata, K. (**1998**) *J. Chem. Soc., Perkin Trans. 1*, 689–98.
52. (a) Sweeney, J. B., Cantrill, A. A., McLaren, A. B. and Thobhani, S. (**2006**) *Tetrahedron*, **62**, 3681–93; (b) Sweeney, J. B., Cantrill, A. A., Drew, M. G. B., McLaren, A. B. and Thobhani, S. (**2006**) *Tetrahedron*, **62**, 3694–703.
53. Hanessian, S. and Cantin, L. D. (**2000**) *Tetrahedron Lett.*, **41**, 787–90.
54. Williams, A. L. and Johnston, J. N. (**2004**) *J. Am. Chem. Soc.*, **126**, 1612–13.
55. Hansen, K. B., Finney, N. S. and Jacobsen, E. N. (**1995**) *Angew. Chem. Int. Ed. Engl.*, **34**, 676–78.
56. Rasmussen, K. G. and Jørgensen, K. A. (**1997**) *J. Chem. Soc., Perkin Trans. 1*, 1287–91.
57. Karsten, J., Hazell, R. G. and Jørgensen, K. A. (**1997**) *J. Chem. Soc., Perkin Trans. 1*, 2293–97.
58. Antilla, J. C. and Wulff, D. W. (**1999**) *J. Am. Chem. Soc.*, **121**, 5099–100.
59. Antilla, J. C. and Wulff, D. W. (**2000**) *Angew. Chem. Int. Ed. Engl.*, **39**, 4518–21.
60. Garcia Ruano, Jl., Fernandez, I., del Prado Catalina, M. and Cruz, A. A. (**1996**) *Tetrahedron: Asymmetry*, **7**, 3407–14.

61 Khiar, N., Fernandez, I. and Alcudia, F. (**1994**) *Tetrahedron Lett.*, **35**, 5719.
62 Morton, D., Pearson, D., Field, R. A. and Stockman, R. A. (**2004**) *Org. Lett.*, **6**, 2377–80.
63 Li, A. H., Zhou, Y. G., Dai, L. X., Hou, X. L., Xia, L. J. and Lin, L. (**1998**) *J. Org. Chem.*, **63**, 4338–48.
64 Saito, T., Sakairi, M. and Akiba, D. (**2001**) *Tetrahedron Lett.*, **42**, 5451–54.
65 Solladiè-Cavallo, A., Roje, M., Welter, R. and Sunjic, V. (**2004**) *J. Org. Chem.*, **69**, 1409–12.
66 Iska, V. B. R., Gais, H. J., Tiwari, S. K., Babu, G. S. and Adrien, A. (**2007**) *Tetrahedron Lett.*, **48**, 7102–7.
67 Aggarwal, V. K., Thompson, A., Jones, R. V. H. and Standen, M. C. H. (**1996**) *J. Org. Chem.*, **61**, 8368–69.
68 (a) Aggarwal, V. K., Ferrara, M., O'Brien, C. J., Thompson, A., Jones, R. V. H. and Fieldhouse, R. (**2001**) *J. Chem. Soc., Perkin Trans. 1*, 1635–43; (b) Aggarwal, V. K., Charmant, J. P. H., Ciampi, C., Hornby, J. M., O'Brien, C. J., Hynd, G. and Parsons, R. (**2001**) *J. Chem. Soc., Perkin Trans. 1*, 3159–66.
69 (a) Aggarwal, V. K., Alonso, E., Fang, G., Ferrara, M., Hynd, G. and Porcelloni, M. (**2001**) *Angew. Chem. Int. Ed. Engl.*, **40**, 1433–36; (b) Aggarwal, V. K. and Vasse, J. L. (**2003**) *Org. Lett.*, **5**, 3987–90.
70 (a) Staudinger, H. and Meyer, J. (**1919**) *Helv. Chim. Acta*, **2**, 635; (b) Gololobov, Y. G., Zhmurova, I. N. and Kasukhin, L. F. (**1981**) *Tetrahedron*, **37**, 437–72.
71 Legters, J., Thijs, L. and Zwanenburg, B. (**1989**) *Tetrahedron Lett.*, **30**, 4881–84.
72 (a) Legters, J., Thijs, L. and Zwanenburg, B. (**1991**) *Tetrahedron*, **47**, 5287–94; (b) Tanner, D., Birgersson, C. and Dhaliwal, H. K. (**1990**) *Tetrahedron Lett.*, **31**, 1903–8.
73 Tanner, D. and Somfai, P. (**1988**) *Tetrahedron*, **44**, 619–24.
74 Zamboni, R. and Rokach, J. (**1983**) *Tetrahedron Lett.*, **24**, 331–34.
75 Dubois, L. and Dodd, R. H. (**1993**) *Tetrahedron*, **49**, 901–10.
76 Benfatti, F., Cardillo, G., Gentilucci, L., Perciaccante, R., Tolomelli, A. and Catapano, A. (**2006**) *J. Org. Chem.*, **71**, 9229–32.
77 Tsuchiya, Y., Kumamoto, T. and Tsutomu, I. (**2004**) *J. Org. Chem.*, **69**, 8504–5.
78 Larrow, J. F., Schaus, S. E. and Jacobsen, E. N. (**1996**) *J. Am. Chem. Soc.*, **118**, 7420–21.
79 Kim, S. K. and Jacobsen, E. N. (**2004**) *Angew. Chem. Int. Ed. Engl.*, **43**, 3952–54.
80 Ishikawa, T., Kudoh, T., Yoshida, J., Yasuhara, A., Shinobu, M. and Saito, S. (**2002**) *Org. Lett.*, **4**, 1907–10.
81 Mori, K. and Toda, F. (**1990**) *Tetrahedron: Asymmetry*, **1**, 281–82.
82 Bucciarelli, M., Forni, A., Moretti, I., Prati, F. and Torre, G. (**1993**) *J. Chem. Soc., Perkin Trans. 1*, 3041–45.
83 Sakai, T., Liu, Y., Ohta, H., Korenaga, T. and Ema, T. (**2005**) *J. Org. Chem.*, **70**, 1369–75.
84 Sakai, T. (**2004**) *Tetrahedron: Asymmetry*, **15**, 2749–56.
85 Moràn-Ramallal, R., Liz, R. and Gotor, V. (**2007**) *Org. Lett.*, **9**, 521–24.
86 Wang, J.-Y., Wang, D.-X., Pan, J., Hyuang, Z.-T.-. and Wang, M.-X. (**2007**) *J. Org. Chem.*, **72**, 9391–94.
87 (a) Neber, P. W. and Burgard, A. (**1932**) *Justus Liebigs Ann. Chem.*, **493**, 281–94; (b) Neber, P. W. and Huh, G. (**1935**) *Justus Liebigs Ann. Chem.*, **515**, 283–96.
88 Pinho e Melo, T. M. V. D. and Rocha Gonsalves, A. Md. A. (**2004**) *Curr. Org. Synth.*, **1**, 275.
89 (a) Piskunova, I. P., Eremeev, A. V., Mishnev, A. F. and Vosekalna, I. A. (**1993**) *Tetrahedron*, **49**, 4671–76; (b) Palacios, F., Ochoa de Retana, A. M., de Marigorta, E. M. and de los Santos, J. M. (**2001**) *Eur. J. Org. Chem.*, 2401–14.
90 Verstappen, M. M. H., Ariaans, G. J. A. and Zwanenburg, B. (**1996**) *J. Am. Chem. Soc.*, **118**, 8491–92.
91 Palacios, F., Ochoa de Retana, A. M., Gil, J. I. and Ezpeleta, J. M. (**2000**) *J. Org. Chem.*, **65**, 3213–17.

92 Palacios, F., Aparicio, D., de Retana, A. M. O., de los Santos, J. M., Gil, J. I. and Lopez de Munain, R. (**2003**) *Tetrahedron: Asymmetry*, **14**, 689–700.

93 (a) Bucher, C. B. and Heimgartner, H. (**1996**) *Helv. Chim. Acta*, **79**, 1903–15; (b) Bucher, C. B., Linden, A., Heimgartner, H. (**1995**) *Helv. Chim. Acta*, **78**, 935–46.

94 Henriet, M., Houtekie, M., Techy, B., Touillaux, R. and Ghosez, L. (**1980**) *Tetrahedron Lett.*, **21**, 223–26.

95 Villalgordo, J. M. and Heimgartner, H. (**1993**) *Helv. Chim. Acta*, **76**, 2830–37.

96 Villalgordo, J. M., Enderli, A., Linden, A. and Heimgartner, H. (**1995**) *Helv. Chim. Acta*, **78**, 1983–98.

97 Legters, J., Thijs, L. and Zwanenburg, B. (**1992**) *Recl. Trav. Chim. Pays-Bas*, **111**, 75–78.

98 (a) Davis, F. A., Reddy, G. V. and Liu, H. (**1995**) *J. Am. Chem. Soc.*, **117**, 3651–52; (b) Davis, F. A., Liu, H., Liang, C.-H., Reddy, G. V., Zhang, Y., Fang, T. and Titus, D. D. (**1999**) *J. Org. Chem.*, **64**, 8929–35; (c) Davis, F. A., Liu, H., Zhou, P., Fang, T., Reddy, G. V. and Zhang, Y. (**1999**) *J. Org. Chem.*, **64**, 7559–67; (d) Davis, F. A., Zhou, P. and Reddy, G. V. (**1994**) *J. Org. Chem.*, **59**, 3243–45; (e) Davis, F. A., Liang, C.-H. and Liu, H. (**1997**) *J. Org. Chem.*, **62**, 3796–97.

99 Davis, F. A. and Deng, J. (**2007**) *Org. Lett.*, **9**, 1707–10.

100 Sakai, T., Kawabata, I., Kishimoto, T., Ema, T. and Utaka, M. (**1997**) *J. Org. Chem.*, **62**, 4906–7.

101 Gentilucci, L., Grijzen, Y., Thijs, L. and Zwanenburg, B. (**1995**) *Tetrahedron Lett.*, **36**, 4665–68.

102 Davis, F. A., Wu, Y., Yan, H., Prasad, K. R. and McCoull, W. (**2002**) *Org. Lett.*, **4**, 655–58.

103 (a) Alcaide, B., Almendros, P. and Aragoncillo, C. (**2007**) *Chem. Rev.*, **107**, 4437–92; (b) Coates, C., Kabir, J. and Turos, E. (**2005**) *Sci. Synth.*, **21**, 609–46; (c) Singh, G. S. (**2003**) *Tetrahedron*, **59**, 7631–49; (d) Bateson, J. H. (**1991**) *Prog. Heterocycl. Chem.*, **3**, 1–20.

104 (a) Suzuki, S., Isono, K., Nagatsu, J., Mizutani, T., Kawashima, Y. and Mizuno, T. (**1965**) *J. Antibiot. Ser. A*, **18**, 131; (b) Isono, K., Funayama, S. and Suhadolnik, R. (**1975**) *J. Biochem.*, **14**, 2992.

105 Couty, F., Evano, G. and Rabasso, N. (**2003**) *Tetrahedron: Asymmetry*, **14**, 2407–12.

106 Barluenga, J., Fernandez-Marì, F., Viado, A. L., Aguilar, E. and Olano, B. (**1996**) *J. Org. Chem.*, **61**, 5659–62.

107 (a) Knapp, S. and Dong, Y. (**1997**) *Tetrahedron Lett.*, **38**, 3813–16; (b) Obika, S., Andoh, J., Onoda, M., Nakagawa, O., Hiroto, A., Sugimoto, T., Imanishi, T. (**2003**) *Tetrahedron Lett.*, **44**, 5267–70.

108 Fernandez-Megia, E., Montaos, M. A. and Sardina, F. J. (**2000**) *J. Org. Chem.*, **65**, 6780–83.

109 (a) Liu, D. G. and Lin, G. Q. (**1999**) *Tetrahedron Lett.*, **40**, 337–40; (b) Takikawa, H., Maeda, T. and Mori, K. (**1995**) *Tetrahedron Lett.*, **36**, 7689–92; (c) Yoda, H., Uemura, T. and Takanabe, K. (**2003**) *Tetrahedron Lett.*, **44**, 977–79.

110 Ohno, H., Hamaguchi, H. and Tanaka, T. (**2001**) *J. Org. Chem.*, **66**, 1867–75.

111 Marinetti, A., Hubert, P. and Gent, J. P. (**2000**) *Eur. J. Org. Chem.*, 1815–20.

112 Concellon, J. M., Bernad, P. L. and Perez-Andres, J. A. (**2000**) *Tetrahedron Lett.*, **41**, 1231–34.

113 Ohno, H., Anzai, M., Toda, A., Ohishi, S., Fujii, N., Tanaka, T., Takemoto, Y. and Ibuka, T. (**2001**) *J. Org. Chem.*, **66**, 4904–14.

114 (a) Agami, C., Couty, F. and Rabasso, N. (**2002**) *Tetrahedron Lett.*, **43**, 4633–36; (b) Agami, C., Couty, F. and Evano, G. (**2002**) *Tetrahedron: Asymmetry*, **13**, 297–302.

115 Carlin-Sinclair, A., Couty, F. and Rabasso, N. (**2003**) *Synlett*, 726–28.

116 (a) Rodebangh, R. M. and Cromwell, N. H. (**1969**) *J. Heterocycl. Chem.*, **6**, 993; (b) Rama Rao, A. V., Gurjar, M. K. and Kaiwar, V. (**1992**) *Tetrahedron: Asymmetry*, **3**, 859.

117 (a) Hoshino, J., Hiraoka, J., Hata, Y., Sawada, S. and Yamamoto, Y. (**1995**) *J. Chem. Soc., Perkin Trans. 1*, 693–97; (b) Shi, M. and Jiang, J. K. (**1999**) *Tetrahedron: Asymmetry*, **10**, 1673–79; (c) Wilken, J., Erny, S., Wassmann, S. and Martens, J. (**2000**) *Tetrahedron: Asymmetry*, **11**, 2143–48.

118 Behnen, W., Mehler, T. and Martens, J. (**1993**) *Tetrahedron: Asymmetry*, **4**, 1413–16.

119 Kozikowski, A. P., Tuckmantel, W., Liao, Y., Manev, H., Ikonomovic, S. and Wroblewski, J. T. (**1993**) *J. Med. Chem.*, **36**, 2706–8.

120 Guanti, G. and Riva, R. (**2001**) *Tetrahedron: Asymmetry*, **12**, 605–18.

121 (a) Starmans, W. A. J., Doppen, R. G., Thijs, L. and Zwanenburg, B. (**1998**) *Tetrahedron: Asymmetry*, **9**, 429–35; (b) Hermsen, P. J., Cremers, J. G. O., Thijs, L. and Zwanenburg, B. (**2001**) *Tetrahedron Lett.*, **42**, 4243–45.

122 (a) De Smaele, D., Dejaegher, Y., Duvey, G. and De Kimpe, N. (**2001**) *Tetrahedron Lett.*, **42**, 2373–75; (b) Salgado, A., Dejaegher, Y., Verniest, G., Boeykens, M., Gauthier, C., Lopin, C., Therani, K. A. and De Kimpe, N. (**2003**) *Tetrahedron*, **59**, 2231–39.

123 (a) Ojima, I., Zhao, M., Yamato, T., Nakahashi, K., Yamashita, M. and Abe, R. (**1991**) *J. Org. Chem.*, **56**, 5263–77; (b) Alcaide, B., Almendros, P., Aragoncillo, C. and Salgado, N. R. (**1999**) *J. Org. Chem.*, **64**, 9596–04.

124 Hanessian, S., Fu, J. M., Chiara, J. L. and Di Fabio, R. (**1993**) *Tetrahedron Lett.*, **34**, 4157–60.

2
Asymmetric Synthesis of Five-Membered Ring Heterocycles

Pei-Qiang Huang

2.1
Monocyclic Pyrrolidines and Pyrrolidinones

2.1.1
Generalities

The pyrrolidine structural unit is one of the most commonly occurring structural cores in a large number of biologically active alkaloids [1]. For example, since its first isolation in 1954 [2a], anisomycin (**1**) has become a valuable tool in molecular biology, having been used for the treatment of *Trichomonas vaginitis* and amoebic dysentery, and also as an agricultural fungicide. Recent investigation showed that anisomycin and its higher homolog 3097-B1 (**1a**) as well as 3097-B2 and 3097-C2 exhibited *in vitro* anticytotoxic activity (Figure 2.1) [2b]. The role of the naturally occurring excitatory amino acid (*R*)-α-kainic acid (**2**) in mediating synaptic responses has made it an important reagent for investigations into Alzheimer's disease, epilepsy, and other neurological disorders [3]. It has also been used as an antiworming agent to eliminate parasites from humans and animals. (+)-Lactacystin (**3**) is a potent and selective proteasome inhibitor that was isolated from *Streptomyces* sp [4]. Recently, a more potent proteasome inhibitor, (−)-salinosporamide A (**4**), was isolated from a marine actinomycete [5] and it inhibits proteasomal proteolytic activity with an IC_{50} value of 1.3 nM. Moreover, compound **4** has potent *in vitro* cytotoxicity ($LC_{50} < 10$ nM against four different cancer cell lines).

On the other hand, pyrrolidine is also a motif for designing pharmaceuticals. For example, nemonapride (YM-09151-2) (**5**) was marketed in 1991 as a potent antipsychotic drug [6]. Recently, A-315675 (**6**) was developed by scientists at Abbott laboratories as a potent influenza neuraminidase inhibitor and is a candidate for development as an anti-influenza drug [7].

In addition to their underexplored medicinal importance, functionalized pyrrolidines have found wide applications as chiral ligands [8], chiral auxiliaries [9], and organocatalysts [10] in asymmetric synthesis.

Asymmetric Synthesis of Nitrogen Heterocycles. Edited by Jacques Royer
Copyright © 2009 WILEY-VCH Verlag GmbH & Co. KGaA, Weinheim
ISBN: 978-3-527-32036-3

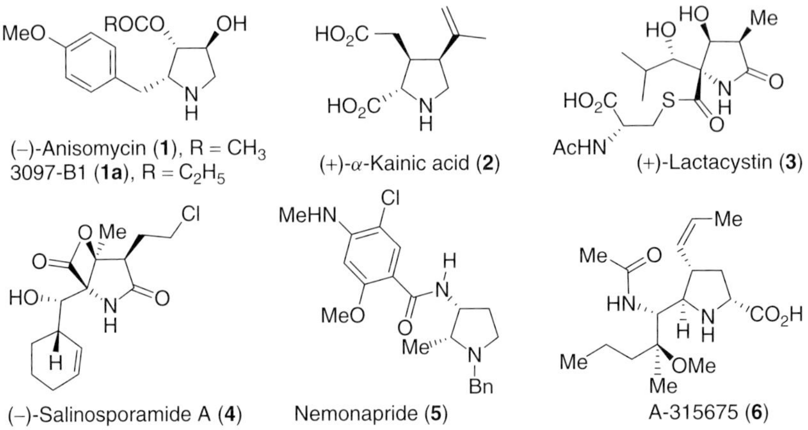

2 Asymmetric Synthesis of Five-Membered Ring Heterocycles

(−)-Anisomycin (**1**), R = CH₃
3097-B1 (**1a**), R = C₂H₅

(+)-α-Kainic acid (**2**)

(+)-Lactacystin (**3**)

(−)-Salinosporamide A (**4**)

Nemonapride (**5**)

A-315675 (**6**)

Fig. 2.1 Selected examples of bioactive pyrrolidines/pyrrolidin-2-ones.

Consequently, significant efforts have been devoted to the development of efficient routes to substituted pyrrolidines.

2.1.2
Cyclization Methods

Formation of the pyrrolidine/pyrrolidinone ring by cyclization is a widely used approach. Almost all disconnections around pyrrolidine/pyrrolidine-2-one ring are viable, including C_1/C_5–N bond, C_2–C_3 bond, and C_3–C_4 bond disconnections, as well as C_1–N/C_5–N double disconnections.

2.1.2.1 Cyclization via C_1/C_5–N Bond Formation
Many methods have been reported for the synthesis of pyrrolidines and pyrrolidin-2-ones by intramolecular amination or amidation of chiral nonracemic alicyclic amines or amides. They will be classified according to the methods used to synthesize the chiral nonracemic alicyclic precursors.

Naturally Occurring Chiron Approach Enantiomerically pure α-amino acids are cheap and easily available chirons [11]. A three-step synthesis of 5-alkyltetramic acids **7**, developed by Jouin and Castro [12], provides a flexible approach to cis-4-hydroxy-5-alkylpyrrolidin-2-ones **8** after a highly stereoselective reduction with NaBH₄, and to 2,5-disubstituted pyrrolidin-4-ols after further transformations (Scheme 2.1).

Sugars provide a rich source of chirons for the synthesis of enantiomeric pure hydroxylated pyrrolidines/pyrrolidin-2-ones. Nicotra developed a four-step method for the synthesis of 4,5-cis-substituted pyrrolidine-2-ones (**11**) and pyrrolidines (**12**) starting from 2,3,5-tri-O-benzyl-D-arabinofuranose (**9**) (Scheme 2.2) [13]. This provided a basis for the asymmetric synthesis of antibiotic anisomycin (**1**) [14a] and the antifungal agent (+)-preussin (**15**) [14b], as well as pyrrolizidine alkaloid

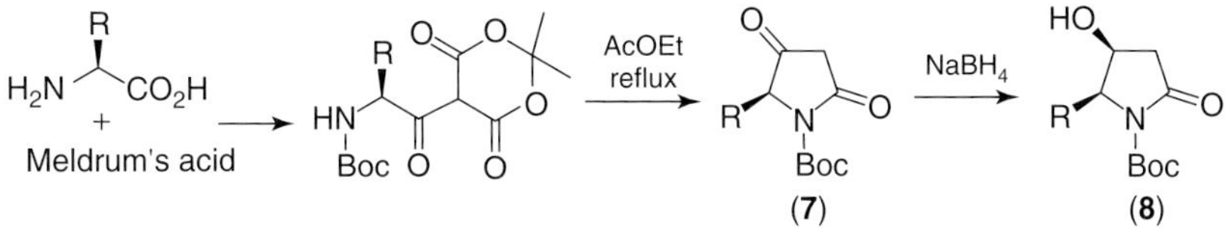

2.1 Monocyclic Pyrrolidines and Pyrrolidinones

Scheme 2.1 An α-amino acid–based approach to tetramic acids and cis-4-hydroxy-5-alkylpyrrolidin-2-ones.

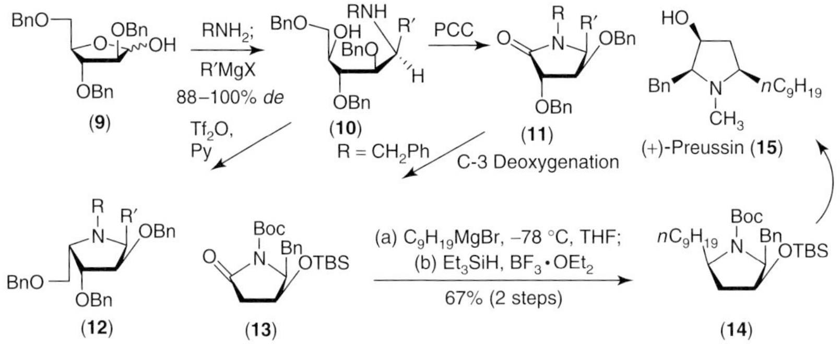

Scheme 2.2 Nicotra's approach to hydroxylated pyrrolidin-2-ones/pyrrolidines (TBS = tert-butyl-dimethylsilyl).

(+)-alexine [14c]. To introduce the C-5 side chain of (+)-preussin, Yoda et al. developed a reductive alkylation method (**13 → 14**), which allows the establishment of C_2/C_5 cis stereochemistry with high diastereoselectivity [14d,e].

Scheme 2.3 shows three approaches for the rapid transformation of sugars into pyrrolidine/1-pyrroline derivatives [15–17]. In Ganem's approach (Scheme 2.3a) [15], upon treatment of bromobenzoate **16** with Zn-NaBH$_3$CN, reductive ring opening and reductive amination occurred to give **17**, which was followed by a spontaneous intramolecular displacement of the benzoate in **17**, leading to pyrrolidine **18** in 70% yield from **16**.

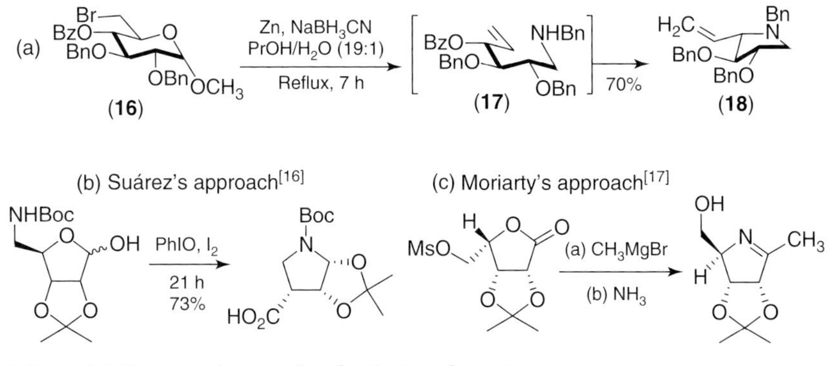

Scheme 2.3 Three rapid approaches for the transformation of sugars to pyrrolidine/1-pyrroline derivatives.

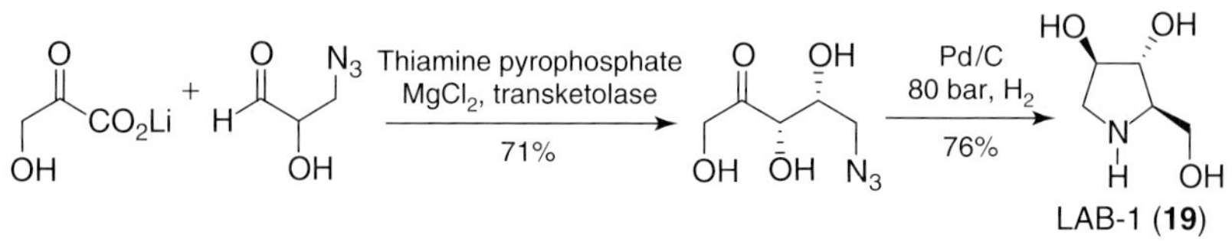

Scheme 2.4 Enzymatic synthesis of azasugar LAB-1.

Enzymatic Approach Among various routes to azasugars, an enzymatic route should be quite efficient, because such approaches are generally free of deprotection–protection procedures. Effenberger et al. developed an enzymatic synthesis of 5-azido-5-deoxy-D-xylose, which led to the synthesis of azasugar LAB-1 (**19**, Scheme 2.4) [18]. Wong et al. undertook a systematic study on chemo-enzymatic synthesis, which led to the synthesis of many polyhydroxylated pyrrolidines [19].

Catalytic Asymmetric and Chiral Auxiliary-Induced Approaches The efficient enantioselective synthesis of polyhydroxypyrrolidine **21** shown in Scheme 2.5 involves a sequential Sharpless asymmetric dihydroxylation and asymmetric epoxidation in the presence of the Mizuno's dinuclear peroxotungstate catalyst. Final treatment of **20** with 10% NH_3 in water gave azasugar **21** in 88% yield [20] via a tandem nucleophilic substitution–intramolecular ring opening of the epoxide.

A recent synthesis of (−)-codonopsinine (**22**) illustrates a new approach to substituted pyrrolidines. It involves a highly *trans*-diastereoselective intramolecular S_N1 reaction via a resonance-stabilized benzylic carbacation intermediate derived from **23**, which was synthesized via the Sharpless asymmetric dihydroxylation reaction (dr = 91:9) (Scheme 2.6) [21].

On the basis of iterative asymmetric Ti-mediated allylations, Cossy and coworkers developed an asymmetric approach to (+)-preussin (**15**). The two titanium-complex-

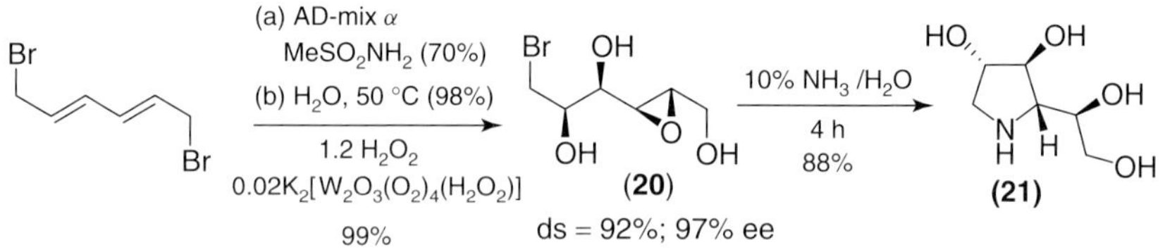

Scheme 2.5 A concise enantioselective synthesis of polyhydroxypyrrolidine.

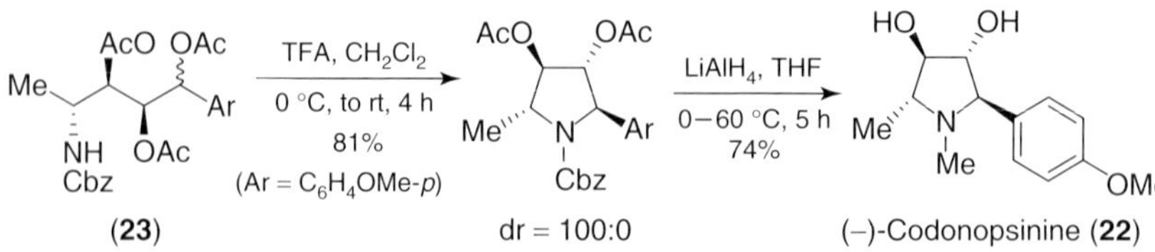

Scheme 2.6 Rao's enantioselective synthesis of polyhydroxypyrrolidine.

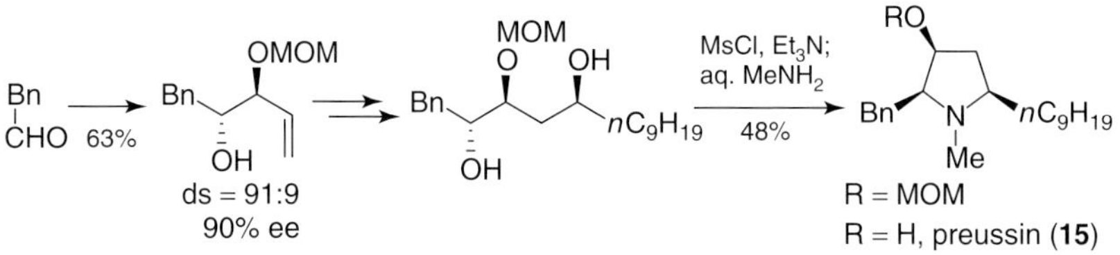

Scheme 2.7 Cossy's Ti-mediated asymmetric allylation approach to (+)-preussin.

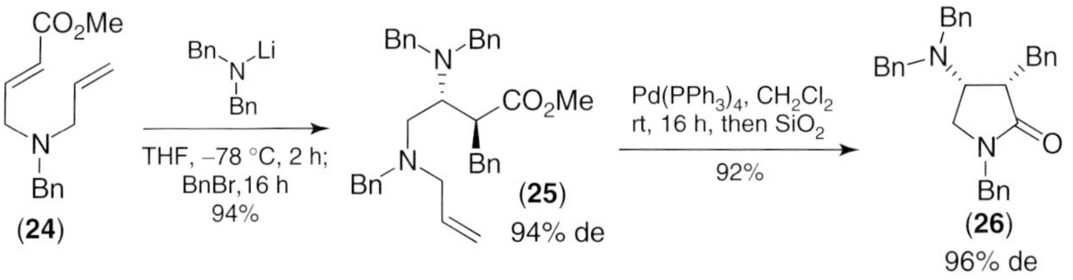

Scheme 2.8 Davies' diastereoselective approach to 3-substituted 4-aminopyrrolidines.

catalyzed asymmetric allylation reactions proceeded with 91:9 and 96:4 diastereoselectivities respectively, which established all three stereogenic centers of the target molecule (Scheme 2.7) [22].

As an application of their chiral lithium amide conjugate addition methodology (**24** → **25**), Davies et al. developed a highly diastereoselective and enantioselective approach to 3,4-*trans*- and 3,4-*cis*-substituted aminopyrrolidines (Scheme 2.8) [23].

Catalytic enantioselective and diastereoselective addition of enecarbamates with α-oxo aldehydes provides a one-step synthesis of γ-amino acid derivative **27**, which was converted subsequently into substituted 3-hydroxypyrrolidin-2-one **28** and 3-hydroxy-4-methylproline (HMP, **29**, Scheme 2.9) [24].

Recently, two catalytic enantioselective Michael addition–based approaches to β-aryl-γ-lactams **30** have been developed by Kanemasan [25] and Barnes [26] (Scheme 2.10a and b respectively), with the latter being used in the synthesis of the antidepressant rolipram. The same product can also be obtained via Ley's butane-2,3-diacetal (BDA) derivative **31**-based Michael addition (Scheme 2.10c) [27].

In Shibasaki's catalytic asymmetric total synthesis of (+)-lactacystin, the lactam intermediate **32** was made by subsequent oxidative cleavage, oxidation, and lactamization of the chiral cyclopentene derivative **33**, obtained via an enantioselective Strecker reaction of ketoimine **34** (Scheme 2.11) [28].

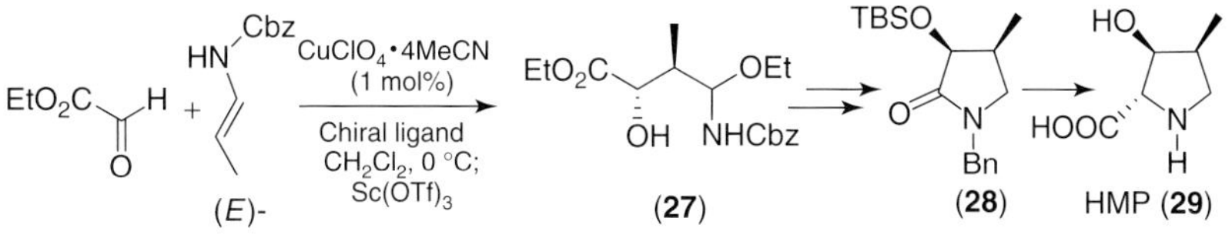

Scheme 2.9 Enantioselective Stork-type approach to substituted 3-hydroxypyrrolidin-2-one.

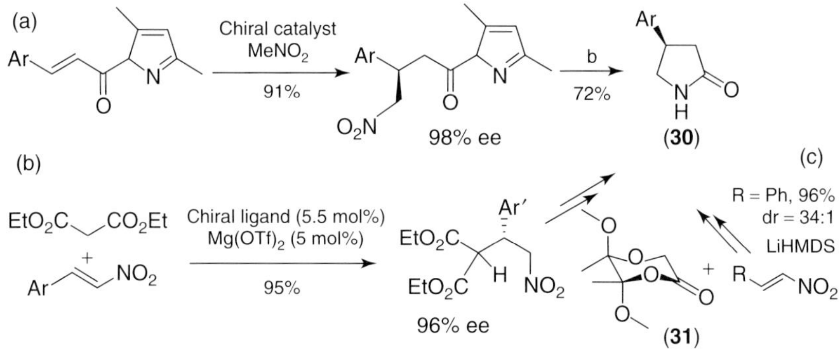

Scheme 2.10 Three enantioselective approaches to β-aryl-γ-lactams.

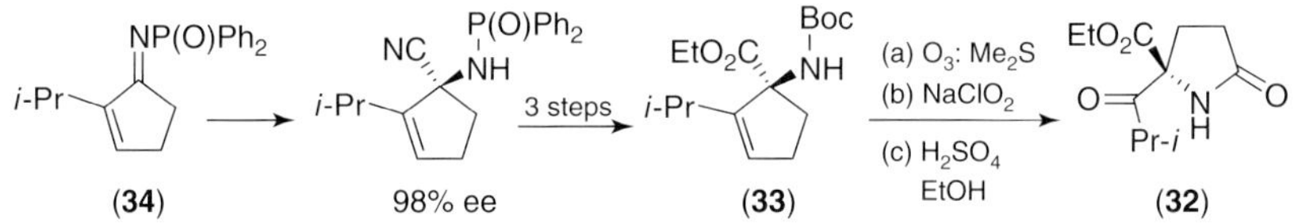

Scheme 2.11 Shibasaki's catalytic asymmetric approach to (+)-lactacystin.

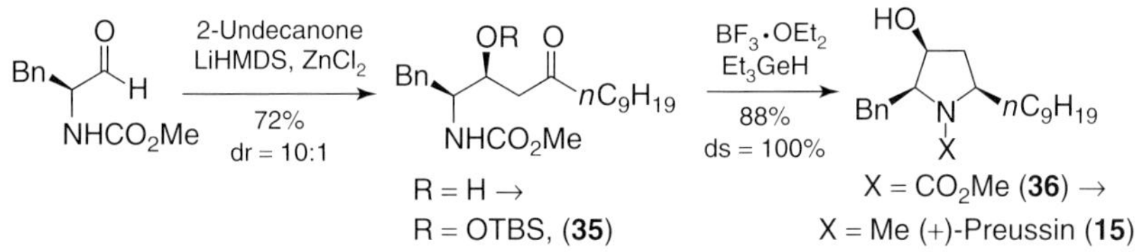

Scheme 2.12 Kitahara's intramolecular reductive amidoalkylation approach to (+)-preussin.

A combination of chelation-controlled *syn*-selective aldol reaction of α-amino aldehyde and zinc enolate of 2-undecanone, with Lewis acid-promoted intramolecular reductive amidoalkylation (**35** → **36**), establishes a concise and highly diastereoselective approach to (+)-preussin (**15**) (Scheme 2.12) [29].

Synthesis of substituted pyrrolidines by both substrate [30] and chiral auxiliary [31]-induced asymmetric intramolecular aza-Michael addition was reported. As shown in Scheme 2.13, reaction of hemiaminal **37** with the carbanion generated from trimethyl phosphonoacetate led, via a tandem Wittig-aza-Michael reaction, to the formation of all-*cis*-trisubstituted pyrrolidine **38** in 5.6:1 diastereoselectivity [30]. Remarkably, starting from methylated hemiaminal **40**, all-*cis*-tetrasubstituted pyrrolidine **41** was formed as a single diastereomer in 75% yield.

Metal and halogen-mediated intramolecular cyclizations of amido/amino-olefins/ allenes/alkynes constitute another versatile strategy for pyrrolidine formation. Pd(0) complex, Hg(II) salt, and Ag(I) salt are the most frequently used metal catalysts.

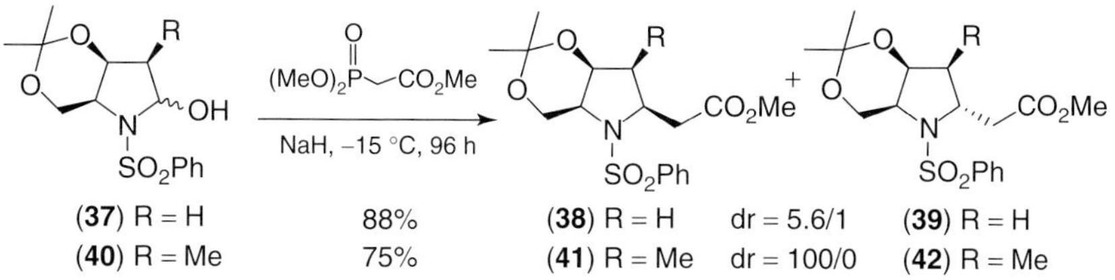

2.1 Monocyclic Pyrrolidines and Pyrrolidinones

Scheme 2.13 Synthesis of polysubstituted pyrrolidines by intramolecular aza-Michael addition.

(37) R = H	88%	(38) R = H dr = 5.6/1 (39) R = H
(40) R = Me	75%	(41) R = Me dr = 100/0 (42) R = Me

The seminal work of Harding established that pyrrolidine ring systems can be formed via intramolecular amidomercuration, and 2,5-*trans*-disubstituted pyrrolidines can be obtained under kinetically controlled conditions, while thermodynamically controlled conditions allow access to 2,5-*cis*-disubstituted pyrrolidines [32]. Starting from this reaction, Momose and coworkers accomplished the asymmetric synthesis of several alkaloids [33]. The mercury(II)-mediated amidomercuration-oxidative demercuration reaction of the δ-alkenylamide **43**, prepared from achiral 4-(*p*-methoxy)phenoxylated α,β-unsaturated ester via a regio- and enantioselective Sharpless aminohydroxylation, was shown to be highly *trans*-diastereoselective. On the basis of these reactions, a six-step enantioselective synthesis of polyhydroxylated pyrrolidines **44** was developed (Scheme 2.14) [34].

The concise five-step synthesis of the antifungal agent (+)-preussin (**15**) is an excellent demonstration of the power of Hg(II)-mediated ring closure of ynone **45** (Scheme 2.15) [35].

The advantages of Pd-catalyzed amination of alkenes/alkynes reside on the usefulness of the Pd-intermediate formed, which in turn can be used for carboxylation or arylation. This was demonstrated by the concise syntheses of (+)-preussin, 3-*epi*-preussin, and its analogs shown partially in Scheme 2.16a [36]. In addition,

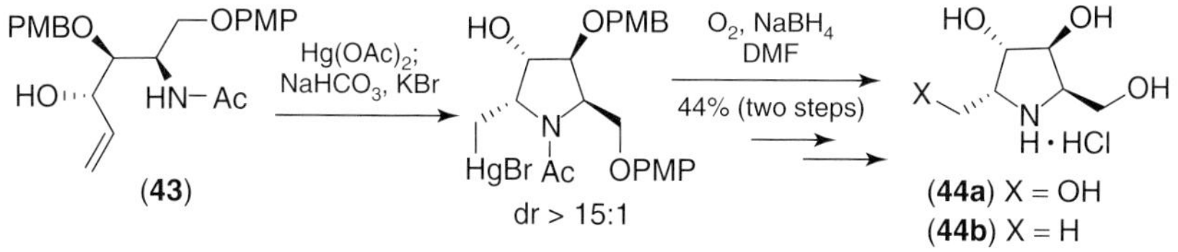

Scheme 2.14 Han's asymmetric synthesis of azasugars by mercury(II)-mediated amidomercuration.

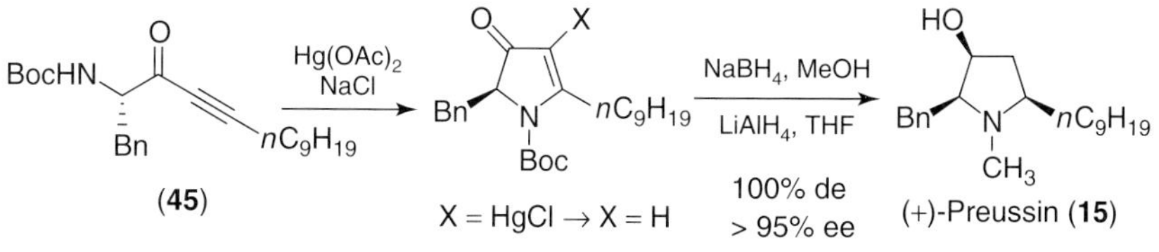

Scheme 2.15 Hecht's Hg(II)-mediated ring closure approach to (+)-preussin.

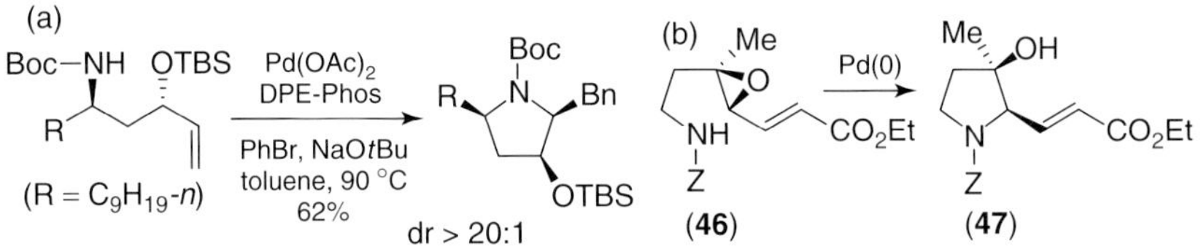

Scheme 2.16 Wolfe's Pd-catalyzed carboamination approach to preussin and Pd(0)-catalyzed amidation of allylic epoxide.

Pd(0)-catalyzed ring opening of allylic epoxide **46** yielded **47** and its diastereomer in 90:10 ratio (Scheme 2.16b) [37].

The intramolecular iodosulfonamidation of **48** was achieved by treating with Na_2CO_3 and I_2 under biphasic conditions, which afforded 2,5-*trans*-pyrrolidine derivative **50** as the major diastereomer (Scheme 2.17) [30]. This approach complements that by intramolecular aza-Michael reaction (Scheme 2.13) [30] in terms of diastereoselection.

Taking advantages of the 5-*endo*-trig iodoamidation reaction developed by Knight [38], flexible approaches to 2,5-dialkylpyrrolidines [39] and anisomycin (**1**) [40] were developed [40]. In the synthesis of anisomycin (**1**), the 4-iodopyrrolidine **53** was converted efficiently to pyrrolidinediol **54** by a Woodward–Prévost reaction (Scheme 2.18).

The osmium-catalyzed oxidative cyclization of aminoalkenes developed by Donohoe exhibits notable stereospecificity (*syn*-addition across the tethered alkene) and stereoselectivity with exclusive formation of *cis*-2,5-disubstituted pyrrolidines (Scheme 2.19) [41].

Borhan and coworkers developed a tandem aza-Payne/hydroamination reaction-based approach to tetrasubstituted pyrrolidines. The starting material was synthesized by addition of alkynyl Grignard reagent to 2,3-aziridinal, which gave the aziridinol **55** in high *syn*-selectivity. Treatment of the *syn*-aziridinol **55** with base led,

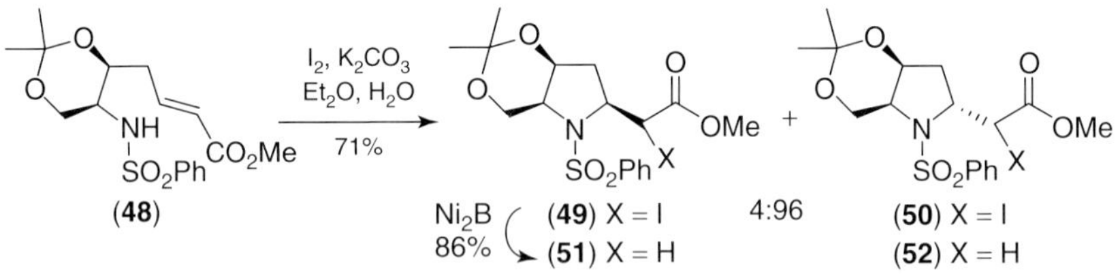

Scheme 2.17 Intramolecular iodosulfonamidation approach to 2,5-*trans*-pyrrolidines.

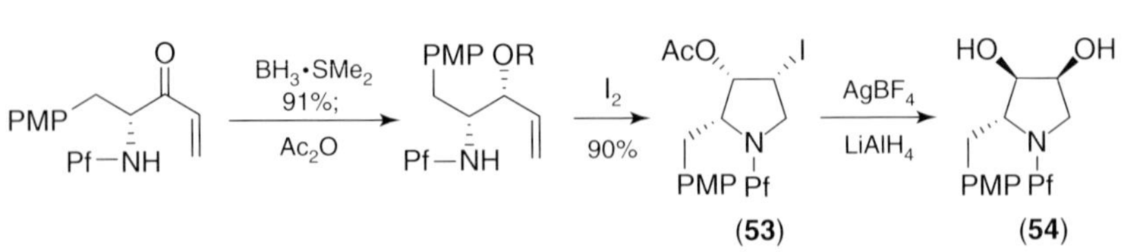

Scheme 2.18 Park's approach to anisomycin (PMP = *para*-methoxyphenyl).

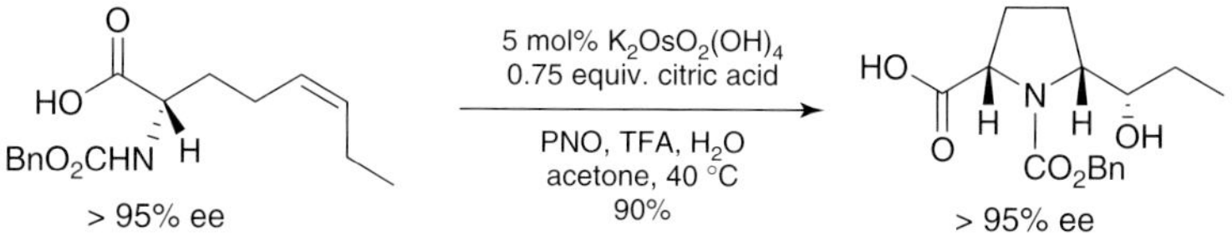

Scheme 2.19 Synthesis of *cis*-2,5-disubstituted pyrrolidines by osmium-catalyzed oxidative cyclization of aminoalkenes.

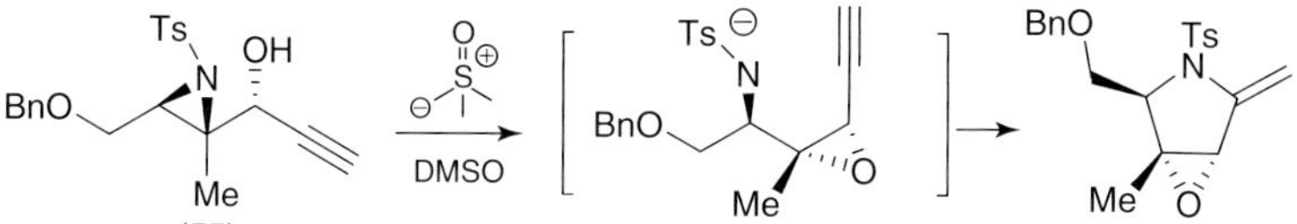

(55) (56)

Scheme 2.20 Borhan's tandem aza-Payne/hydroamination reaction–based approach to tetrasubstituted pyrrolidines.

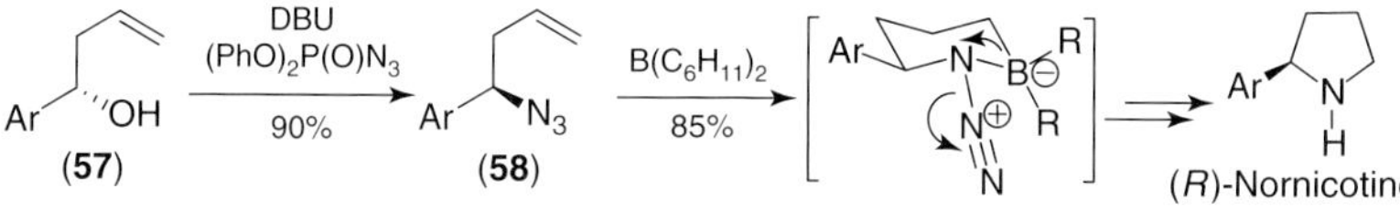

(57) (58)

(Ar = 3-Pyridinyl; R = Cyclohexyl)

Scheme 2.21 Synthesis of pyrrolidines by intramolecular hydroboration–cycloalkylation.

via successive aza-Payne rearrangement and hydroamination, to epoxypyrrolidine **56** (Scheme 2.20) [42].

The intramolecular hydroboration–cycloalkylation of azido-olefins affords an attractive method for the synthesis of pyrrolidines [43, 44] because of the easy availability of the precursor. As shown in Scheme 2.21, the unnatural (*R*)-nicotine was synthesized in four steps with an overall yield of 51%. The starting (*S*)-homoallylic alcohol was obtained in 86% yield with 94% ee by asymmetric allylation.

Scheme 2.22a shows a general method for the synthesis of ketimine-type iminosugars, based on the tandem Grignard reagent addition/cyclization reaction of methanesulfonyl glycononitriles, which are easily available from sugars [45].

In recent years, several catalytic asymmetric hydroamination methods have been developed. For example, chiral phosphoric acid diesters [46] and binaphthylamido ytterbium ate complexes [47] were shown to be effective catalysts for the intramolecular hydroaminations of nonactivated alkenes (Scheme 2.22b).

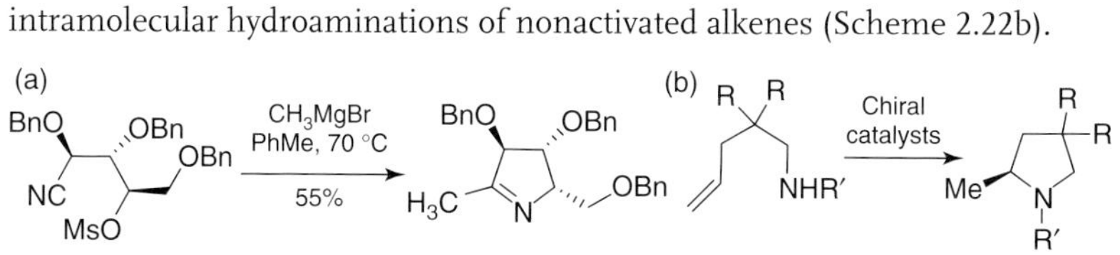

Scheme 2.22 Synthesis of ketimine-type iminosugars and catalytic asymmetric intramolecular hydroamination.

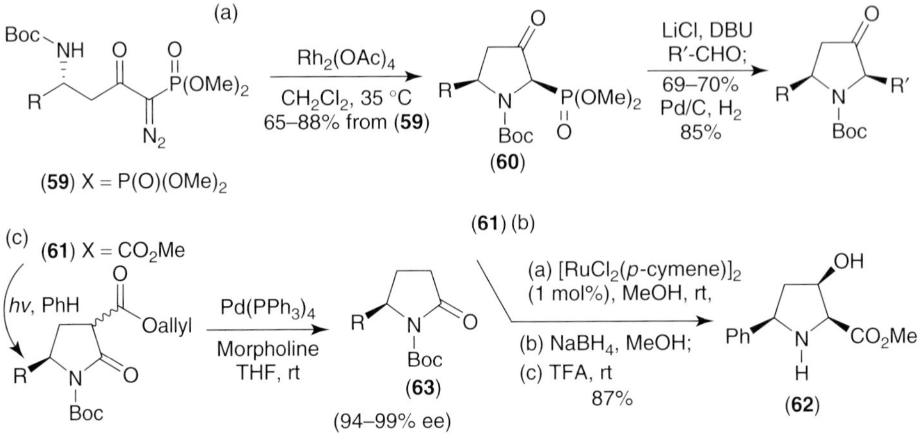

Scheme 2.23 Synthesis of 3-oxopyrrolidines via N–H insertion and 2-oxopyrrolidines via the Wolff rearrangement.

Metal carbenoid N–H insertion constitutes a particular method for the formation of 3-oxopyrrolidine rings (Scheme 2.23). Treatment of the α-diazo compound **59** with 4 mol% of $Rh_2(OAc)_4$ resulted in the formation of 3-oxopyrrolidine phosphate **60**, which can be converted into cis-2,5-disubstituted 3-oxopyrrolidines, and cis-2,5-disubstituted pyrrolidines (Scheme 2.23a) [48]. Che reported that $[RuCl_2(p\text{-cymene})]_2$ is also an effective catalyst for the N–H insertion reaction which, in tandem with sodium borohydride reduction, afforded 5-substituted 3-hydroxyprolinates **62** with good chemical yield and stereoselectivity (Scheme 2.23b) [49]. Wang et al. showed that the reactions of α-diazo compound **61** can lead to either 2-oxo or 3-oxo-pyrrolidine derivatives (Scheme 2.23c) [50]. Under photoinduced conditions, a Wolff rearrangement reaction occurred, leading to the formation of 2-oxopyrrolidine derivatives (**63**) after subsequent decarboxylation.

Divinylcarbinol **64** is a usual prochiral building block for the asymmetric synthesis of pyrrolidines via Sharpless asymmetric epoxidation [51–53]. Aminolysis followed by the reverse-Cope cyclization [54] afford an easy access to substituted pyrrolidine N-oxides **65** (Scheme 2.24) [53].

Denmark developed a chiral auxiliary-induced regioselective asymmetric [4 + 2] cycloaddition of 2-(acetoxy)vinyl ethers with nitroalkenes, which provides an efficient approach to pyrrolidines with a quaternary chiral center (Scheme 2.25) [55a]. Extension of this strategy to asymmetric tandem [4 + 2]/[3 + 2] cycloadditions

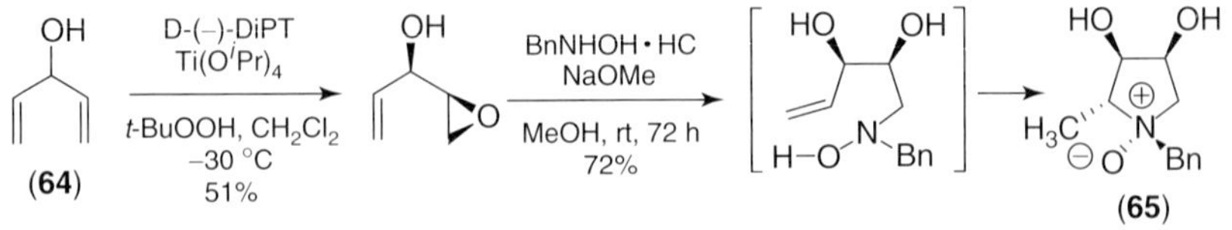

Scheme 2.24 Synthesis of substituted pyrrolidine N-oxides by the reverse-Cope cyclization.

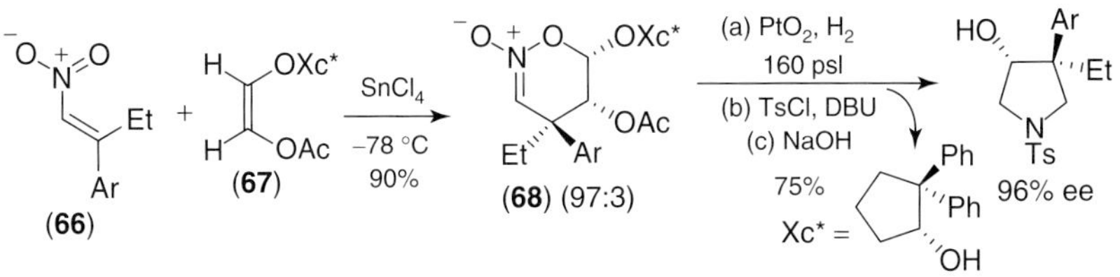

Scheme 2.25 Denmark's approach to 3-hydroxypyrrolidines.
Ar = 3,4-dimethoxyphenyl; Ac = acetyl; Ts = p-toluenesulfonyl.

provides a powerful methodology for the synthesis of pyrrolizidine alkaloids such as (+)-macronecine, (+)-petasinecine, and (−)-hastanecine [55b].

2.1.2.2 Cyclization via C_2–C_3 Bond Formation

Several methods for the construction of pyrrolidine ring by forming the C_2–C_3 bond have been reported. The highly enantioselective S_Ni cyclization reactions reported independently by Kawabata [56] and Kolaczkowski [57] (**69**, Scheme 2.26a) involve the extension of the concept of memory of chirality developed by Fuji and coworkers.

In Tian's asymmetric synthesis of dehydroclausenamide, a highly diastereoselective nucleophilic epoxide-opening reaction was used as the key step. When using a 1:1 CH_2Cl_2:H_2O mixture as the biphase solvent, the reaction of **70** provided the lactam **71** as a single product (Scheme 2.26b) [58].

In Fukuyama's synthesis of (−)-α-kainic acid, a highly stereoselective intramolecular Michael addition reaction was used as the key step (Scheme 2.27a) [59].

The amino-zinc-ene-enolate cyclization developed by Karoyan and coworkers [60] turned out to be a stereoselective method for the synthesis of cis-3-substituted prolinate. The zinc intermediate **73** formed during the reaction can be further

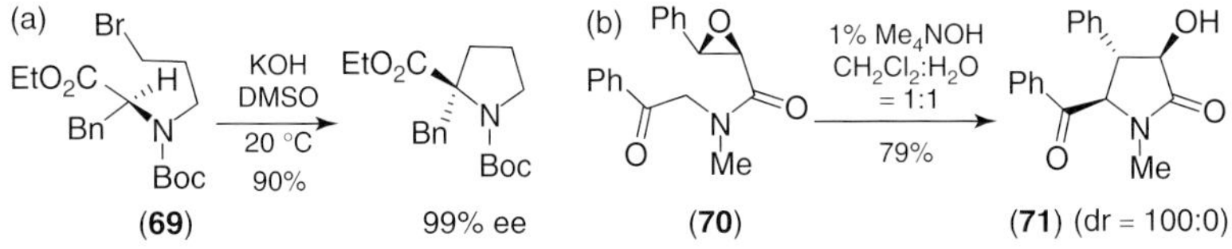

Scheme 2.26 Synthesis of pyrrolidines by S_Ni reactions.

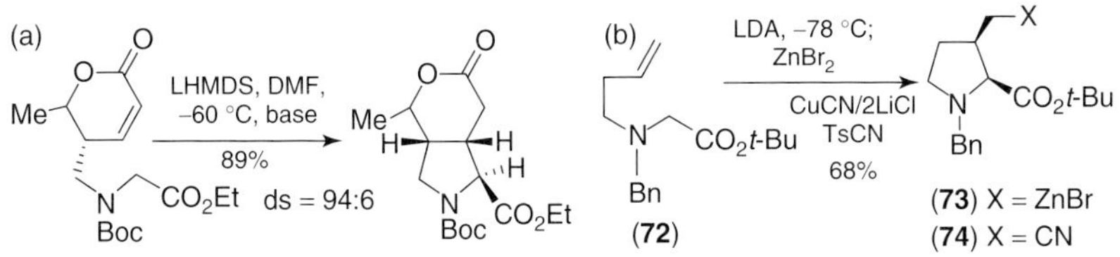

Scheme 2.27 Synthesis of pyrrolidines by intramolecular Michael addition and the amino-zinc-ene-enolate cyclization.

functionalized. For example, tandem cyclization–cyanation of **72** gave **74** in good overall yield (Scheme 2.27b) [60].

Recently, Szymoniak reported a novel method for the construction of the pyrrolidine ring via a tandem hydrozirconation-stereoselective Lewis acid–mediated cyclization [61]. By using N-3-alkenylcarbamates as the starting material and replacing Lewis acid by n-butyllithium the above-mentioned concept was applied to the asymmetric synthesis of pyrrolidin-2-ones (Scheme 2.28).

2.1.2.3 Cyclization Involving C_3–C_4 Bond Formation

Dieckmann reaction of the chiral β-amino ester, prepared by Davies' asymmetric aza-Michael addition method, was shown to give good regioselectivity through the β-aminoenolate. The use of Weinreb amide served as a better electrophilic component. Similarly, intramolecular Thorpe reaction of a chiral dicyanide also gave good regioselectivity (Scheme 2.29a) [62].

In a synthesis of kainoic acid, the reaction of methylvinyl ketone (MVK) with the Michael donor–acceptor α-amino-α,β-unsaturated eater **75** proceeded smoothly to give the tandem Michael addition product **76** in 90% yield (Scheme 2.29b) [63].

In the synthesis of salinosporamide A (**4**), an even more effective proteasome inhibitor than omuralide, Corey and coworkers developed two highly diastereoselective syntheses of the highly substituted lactam intermediates via an internal Kulinkovich reaction (Scheme 2.30a) [64], and an internal Baylis–Hilman–aldol reaction (Scheme 2.30b) [65].

In the synthesis of the amino acid moiety of polyoxypeptins A/B, the SmI_2-mediated Barbier-type cyclization of iodoketone **77** led to the desired diastereomer in 97:3 (Scheme 2.31a) [66].

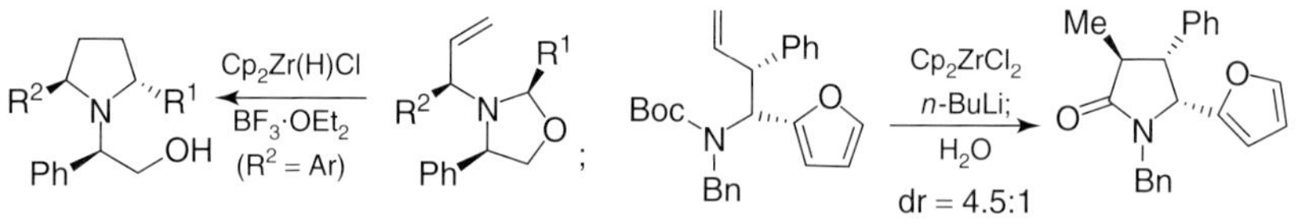

Scheme 2.28 Synthesis of pyrrolidines/pyrrolidin-2-ones by tandem hydrozirconation–cyclization.

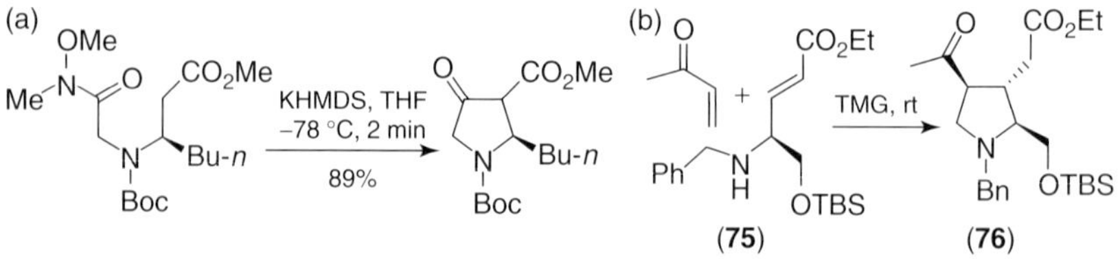

Scheme 2.29 Synthesis of pyrrolidine derivatives by Dieckmann reaction and tandem Michael addition (KHMDS = potassium hexamethyldisilylamide).

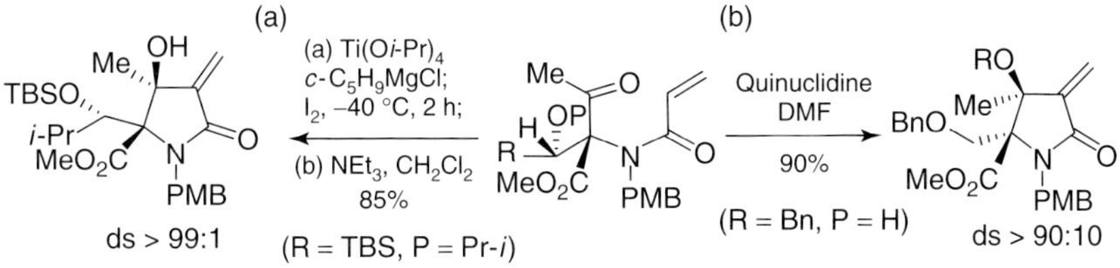

Scheme 2.30 Corey's approaches to the intermediates for the synthesis of salinosporamide A.

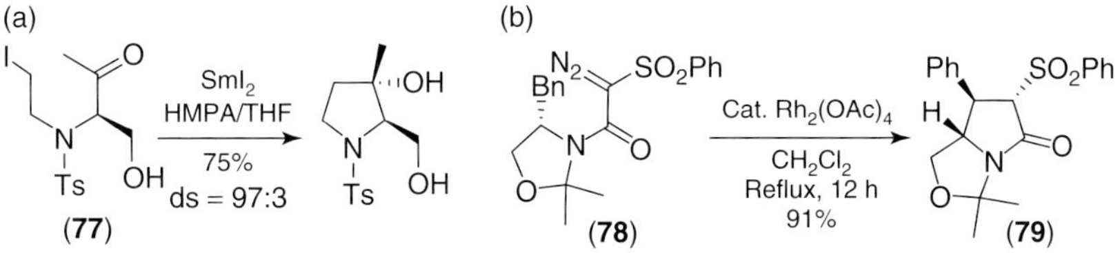

Scheme 2.31 Synthesis of pyrrolidine/pyrrolidin-2-one by radical cyclization and carbene insertion.

Rh-Catalyzed intramolecular carbene C–H insertion of diazo compound **78**, derived from α-amino acid in six steps, gave the chiral γ-lactam **79** with high regio- and diastereoselectivity (Scheme 2.31b) [67].

In association with the synthesis of kainoids, several useful methods have been developed for the construction of pyrrolidine ring by C_3–C_4 bond formation. They include, among others, Oppolzer's intramolecular thermal type I Ene reaction (Scheme 2.32a) [68], Hoppe's (−)-sparteine-mediated asymmetric deprotonation–cycloalkylation (Scheme 2.33b) [69], and Ganem's catalytic enantioselective metallo-ene reaction (Scheme 2.32c) [70].

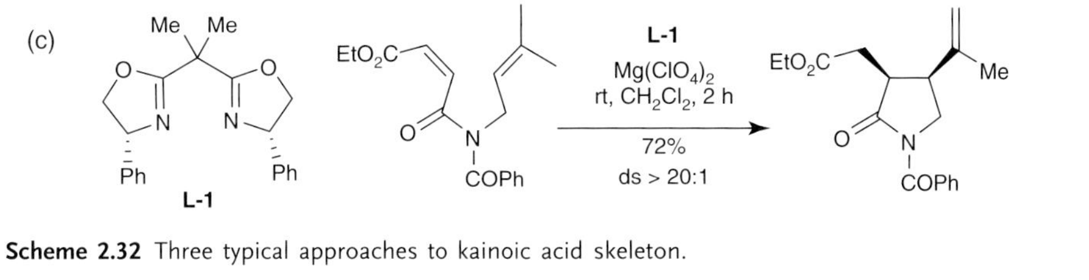

Scheme 2.32 Three typical approaches to kainoic acid skeleton.

Recently, this concept was extended to an organocatalytic version. The protected diphenylprolinol-catalyzed, enantioselective, three-component coupling between aldehydes, dialkyl 2-aminomalonates, and α,β-unsaturated aldehydes gives access to highly substituted pyrrolidines in good yield with >10:1 diastereoselectivity and 90–98% ee (Scheme 2.38) [79].

2.1.3.2 Annulation Approach

Pyrrolidines and pyrrolidine-2-ones can be synthesized by nucleophilic ring opening of optically active aziridines in a [3 + 2] annulation manner [80]. Reaction of configurationally stable alkoxy allenylzinc reagent **85** with the imine derived from (S)-malic acid furnished a route to pyrrolidin-2-one **86** with *anti–anti* stereochemistry (Scheme 2.39) [81].

A novel [4 + 1] annulation method for the efficient synthesis of diastereomerically and enantiomerically pure 2,3-disubstituted pyrrolidines is shown in Scheme 2.40. The method involves a sulfoxoniumylide-based aza-Payne rearrangement of the 2,3-aziridin-1-ols and the subsequent [1 + 4] annulation reaction [82].

Lewis acid–catalyzed [3 + 2] annulation [83] of optically active N-Ts-α-amino aldehydes and 1,3-bis(silyl)propenes opens another one-step access to polysubstituted pyrrolidines in excellent diastereoselectivity (98:2). The conversion of the silyl group into a hydroxyl group by Tamao oxidation allows the total synthesis of polyhydroxylated alkaloids (Scheme 2.41) [84].

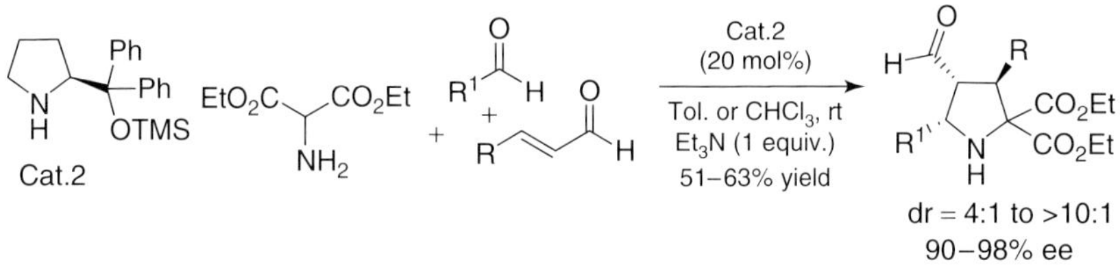

Scheme 2.38 Organocatalytic, three-component synthesis of highly substituted pyrrolidines.

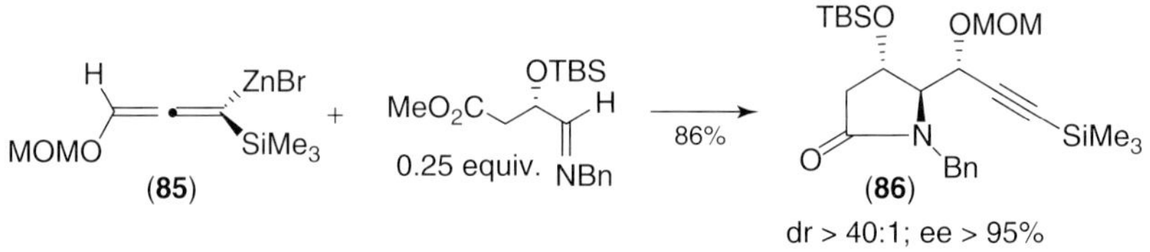

Scheme 2.39 Normant's [3 + 2]-annulation approach to substituted pyrrolidin-2-ones.

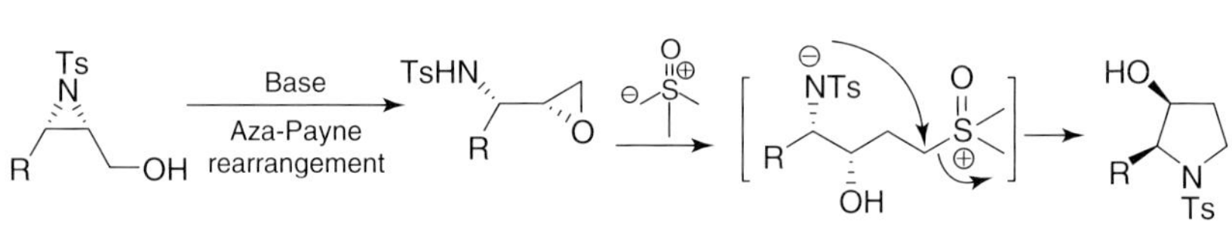

Scheme 2.40 Borhan's [4 + 1] annulation approach to 2,3-disubstituted pyrrolidines.

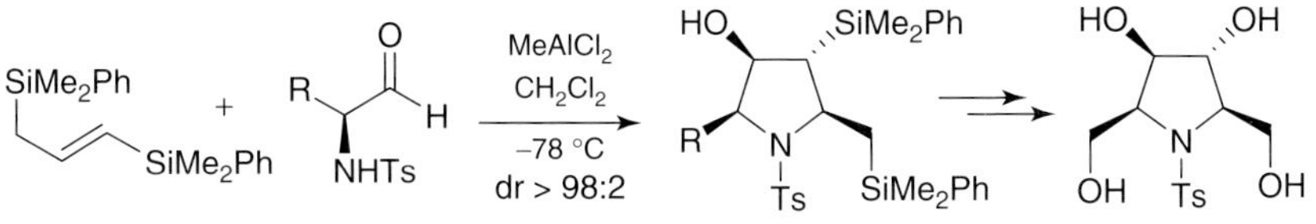

Scheme 2.41 Synthesis of substituted pyrrolidines via a Lewis acid–catalyzed [3 + 2] annulation.

2.1.4
Ring Transformation Methods

2.1.4.1 Ring Expansion Methods

In a formal synthesis of (−)-anisomycin (**1**), Somfai et al. showed that microwave-assisted rearrangement of vinylaziridines provides an easy access to 3-pyrrolines (Scheme 2.42) [85].

The β-lactam skeleton has been recognized as a valuable precursor for the synthesis of pyrrolidine-2-ones via ring expansion [86]. For example, upon treatment of N-unsubstituted 4-(α-aminoalkyl) β-lactam with methanolic HCl at 60 °C for 2–24 h, the expected rearrangement occurred smoothly to give the pyrrolidine-2-one bearing four contiguous chiral centers in almost quantitative yield. The starting β-lactams are easy available via the Staudinger reaction (Scheme 2.43) [87].

Using Tamura's Beckmann reagent, O-(mesitylenesulfonyl) hydroxylamine (MSH), Greene et al. developed a chiral auxiliary-induced, regio-controlled cyclobutanone → pyrrolidinone ring expansion reaction. This, after dechlorination with zinc–copper coupling, led to cis-pyrrolidin-2-ones in high yields (Scheme 2.44). Using this strategy, they accomplished the total syntheses of pyrrolidine alkaloids (+)-preussin (**15**) [88a], (−)-anisomycin (**1**) [88b], and (−)-detoxinine [88c].

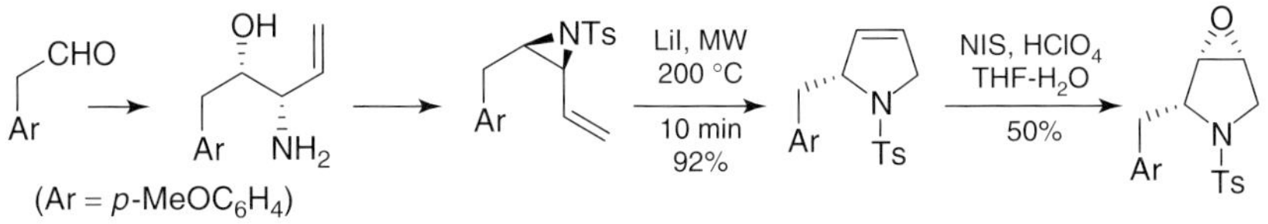

Scheme 2.42 Microwave-assisted rearrangement of vinylaziridines to 3-pyrrolines en route to (−)-anisomycin.

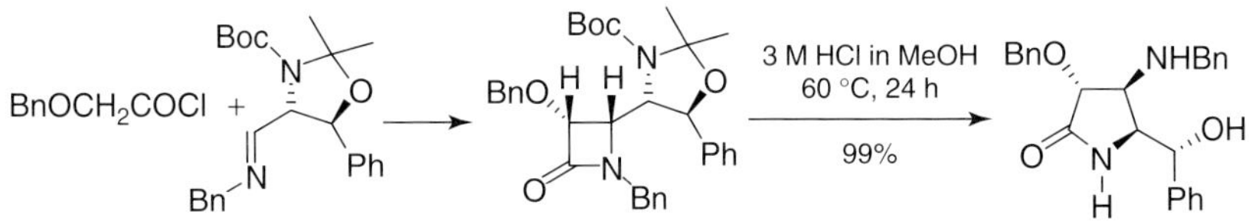

Scheme 2.43 Ring expansion of β-lactam derivatives to pyrrolidine-2-one derivative.

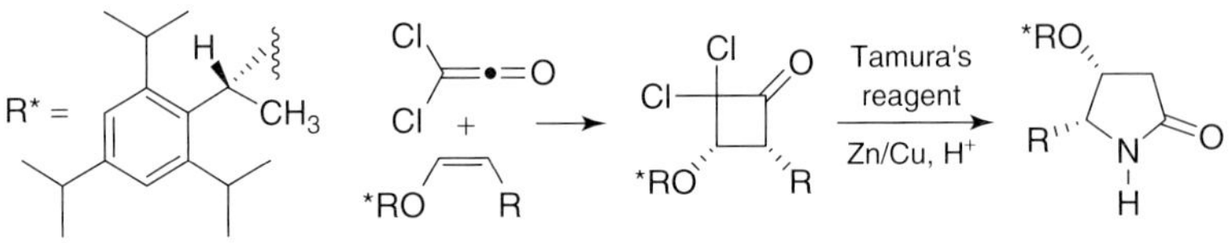

Scheme 2.44 Greene's synthesis of cis-4,5-disubstituted pyrrolidin-2-ones by cyclobutanone ring expansion.

2.1.4.2 Ring Contraction Methods

Although rarely used in pyrrolidine syntheses, the carbene ring contraction–decarboxylation of (2R,3S)- and (2S,3R)-6-diazo-3,4-dimethyl-2-phenyloxazepane-5,7-diones opens an easy regiospecific entrance to (S)- and (R)-1,5-dimethyl-1-4-phenyl-1,5-dihydro-2H-pyrrol-2-ones in enantiomeric pure forms (Scheme 2.45) [89].

Chiral auxiliary-induced, diastereoselective hetero-Diels-Alder reaction of nitroso derivatives and the subsequent reductive ring contraction of the resulting 1,2-oxazines under catalytic hydrogenolytic conditions constitutes a useful approach to dihydroxypyrrolidines (Scheme 2.46) [90].

2.1.5
Substitution of Already Formed Heterocycle

A number of approaches are available for the functionalization of a pyrrolidine/pyrrolidin-2-one ring at C-2 because of its presence in many alkaloids, including pyrrolidines, pyrrolizidines, and indolizidines alkaloids. Hence, the synthesis of 2-alkyl and 2,5-dialkyl pyrrolidines has attracted much attention [91].

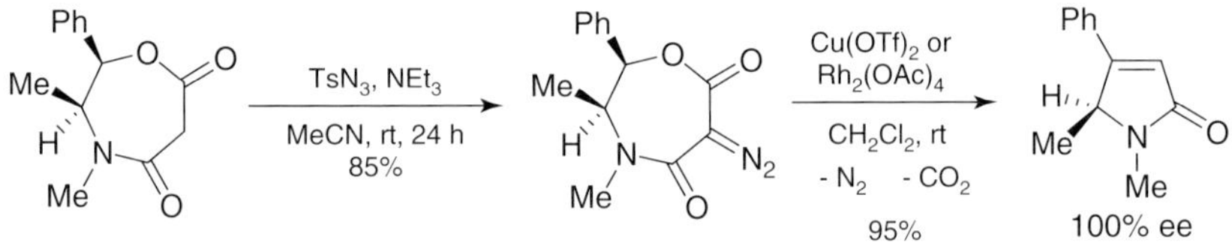

Scheme 2.45 Regiospecific ring contraction of oxazepane-5,7-dione to α,β-unsaturated pyrrolidin-2-one derivatives.

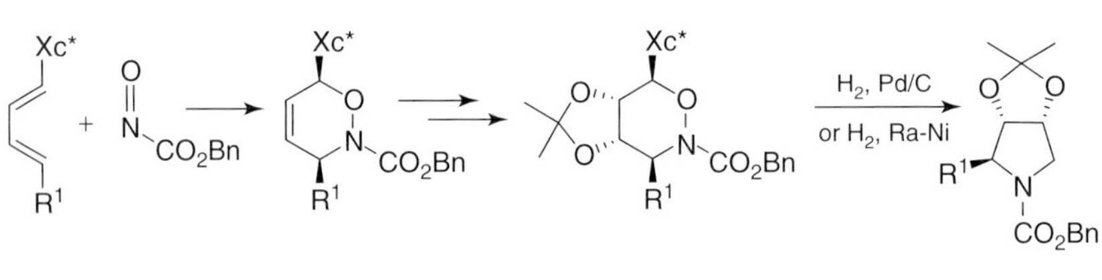

Scheme 2.46 Stereoselective access to dihydroxylated pyrrolidines by reductive ring contraction of 1,2-oxazines.

2.1 Monocyclic Pyrrolidines and Pyrrolidinones | 69

With Seebach's SRS methodology (self-regeneration of stereochemistry), single-enantiomer proline and pyroglutamate may serve as reliable chirons for the generation and reaction of pyrrolidine α-carbanions [92]. It was shown that the alkylation of the enolate proceeds *cis* with respect to the *tert*-butyl group ("*cis* rule"), with overall retention of configuration. For example, in the first total synthesis of kaitocephalin, the condensation of **87** with **89** led only to *cis* adducts **90** (Scheme 2.47) [93]. In contrast, the MgBr$_2$•OEt$_2$-mediated conjugate addition of the silyl enol ether, formed *in situ* from the N,N-acetal **88**, with a nitroolefin resulted in a 10:1 mixture of two *trans* diastereomers **91**. The major diastereomer **91** was used as a key intermediate for the synthesis of the marine alkaloid (−)-amathaspiramide F [94].

tert-Butoxycarbonyl (Boc) is an efficient group for the stabilization of α-amino carbanions. The deprotonation of N-Boc-pyrrolidin-3-ol **92** occurred regioselectively at C-5. The reaction of the resulting dianion with electrophiles yielded a 1:1 diastereomeric mixture in each case tested except with methyl iodide (dr = 5:1) (Scheme 2.48) [95, 96]. The SmI$_2$-mediated reaction of the N,O-diprotected 2-pyridyl 3-pyrrolidinol-2-yl sulfide **94** with carbonyl compounds affords the protected N-α-hydroxyalkyl-3-pyrrolidines **95** with excellent diastereoselectivity at the newly formed chiral center in the pyrrolidine ring [97].

Asymmetric alkylation of the β-hydroxyproline ethyl ester N-carbamate, prepared by enantioselective reduction of **96** with Baker's yeast, was shown to proceed with net retention of configuration. By developing this method, Williams *et al.* achieved a total asymmetric synthesis of paraherquamide A (Scheme 2.49a) [98].

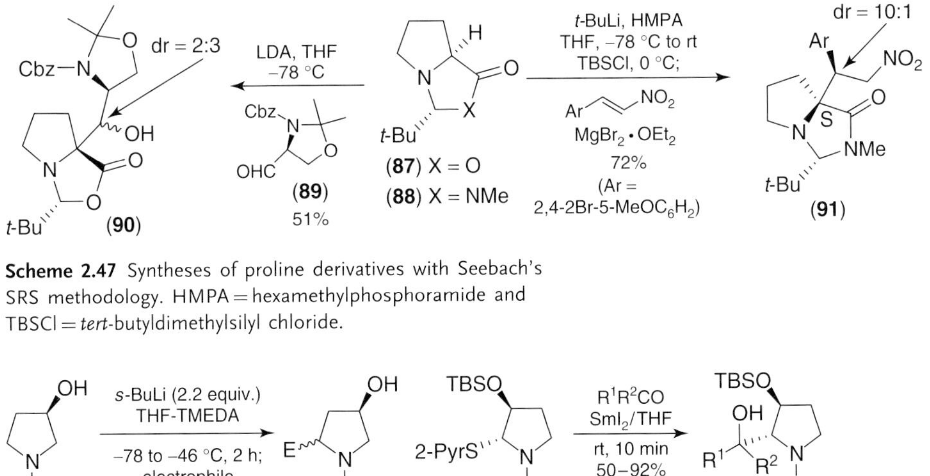

Scheme 2.47 Syntheses of proline derivatives with Seebach's SRS methodology. HMPA = hexamethylphosphoramide and TBSCl = *tert*-butyldimethylsilyl chloride.

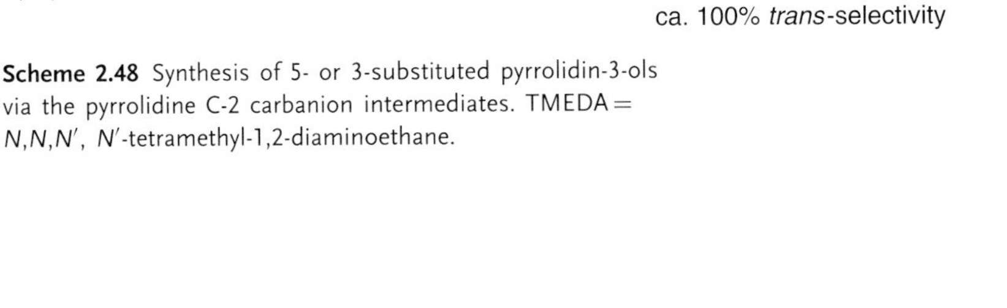

Scheme 2.48 Synthesis of 5- or 3-substituted pyrrolidin-3-ols via the pyrrolidine C-2 carbanion intermediates. TMEDA = N,N,N′,N′-tetramethyl-1,2-diaminoethane.

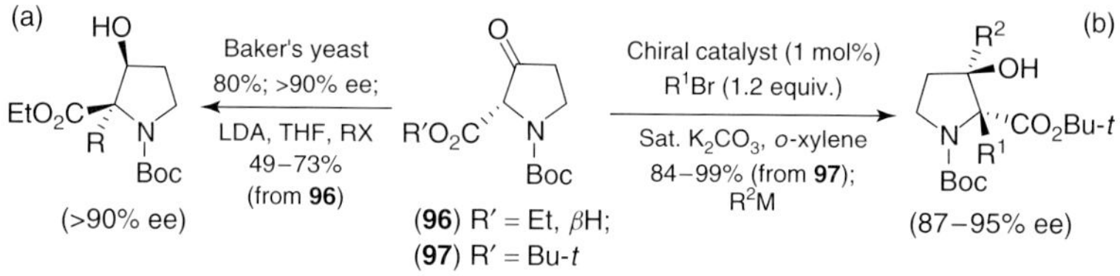

Scheme 2.49 Syntheses of substituted 3-hydroxyproline derivatives by alkylation.

A phase-transfer-catalyzed alkylation strategy developed by Maruoka provides an efficient and high-yielding approach to another diastereomer and its derivatives (Scheme 2.49b) [99].

Beak and coworkers showed that enantioselective deprotonation (lithiation) of N-Boc-pyrrolidine (**98**) can be achieved using sec-BuLi/(−)-sparteine (**L-2**). The carbanion thus generated (**100**, α-Li) reacts with electrophiles to give 2-substituted pyrrolidines with high enantioselectivities (Scheme 2.50) [95, 100]. The unnatural ligand **L-3** developed by O'Brien allows access to another enantiomer of the 2-substituted pyrrolidines [101]. Coldham's ligands **L-4**/**L-5** afford a highly enantioselective dynamic kinetic resolution pathway of the racemic organolithium **100**, derived from **98** or **99** by deprotonation or by tin–lithium exchange [102]. Gawley showed that the tin–lithium exchange of enantioenriched 2-stannylpyrrolidine **99** affords configurationally stable 2-lithiopyrrolidine **100** with retention of configuration [103].

Through the reaction sequence displayed in Scheme 2.51, tetramate derivative **101** turned out to be a valuable synthetic equivalent of the 4-hydroxypyrrolidin-2-one 2-carbanion synthon, which is used for the synthesis of pyrrolidin-2-ones **102** [104].

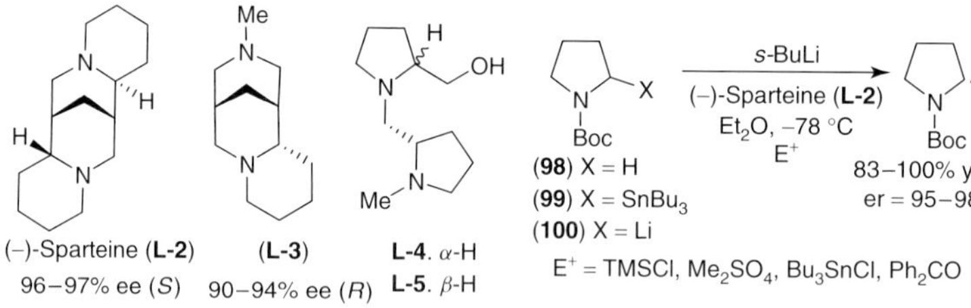

Scheme 2.50 Methods for the generation and reaction of configurationally stable enantioenriched N-Boc 2-lithiopyrrolidine.

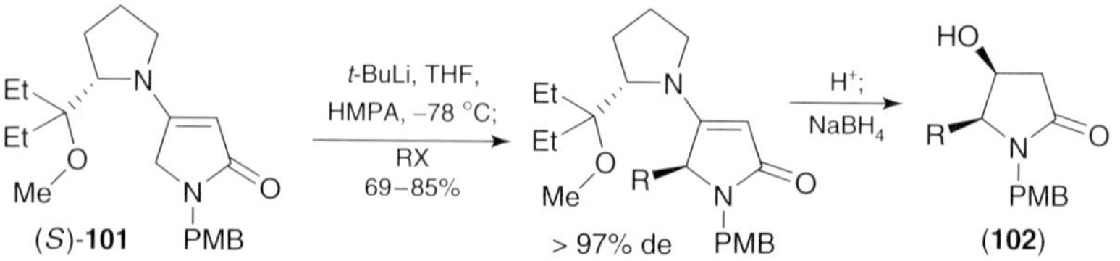

Scheme 2.51 A flexible approach to cis-5-alkyl-4-hydroxypyrrolidin-2-ones.

2.1.5.1 By Nucleophilic Reaction of Pyrrolidinium Ions

Because of its double functionality and easy availability, (S)-pyroglutamic acid is a versatile useful chiron (Scheme 2.52) [105] for the asymmetric synthesis of 2,5-disubstituted pyrrolidines [106], pyrrolizidines [107], and indolizidines [108].

Both the lactam carbonyl and carboxyl groups in pyroglutamic acid can be employed to introduce side chains. Elongation of the carboxyl group can be achieved by reduction to a hydroxymethyl group followed by tosylation and coupling with an organocopper reagent [106, 107], or by transformation into a formyl group, followed by Kocienski's modification of Julia olefin synthesis [108]. The amide carbonyl can be used to install a second alkyl group via the corresponding thiolactam by Eschenmoser sulfide contraction reaction [109], via enamino diester derivative [106, 107], or via thioimidate derivative [106].

A more versatile method for the activation of lactams consists in converting lactams into the corresponding N-carbamates (imides). After their regioselective partial reduction with diisobutyl aluminium hydride (DIBAL-H) or Super hydride, they are converted into N,O-acetals and subjected to Lewis acid–promoted α-amidoalkylation with lower order organocuprates (107 → 112, Scheme 2.53) [110] or with Si-nucleophiles (110 → 113) [111]. The latter reaction can also be achieved by Ley's sulfone chemistry [112] (108 → 112) [113]. Martin's improvements over the stepwise reductive alkylation method [114] provides a general and highly chemo- and diastereoselective reductive alkylation (114 → 115, Scheme 2.54) [115]. It is well recognized that the α-aminoalkylation passes through N-acyliminium ion intermediates [116].

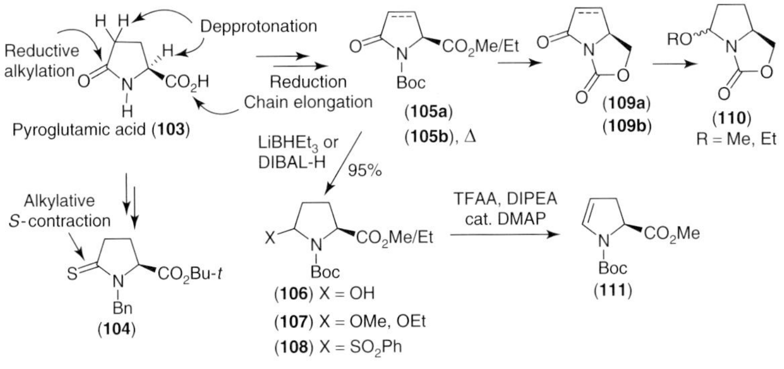

Scheme 2.52 Conversion of pyroglutamic acid into useful synthetic intermediates.

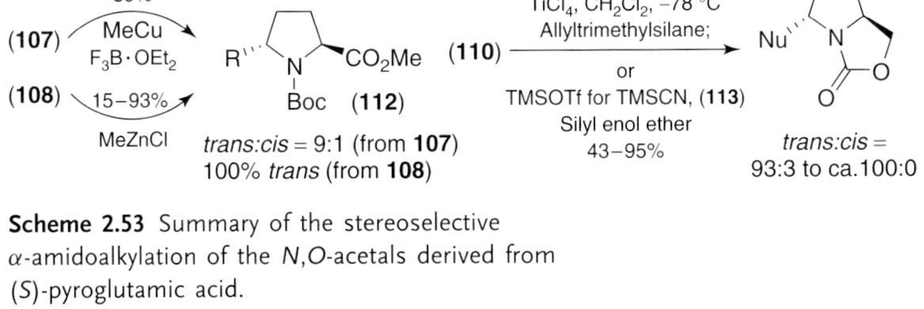

Scheme 2.53 Summary of the stereoselective α-amidoalkylation of the N,O-acetals derived from (S)-pyroglutamic acid.

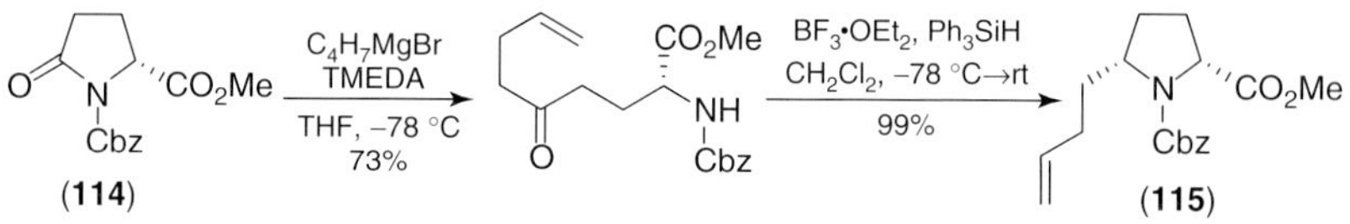

Scheme 2.54 Martin's modified reductive alkylation method.

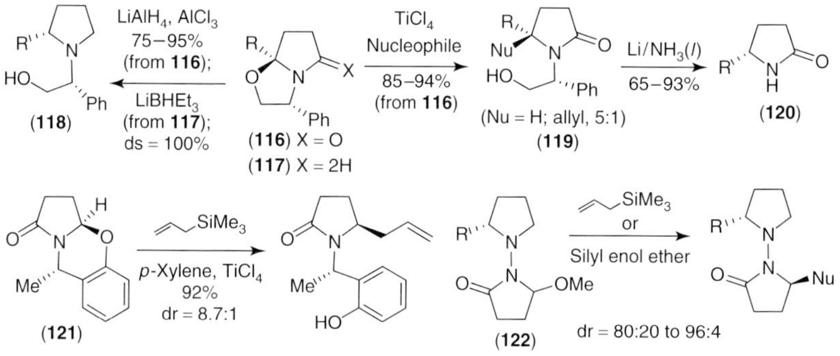

Scheme 2.55 The α-amidoalkylation reactions of Meyers' and Kibayashi's chiral N,O-acetals.

A simple asymmetric synthesis of 2-substituted pyrrolidines (**116** → **118**, Scheme 2.55) [117] and 5-substituted pyrrolidin-2-ones (**116** → **119**) [118] starting form γ-keto acid and (R)-phenylglycinol was developed by Meyers et al. [119]. This method was extended to the (R)-phenylglycinol-derived bicyclic 1,3-oxazolidine **117** [120]. The N,O-acetals bearing easily cleavable chiral auxiliaries **121** [117] and **122** [121] were developed by Kibayashi et al. and were shown to give good stereoselectivities in the α-amidoalkylation. Except one case [122], retention of configuration was observed in the above-mentioned reactions.

The α-amidoalkylation of N,O-acetals **123** [123], **125** [124] and **126** [125] with trimethylallylsilane and silyl enol ether give the corresponding products with excellent trans diastereoselection (Scheme 2.56). By contrast, the reactions of O-TBS-protected N,O-acetals derived from tartaric acid (**124**) [126], malic acid [127], and related compounds [128] were shown to give good-to-excellent cis diastereoselectivities.

The α-amidoalkylation of N,O-acetals **127** [129], **128** [130], and **129** [131] and also the borono-Mannich reaction of **131** [132] with more flexible carbon nucleophiles are trans-diastereoselective (Scheme 2.57). By contrast, reactions of O-acetyl-protected N,O-acetals derived from tartaric acid (**130**) [133] give cis diastereomers as major products.

2.1.5.2 By Nucleophilic Reaction of Cyclic Imides

Results from Yoda's laboratory [134] show that the reductive alkylation of the C_2-symmetrical tartarimides is a better alternative to the α-amidoalkylation reaction displayed in Schemes 2.56 and 2.57 both in terms of chemical yields and in trans-diastereoselectivities (Scheme 2.58a) [134a]. They applied this method to the

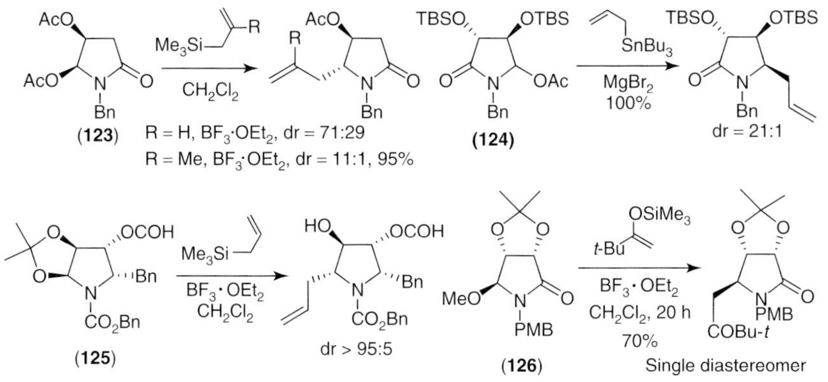

Scheme 2.56 α-Amidoalkylation reactions of the chiral pool-derived N,O-acetals.

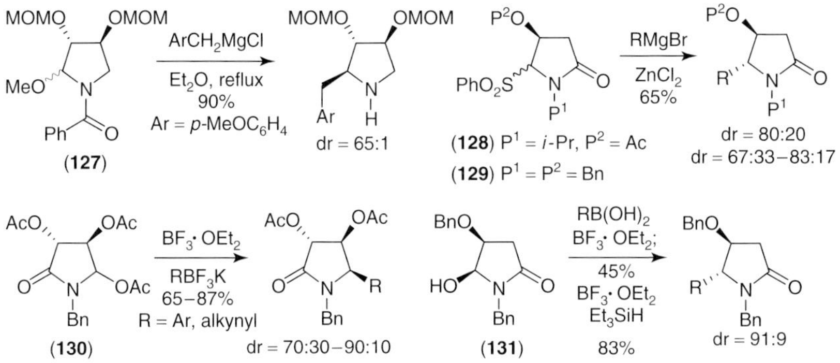

Scheme 2.57 α-Amidoalkylation reactions of chiral pool-derived N,O-acetals with organometallic compounds.

asymmetric synthesis of the pyrrolidine alkaloid codonopsinine (**22**) [134b]. Pilli et al. showed that the Grignard addition products, N,O-acetals **133**, can undergo an α-amidoallylation with allyltributyltin leading to the formation of a quaternary stereocenter [135a].

Regioselectivity in the reductive alkylation of the protected malimides [136] was shown to be dependent on the protecting group [136b]. Moreover, the reaction of the O-benzyl-protected malimides with Grignard reagents gives excellent C-2 regioselectivity [136], while addition of organocerium reagents affords C-5 adducts as the major products (Scheme 2.58b) [135b]. In both cases, subsequent reductive dehydroxylation reactions give excellent trans diastereoselectivities. Unexpectedly, the reaction of malimides with organotitanium reagents gives only modest regioselectivity [137].

The intramolecular Wittig reaction of a malimide derivative followed by stereoselective hydrogenation was used as key steps in a highly regio- and diastereoselective synthesis of the Geissman–Waiss lactone (Scheme 2.59) [138].

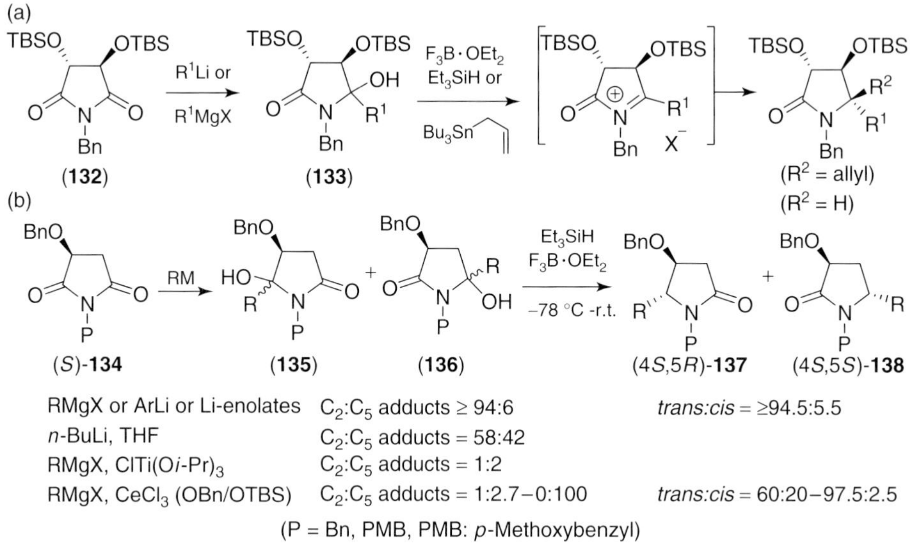

Scheme 2.58 The stepwise reductive alkylation of tartarimides and malimides.

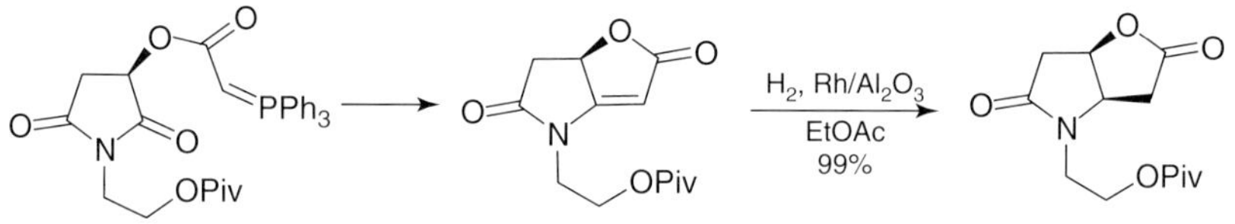

Scheme 2.59 A highly regio- and diastereoselective synthesis of the Geissman–Waiss lactone.

2.1.5.3 By Nucleophilic Addition/Cycloaddition of Pyrrolidine Nitrones

Enantioenriched pyrrolidine nitrones [139] are versatile building blocks for the asymmetric synthesis of pyrrolidines, pyrrolizidines, and indolizidines through nucleophilic addition [140] and 1,3-dipolar cycloaddition [141]. Pyrrolidine nitrones can be prepared by intramolecular condensation of a hydroxylamine with a carbonyl group or by oxidation of cyclic hydroxyamines, amines, and imines available from chiral pools such as tartaric acid, malic acid, and sugars.

For the synthesis of nitrones from unsymmetric N-hydroxypyrrolidines, such as those derived from malic acid, regioselectivity may be a problem. Fortunately, oxidation with yellow HgO was reported to give good regioselectivities (Scheme 2.60a) [142]. The environmentally friendly oxidant, bleach, gives only modest regioselectivities [143]. Catalytic oxidation of preformed imines by methyl trioxorhenium/urea hydrogen peroxide (Scheme 2.60b) eliminates the problem of regioselectivity [144], which is also the case by decarboxylative oxidation of α-amino acids (Scheme 2.60c) [145].

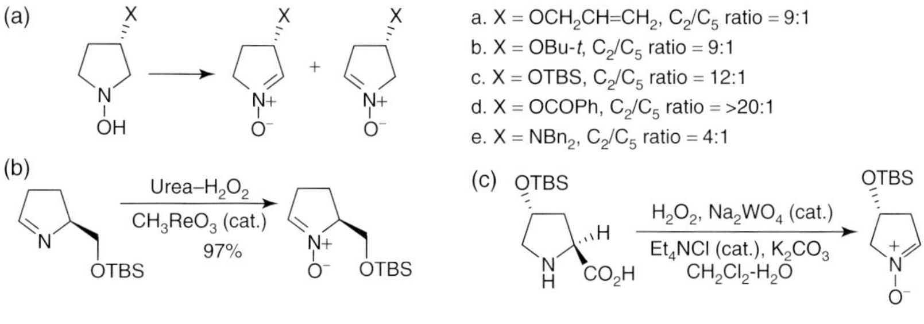
(a)
a. X = OCH2CH=CH2, C2/C5 ratio = 9:1
b. X = OBu-t, C2/C5 ratio = 9:1
c. X = OTBS, C2/C5 ratio = 12:1
d. X = OCOPh, C2/C5 ratio = >20:1
e. X = NBn2, C2/C5 ratio = 4:1
(b)
Urea–H2O2
CH3ReO3 (cat.)
97%
OTBS
(c) OTBS
H2O2, Na2WO4 (cat.)
Et4NCl (cat.), K2CO3
CH2Cl2-H2O
CO2H

Scheme 2.60 Regioselective synthesis of chiral, nonracemic, unsymmetric nitrones.

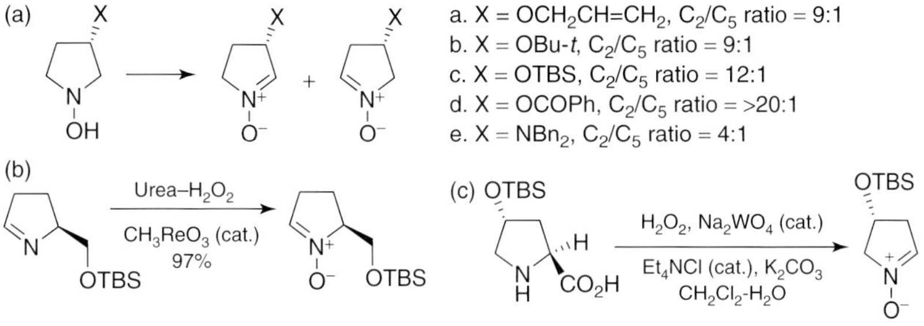
(a)
a. X = OCH2CH=CH2, C2/C5 ratio = 9:1
b. X = OBu-t, C2/C5 ratio = 9:1
c. X = OTBS, C2/C5 ratio = 12:1
d. X = OCOPh, C2/C5 ratio = >20:1
e. X = NBn2, C2/C5 ratio = 4:1
(b)
Urea–H2O2
CH3ReO3 (cat.)
97%
OTBS
(c) OTBS
H2O2, Na2WO4 (cat.)
Et4NCl (cat.), K2CO3
CH2Cl2-H2O
CO2H

Scheme 2.61 Stereoselective Grignard addition to nitrones.

Stereoselection in Grignard additions to nitrones is substrate and reagent dependent. While the addition of benzylic Grignard reagents to nitrone **139** gave a 3:2 *trans/cis* selectivity, a 3:7 ratio was obtained by using magnesium bromide etherate as an additive (Scheme 2.61) [146]. The vinylation of nitrone **140** was shown to be entropy-controlled: a 93:7 *trans/cis* diastereomeric ratio was obtained independent of the reaction temperature [147]. Similarly, ZnI_2-promoted reactions of ketene *t*-butyldimethylsilyl methyl/*t*-butyl acetals with nitrone **141** gave excellent *trans* diastereoselectivities [148].

The structural similarity between nitrone **142** and alkaloids radicamine A (**143**), radicamine B (**144**), as well as codonopsine and codonopsinine (**22**) justifies consideration of the former as the chiron for the synthesis of radicamines A and B. Indeed, starting from nitrone **142**, derived in seven steps from D-xylose, the antipodes of (+)-radicamine A and (+)-radicamine B were synthesized as the sole diastereomers respectively (Scheme 2.62) [149]. Similarly, (−)-codonopsine was synthesized from L-xylose [150].

Nitrones are excellent 1,3-dipoles for cycloaddition and, after cleavage of the *N–O* bond, can be used to introduce side chains at C-2 of pyrrolidine ring. The asymmetric total synthesis of marine natural product crambescidin 359 (**145**) nicely demonstrates the value of nitrone cycloaddition methodology for the asymmetric synthesis of pyrrolidines (Scheme 2.63) [151].

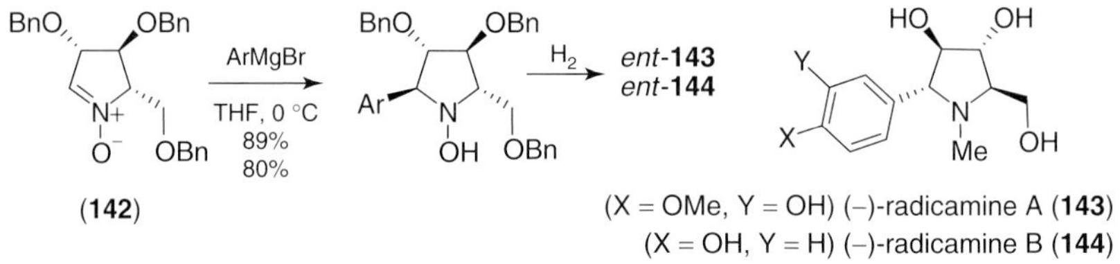

76 | 2 Asymmetric Synthesis of Five-Membered Ring Heterocycles

Scheme 2.62 Asymmetric syntheses of radicamines A and B via nitrone intermediates.

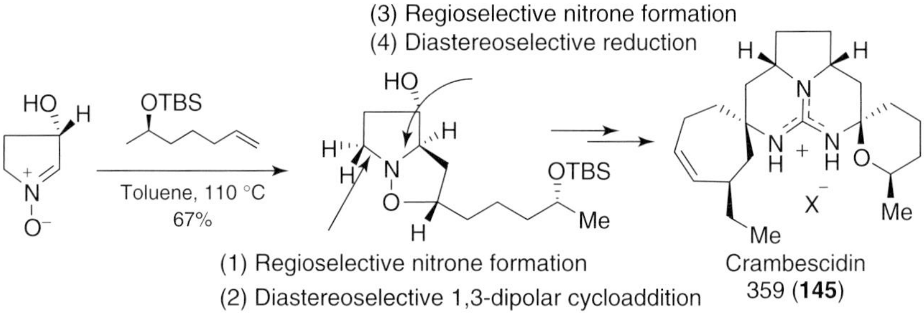

Scheme 2.63 Asymmetric synthesis of crambescidin 359 via iterative selective nitrones formation-1,3-dipolar cycloaddition.

2.1.5.4 By Functionalization of 2-Pyrrolines

Three approaches are available for the functionalization of endocyclic enecarbamates (cf. Scheme 2.79). Ketene [2 + 2] cycloaddition with endocyclic enecarbamate **146**, developed by Correia showed a 1:4 diastereoselectivity. An *meta*-chloroperoxybenzoic acid (CPBA)-mediated, regioselective Baeyer–Villiger oxidation led to the synthesis of the Geissman–Waiss lactone (**147**) (Scheme 2.64a) [152].

By applying the Paternò-Büchi reaction to the chiral 2-substituted 2-pyrroline **148** derived from pyroglutamic acid, Bach *et al.* developed a stereoselective

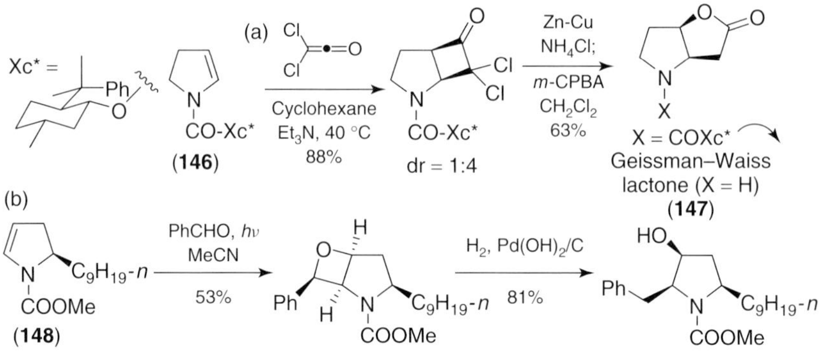

Scheme 2.64 Synthesis of substituted pyrrolidines via [2 + 2] cycloaddition reactions of endocyclic enecarbamates.

2.1 Monocyclic Pyrrolidines and Pyrrolidinones

cis-3-hydroxy-2-benzylation of 2-pyrroline **148**. The key [2 + 2] photocycloaddition was shown to exhibit an unprecedented facial diastereoselectivity. Taking advantage of this reaction, a short total synthesis of (+)-preussin (**15**) was achieved (Scheme 2.64b) [153].

2.1.5.5 By Enantioselective Reactions

Catalytic hydrogenation of substituted pyrrole derivatives is a classical method for the synthesis of all-cis-substituted pyrrolidines. A catalytic enantioselective synthesis version of this method has recently been shown to give high enantioselectivities (Scheme 2.65) [154].

Direct asymmetric α-substitution of pyrrolidine carbamates can be achieved by highly regio-, diastereo-, and enantioselective C–H insertion of methyl aryldiazoacetates. Using this method, both monosubstituted and C_2-symmetric 2,5-trans-disubstituted pyrrolidines can be synthesized with high enantioselectivities (Scheme 2.66) [155a]. A Rh(II)–Carbenoid mediated, intramolecular C–H insertion reaction affords a short approach to both Geissman–Waiss lactone and the necine base (−)-turneforcidine [155b].

Katsuki found that asymmetric desymmetrization of N-t-butoxycarbonyl-meso-3,4-isopropylidenedioxy-pyrrolidine **149** can be effected through an enantiotropic selective oxidation using an (R,R)-(salen)manganese(III) complex as catalyst (Scheme 2.67) [156].

2.1.5.6 By Functionalization at C_3/C_4 Positions of Pyrrolidines

As shown in Scheme 2.52, pyroglutamic acid is not only suitable for the introduction of substituents at C_2 and C_5 positions but it has also been widely used to synthesize 3/4-substituted pyrrolidines/pyrrolidin-2-ones [105]. This can be done by deprotonation–substitution at α-position of the lactam carbonyl group [157]. Through the formation of an α,β-unsaturated lactam system (**105b** and **109b**, Scheme 2.52), a number of reactions can be realized in the olefin portion, including

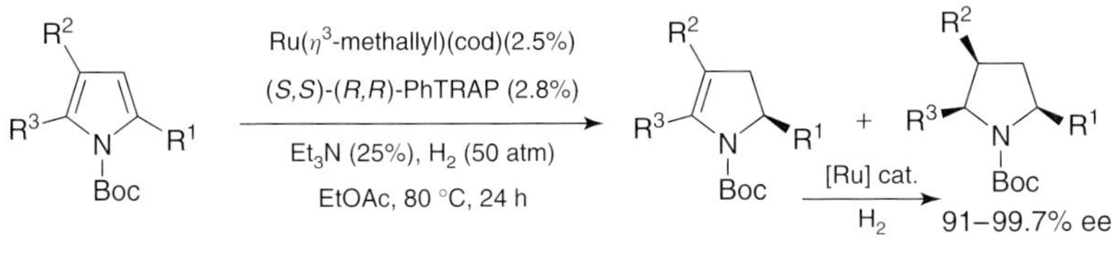

Scheme 2.65 Synthesis of all-cis-substituted pyrrolidines by catalytic enantioselective hydrogenation.

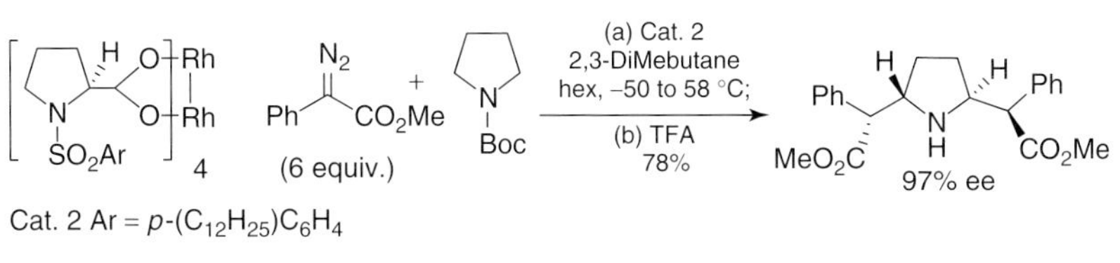

Scheme 2.66 α-Substitution of pyrrolidine carbamates by enantioselective C–H insertion.

conjugate addition [158], tandem conjugate addition-trapping of enolate intermediates [159], epoxidation [160], and [2 + 3] dipolar cycloaddition reactions [160].

Trans-4-Hydroxyproline (**152**) constitutes another useful chiron for the synthesis of 3,4-dihydroxyprolinol derivatives. For example, it was converted into **153** [161], **154** [162], **155** [163], and **156** [164] (Scheme 2.68). Compounds **155** and **156** were used in a total synthesis of α–kainic acid [162, 163].

Meyers' chiral bicyclic lactams **158** are versatile building blocks for highly *endo*-selective α-alkylations, α,α-dialkylations, cyclopropanations, and so forth (Scheme 2.69) [119].

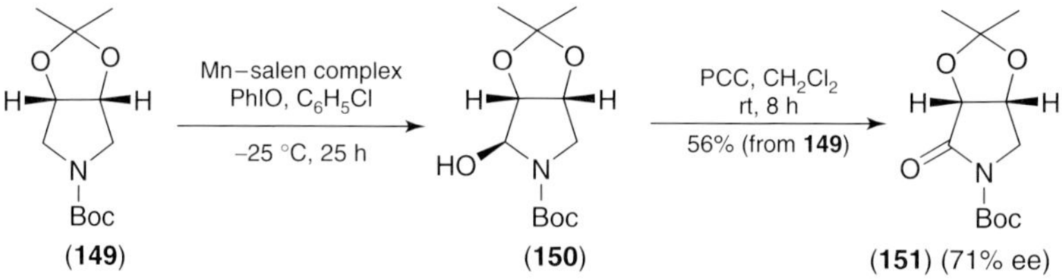

Scheme 2.67 Enantioselective α-oxidation of pyrrolidine carbamates.

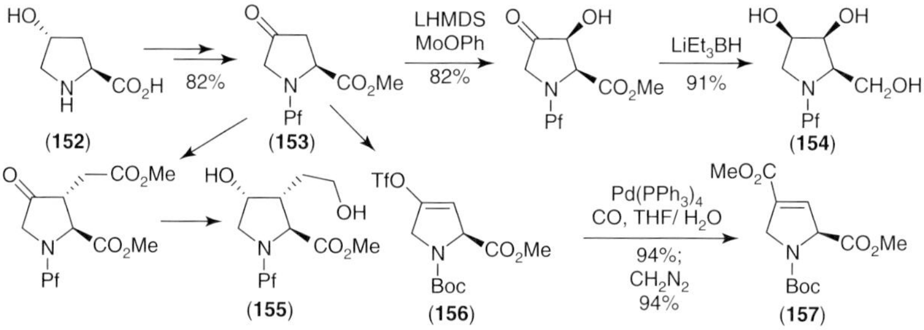

Scheme 2.68 Some pyrrolidine derivatives derived from *trans*-4-hydroxyproline.

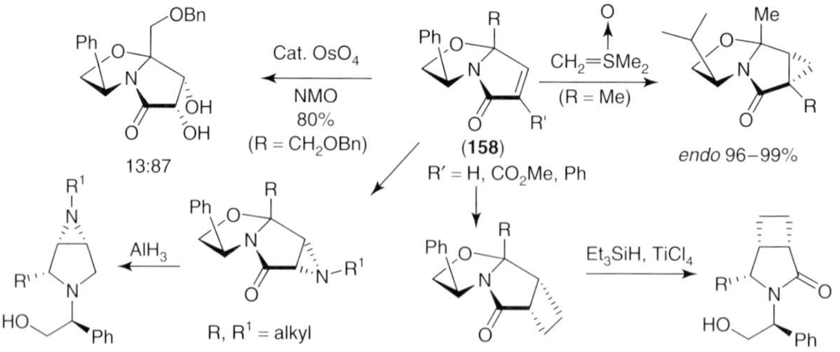

Scheme 2.69 Selected transformations of the Meyers' chiral bicyclic lactams.

2.2 Pyrrolines

2.2.1 Synthesis of Pyrrolines by Cyclization and Annulation Reactions

Several "2 + 3" annulation approaches to 2-pyrrolines have been reported. Goré [164] and Reißig [165] independently developed a one-pot "3 + 2" annulation approach to 2-pyrroline derivatives (Scheme 2.70). The chiral inducer may be a part of the molecule or introduced as a chiral auxiliary on either imines or alkoxyallenes. Using this method, a straightforward synthesis of (−)-detoxinine was achieved from the adduct **162** in five steps.

Recently, a catalytic enantioselective three-component synthesis of 4-arylated dehydroprolines was reported. In the presence of 10 mol% of [Cu(MeCN)$_4$]ClO$_4$/(R)-Ti(BINAP) (cat. 3), the "3 + 2" annulation of allenylstannane with an α-imino ester, followed by a tandem Stille coupling of the resulting 4-stannyl dehydroprolinate, gave 4-arylated dehydroprolinates (Scheme 2.71) [166].

A tandem aza-Michael-type-reaction–Wittig reaction between α-amide ketones and vinylphosphonium salts was reported to give 3-pyrrolines **164** and pyrrolizidines in high yields. The stereochemistry of the starting α-amido ketone is retained during this process (Scheme 2.71) [167].

Nucleophilic addition of lithio propiolate to nitrones derived from D-glyceraldehyde and α-amino aldehydes has been shown to give excellent diastereoselectivity. This opens a ready access to enantiopure 5-substituted-3-pyrrolin-2-ones [168]. After further refinement, Hanessian and coworkers applied this methodology to the total synthesis of A-315675 (**6**), a potent neuraminidase inhibitor (Scheme 2.72) [169].

Several ring-closing olefin metathesis (RCM) reaction based approaches to pyrrolines have been reported [170]. Sharpless asymmetric epoxidation [171],

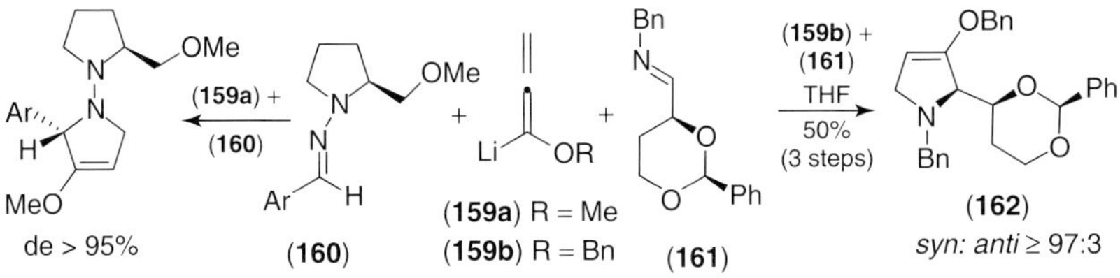

Scheme 2.70 Goré's and Reißig's one-pot "3 + 2" annulation approaches to 3-pyrrolines.

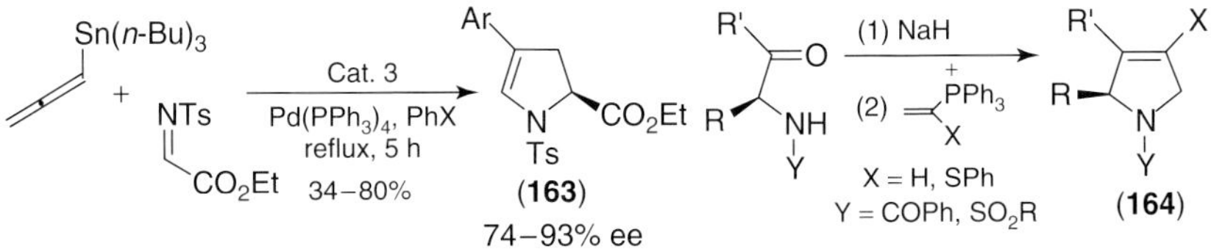

Scheme 2.71 Two "3 + 2" annulation approaches to pyrrolines.

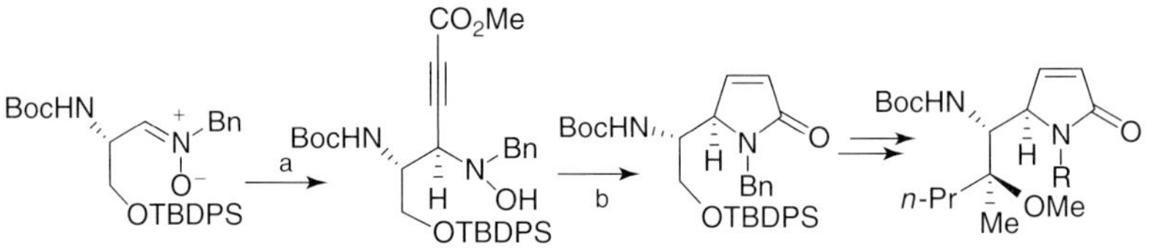

Scheme 2.72 Hanessian's synthesis of 5-substituted-3-pyrrolin-2-ones.

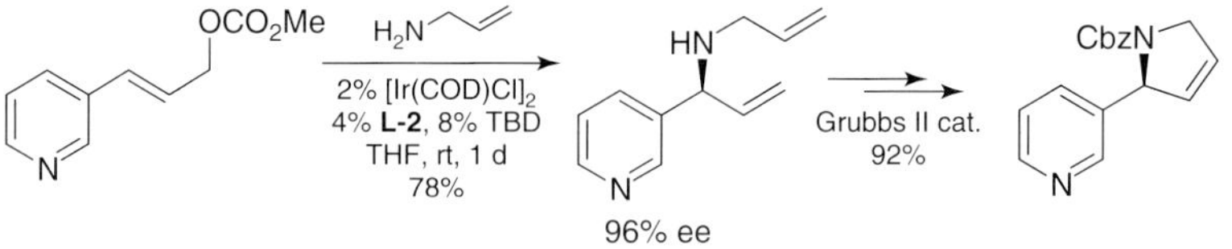

Scheme 2.73 Helmchen's enantioselective synthesis of (R)- and (S)-nicotine.

regioselective rhodium-catalyzed allylic amination [172], palladium-catalyzed dynamic kinetic asymmetric transformation (DYKAT) of vinyl epoxides [173], and Ir-catalyzed enantioselective allylic amination [174] (Scheme 2.73) have all been used to form the bis-allylic alicyclic precursors.

Hayes et al. developed a KHMDS-promoted stereospecific alkylidene carbene 1,5-CH insertion for the synthesis of substituted 3-pyrrolines (Scheme 2.74). This was used as a key step in an enantioselective total syntheses of omuralide, of its C_7-epimer, and of (+)-lactacystin (**3**) [175].

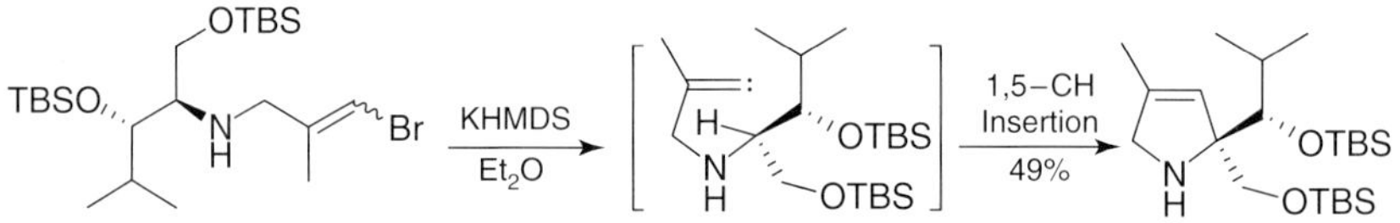

Scheme 2.74 Synthesis of substituted 3-pyrrolines by base-promoted stereospecific alkylidene carbene 1,5-CH insertion.

2.2.2
Synthesis of Pyrrolines by Substitution of Already Formed Heterocycles

Chiral auxiliary-induced, tandem Birch-type reduction–protonation/alkylation of N-Boc-pyrrole carboxylic ester/amide [176a,b] **165** gives rise to nonalkylated/alkylated 3,4-dehydroprolines with good diastereoselectivities (Scheme 2.75) [176]. More interestingly, and through chiral protonation, enantioselective partial reduction of pyrrole 2,5-dicarboxylate can be achieved with good diastereo- and enantioselectivities following recrystallization [177].

Alkylation of a chiral, nonracemic formamidine developed by Meyers et al. affords a highly enantioselective entrance to 2-alkyl-3-pyrrolines (95–96% ee, Scheme 2.76a) [178], while Royer et al. showed that chiral nonracemic α,β-

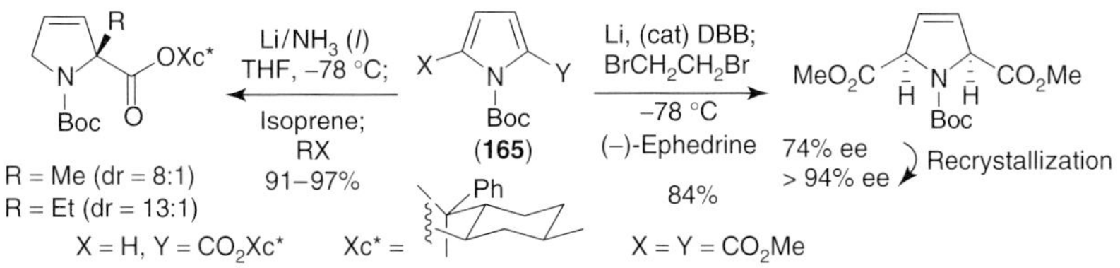

2.2 Pyrrolines

Scheme 2.75 Donohoe's approaches to proline derivatives.

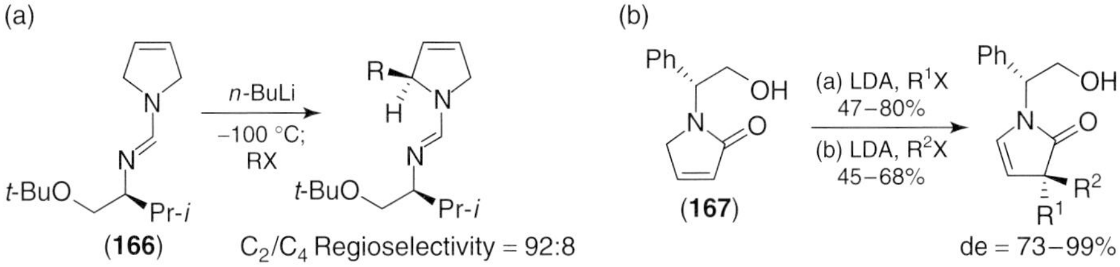

Scheme 2.76 Diastereoselective approaches to 2-alkyl-3-pyrrolines and to β,γ-unsaturated lactams.

unsaturated γ-lactam **167** is a useful building block for the construction of β,γ-unsaturated lactams bearing a chiral quaternary center (Scheme 2.76b) [179].

The vinylogous Mukaiyama-type aldol addition methodology, originally developed by Rassu/Casiraghi et al., affords a versatile approach to chiral nonracemic α,β-unsaturated γ-lactams [180]. For example, in the presence of SnCl$_4$, the reaction of 2,3-O-isopropylidene-D-glyceraldehyde (**169**) with N-(tert-butoxy carbonyl)-2-(tert-butyldimethylsiloxy) pyrrole (TBSOP) **168** gave D-arabino α,β-unsaturated γ-lactam **170** as the sole product (Scheme 2.77) [180a]. However, BF$_3$ etherate–mediated reaction led to D-ribo-epimer **171**. Moreover, a clean and almost quantitative epimerization at C-5 can be achieved by stirring lactam **170** with Et$_3$N/DMAP, which provides a good alternative preparation of the thermodynamically more stable D-ribo-lactam **171**.

The reactions of chiral 2-(tert-butyldimethylsiloxy) pyrroles **172** and **173** with achiral aldehydes were investigated independently by Baldwin and Royer and used in the syntheses of (+)-lactacystin (**3**) [181] and cephalotaxine [182] respectively.

The Rassu/Casiraghi methodology is applicable to imines, which constitutes a vinylogous Mannich-type reaction [183]. The reaction of TBSOP (**168**) with chiral imines was used as a key step in the total synthesis of an influenza

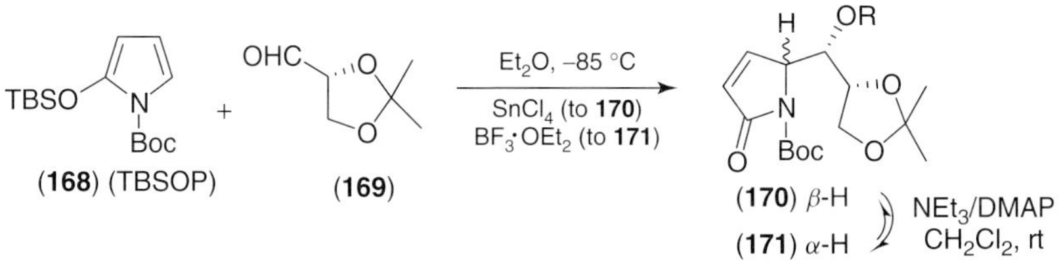

Scheme 2.77 Rassu/Casiraghi vinylogous Mukaiyama-type aldol addition methodology.

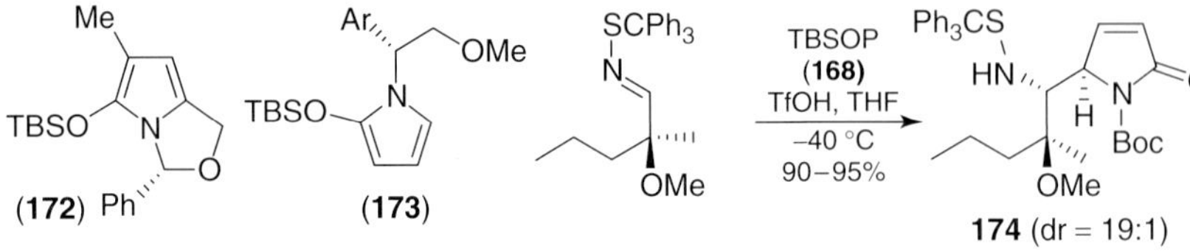

Scheme 2.78 Two chiral 2-(*tert*-butyldimethylsiloxy) pyrroles and a vinylogous Mannich-type reaction.

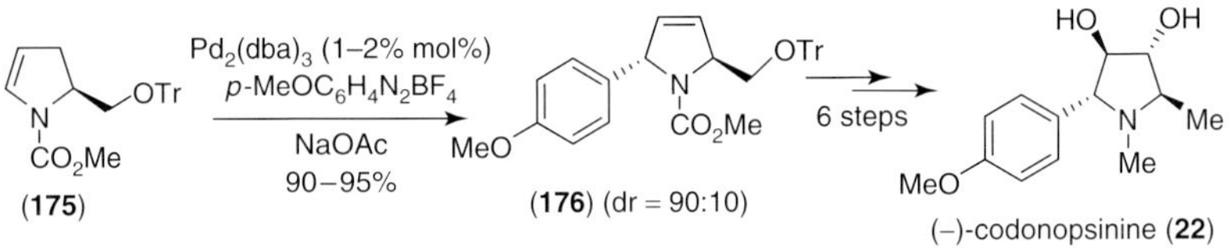

Scheme 2.79 Correia's synthesis of (−)-codonopsinine.

neuraminidase inhibitor. It was found that the use of TfOH as a catalyst gave excellent diastereoselectivity and chemical yield (Scheme 2.78) [184].

Heck reaction of enecarbamates **175** derived from pyroglutamic acid with *p*-methoxybenzenediazonium tetrafluoroborate was shown to give 2,5-disubstituted 3-pyrroline **176** with good regio- and *trans*-diastereoselectivity. The product was converted into (−)-codonopsinine (**22**) in six steps (Scheme 2.79) [185].

2.3
Fused Bicyclic Systems with Bridgehead Nitrogen

2.3.1
Pyrrolizidines

2.3.1.1 Through Extension of Methods for the Synthesis of Pyrrolidines

Because a pyrrolizidine can be viewed as a fused pyrrolidine, many methods developed for the asymmetric synthesis of pyrrolidines can be extended to pyrrolizidines [167,186–189]. Thus methods employed for the formation of pyrrolidine rings can be used to construct the pyrrolizidine ring system starting from pyrrolidine derivatives.

For example, the palladium-catalyzed cyclization of proline-derived allylic carbonates, initially developed for the synthesis of 4,5-*cis*-disubstituted γ-lactams, was extended to the synthesis of pyrrolizidinone, which led to a stereoselective synthesis of (−)-trachelanthamidie (**177**) (Scheme 2.80) [190].

Intramolecular reductive alkylation is a useful method widely used for the synthesis of pyrrolidines [116], and has also found widespread applications in the synthesis of pyrrolizidines and indolizidines. In the first enantioselective synthesis

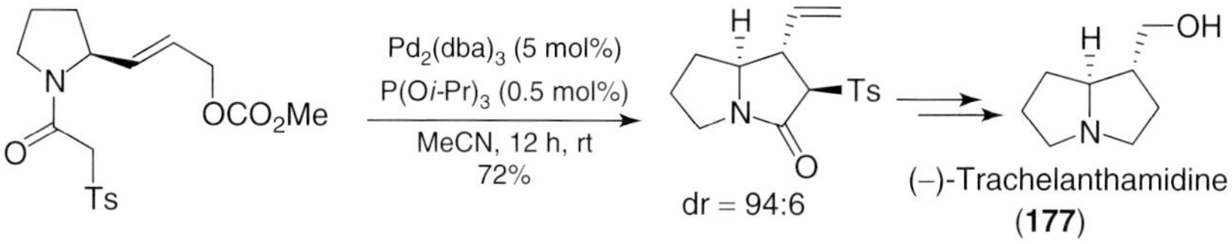

Scheme 2.80 Synthesis of pyrrolizidinones by palladium-catalyzed cyclization of proline-derived allylic carbonate.

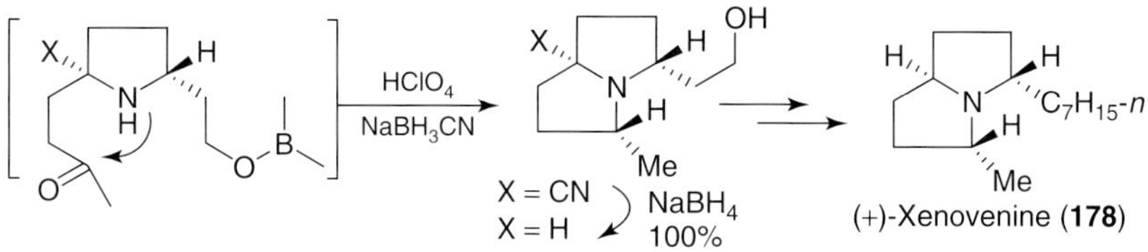

Scheme 2.81 Takano's intramolecular reductive alkylation approach to xenovenine.

of the ant venom alkaloid xenovenine (**178**, Scheme 2.81) [191], double intramolecular reductive alkylation was employed to form the pyrrolizidine ring in high stereoselectivity. Later, this became a standard procedure for the synthesis of this alkaloid [192], as well as of other pyrrolizidine alkaloids such as the pyrrolizidines 239K′, 265H′, 267H′ [193], epohelmin B [194a], as well as hyacinthacines A_2 and A_3 [194b].

Husson's synthesis of *trans*-2,5-dialkylpyrrolidines [192a] and xenovenine (**178**) [192b] is another demonstration of such an approach (Scheme 2.82) [192c]. Similarly, Greene's methodology [88] was extended to the synthesis of (+)-hyacinthacine A_1 [189a], (+)-amphorogynines A and D, and (+)-retronecine [189b].

The cumulated ylide Ph_3PCCO is a useful $C_2^{a,d}$ component. Its reaction with α-amino esters can lead to tetramates [195a]. The reaction with (R)-prolinate gave, via a domino addition–Wittig reaction, bicyclic tetramate, which was converted into (−)-pyrrolam A (**179**) in four steps from prolinate (Scheme 2.83) [195b].

2.3.1.2 Other Methods for the Synthesis of Pyrrolizidines

Formation of the pyrrolizidine ring system by intramolecular α-amidoalkylation is a widely adopted strategy [116]. Ketene dithioacetal (**180**, Scheme 2.84) [196] and allyl tin (**181**) [197] were shown to be excellent nucleophiles for intramolecular α-amidoalkylation. Using acetoxy-directed N-acyliminium ion-ketene dithioacetal cationic cyclization (of **180**) as the key step, Chamberlin et al. established a versatile

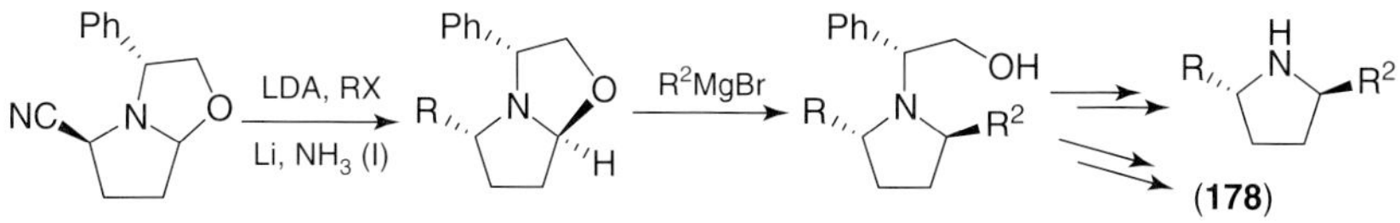

Scheme 2.82 Husson's approach to pyrrolidines and pyrrolizidine.

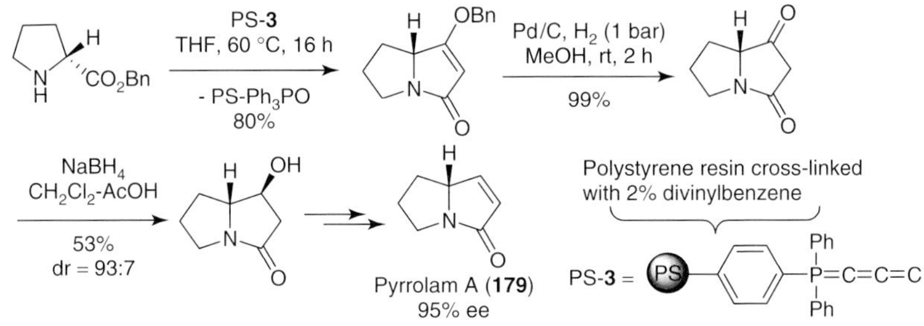

Scheme 2.83 Schobert's synthesis of (−)-pyrrolam A.

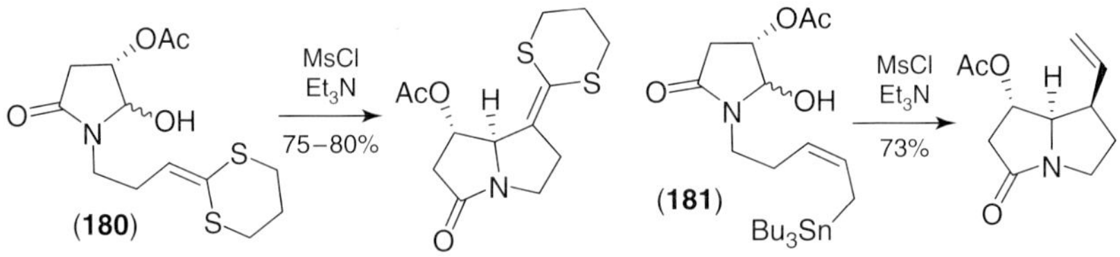

Scheme 2.84 Synthesis of pyrrolizidinones by cyclization involving N-acyliminium ion intermediates.

entrance to six pyrrolizidine alkaloids [196], while cyclization of **181** led to the synthesis of (+)-heliotridine [197].

SmI_2-mediated intramolecular reductive alkylation of tartarimide derivative **182** affords an easy approach to dihydroxylated pyrrolizidinone (Scheme 2.85) [198]. A similar product can be obtained through silylative Dieckmann-like cyclizations of ester-imides **183** (Scheme 2.85) [199].

A general 1,3-thiazolidine-2-thione-induced asymmetric α-amidoalkylation-type reaction affords a two-step approach to (−)-trachelantamidine (**177**, Scheme 2.86) [200]. Using this strategy, Pilli et al. established a synthesis of (+)-hastanecine [201].

"Cp_2Zr"-Mediated ring contraction of vinylmorpholine derivatives **184** was used to synthesize pyrrolidines and pyrrolizidine derivative **185**. The latter was further transformed into (−)-macronocine (Scheme 2.87a) [202]. In White's total synthesis of (+)-australine (**186**), a transannular cyclization was used to build the pyrrolizidine ring (Scheme 2.87b) [203].

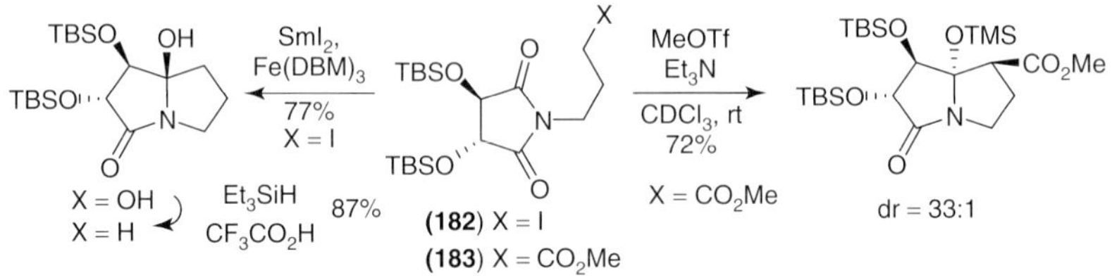

Scheme 2.85 Nucleophilic cyclizations leading to dihydroxylated pyrrolizidinone derivatives. $(Fe(DBM)_3) = tris(dibenzoylmethido)iron(III)$.

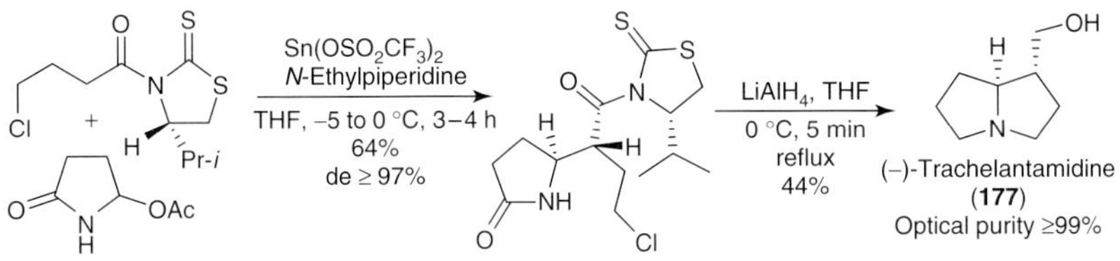

Scheme 2.86 Nagao's approach to (−)-trachelantamidine.

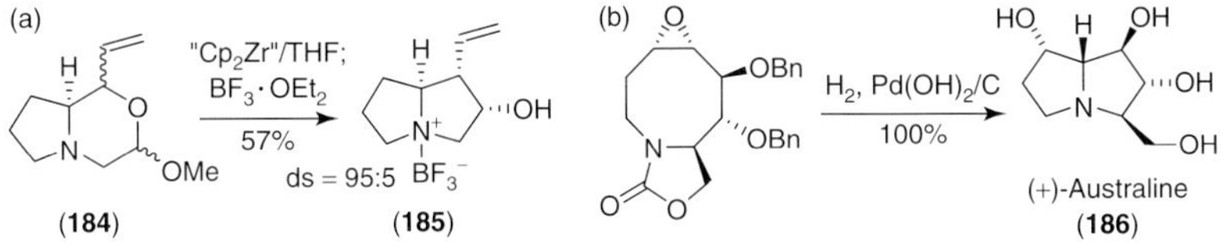

Scheme 2.87 Synthesis of pyrrolizidine by ring contraction and transannular cyclization reactions.

2.3.1.3 Asymmetric Synthesis of Polyhydroxylated Pyrrolizidines

By a combination of Nicotra's highly stereoselective vinylation, Ganem's pyrrolidine formation procedure [15], and the RCM reaction, Martin achieved the first synthesis of (+)-hyacinthacine A$_2$ (**187**) in six steps starting from protected D-arabinofuranose (**9**) (Scheme 2.88) [204].

Reductive double cyclization is an efficient strategy for the synthesis of polyhydroxylated pyrrolizidines [53, 205, 206], as was demonstrated by Pearson et al. in their synthesis of (+)-australine (**186**) and (−)-7-epi-alexine (Scheme 2.89a) [205]. Another nice demonstration of the efficiency of the reductive double cyclization strategy was made by Wong and coworkers, in which a skillful combination of an improved Sharpless asymmetric epoxidation of divinylcarbinol (**64**) with Payne rearrangement, an enzymatic aldol reaction, and an intramolecular bis-reductive amination led to a concise synthesis of australine (**186**) (Scheme 2.89b) [53].

The Cope–House cyclization is a useful method for the construction of pyrrolidine rings. This strategy was used for the synthesis of pyrrolizidines 5-epi-hyacinthacine A$_3$ and 5-epi-hyacinthacine A$_5$ (Scheme 2.90) [207].

Besides serving as electrophiles in the nucleophilic addition with Grignard reagents, the umpolung of nitrones can be achieved by a samarium diiodide–induced coupling of nitrones with electron-deficient olefins. This paved the route for a very concise synthesis of (+)-hyacinthacine A$_2$ (**187**) (Scheme 2.91) [208].

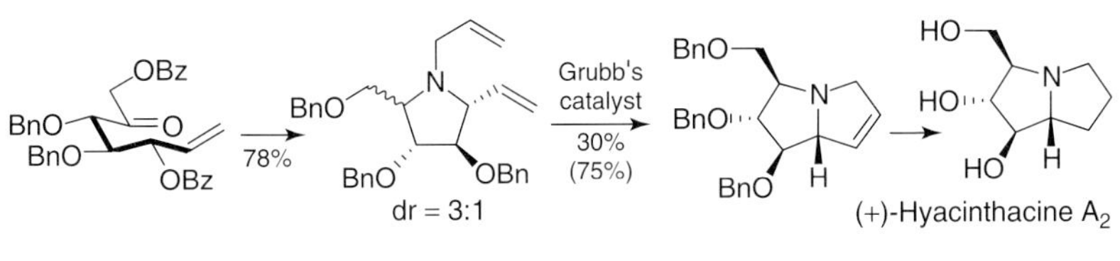

Scheme 2.88 Martin's synthesis of (+)-hyacinthacine A$_2$.

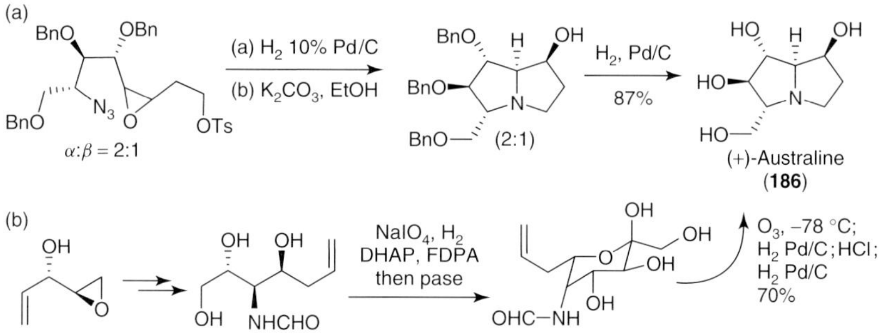

Scheme 2.89 Pearson's and Wong's syntheses of (+)-australine. DHAP = dihydroxyacetone phosphate; FDPA = fructose-1,6-diphosphate aldolase; and Pase = acid phosphatase.

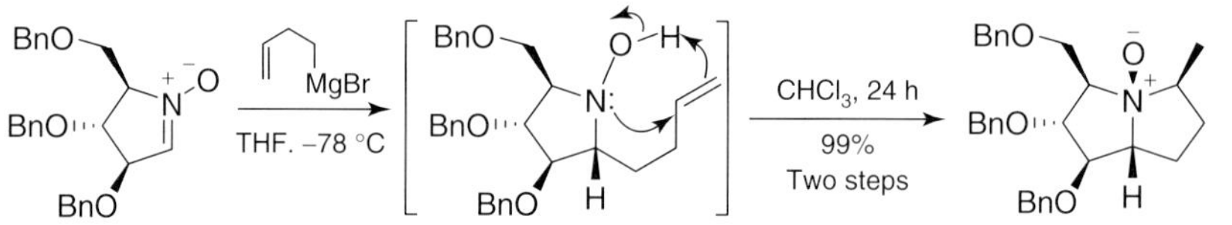

Scheme 2.90 Py's synthesis of (+)-hyacinthacine A_2.

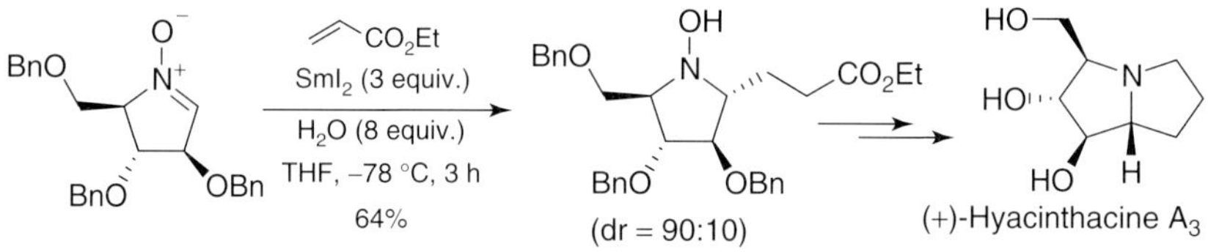

Scheme 2.91 Py's synthesis of (+)-hyacinthacine A_3.

By regio- and diastereoselective 1,3-dipolar cycloaddition of functionalized dipolarphiles with pyrrolidine nitrones, several pyrrolizidines and indolizidines have been synthesized. For example, the 1,3-dipolar cycloaddition of nitrone **188** with ethyl 4-bromocrotonate provided a highly selective and concise synthesis of (+)-heliotridine (Scheme 2.92) [209]. A similar strategy was also used in the synthesis of hyacinthacine A_2 and 7-deoxycasuarine [210].

Blechert's enantiospecific synthesis of (+)-hyacinthacine A_2 (**187**) exhibits the efficiency of a convergent synthesis. Indeed, after synthesizing fragments A and B respectively, it needs only three steps to combine the two fragments and convert the coupled product into the target molecule hyacinthacine A_2, in which two rings and three chiral centers have been formed stereoselectively (Scheme 2.93) [211].

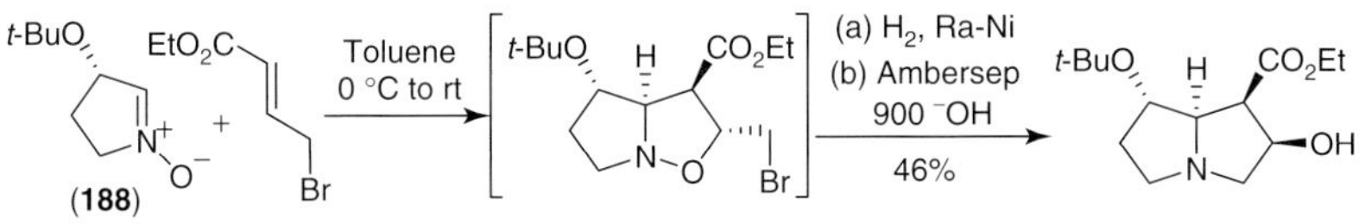

Scheme 2.92 Brandi's synthesis of (+)-heliotridine.

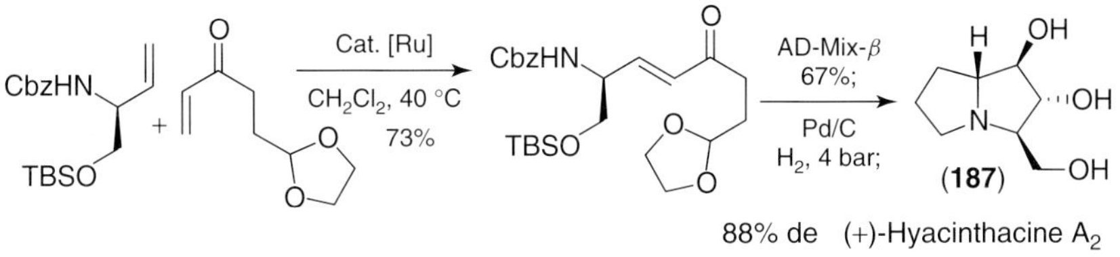

Scheme 2.93 Blechert's enantiospecific convergent synthesis (+)-hyacinthacine A_2.

2.4
Acknowledgments

The author thanks Ms Yan-Jiao Gao and Ms Jing-Wei Chen for technical assistance in preparing this manuscript. We are indebted to Professor G. Michael Blackburn for valuable discussions and help during the preparation of this manuscript.

References

1. (a) Liddell, J.R. (**1999**) *Nat. Prod. Rep.*, **16**, 499–507; (b) O'Hagan, D. (**2000**) *Nat. Prod. Rep.*, **17**, 435–46; (c) Burgess, K. and Henderson, I. (**1992**) *Tetrahedron*, **48**, 4045–66; (d) Michael, J.P. (**2005**) *Nat. Prod. Rep.*, **22**, 603–26.
2. (a) Sobin, B.A. and Tanner, F.W. (**1954**) *J. Am. Chem. Soc.*, **76**, 4053–54; (b) Hosoya, Y., Kameyama, T., Naganawa, H., Okami, Y. and Takeuchi, T. (**1993**) *J. Antibiot.*, **46**, 1300–2.
3. Hollmann, M. and Heinemann, S. (**1994**) *Annu. Rev. Neurosci.*, **17**, 31–108.
4. (a) Omura, S., Fujimoto, T., Otoguro, K., Matsuzaki, K., Moriguchi, R., Tanaka, H. and Sasaki, Y. (**1991**) *J. Antibiot.*, **44**, 113–16; (b) Omura, S., Natsuzaki, K., Fujimoto, T., Kosuge, K., Furuya, T., Fujita, S. and Nakagawa, A. (**1991**) *J. Antibiot.*, **44**, 117–18.
5. Feling, R.H., Buchanan, G.O., Mincer, T.J., Kauffman, C.A., Jensen, P.R. and Fenical, W. (**2003**) *Angew. Chem., Int. Ed.*, **42**, 355–57.
6. Iwanami, S., Takashima, M., Hirata, Y., Hasegawa, O. and Usuda, S. (**1981**) *J. Med. Chem.*, **24**, 1224–30.
7. Maring, C., McDaniel, K., Krueger, A., Zhao, C., Sun, M., Madigan, D., DeGoey, D., Chen, H.-J., Yeung, M.C., Flosi, W., Grampovnik, D., Kati, W., Klein, L., Stewart, K., Stoll, V., Saldivar, A., Montgomery, D., Carrick, R., Steffy, K., Kempf, D., Molla, A., Kohlbrenner, W., Kennedy, A., Herrin, T., Xu, Y. and Laver, W.G. (**2001**) *Antiviral Res.*, **50**, A76, Abstract 129.
8. (a) Sweet, J.A., Cavallari, J.M., Price, W.A., Ziller, J.W. and McGrath, D.V. (**1997**) *Tetrahedron: Asymmetry*, **8**, 207–11; (b) Fache, F., Schulz, E., Tommasino, M.L. and Lemaire, M. (**2000**) *Chem. Rev.*, **100**, 2159–231.

9 (a) Mukaiyama, T., Sakito, Y. and Asami, M. (1979) *Chem. Lett.*, 705–8; (b) Enders, D. and Thiebes, C. (2001) *Pure Appl. Chem.*, **73**, 573–78.
10 Dalko, P.I. and Moisan, L. (2004) *Angew. Chem., Int. Ed.*, **43**, 5138–75.
11 Sardina, F.J. and Rapoport, H. (1996) *Chem. Rev.*, **96**, 1825–72.
12 Jouin, P. and Castro, B. (1987) *J. Chem. Soc., Perkin Trans. 1*, 1177–82.
13 Lay, L., Nicotra, F., Paganini, A., Pangrazio, C. and Panza, L. (1993) *Tetrahedron Lett.*, **34**, 4555–58.
14 (a) Yoda, H., Yamazaki, H. and Takabe, K. (1996) *Tetrahedron: Asymmetry*, **7**, 373–74; (b) Yoda, H., Nakajima, T., Yamazaki, H. and Takabe, K. (1995) *Heterocycles*, **41**, 2423–26; (c) Yoda, H., Yamazaki, H., Kawauchi, M. and Takabe, K. (1995) *Tetrahedron: Asymmetry*, **6**, 2669–72; (d) Yoda, H., Katoh, H. and Takabe, K. (2000) *Tetrahedron Lett.*, **41**, 7661–65; (e) Yoda, H. (2002) *Curr. Org. Chem.*, **6**, 223–43.
15 Liotta, L.J. and Ganem, B. (1990) *Synlett*, 503–4.
16 Francisco, C.G., Freire, R., Gonzalez, C.C., Leon, E.I., Riesco-Fagundo, C. and Suarez, E. (2001) *J. Org. Chem.*, **66**, 1861–66.
17 Moriarty, R.M., Mitan, C.I., Branza-Nichita, N., Phares, K.R. and Parrish, D. (2006) *Org. Lett.*, **8**, 3465–67.
18 Ziegler, T., Straub, A. and Effenberger, F. (1988) *Angew. Chem., Int. Ed. Engl.*, **27**, 716–17.
19 Takaoka, Y., Kajimoto, T. and Wong, C.H. (1993) *J. Org. Chem.*, **58**, 4809–12.
20 Lindstrom, U.M., Ding, R. and Hidestal, O. (2005) *Chem. Commun.*, 1773–74.
21 Reddy, J.S. and Rao, B.V. (2007) *J. Org. Chem.*, **72**, 2224–27.
22 Canova, S., Bellosta, V. and Cossy, J. (2004) *Synlett*, 1811–13.
23 Davies, S.G., Garner, A.C., Goddard, E.C., Kruchinin, D., Roberts, P.M., Smith, A.D., Rodriguez-Solla, H., Thomson, J.E. and Toms, S.M. (2007) *Org. Biomol. Chem.*, **5**, 1961–69.
24 Matsubara, R., Kawai, N. and Kobayashi, S. (2006) *Angew. Chem., Int. Ed.*, **45**, 3814–16.
25 Itoh, K. and Kanemasa, S. (2002) *J. Am. Chem. Soc.*, **124**, 13394–95.
26 Barnes, D.M., Ji, J.G., Fickes, M.G., Fizgerald, M.A., King, S.A., Morton, H.E., Plagge, F.A., Preskill, M., Wagwa, S.H., Wittenberger, S.J. and Zhang, J. (2002) *J. Am. Chem. Soc.*, **124**, 13097–105.
27 Dixon, D.J., Ley, S.V. and Rodriguez, F. (2001) *Org. Lett.*, **3**, 3753–55.
28 Fukuda, N., Sasaki, K., Sastry, T.V.R.S., Kanai, M. and Shibasaki, M. (2006) *J. Org. Chem.*, **71**, 1220–25.
29 Okue, M., Watanabe, H. and Kitahara, T. (2001) *Tetrahedron*, **57**, 4107–10.
30 Verma, S.K., Atanes, M.N., Busto, J.H., Thai, D.L. and Rapoport, H. (2002) *J. Org. Chem.*, **67**, 1314–18.
31 Wakabayashi, T. and Saito, M. (1977) *Tetrahedron Lett.*, **18**, 93–96.
32 Harding, K.E. and Marman, T.H. (1984) *J. Org. Chem.*, **49**, 2838–40.
33 Takahata, H., Banba, Y., Tajima, M. and Momose, T. (1991) *J. Org. Chem.*, **56**, 240–45.
34 Singh, S., Chikkanna, D., Singh, O.V. and Han, H. (2003) *Synlett*, 1279–982.
35 Overhand, M. and Hecht, S.M. (1994) *J. Org. Chem.*, **59**, 4721–22.
36 Bertrand, M.B. and Wolfe, J.P. (2006) *Org. Lett.*, **8**, 2353–56.
37 Noguchi, Y., Uchiro, H., Yamada, T. and Kobayashi, S. (2001) *Tetrahedron Lett.*, **42**, 5253–56.
38 Jones, A.D., Knight, D.W. and Hibbs, D.E. (2001) *J. Chem. Soc., Perkin Trans. 1*, 1182–203.
39 Davis, F.A., Song, M. and Augustine, A. (2006) *J. Org. Chem.*, **71**, 2779–86.
40 Kim, J.H., Curtis-Long, M.J., Seo, W.D., Ryu, Y.B., Yang, M.S. and Park, K.H. (2005) *J. Org. Chem.*, **70**, 4082–87.
41 Donohoe, T.J., Wheelhouse, K.M.P., Lindsay-Scott, P.J., Glossop, P.A.,

Nash, I.A. and Parker, J.S. (2008) *Angew. Chem., Int. Ed.*, **47**, 2872–75.
42. Schomaker, J.M., Geiser, A.R., Huang, R. and Borhan, B. (2007) *J. Am. Chem. Soc.*, **129**, 3794–95.
43. Salmon, A. and Carboni, B. (1998) *J. Organomet. Chem.*, **567**, 31–37.
44. Girard, S., Robins, R.J., Villieras, J. and Lebreton, J. (2000) *Tetrahedron Lett.*, **41**, 9245–49.
45. Behr, J.-B., Kalla, A., Harakat, D. and Plantier-Royon, R. (2008) *J. Org. Chem.*, **73**, 3612–15.
46. Ackermann, L. and Althammer, A. (2008) *Synlett*, 995–98.
47. Aillaud, I., Collin, J., Duhayon, C., Guillot, R., Lyubov, D., Schulz, E. and Trifonov, A. (2008) *Chem. Eur. J.*, **14**, 2189–200.
48. Davis, F.A., Xu, H., Wu, Y.Z. and Zhang, J.Y. (2006) *Org. Lett.*, **8**, 2273–76.
49. Deng, Q.H., Xu, H.W., Yuen, A.W.H., Xu, Z.J. and Che, C.M. (2008) *Org. Lett.*, **10**, 1529–32.
50. Dong, C.Q., Mo, F. and Wang, J.B. (2008) *J. Org. Chem.*, **73**, 1971–74.
51. Shi, Z.-C. and Lin, G.-Q. (1995) *Tetrahedron: Asymmetry*, **6**, 2907–10.
52. O'Neil, I.A., Cleator, E., Southern, J.M., Hone, N. and Tapolczay, D.J. (2000) *Synlett*, 1408–10.
53. Romero, A. and Wong, C.H. (2000) *J. Org. Chem.*, **65**, 8264–68.
54. Hanrahan, J.R., Knight, D.W. and Salter, R. (2001) *Synlett*, 1587–89.
55. (a) Denmark, S.E. and Schnute, M.E. (1994) *J. Org. Chem.*, **59**, 4576–95; (b) Denmark, S.E. and Hurd, A.R. (1998) *J. Org. Chem.*, **63**, 3045–50.
56. Kawabata, T., Moriyama, K., Kawakami, S. and Tsubaki, K. (2008) *J. Am. Chem. Soc.*, **130**, 4153–57.
57. Kolaczkowski, L. and Barnes, D.M. (2007) *Org. Lett.*, **9**, 3029–32.
58. Yan, Z.H., Wang, J.Q. and Tian, W.S. (2003) *Tetrahedron Lett.*, **44**, 9383–84.
59. Sakaguchi, H., Tokuyama, H. and Fukuyama, T. (2007) *Org. Lett.*, **9**, 1635–38.
60. Quancard, J., Labonne, A., Jacquot, Y., Chassaing, G., Lavielle, S. and Karoyan, P. (2004) *J. Org. Chem.*, **69**, 7940–48.
61. (a) Ahari, M., Joosten, A., Vasse, J.L. and Szymoniak, J. (2008) *Synthesis*, 61–68; (b) Denhez, C., Vasse, J.L., Harakat, D. and Szymoniak, J. (2007) *Tetrahedron: Asymmetry*, **18**, 424–34.
62. Pinto, A.C., Abdala, R.V. and Costa, P.R.R. (2000) *Tetrahedron: Asymmetry*, **11**, 4239–43.
63. Barco, A., Benetti, S. and Spalluto, G. (1992) *J. Org. Chem.*, **57**, 6279–86.
64. Reddy, L.R., Fournier, J.F., Reddy, B.V.S. and Corey, E.J. (2005) *J. Am. Chem. Soc.*, **127**, 8974–76.
65. Reddy, L.R., Saravanan, R. and Corey, E.J. (2004) *J. Am. Chem. Soc.*, **126**, 6230–31.
66. Makino, K., Kondoh, A. and Hamada, Y. (2002) *Tetrahedron Lett.*, **43**, 4695–98.
67. Yoon, C.H., Flanigan, D.L., Chong, B.D. and Jung, K.W. (2002) *J. Org. Chem.*, **67**, 6582–84.
68. Oppolzer, W. and Thirring, K. (1982) *J. Am. Chem. Soc.*, **104**, 4978–79.
69. Martinez, M.M. and Hoppe, D. (2004) *Org. Lett.*, **6**, 3743–46.
70. Xia, Q. and Ganem, B. (2001) *Org. Lett.*, **3**, 485–87.
71. Mcgrane, P.L. and Livinghouse, T. (1993) *J. Am. Chem. Soc.*, **115**, 11485–89.
72. Scott, M.E. and Lautens, M. (2005) *Org. Lett.*, **7**, 3045–47.
73. Jackson, S.K., karadeolian, A., Driega, A.B. and Kerr, M.A. (2008) *J. Am. Chem. Soc.*, **130**, 4196–201.
74. (a) Pandey, G., Banerjee, P. and Gadre, S.R. (2006) *Chem. Rev.*, **106**, 4484–517; (b) Coldham, I. and Hufton, R. (2005) *Chem. Rev.*, **105**, 2765–809.
75. Garber, P., Dogan, O., Youngs, W.J., Kennedy, V.O., Protasiewicz, J. and Zaniewski, R. (2001) *Tetrahedron*, **57**, 71–85.
76. Agbodjan, A.A., Cooley, B.E., Copley, R.C.B., Corfield, J.A., Flanagan, R.C., Glover, B.N., Guidetti, R., Haigh, D., Howes, P.D., Jackson, M.M., Matsuoka, R.T., Medhurst, K.J., Millar, A., Sharp, M.J., Slater,

M.J., Toczko, J.F. and Xie, S. (**2008**) *J. Org. Chem.*, **73**, 3094–102.

77 Garner, P., Hu, J.Y., Parker, C.G., Youngs, W.J. and Medvetz, D. (**2007**) *Tetrahedron Lett.*, **48**, 3867–70.

78 Garner, P., Kaniskan, H.U., Hu, J., Youngs, W.J. and Panzner, M. (**2006**) *Org. Lett.*, **8**, 3647–50.

79 (a) Ibrahem, I., Rios, R., Vesely, J. and Cordova, A. (**2007**) *Tetrahedron Lett.*, **48**, 6252–57; (b) Cabrera, S., Arrayás, R.G., Martín-Matute, B., Cossío, F.P. and Carretero, J.C. (**2007**) *Tetrahedron*, **63**, 6587–602.

80 (a) Vicario, J.L., Badia, D. and Carrillo, L. (**2001**) *J. Org. Chem.*, **66**, 5801–7; (b) Akiyama, T., Ishida, Y. and Kagoshima, H. (**1999**) *Tetrahedron Lett.*, **40**, 4219–22.

81 Poisson, J.F. and Normant, J.F. (**2001**) *Org. Lett.*, **3**, 1889–91.

82 Schomaker, J.M., Bhattacharjee, S., Yan, J. and Borhan, B. (**2007**) *J. Am. Chem. Soc.*, **129**, 1996–2003.

83 Reggelin, M. and Heinrich, T. (**1998**) *Angew. Chem., Int. Ed.*, **37**, 2883–86.

84 Restorp, P., Fischer, A. and Somfai, P. (**2006**) *J. Am. Chem. Soc.*, **128**, 12646–47.

85 Hirner, S. and Somfai, P. (**2005**) *Synlett*, 3099–102.

86 Alcaide, B., Almendros, P. and Aragoncillo, C. (**2007**) *Chem. Rev.*, **107**, 4437–92.

87 (a) Palomo, C., Cossio, F.P., Cuevas, C., Odriozola, J.M. and Ontoria, J.M. (**1992**) *Tetrahedron Lett.*, **33**, 4827–30; (b) Jayaraman, M., Puranik, V.G. and Bhawal, B.M. (**1996**) *Tetrahedron Lett.*, **52**, 9005–16.

88 (a) Kanazawa, A., Gillet, S., Delair, P. and Greene, A.E. (**1998**) *J. Org. Chem.*, **63**, 4660–63; (b) Delair, P., Brot, E., Kanazawa, A. and Greene, A.E. (**1999**) *J. Org. Chem.*, **64**, 1383–86; (c) Ceccon, J., Poisson, J.-F. and Greene, A.E. (**2005**) *Synlett*, 1413–16.

89 Chelucci, G. and Saba, A. (**1995**) *Angew. Chem. Int. Ed. Engl.*, **34**, 78–79.

90 Streith, J. and Defoin, A. (**1996**) *Synlett*, 189–200.

91 Pichon, M. and Figadère, B. (**1996**) *Tetrahedron: Asymmetry*, **7**, 927–64.

92 Seebach, D., Sting, A.R. and Hoffmann, M. (**1996**) *Angew. Chem., Int. Ed. Engl.*, **35**, 2709–48.

93 Watanabe, H., Okue, M., Kobayashi, H. and Kitahara, T. (**2002**) *Tetrahedron Lett.*, **43**, 861–64.

94 Hughes, C.C. and Trauner, D. (**2002**) *Angew. Chem., Int. Ed.*, **41**, 4556–59.

95 (a) Beak, P., Basu, A., Gallagher, D.J., Park, Y.S. and Thayumanavan, S. (**1996**) *Acc. Chem. Res.*, **29**, 552–60; (b) Kizirian, J.C. (**2008**) *Chem. Rev.*, **108**, 140–205.

96 Sunose, M., Peakman, T.M., Charmant, J.P.H., Gallagher, T. and Macdonald, S.J.F. (**1998**) *Chem. Commun.*, 1723–24.

97 Zheng, X., Feng, C.-G. and Huang, P.-Q. (**2005**) *Org. Lett.*, **7**, 553–56.

98 Williams, R.M., Cao, J. and Tsujishima, H. (**2000**) *Angew. Chem., Int. Ed.*, **39**, 2541–44.

99 Ooi, T., Miki, T. and Maruoka, K. (**2005**) *Org. Lett.*, **7**, 191–93.

100 Beak, P., Kerrick, S.T., Wu, S. and Chu, J. (**1994**) *J. Am. Chem. Soc.*, **116**, 3231–39.

101 Hermet, J.-P.R., Porter, D.W., Dearden, M.J., Harrison, J.R., Koplin, T., O'Brien, P., Parmene, J., Tyurin, V., Whitwood, A.C., Gilday, J. and Smith, N.M. (**2003**) *Org. Biomol. Chem.*, **1**, 3977–88.

102 Coldham, I., Dufour, S., Haxell, T.F.N., Patel, J.J. and Sanchez-Jimenez, G. (**2006**) *J. Am. Chem. Soc.*, **128**, 10943–51.

103 Gawley, R.E., Zhang, Q.H. and Campagna, S. (**1995**) *J. Am. Chem. Soc.*, **117**, 11817–18.

104 Huang, P.Q., Wu, T.J. and Ruan, Y.P. (**2003**) *Org. Lett.*, **5**, 4341–44.

105 (a) Nájera, C. and Yus, M. (**1999**) *Tetrahedron: Asymmetry*, **10**, 2245–303; (b) Ager, D.J., Prakash, I. and Schaad, D.R. (**1996**) *Chem. Rev.*, **96**, 835–76.

106 Rosset, S., Célérier, J.P. and Lhommet, G. (**1991**) *Tetrahedron Lett.*, **32**, 7521–24.

107 Provot, O., Célérier, J.P., Petit, H. and Lhommet, G. (**1992**) *J. Org. Chem.*, **57**, 2163–66.

108 Potts, D., Stevenson, P.J. and Thompson, N. (**2000**) *Tetrahedron Lett.*, **41**, 275–78.

109 Shiosaki, K. and Rapoport, H. (**1985**) *J. Org. Chem.*, **50**, 1229–39.

110 Cuny, G.D. and Buchwald, S.L. (**1995**) *Synlett*, 519–22.

111 Dhimanea, H., Vanucci-Bacqué, C., Hamonb, L. and Lhommet, G. (**1998**) *Eur. J. Org. Chem.*, 1955–63.

112 Brown, D.S., Charreau, P., Hansson, T. and Ley, S.V. (**1991**) *Tetrahedron*, **47**, 1311–28.

113 Moloney, M.G., Panchal, T. and Pike, R. (**2006**) *Org. Biomol. Chem.*, **4**, 3894–97.

114 Giovannini, A., Savoia, D. and Umani-Ronchi, A. (**1989**) *J. Org. Chem.*, **54**, 228–34.

115 Brenneman, J.B., Machauer, R. and Martin, S.F. (**2004**) *Tetrahedron*, **60**, 7301–14.

116 (a) Speckamp, W.N. and Moolenaar, M.J. (**2000**) *Tetrahedron*, **56**, 3817–56; (b) Marson, C.M. (**2001**) *Arkivoc*, part 1, 1–16, at *www.arkat-usa.org*; (c) Maryanoff, B.E., Zhang, H.-C., Cohen, J.H., Turchi, I.J. and Maryanoff, C.A. (**2004**) *Chem. Rev.*, **104**, 1431–628; (d) Royer, J. (**2004**) *Chem. Rev.*, **104**, 2311–52.

117 Yamazaki, N., Ito, T. and Kibayashi, C. (**2000**) *Org. Lett.*, **2**, 465–67.

118 (a) Burgess, L.E. and Meyers, A.I. (**1991**) *J. Am. Chem. Soc.*, **113**, 9858–59; (b) Burgess, L.E. and Meyers, A.I. (**1992**) *J. Org. Chem.*, **57**, 1656–62; (c) Maury, C., Wang, Q., Gharbaoui, T., Chiadmi, M., Tomas, A., Royer, J. and Husson, H.-P. (**1997**) *Tetrahedron*, **53**, 3627–36.

119 Groaning, M.D. and Meyers, A.I. (**2000**) *Tetrahedron*, **56**, 9843–73.

120 Andres, J.M., Herraiz-Sierra, I., Pedrosa, R. and Perez-Encabo, A. (**2000**) *Eur. J. Org. Chem.*, 1719–26.

121 Suzuki, H., Aoyagi, S. and Kibayashi, C. (**1994**) *Tetrahedron Lett.*, **35**, 6119–22.

122 Meyers, A.I. and Burgess, L.E. (**1991**) *J. Org. Chem.*, **56**, 2294–96.

123 (a) Koot, W.J., Ginkel, R.V., Kranenburg, M., Hiemstra, H., Louwrier, S., Moolenaar, M.J. and Speckamp, W.N. (**1991**) *Tetrahedron Lett.*, **32**, 401–4; (b) Louwrier, S., Ostendorf, M., Boom, A., Hiemstra, H. and Speckamp, W.N. (**1996**) *Tetrahedron*, **52**, 2603–28.

124 Armas, P.D., Garcia-Tellado, F., Marrero-Tellado, J.J. and Robles, J. (**1998**) *Tetrahedron Lett.*, **39**, 131–34.

125 Smith, A.B., Saivatore, B.A., Hull, K.G. and Duan, J.J.W. (**1991**) *Tetrahedron Lett.*, **32**, 4859–62.

126 Ryu, Y. and Kim, G. (**1995**) *J. Org. Chem.*, **60**, 103–8.

127 Keum, G. and Kim, G. (**1994**) *Bull. Korean Chem. Soc.*, **15**, 278–79.

128 (a) Taning, M. and Wistrand, L.G. (**1990**) *J. Org. Chem.*, **55**, 1406–8; (b) Lennartz, M., Sadakane, M. and Steckhan, E. (**1999**) *Tetrahedron*, **55**, 14407–20.

129 Han, G., LaPorte, M.G., McIntosh, M.C., Weinreb, S.M. and Parvez, M. (**1996**) *J. Org. Chem.*, **61**, 9483–93.

130 Washburn, D.G., Heidebrecht, R.W. Jr. and Martin, S. (**2003**) *Org. Lett.*, **5**, 3523–25.

131 Huang, P.-Q., Xu, T. and Chen, A.-Q. (**2000**) *Synth. Commun.*, **30**, 2259–68.

132 Morgan, I.R., Yazici, A. and Pyne, S.G. (**2008**) *Tetrahedron*, **64**, 1409–19.

133 Vieira, A.S., Ferreira, F.P., fiorante, P.F., Guadagnin, R.C. and Stefanni, H.A. (**2008**) *Tetrahedron*, **64**, 3306–14.

134 (a) Yoda, H., Kitayama, H., Yamada, W., Katagiri, T. and Takabe, K. (**1993**) *Tetrahedron: Asymmetry*, **4**, 1451–55; (b) Yoda, H., Nakajima, T. and Takabe, K. (**1996**) *Tetrahedron Lett.*, **37**, 5531–34.

135 (a) Schuch, C.M. and Pilli, R.A. (**2000**) *Tetrahedron: Asymmetry*, **11**, 753–64; (b) Schuch, C.M. and Pilli, R.A. (**2002**) *Tetrahedron: Asymmetry*, **13**, 1973–80.

136 (a) Huang, P.-Q., Wang, S.L., Zheng, H. and Fei, X.-S. (**1997**) *Tetrahedron Lett.*, **38**, 271–72; (b) He, B.-Y., Wu, T.-J., Yu, X.-Y. and Huang,

P.-Q. (2003) *Tetrahedron: Asymmetry*, **14**, 2101–8; (c) Huang, P.-Q. (2005) Recent advances on the asymmetric synthesis of bioactive 2-pyrrolidinone-related compounds starting from enantiomeric malic acid, in *New Methods for the Asymmetric Synthesis of Nitrogen Heterocycles* (eds J.L. Vicario, D. Badia and L. Carrillo), Research Signpost, Kerala, pp. 197–222; (d) Huang, P.-Q. (2006) *Synlett*, 1133–47.

137 Kim, S.H., Park, Y., Choo, H. and Cha, J.K. (2002) *Tetrahedron Lett.*, **43**, 6657–60.

138 Niwa, H., Miyachi, Y., Okamoto, O., Uosaki, Y., Kuroda, A., Ishiwata, H. and Yamada, K. (1992) *Tetrahedron*, **48**, 393–412.

139 Revuelta, J., Cicchi, S., Goti, A. and Brandi, A. (2007) *Synthesis*, 485–504.

140 Merino, P.C. (2005) *Comptes Rendues Chim.*, **8**, 775–88.

141 Gothelf, K.V. and Jorgensen, K.A. (2000) *Chem. Commun.*, 1449–58.

142 (a) Cicchi, S., Goti, A. and Brandi, A. (1995) *J. Org. Chem.*, **60**, 4743–48; (b) Goti, A., Cicchi, S., Fedi, V., Nannelli, L. and Brandi, A. (1997) *J. Org. Chem.*, **62**, 3119–25.

143 Cicchi, S., Corsi, M. and Goti, A. (1999) *J. Org. Chem.*, **64**, 7243–45.

144 Soldaini, G., Cardona, F. and Goti, A. (2007) *Org. Lett.*, **9**, 473–76.

145 Murahashi, S.I., Imada, Y. and Ohtake, H. (1994) *J. Org. Chem.*, **59**, 6170–72.

146 Ballini, R., Marcantoni, E. and Petrini, M. (1992) *J. Org. Chem.*, **57**, 1316–18.

147 Shen, J.W., Qin, D.G., Zhang, H.W. and Yao, Z.J. (2003) *J. Org. Chem.*, **68**, 7479–84.

148 Ohtake, H., Imada, Y. and Murahashi, S.I. (1999) *Bull. Chem. Soc. Jpn.*, **72**, 2737–54.

149 Yu, C.-Y. and Huang, M.-H. (2006) *Org. Lett.*, **8**, 3021–24.

150 Toyao, A., Tamura, O., Takagi, H. and Ishibashi, H. (2003) *Synlett*, 35–38.

151 Nagasawa, K., Georgieva, A., Koshino, H., Nakata, T., Kita, T. and Hashimoto, Y. (2002) *Org. Lett.*, **4**, 177–80.

152 Miranda, P.C.M.L. and Correia, C.R.D. (1999) *Tetrahedron Lett.*, **40**, 7735–38.

153 Bach, T., Brummerhop, H. and Harms, K. (2000) *Chem. Eur. J.*, **6**, 3838–48.

154 Kuwano, R., Kashiwabara, M., Ohsumi, M. and Kusano, H. (2008) *J. Am. Chem. Soc.*, **130**, 808–9.

155 (a) Davies, H.M.L., Hansen, T., Hopper, D.W. and Panaro, S.A. (1999) *J. Am. Chem. Soc.*, **121**, 6509–10; (b) Wee, A.G.H. (2001) *J. Org. Chem.*, **66**, 8513–17.

156 Punniyamurthy, T., Irie, R. and Katsuki, T. (2000) *Chirality*, **12**, 464–68.

157 (a) Ezquerra, J., Pedregal, C., Rubio, A., Yruretagoyena, B., Escribano, A. and Sánchez-Ferrando, F. (1993) *Tetrahedron*, **49**, 8665–78; (b) Dikshit, D.K. and Panday, S.K. (1992) *J. Org. Chem.*, **57**, 1920–24; (c) Breña-Valle, L.J., Sánchez, R.C. and Cruz-Almanza, R. (1996) *Tetrahedron: Asymmetry*, **7**, 1019–26.

158 (a) Herdeis, C. and Hubmann, P. (1992) *Tetrahedron: Asymmetry*, **3**, 1213–21; (b) Langlois, N., Calvez, O. and Radom, M.O. (1997) *Tetrahedron Lett.*, **38**, 8037–40.

159 (a) Hanessian, S. and Ratovelomanana, V. (1990) *Synlett*, 501–3; (b) Baldwin, J.E., Moloney, M.G. and Shim, S.B. (1991) *Tetrahedron Lett.*, **32**, 1379–80.

160 Langlois, N. and Rakaotondradany, F. (2000) *Tetrahedron*, **56**, 2437–48.

161 Blanco, M.J. and Sardina, F.J. (1999) *J. Org. Chem.*, **64**, 4748–55.

162 Pandey, S.K., Orellana, A., Greene, A. and Poisson, J.F. (2006) *Org. Lett.*, **8**, 5665–68.

163 Poisson, J.F., Orellana, A. and Greene, A.E. (2005) *J. Org. Chem.*, **70**, 10860–63.

164 Breuil-Desvergnes, B. and Goré, J. (2000) *Tetrahedron*, **56**, 1951–60.

165 Flogel, O., Amombo, M.G.O., Reibig, H.U., Zahn, G., Brudgam, I. and Hartl, H. (2003) *Chem. Eur. J.*, **9**, 1405–14.

166 Fuchibe, K., Hatemata, R. and Akiyama, T. (**2005**) *Tetrahedron Lett.*, **46**, 8563–66.
167 Boynton, C.M., Hewson, A.T. and Mitchell, D. (**2000**) *J. Chem. Soc., Perkin Trans. 1*, 3599–602.
168 Gawley, R.E., Barolli, G., Madan, S., Saverin, M. and O'Connor, S. (**2004**) *Tetrahedron Lett.*, **45**, 1759–61.
169 Hanessian, S., Bayrakdarian, M. and Luo, X.H. (**2002**) *J. Am. Chem. Soc.*, **124**, 4716–21.
170 Felpin, F.-X. and Lebreton, J. (**2003**) *Eur. J. Org. Chem.*, 3693–712.
171 Murruzzu, C. and Riera, A. (**2007**) *Tetrahedron: Asymmetry*, **18**, 149–54.
172 Evans, P.A. and Robinson, J.E. (**1999**) *Org. Lett.*, **1**, 1929–31.
173 Trost, B.M., Horne, D.B. and Woltering, M.J. (**2003**) *Angew. Chem., Int. Ed.*, **42**, 5987–90.
174 Welter, C., Moreno, R.M., Streiff, S. and Helmchen, G. (**2005**) *Org. Biomol. Chem.*, **3**, 3266–68.
175 Hayes, C.J., Sherlock, A.E., Green, M.P., Wilson, C., Blake, A.J., Selby, M.D. and Prodger, J.C. (**2008**) *J. Org. Chem.*, **73**, 2041–51.
176 (a) Donohoe, T.J., Guyo, P.M. and Helliwell, M. (**1999**) *Tetrahedron Lett.*, **40**, 435–38; (b) Schafer, A. and Schafer, B. (**1999**) *Tetrahedron*, **55**, 12309–12. (c) Donohoe, T.J. and Thomas, R.E. (**2007**) *Chem. Rec.*, **7**, 180–90.
177 Donohoe, T.J., Freestone, G.C., Headley, C.E., Rigby, C.L., Cousins, R.C.C. and Bhalay, G. (**2004**) *Org. Lett.*, **6**, 3055–58.
178 (a) Meyers, A.I., Dickman, D.A. and Bailey, T.R. (**1985**) *J. Am. Chem. Soc.*, **107**, 7974–78; (b) Meyers, A.I. (**1996**) *Tetrahedron*, **52**, 2589–612.
179 Baussanne, I., Chiaroni, A., Husson, H.P., Eiche, C. and Royer, J. (**1994**) *Tetrahedron Lett.*, **35**, 3931–34.
180 (a) Casiraghi, G., Rassu, G., Spanu, P. and Pinna, L. (**1992**) *J. Org. Chem.*, **57**, 3760–63; (b) Rassu, G., Zanardi, F., Battistini, L. and Casiraghi, G. (**2000**) *Chem. Soc. Rev.*, **29**, 109–18; (c) Casiraghi, G., Zanardi, F., Appendino, G. and Rassu, G. (**2000**) *Chem. Rev.*, **100**, 1929–72.
181 Uno, H., Baldwin, J.E. and Russell, A.T. (**1994**) *J. Am. Chem. Soc.*, **116**, 2139–40.
182 Planas, L., Perard-Viret, J. and Royer, J. (**2004**) *J. Org. Chem.*, **69**, 3087–92.
183 Bur, S.K. and Martin, S.F. (**2001**) *Tetrahedron*, **57**, 3221–42.
184 Barnes, D.M., Bhagavatula, L., DeMattei, J., Gupta, A., Hill, D.R., Manna, S., McLaughlin, M.A., Nichols, P., Premchandran, R., Rasmussen, M.W., Tian, Z.P. and Wittenberger, S.J. (**2003**) *Tetrahedron: Asymmetry*, **14**, 3541–51.
185 Severino, E.A. and Correia, C.R.D. (**2000**) *Org. Lett.*, **2**, 3039–42.
186 David, O., Blot, J., Bellec, C., Fargeau-Bellassoued, M.C., Haviari, G., Célérier, J.P., Lhommet, G., Gramain, J.C. and Gardette, D. (**1999**) *J. Org. Chem.*, **64**, 3122–31.
187 Watson, R.T., Gore, V.K., Chandupatla, K.R., Dieter, R.K. and Snyder, J.P. (**2004**) *J. Org. Chem.*, **69**, 6105–14.
188 Tang, T., Ruan, Y.-P., Ye, J.-L. and Huang, P.-Q. (**2005**) *Synlett*, 231–34.
189 (a) Reddy, P.V., Veyron, A., Koos, P., Bayle, A., Greene, A.E. and Delair, P. (**2008**) *Org. Biomol. Chem.*, **6**, 1170–72; (b) Roche, C., Kadlecikova, K., Veyron, A., Delair, P., Philouze, C., Greene, A.E., Flot, D. and Burghammer, M. (**2005**) *J. Org. Chem.*, **70**, 8352–63.
190 Craig, D., Hyland, C.J.T. and Ward, S.E. (**2006**) *Synlett*, 2142–44.
191 Takano, S., Otaki, S. and Ogasawara, K. (**1983**) *J. Chem. Soc., Chem. Commun.*, 1172–74.
192 (a) Huang, P.-Q., Arseniyadis, S. and Husson, H.-P. (**1987**) *Tetrahedron Lett.*, **28**, 547–50; (b) Arseniyadis, S., Huang, P.-Q. and Husson, H.-P. (**1988**) *Tetrahedron Lett.*, **29**, 1391–94; (c) Husson, H.P. and Royer, J. (**1999**) *Chem. Soc. Rev.*, **28**, 383–94.
193 Takahato, H., Takahashi, S., Azer, N., Eldefrawi, A.T. and Eldefrawi, M.E. (**2000**) *Bioorg. Med. Chem. Lett.*, **10**, 1293–95.
194 (a) Snider, B.B. and Gao, X.L. (**2005**) *Org. Lett.*, **7**, 4419–22; (b) Izquierdo, I., Plaza, M.T.

and Franco, F. (**2002**) *Tetrahedron: Asymmetry*, **13**, 1581–85.

195 (a) Schobert, R., Jagusch, C., Melanophy, C. and Mullen, G. (**2004**) *Org. Biomol. Chem.*, **2**, 3524–29; (b) Schobert, R. and Wicklein, A. (**2007**) *Synthesis*, 1499–502.

196 Chamberlin, A.R. and Chung, Y.L. (**1985**) *J. Org. Chem.*, **50**, 4425–31.

197 Keck, G.E., Cressman, E.N.K. and Enholm, E.J. (**1989**) *J. Org. Chem.*, **54**, 4345–49.

198 Ha, D.-C., Yun, C.-S. and Lee, Y. (**2000**) *J. Org. Chem.*, **65**, 621–23.

199 Hoye, T.R., Dvornikovs, V. and Sizova, E. (**2006**) *Org. Lett.*, **8**, 5191–94.

200 Nagao, Y., Dai, W.M., Ochiai, M., Tsukagoshi, S. and Fujita, E. (**1988**) *J. Am. Chem. Soc.*, **110**, 289–91.

201 Pilli, R.A. and Russowskym, D. (**1996**) *J. Org. Chem.*, **61**, 3187–90.

202 Ito, H., Ikeuchi, Y., Taguchi, T. and Hanzawa, Y. (**1994**) *J. Am. Chem. Soc.*, **116**, 5469–70.

203 White, J.D. and Hrnciar, P. (**2000**) *J. Org. Chem.*, **65**, 9129–42.

204 Rambaud, L., Compain, P. and Martin, O.R. (**2001**) *Tetrahedron: Asymmetry*, **12**, 1807–9.

205 Pearson, W.H. and Hines, J.V. (**2000**) *J. Org. Chem.*, **65**, 5785–93.

206 Ribes, C., Falomir, E., Carda, M. and Marco, J.A. (**2007**) *Org. Lett.*, **9**, 77–80.

207 Kaliappan, K.P. and Das, P. (**2008**) *Synlett*, 841–44.

208 Desvergnes, S., Py, S. and Vallée, Y. (**2005**) *J. Org. Chem.*, **70**, 1459–62.

209 Pisaneschi, F., Cordero, F.M. and Brandi, A. (**2003**) *Eur. J. Org. Chem.*, 4373–75.

210 Cardona, F., Faggti, E., Liguori, F., Cacciarini, M. and Goti, A. (**2003**) *Tetrahedron Lett.*, **44**, 2315–18.

211 Dewi-Wulfing, P. and Blechert, S. (**2006**) *Eur. J. Org. Chem.*, 1852–56.

3
Asymmetric Synthesis of Six-Membered Ring Heterocycles

Naoki Toyooka

3.1
Introduction

The functionalized six-membered heterocycles with one nitrogen (piperidine ring systems) have been found in many biologically and medicinally important natural and nonnatural products. Among the nitrogen heterocycles, the piperidine ring probably represents the more frequently encountered substructure of natural or unnatural products. Indeed, many methodologies for the asymmetric construction of this ring system have been developed. This chapter gathers the most recent methods described so far for the asymmetric access of these products, and is divided into four sections: dihydropyridines, tetrahydropyridines, monocyclic and carbocyclic fused piperidines and finally fused tri- or bicyclic systems with bridgehead nitrogen. The first two sections describe the preparation of unsaturated systems that are often used as intermediates in the syntheses of piperidines, thus the methods reported in these two sections may also be useful for the synthesis of piperidines. Owing to the very important literature and several available reviews [1] on this topic, only the recent literature from 2000 to 2007 is reviewed.

3.2
Dihydropyridines

The main interest of dihydropyridine compounds is their reactivity. They can be transformed to yield highly functionalized and substituted piperidines. Indeed, there are quite unstable products and, of course, there are, in most of the cases, intermediates in the asymmetric synthesis of piperidine derivatives. Thus, the preparation of dihydropyridines should be considered as supplementary methods to attain the saturated piperidine compounds.

Direct addition of nucleophile on pyridine ring in an enantioselective manner is one of the most effective methods for the construction of chiral six-membered

Asymmetric Synthesis of Nitrogen Heterocycles. Edited by Jacques Royer
Copyright © 2009 WILEY-VCH Verlag GmbH & Co. KGaA, Weinheim
ISBN: 978-3-527-32036-3

ring system. The pioneering work on this area was reported from Comins' group [2]. This was the method that Charette used to elaborate a practical and highly regio- and stereoselective synthesis of 2-substituted dihydropyridine. The strategy involved the addition of Grignard reagents to chiral pyridinium salt prepared from **1** and pyridine in the presence of Tf_2O to yield the 2-substituted dihydropyridines. These compounds are valuable synthetic intermediates since an enantioselective synthesis of (−)-coniine was obtained by this methodology in very few steps [3]. Owing to the other application of this method to natural product synthesis, the expedient asymmetric synthesis of (+)-julifloridine in four steps was achieved from **2** (R=$C_{12}H_{24}OBn$) by monohydrogenation followed by a one-pot, highly diastereoselective epoxidation–nucleophilic addition of Me_2Zn. On the other hand, synthesis of the nonracemic 2,5-cis-disubstituted piperidine **3** has also been accomplished (Scheme 3.1) [4].

The first catalytic asymmetric addition reaction of nucleophile to N-acylpyridinium ion was reported by Shibasaki and Kanai using Lewis acid–Lewis base bifunctional asymmetric catalyst. Thus, the catalytic enantioselective Reissert reaction of pyridine was achieved in the presence of chiral ligand **4** (10 mol%) and Et_2AlCl (5 mol%) to yield the adduct **5** (98% yield with 91% ee). Interestingly, the use of **4**, which is not C_2 symmetric, showed best regio- (1,6- vs. 1,2-; 12:1–50:1) and enantioselectivity (up to 96% ee) on this reaction. Adduct **5** was converted to the key intermediate **6** for the synthesis of dopamine D_4 receptor-selective antagonist CP-293,019 (Scheme 3.2) [5].

Ma's group also reported the catalytic enantioselective addition of activated terminal alkynes to 1-acylpyridinium salts using Cu-bis(oxazoline) complexes. Among the chiral bis(oxazoline) catalysts, use of less bulky ligand **7** showed best results, and two adducts **8** and **9** were obtained with 94% and 91% ee, respectively. Both adducts were converted to poison-frog alkaloids **167B** and **223AB**, respectively, as shown in Scheme 3.3 [6].

Yamada reported the synthesis of chiral 1,4-dihydropyridines by face-selective addition to a cation-π complex of a pyridinium salt. The addition reaction of ketene

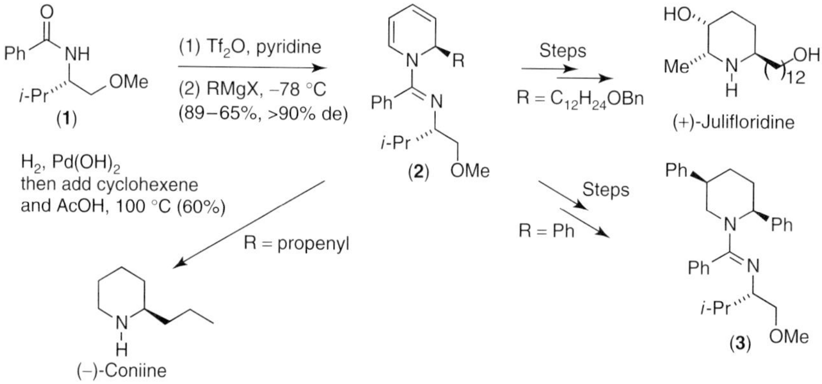

Scheme 3.1

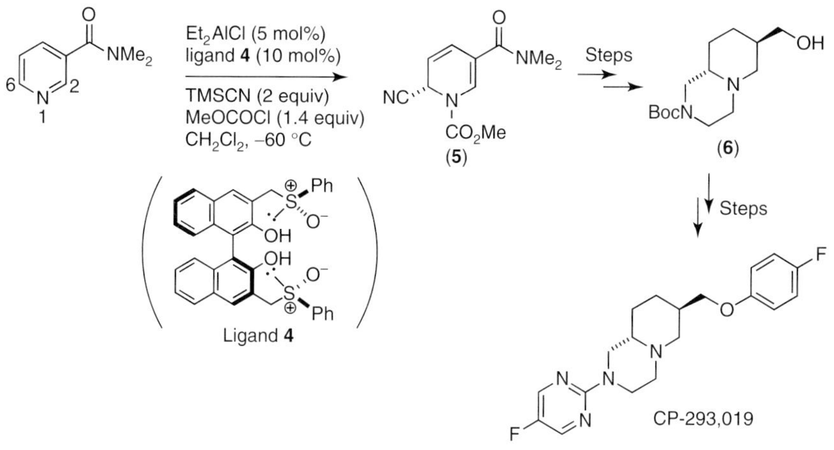

3.2 Dihydropyridines

Scheme 3.2

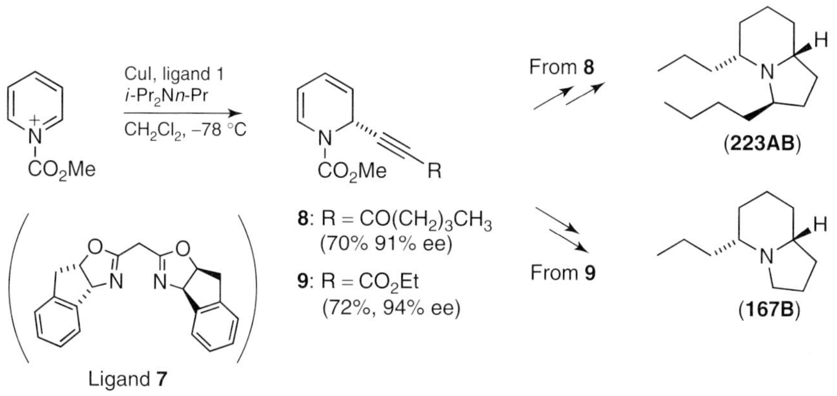

Scheme 3.3

silyl acetals to the pyridinium salt of nicotinic amide derivatives proceeded in good diastereoselectivity to yield the 1,4-dihydropyridine derivatives. The nucleophile attacked on more stable conformer **10** (from *ab initio* calculations) from less hindered top face to yield the product (Scheme 3.4) [7].

The strategy of Comins was broadened by his author with the highly regio- and diastereoselective addition of several nucleophiles such as Grignard reagents, lithium acetylide, and metallo enolates to chiral *N*-acylpyridinium salts to give enantiopure *N*-acyl-2,3-dihydro-4-pyridones. According to this methodology, the first asymmetric synthesis of natural (+)-cannabisativine [8] from **11**, a concise asymmetric synthesis of (+)-allopumiliotoxin **267A** [9] from **12**, the asymmetric synthesis of (+)-β-conhydrine [10] from **13**, and synthetic studies toward (−)-FR901483 [11] from **14** have been reported as shown in Scheme 3.5.

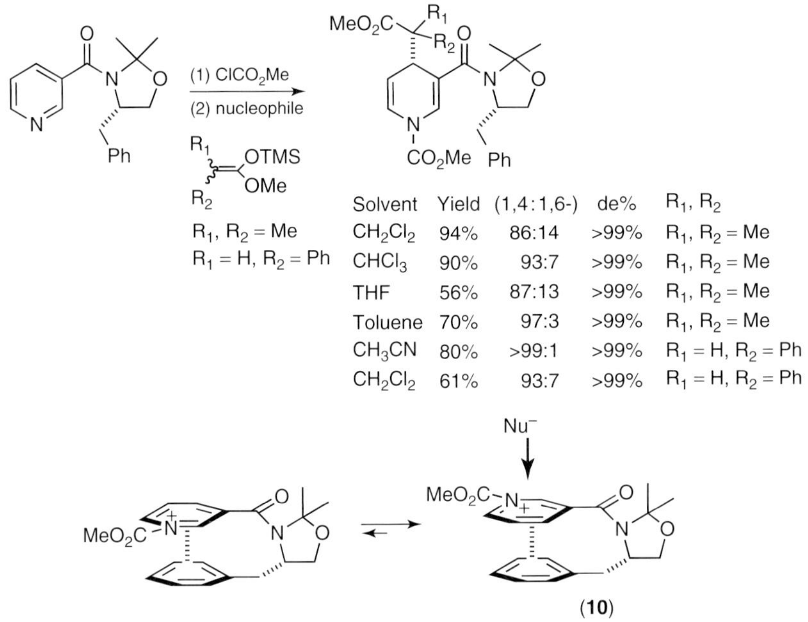

Scheme 3.4

Solvent	Yield	(1,4:1,6-)	de%	R_1, R_2
CH_2Cl_2	94%	86:14	>99%	$R_1, R_2 = Me$
$CHCl_3$	90%	93:7	>99%	$R_1, R_2 = Me$
THF	56%	87:13	>99%	$R_1, R_2 = Me$
Toluene	70%	97:3	>99%	$R_1, R_2 = Me$
CH_3CN	80%	>99:1	>99%	$R_1 = H, R_2 = Ph$
CH_2Cl_2	61%	93:7	>99%	$R_1 = H, R_2 = Ph$

3.3
Tetrahydropyridines

Tetrahydropyridines are more stable compounds than dihydropyridines, but, as dihydropyridines, are mainly prepared as intermediates in the synthesis of piperidines. The position where the unsaturation takes place is important: the 1,2,3,4-tetrahydropyridines are enamines that can react at the C-3 position with electrophiles giving access to more substituted compounds.

Though several methods were described for the construction of substituted tetrahydropyridines, the most versatile methodology is probably the ring-closing metathesis (RCM), which has emerged as one of the most powerful tools in organic synthesis. The total synthesis of piperidine and pyrrolidine natural alkaloids with RCM as a key step has been reviewed by Lebreton in 2003 [12].

3.3.1
Ring-Closing Metathesis (RCM)

Davis reported the asymmetric synthesis of (+)-CP-99,994 by using the RCM reaction of diene **19**, derived from sulfinimine-derived 2,3-diamino ester **17**. Thus, the addition reaction of enolate of **15** to chiral sulfinamide **16** yielded the differentially N-protected diamino ester **17**. The latter was converted to alcohol **18**, which was then transformed into **19**. The RCM reaction of **19** using Grubbs–Hoveyda catalyst

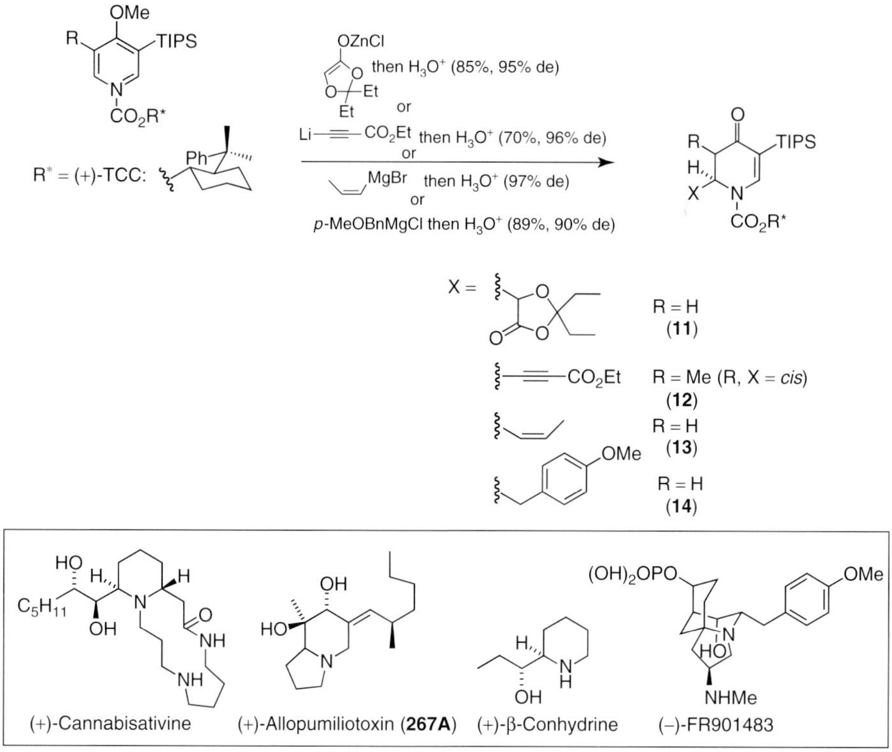

3.3 Tetrahydropyridines

Scheme 3.5

yielded tetrahydropyridine **20** in high yield, which was converted to (+)-CP-99,994 in four-step sequence (Scheme 3.6) [13].

Lebreton achieved the chiral synthesis of (−)-3-epi-deoxoprosopinine by using chiral imino alcohol **21** and compound **23**. The diastereoselective allylation of **21** yielded the *trans*-adduct **22** as the major product and subsequent RCM reaction of

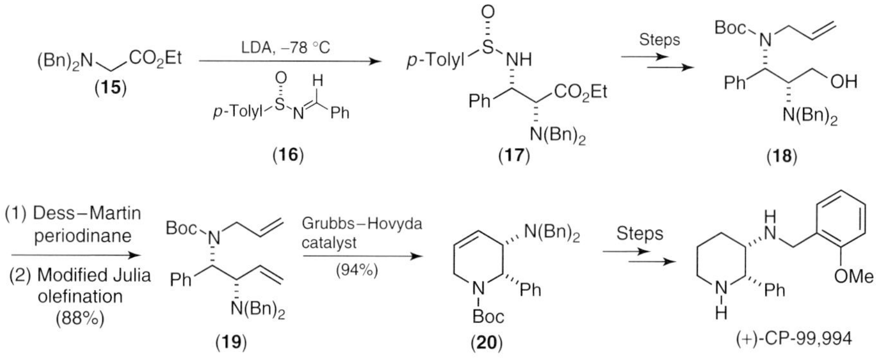

Scheme 3.6

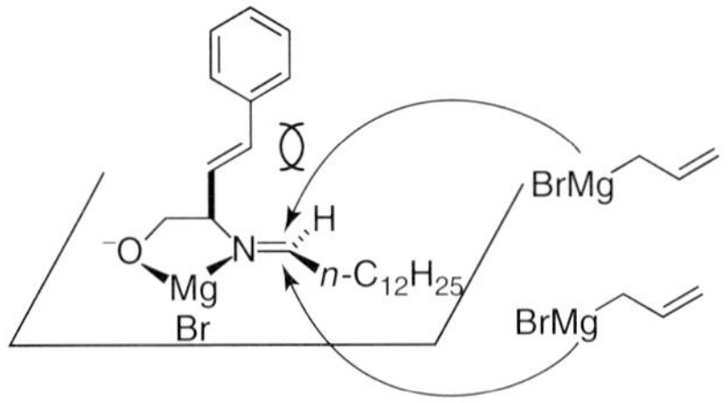

Fig. 3.1 Stereochemical course of Grignard addition to **21**.

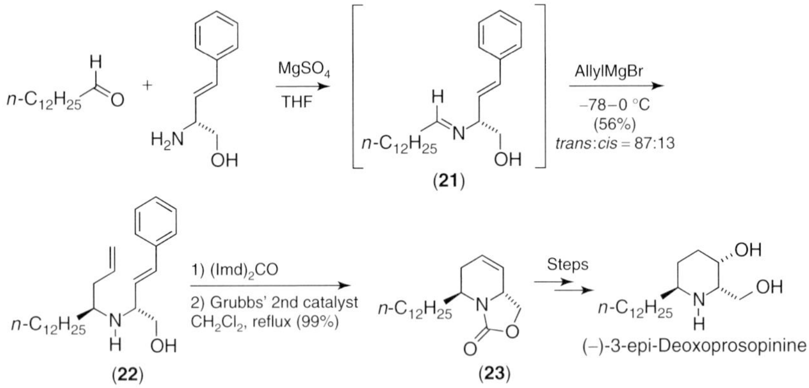

Scheme 3.7

oxazolizinone yielded **23**. The diastereoselectivity of the above addition is explained by the chelation-controlled model shown in Figure 3.1. Epoxidation of **23**, reductive opening of the epoxide followed by alkaline hydrolysis of the oxazolizinone ring, yielded (−)-3-epi-deoxoprosopinine (Scheme 3.7) [14].

In the Lee's chiral synthesis of tetrahydropyridines, the dienes required for the RCM were prepared via the regioselective opening of the chiral aziridine ring systems as the key step. Both chiral aziridines **24** and **25** were treated with PhSH to yield the amino alcohol **26** or **27** in highly regioselective manner, which was converted to tetrahydropyridines **30** or **31** via dienes **28** or **29**, respectively (Scheme 3.8) [15].

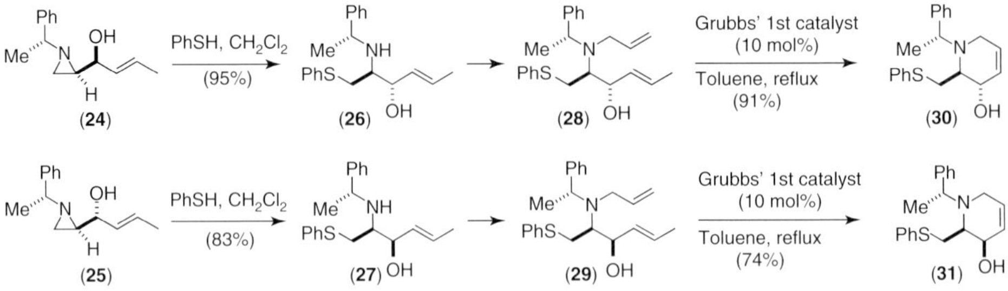

Scheme 3.8

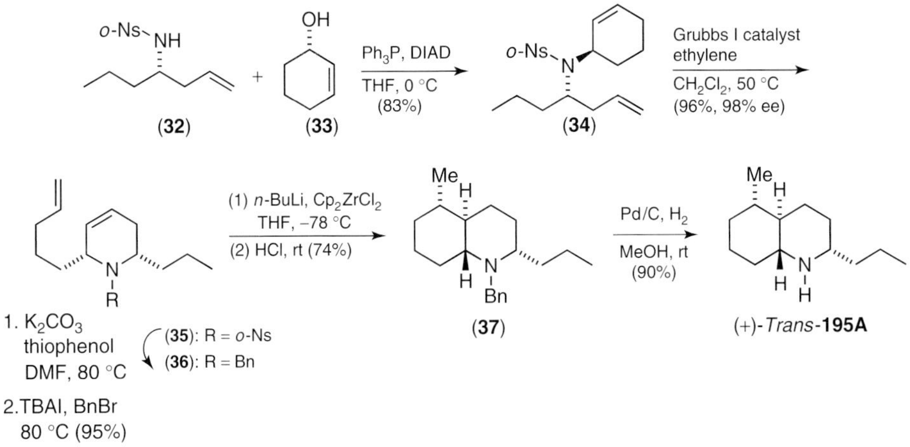

Scheme 3.9

An interesting ring-rearrangement metathesis was developed by Blechert in the total synthesis of poison-frog alkaloid (+)-*trans*-195A. Thus, the Mitsunobu reaction of the chiral sulfonamide **32** with chiral cyclohexenol **33** yielded the secondary amide **34**, which was subjected to ring-rearrangement metathesis using Grubbs I catalyst to yield **35**. The protecting group on nitrogen in **35** was crucial to achieve the next Negishi-coupling reaction. Negishi-coupling reaction of the *N*-benzyl derivative **36** proceeded smoothly to yield **37**, which was converted to (+)-*trans*-195A (Scheme 3.9) [16].

3.3.2
Reduction of Pyridine Derivatives

The chiral Brønsted acid-catalyzed reduction of pyridine derivatives is also an efficient method to prepare tetrahydropyridines or octahydroquinolinones in good ee. Thus, the reduction of several pyridine derivatives, shown below, in the presence of chiral Brønsted acid **38** (5 mol%) and Hantzsch diethyl ester (4 equiv) as the hydride source yielded the reduction products in good ee. This methodology was applied to formal enantioselective synthesis of di-epi-pumiliotoxin C (Scheme 3.10) [17].

3.3.3
Deracemization Processes

Takahata reported the asymmetric synthesis of tetrahydropiperidinol chiral building block **39** using the palladium-catalyzed deracemization of corresponding carbonate (Scheme 3.11) [18]. Thus, the reaction was performed using 8 mol% of phosphine ligand ((*R*)-BPA) developed by Trost and 2 mol% of $Pd_2(dba)_3$ in CH_2Cl_2/H_2O (9:1) to yield **39** (88–94% yield and 87–99% ee), which is the key chiral building block for the synthesis of biologically important alkaloids such as 1-azasugars [19].

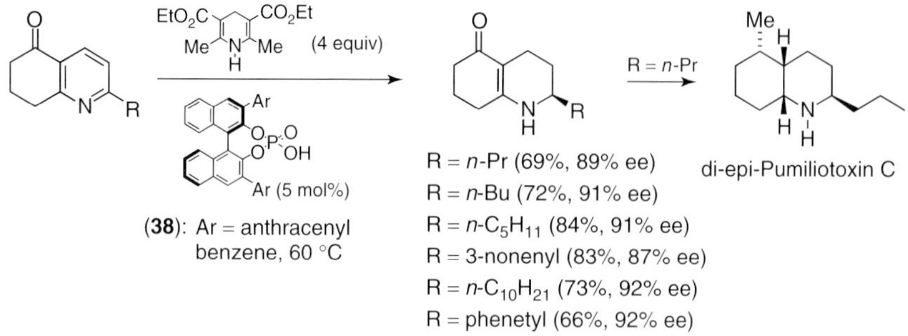
EtO2C CO2Et
Me N Me (4 equiv)
H
Ar
O P O
O OH
Ar (5 mol%)
(38): Ar = anthracenyl
benzene, 60 °C
R = n-Pr
R = n-Pr (69%, 89% ee)
R = n-Bu (72%, 91% ee)
R = n-C5H11 (84%, 91% ee)
R = 3-nonenyl (83%, 87% ee)
R = n-C10H21 (73%, 92% ee)
R = phenetyl (66%, 92% ee)
di-epi-Pumiliotoxin C

Scheme 3.10

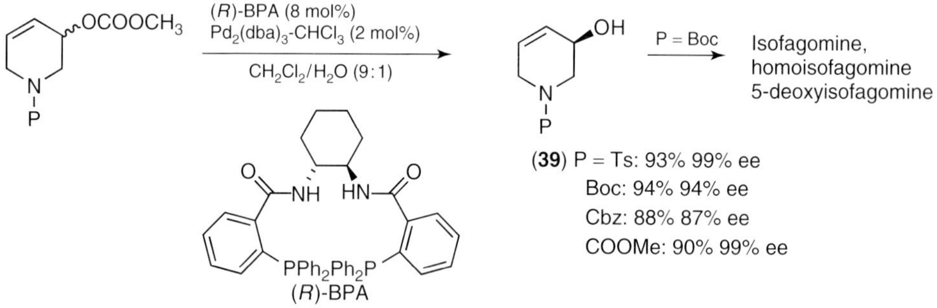
OCOOCH3
(R)-BPA (8 mol%)
Pd2(dba)3-CHCl3 (2 mol%)
CH2Cl2/H2O (9:1)
OH
P = Boc
Isofagomine,
homoisofagomine
5-deoxyisofagomine
(39) P = Ts: 93% 99% ee
Boc: 94% 94% ee
Cbz: 88% 87% ee
COOMe: 90% 99% ee
PPh2Ph2P
(R)-BPA

Scheme 3.11

3.3.4 Michael Addition Followed by Elimination

The synthesis of tetrahydropyridines using the diastereoselective intramolecular Michael cyclization of vinylsulfinyl-containing amino alcohol as the key step, as shown in Scheme 3.12, was described by Delgado. The authors designed a "one-pot" procedure based on the removal of *N*-Boc followed by neutralization with excess Et_3N and "in situ" cyclization under high-dilution conditions at 50 °C to yield the piperidine **40** or indolizidine **41** with high diastereoselectivity. Elimination of the sulfoxide group yielded the tetrahydropyridines **42** or **43**, respectively [20].

Davis used the intramolecular Michael-type addition reaction followed by retro-Michael elimination reaction of the *N*-sulfinyl δ -amino β-keto phosphonate to yield dihydropyridone **44** (Scheme 3.13). The conjugate addition reaction of **44** resulted in the selective formation of 2,6-*trans*-disubstituted tetrahydropyridine **45**, which was used for asymmetric synthesis of (−)-myrtine [21].

3.3.5 Enamine Reaction

The intramolecular enamine reaction in the electrophilic center was used by several groups to prepare the piperidine ring through the C3–C4 bond formation. In the

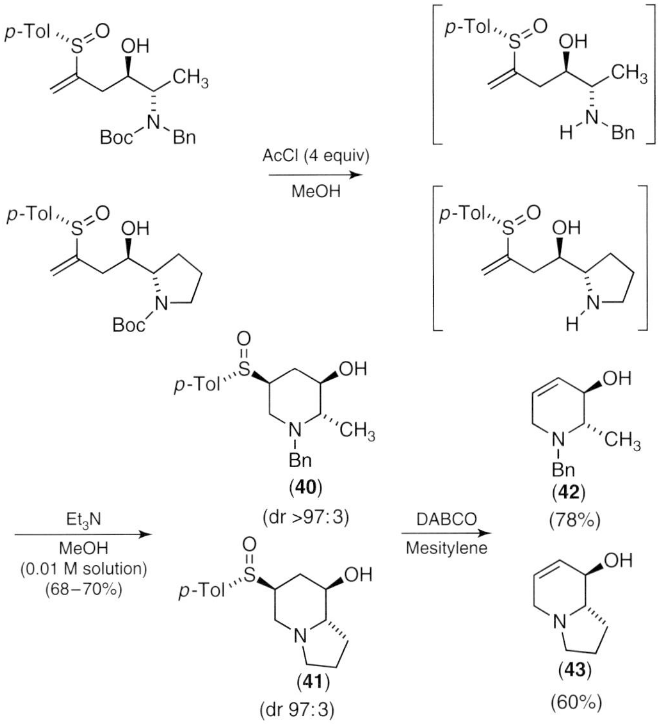

Scheme 3.12

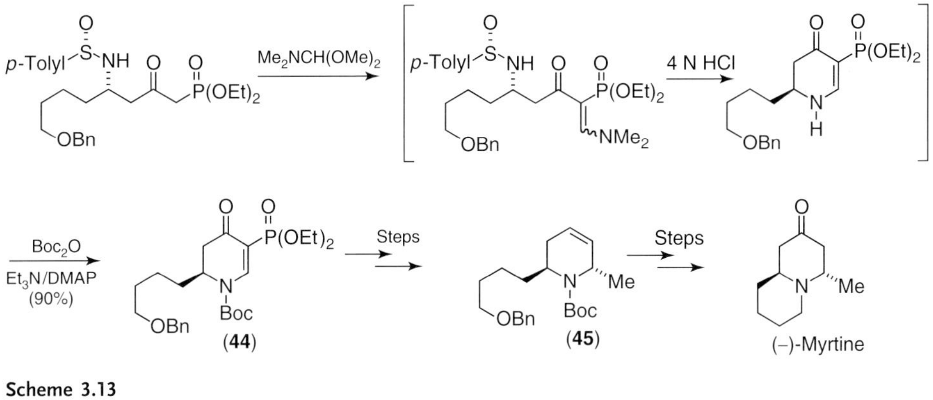

Scheme 3.13

synthesis of indolizidine alkaloid, Ma reported the efficient construction of the functionalized cyclic enamines by the Michael addition of enantiopure amino alcohols to alkynones followed by intramolecular cyclization reaction. On the basis of this methodology, the asymmetric syntheses of poison-frog alkaloid (−)-**223A** and its 6-epimer [22], (−)-deoxoprosophylline [23], and (−)-deoxocassine and (+)-azimic acid [24] have been achieved. The enantiopure amino alcohols **46** and **47** were prepared by Davies' method [25], and were added to alkynone **48** by Michael

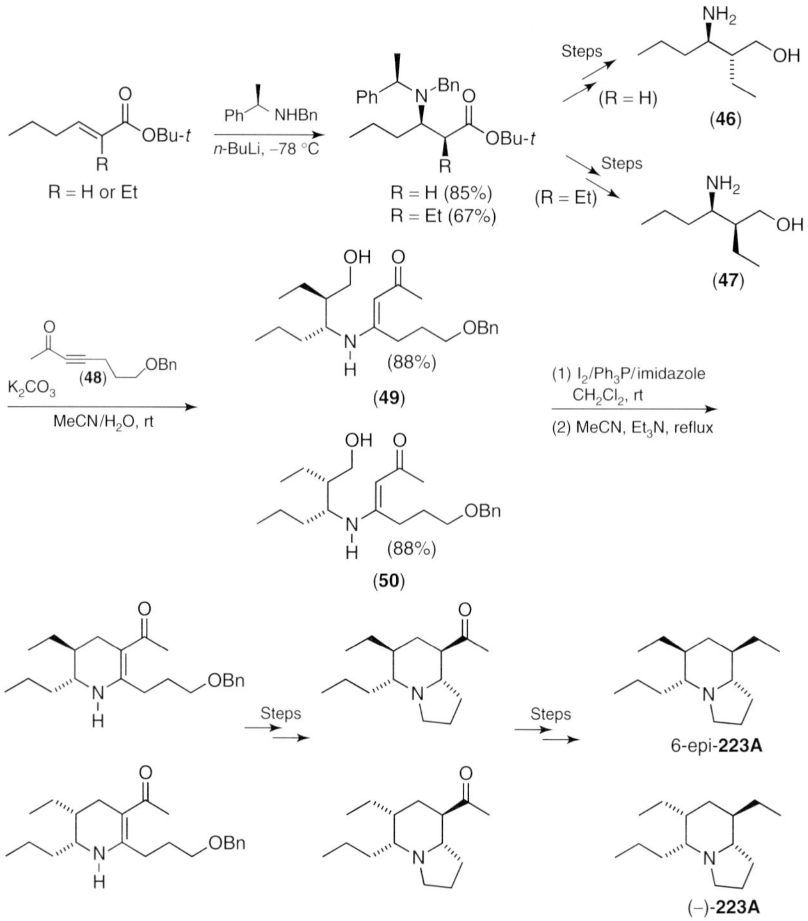

Scheme 3.14

addition reaction to yield the enamines **49** and **50**, respectively. The key two-step cyclization of **49** or **50** yielded the cyclic enaminones, which were converted to 6-epi-**223A** (originally proposed structure for natural **223A**) [26] and (−)-**223A** by modifications on the side chain, reduction of the ketone moiety, and indolizidine formation reaction (Scheme 3.14).

This methodology was developed to one-pot procedure, and then the enantioselective total synthesis of the poison-frog alkaloid (−)-**209I** has been achieved. The key one-pot reaction was performed by the reaction of chloride **51** with conjugated alkynone to yield the piperidine (**52**) after hydrogenation and epimerization at the 3-position, which was converted to (−)-**209I** in some steps (Scheme 3.15) [27].

The total synthesis of lepadins B, D, E, and H and determination of the absolute configuration of lepadins D, E, and H were also achieved by Ma using the same methodology. The key feature of Ma's synthesis was the construction of octahydro

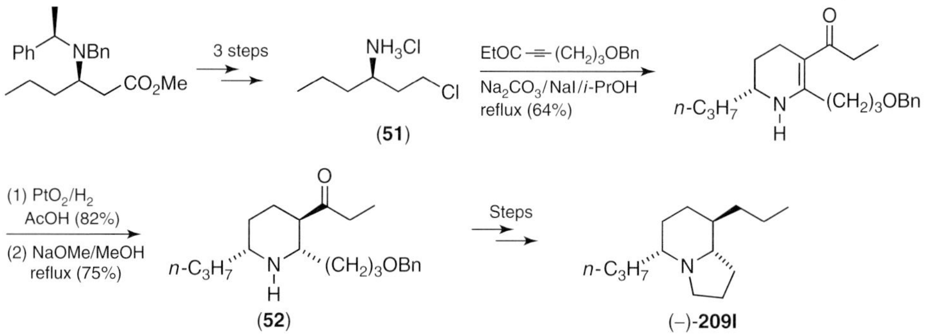

3.3 Tetrahydropyridines

Scheme 3.15

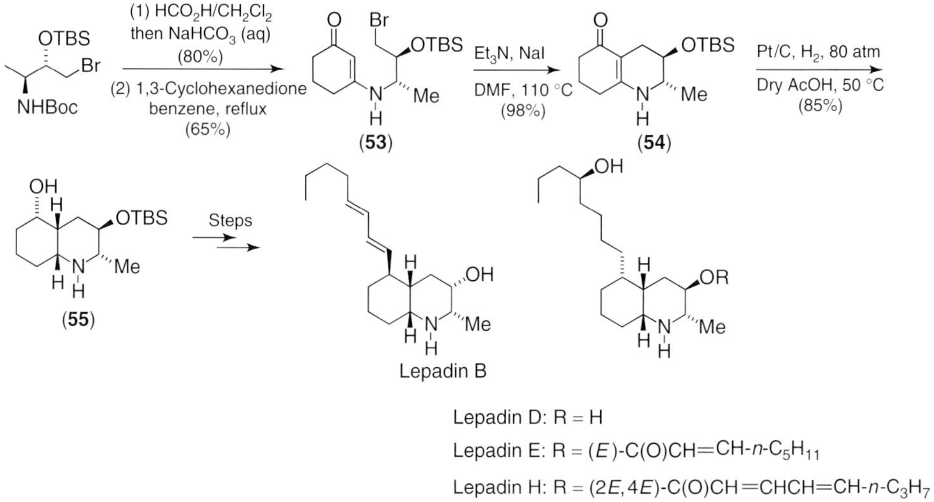

Scheme 3.16

quinolinone ring system by alkylative cyclization of **53**. The diastereoselective hydrogenation of **54** was performed in the presence of Pt/C catalyst by a stereoelectronically controlled axial addition of hydrogen to yield the common intermediate **55** for the synthesis of lepadins after Boc protection and oxidation of the hydroxyl group of the resulting amino alcohol. Elaborations on **55** yielded lepadins B, D, E, and H. The absolute configuration of later three alkaloids was determined by these total syntheses (Scheme 3.16) [28, 29].

3.3.6
Electrocyclization

Katsumura reported the asymmetric 6π-azaelectrocyclization utilizing the 7-alkyl substituted *cis*-1-amino-2-indanols **56** as the novel chiral auxiliary. Then the method

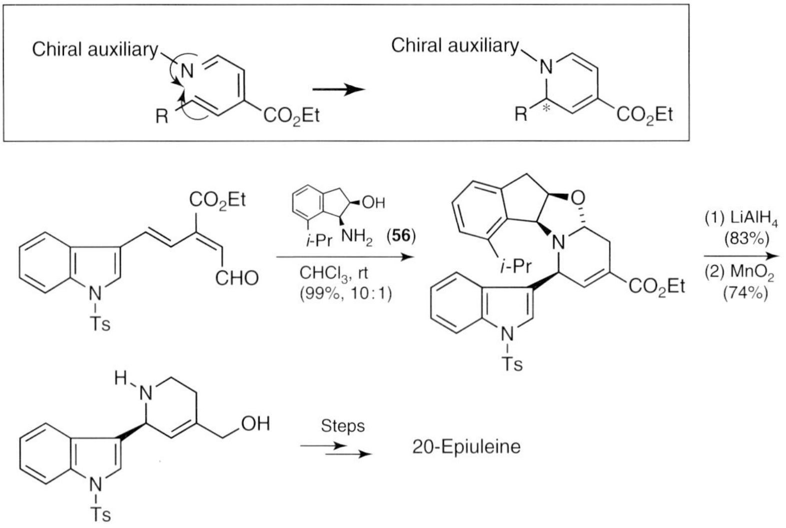

Scheme 3.17

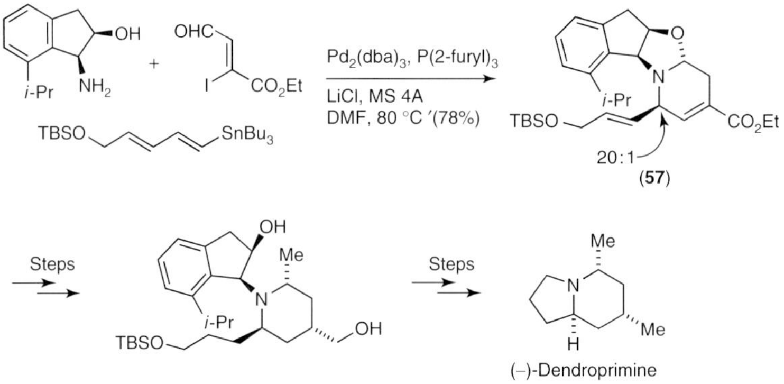

Scheme 3.18

was successfully applied to the asymmetric formal synthesis of 20-epiuleine (Scheme 3.17) [30].

Furthermore, efficient one-pot process of the above reaction has been developed by the same author [31]. Thus, the unsaturated ester **57** was synthesized by this method, and the asymmetric synthesis of (−)-dendroprimine was achieved via the stereoselective opening of the N, O-acetal as shown in Scheme 3.18 [32].

Fu achieved the elegant catalytic asymmetric [4 + 2] annulation of imines with allenes in the presence of the catalytic amounts of binaphthyl-based C_2-symmetric phosphine **58** (5–10 mol%) to yield 2,6-cis-disubstituted piperidine derivatives in good ee (Scheme 3.19). This methodology was applied to the synthesis of a framework (**59**) common to an array of important natural products such as the indole alkaloids of 6-oxoalstophyllal and 6-oxoalstophylline [33].

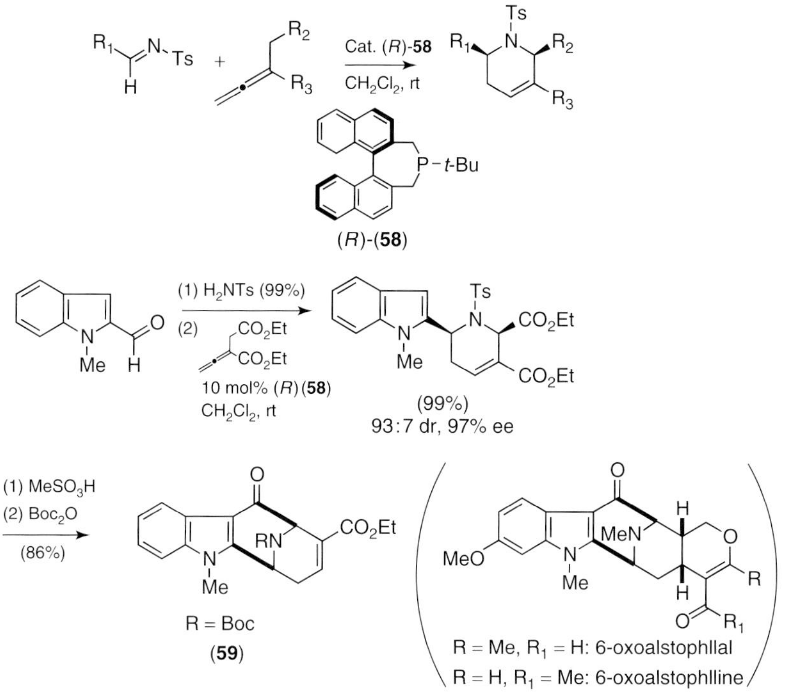

Scheme 3.19

3.4
Monocyclic Piperidines and Carbocyclic Fused Systems

3.4.1
Generalities

In this section, the asymmetric synthesis of piperidines and their corresponding carbocyclic fused systems are described. Of course, most of the methods described above for the preparation of di- and tetrahydropyridines could be applied to piperidines after functionalization or reduction.

The following section deals with different cyclization methods, cycloaddition, ring transformation, and substitution of already formed heterocycles.

3.4.2
Cyclization Methods

Several types of substituted monocyclic piperidines have been synthesized. The more classic cyclization method and more often used is nitrogen intramolecular substitution reaction of appropriate precursors to form the cyclized products in chiral form. Nevertheless, other methods forming the C–C bonds are also reported and described below.

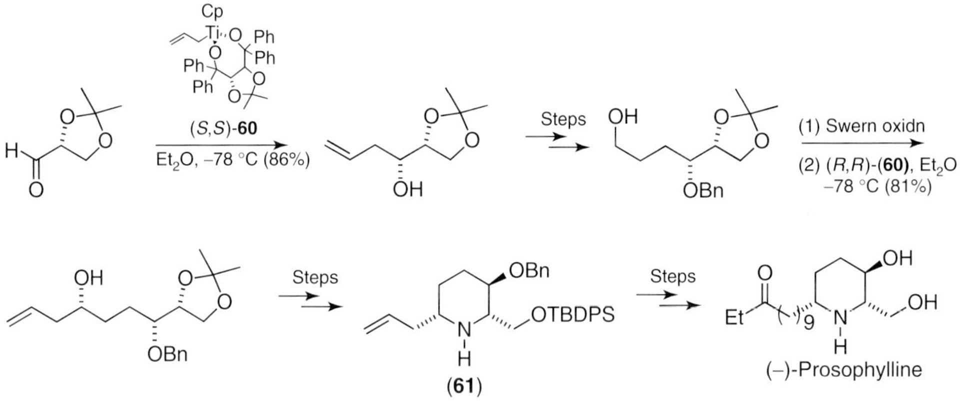

Scheme 3.20

3.4.2.1 Nitrogen as a Nucleophile

S$_N$2 Reactions The nucleophilic substitution with nitrogen as the nucleophile is probably the most obvious method to achieve cyclization; the leaving group may be an activated ester (tosylate), epoxide, or a halogen as illustrated in the following examples. Cossy reported the sequential use of enantioselective allyltitanation using the both enantiomers of **60** to form the chiral homo-allylalcohol, which was converted to the chiral 2,3,6-trisubstituted piperidine **61** by intramolecular cyclization reaction of the chiral allylamine. Construction of the side chain by Grubbs 2nd catalyst-catalyzed cross-metathesis reaction yielded (−)-prosophylline (Scheme 3.20) [34].

The double nucleophilic attack of nitrogen on a bis-tosylate or a bis-epoxide was also described. Takahata reported the synthesis of a novel C$_2$-symmetric 2,6-diallylpiperidine building block for the synthesis of several alkaloids using the double asymmetric allylation under the Brown procedure and cyclization by inter- and intramolecular substitution with benzylamine to yield the key piperidine **62**. Desymmetrization of **62**, crucial step for further elaboration of this building block, was achieved by the intramolecular iodocarbamation to yield the oxazolizinone **63** in high yield. This oxazolizinone is a versatile chiral building block for the synthesis of several alkaloids such as (−)-solenopsin A and (−)-porantheridine, as shown in Scheme 3.21 [35].

Concellon and Rivero achieved the chiral synthesis of 3,4,5-trisubstituted piperidine by sequential inter- and intramolecular ring opening reaction of epoxyaziridine with primary amines as the key step (Scheme 3.22). The aminoketone **64**, derived from L-serine, was converted to aziridine **65** in three steps. The aziridine was transformed into desired epoxyaziridine **66** in three-step sequence, and the key opening reaction of **66** with several primary amines in the presence of lithium perchlorate yielded the 3,4,5-trisubstituted piperidines as the sole product [36].

Gallagher reported [3 + 3] annulation using the 1,3-cyclic sulfate as a double electrophile in the asymmetric synthesis of piperidine. The chiral 1,3-diol was converted to cyclic sulfate, which was treated with dianion of **67** to yield the

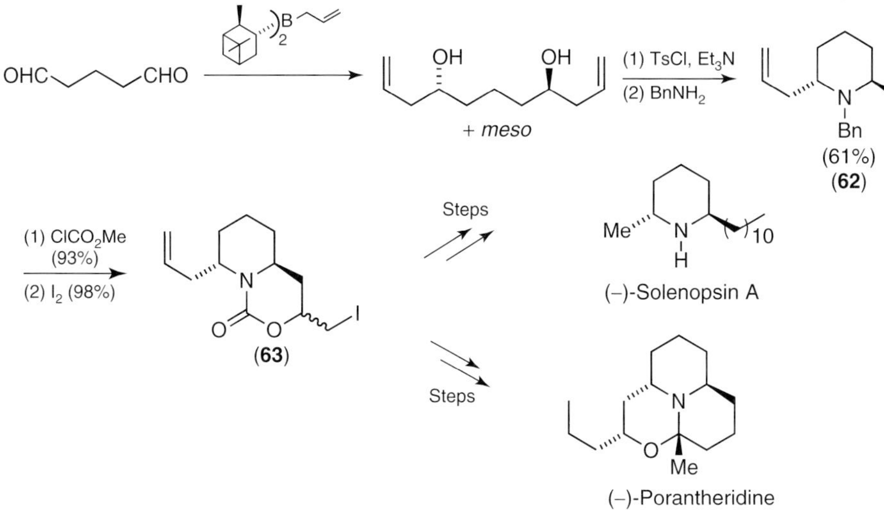

3.4 Monocyclic Piperidines and Carbocyclic Fused Systems

Scheme 3.21

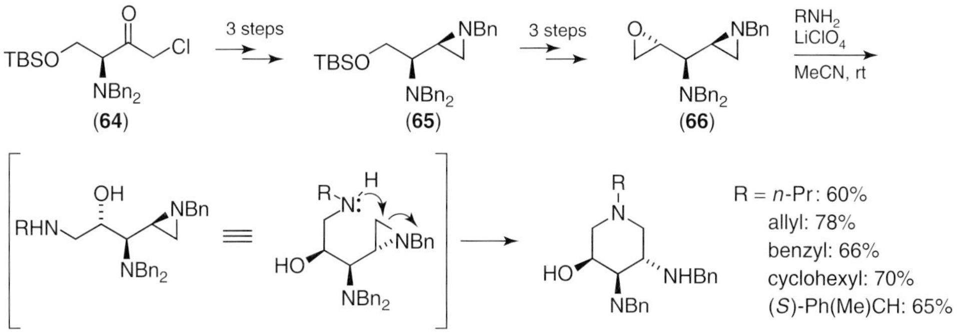

Scheme 3.22

piperidone **69** directly or via sulfate **68**. This piperidone was transformed into (S)-coniine in several steps (Scheme 3.23) [37].

Reductive Alkylation Efficient cyclization reactions via the formation of iminium ion followed by its reduction to form the substituted piperidines have been achieved. Mori described the chiral synthesis of (+)-carpamic acid by stereoselective hydrogenation of the iminium, generated from an acyclic amino ketone, to give 2,6-cis-disubstituted 3-piperidinol derivative. The key substrate **70** was synthesized from (S)-alanine methyl ester involving the diastereoselective addition of the acetylene unit to the corresponding aldehyde. Treatment of **70** with Pearlman's Pd(OH)$_2$ in methanol yielded the piperidine **71**, which was converted to (+)-carpamic acid (Scheme 3.24) [38].

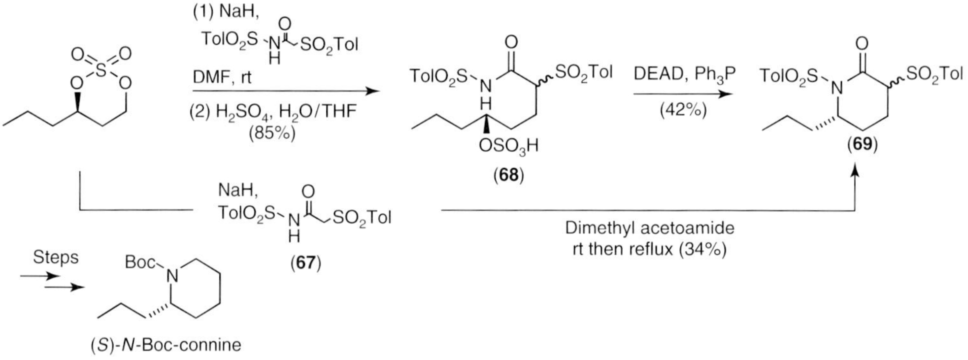

Scheme 3.23

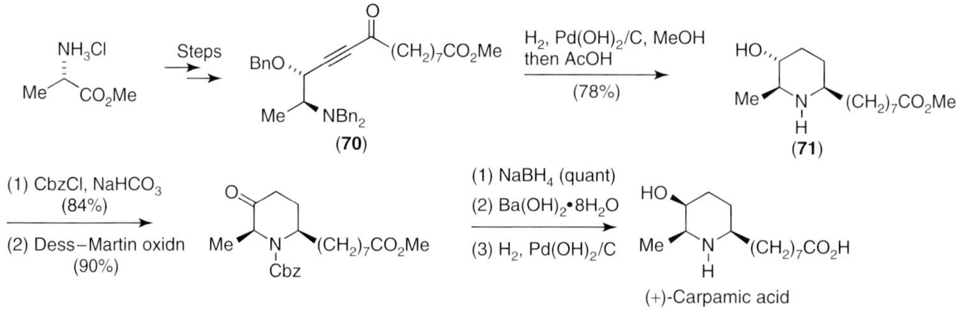

Scheme 3.24

Kumar's [39], Ghosh's [40], and Hum's [41] groups also used the similar iminium cyclization–reduction process to yield 2,6-*cis*-disubstituted 3-piperidinol derivatives for the chiral synthesis of (−)-deoxocassine, (+)-carpamic acid, and (+)-spectaline.

Enders' group reported the asymmetric total synthesis of (+)-2-epi-deoxoprosopinine by using his original SAMP/RAMP hydrazone method as the key step. The α-alkylation of SAMP hydrazone **72** with iodide **73**, which was also prepared by SAMP method, yielded the product **74** with high diastereoselectivity (de = 95%). Hydrolytic cleavage of the auxiliary followed by hydrogenolysis of Cbz-group and subsequent reductive amination resulted in an attack of hydrogen from sterically less hindered *si* face of resulting imine **75** to yield the piperidine **76**. Finally, deprotection of acetonide group yielded (+)-2-epi-deoxyprosopinine, as shown in Scheme 3.25 [42].

π-Allyl Substitution The metal-catalyzed cyclization reactions involving a π-allyl metal have also been used by several groups to form the six-membered nitrogen heterocycles. For instance, Hirai achieved the enantioselective synthesis of azasugars of 1-deoxymannojirimycin [43] and fagomine [44] using his original Pd(II)-catalyzed cyclization of urethanes **77** to form the functionalized piperidine ring systems **78** as the key step. Interestingly, the use of the bulky pivaloyl ester

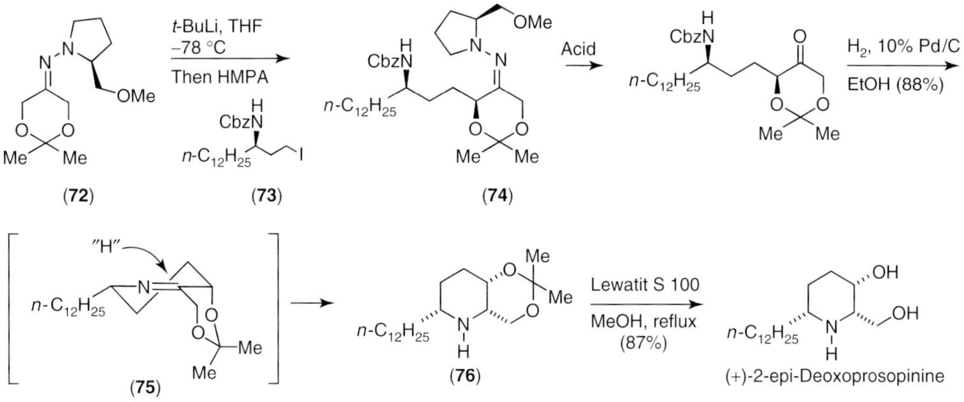

3.4 Monocyclic Piperidines and Carbocyclic Fused Systems | 111

Scheme 3.25

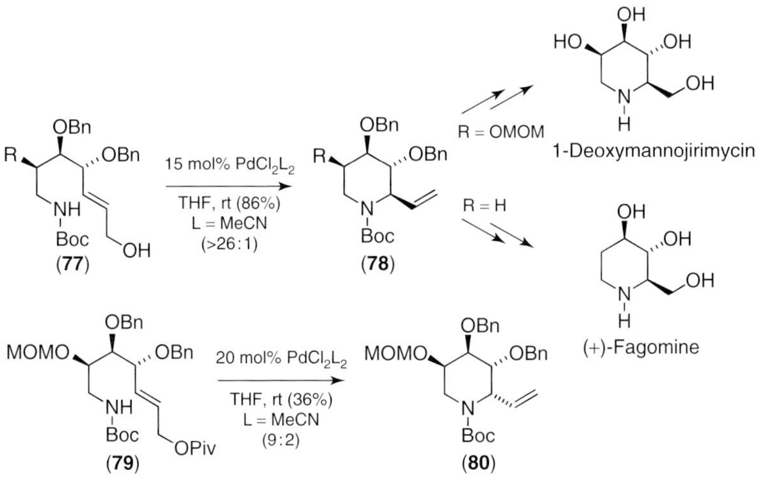

Scheme 3.26

79, instead of allylic alcohol **77**, for the cyclization reaction resulted in the selective formation of the diastereomer **80** (9:2) (Scheme 3.26).

Makabe used the Pd(II)-catalyzed cyclization of amino allylic alcohol, originally developed by Hirai [45], to obtain 2,6-*cis*-disubstituted 3-piperidinol derivative (Scheme 3.27). Thus, the treatment of allylic alcohol **81** with 5 mol% of PdCl$_2$ yielded the 2,6-*cis*-disubstituted 3-piperidinol **82** in a highly diastereoselective manner, which was transformed into (−)-cassine [46].

Palladium was not the only used metal for such transformation; Helmchen's group developed the highly enantioselective synthesis of α-vinyl substituted piperidine **83** by Ir-catalyzed allylic amination. This methodology was extended to sequential reaction to yield the C$_2$-symmetric 2,6-*trans*-divinylpiperidine **84** with an excellent ee (Scheme 3.28) [47].

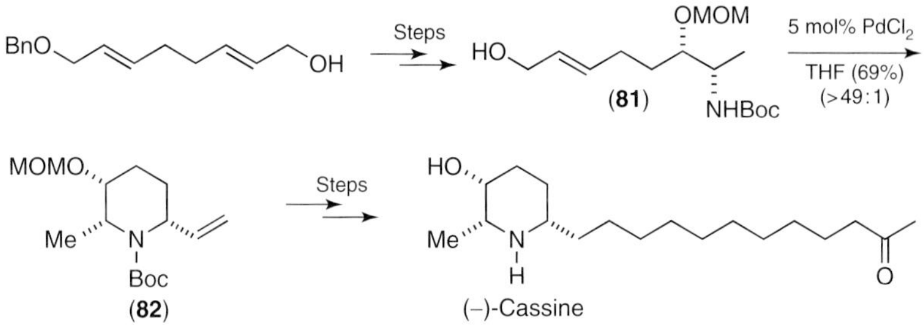

Scheme 3.27

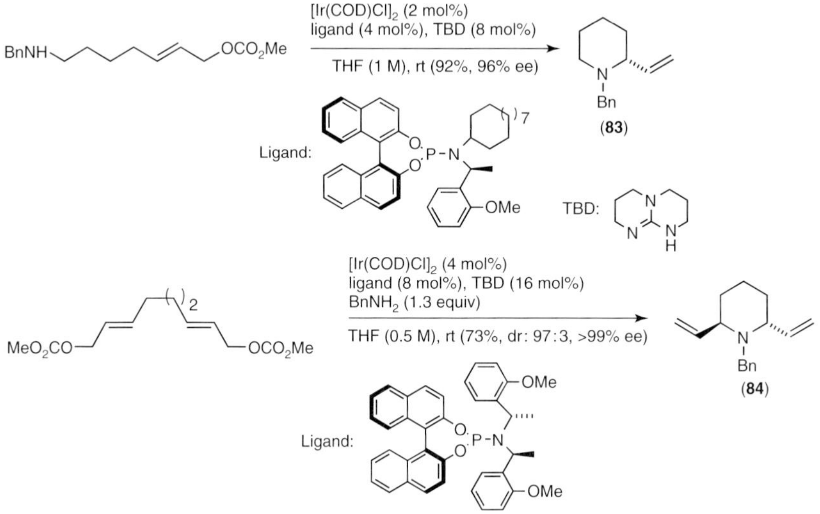

Scheme 3.28

Hydroamination of Allenes The gold(I)-catalyzed intramolecular hydroamination of allenes **85** was used by Toste for the preparation of chiral piperidines. Chiral ligands allowed the reaction to occur with good enantioselectivity (74–98% ee), as shown in Scheme 3.29. A dramatic amplification of enantioselectivity was observed when benzoate counterions were employed in this cyclization reaction [48].

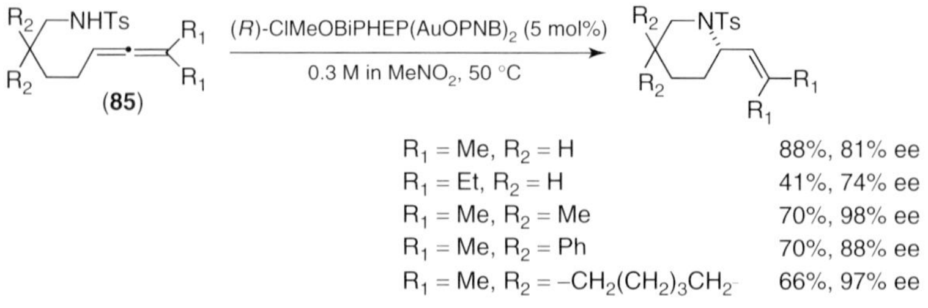

R_1 = Me, R_2 = H   88%, 81% ee
R_1 = Et, R_2 = H   41%, 74% ee
R_1 = Me, R_2 = Me   70%, 98% ee
R_1 = Me, R_2 = Ph   70%, 88% ee
R_1 = Me, R_2 = –CH$_2$(CH$_2$)$_3$CH$_2$–   66%, 97% ee

Scheme 3.29

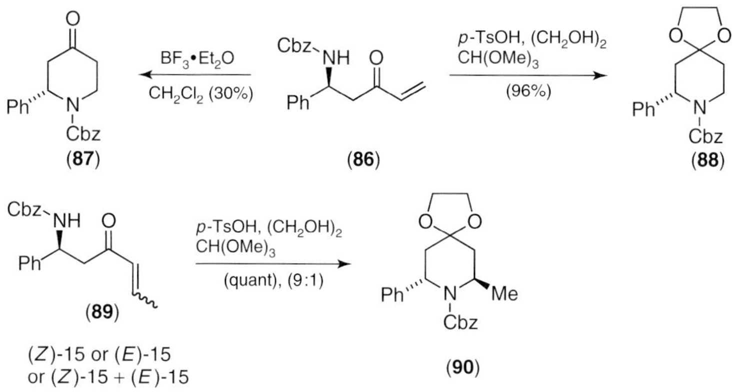

3.4 Monocyclic Piperidines and Carbocyclic Fused Systems

Scheme 3.30

Michael Addition Chalard and Troin studied the 6-*endo*-Michael-type cyclization to yield 2-mono- and 2,6-disubstituted piperidines with 2,6-*trans* selectivity. The intramolecular Michael reaction of **86** in the presence of $BF_3 \cdot Et_2O$ yielded the product **87** (only 30% yield). On the other hand, this type of cyclization reaction was achieved by using *p*-TsOH and ethylene glycol in the presence of $CH(OMe)_3$ to yield the corresponding acetal **88** from **86** and **90** from **89** in 96% and quantitative yield, respectively. The 2,6-*trans* adduct **90** was obtained as the major isomer (9:1) starting from either *E*- or *Z*-olefins (Scheme 3.30) [49].

3.4.2.2 C–C Bond Formation

Nitrogen α–Carbanion Kawabata's group succeeded in the asymmetric cyclization of α-amino acids to yield the six-membered cyclic amino acid with a quaternary stereocenter on the basis of his original concept of memory of chirality (Scheme 3.31). Thus, the amino acid bearing the bromobutyl on the nitrogen was treated with potassium hexa-methyldisilazane (KHMDs) in dimethylformamide (DMF) at −60 °C to yield the cyclic compound **91** (84% yield with high enantioselectivity 97% ee). This methodology can be applied to construct the five-membered ring system too [50].

Iminium Addition The intramolecular Mannich-type cyclization is also a powerful tool for the construction of functionalized piperidine ring system.

Davis reported the efficient asymmetric synthesis of β-amino ketones from sulfinimines followed by the Mannich-type cyclization to give functionalized 4-piperidone ring systems. These piperidones are very versatile building blocks for

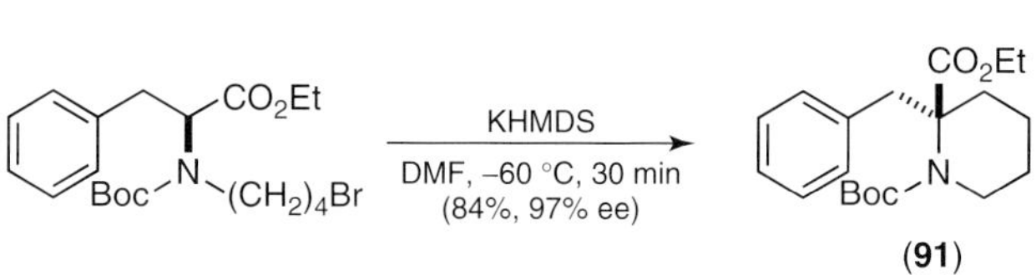

Scheme 3.31

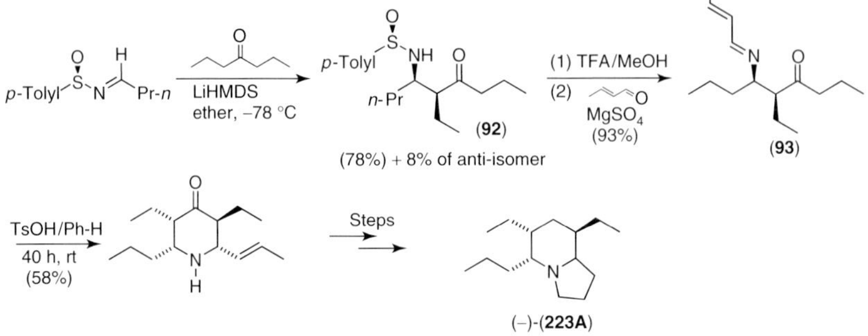

Scheme 3.32

the synthesis of several alkaloids, and the asymmetric syntheses of (−)-**223A** [51], (−)-nupharamine [52], and (+)-**241D** [53] have been achieved. The first key reaction was the *syn*-selective addition of the enolate to chiral sulfinimine to yield the α-substituted β-amino ketone **92**. The second key step was the acid-catalyzed Mannich cyclization of **93**, derived from **92**, to yield the functionalized 4-piperidone, which was converted to (−)-**223A** by indolizidine formation followed by the reduction of the ketone moiety (Scheme 3.32).

The intramolecular attack of allylsilane on iminium ion or imine is also effective to construct the chiral piperidine ring system by cyclization reaction.

Remuson described a new asymmetric synthesis of 2,6-*cis*-disubstituted 4-methylenepiperidine **95** based upon a Mannich-type intramolecular cyclization of an allylsilane on an iminium ion **94**. According to this strategy, the poison-frog alkaloid **241D** was synthesized, as shown in Scheme 3.33 [54].

Hiemstra and Rutjes investigated the amidopalladation of alkoxyallenes to yield the α-alkoxycarbamates or amides. The formal enantioselective synthesis of poison-frog alkaloid **233A** has been achieved by using the above key reaction followed by acyliminium cyclization of the resulting α-alkoxycarbamate. Treatment of **96** with Hoveyda–Grubbs catalyst (5 mol%) resulted in unusual double bond

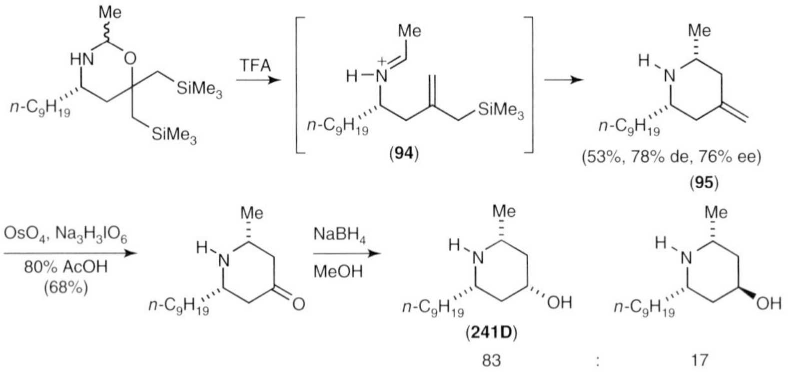

Scheme 3.33

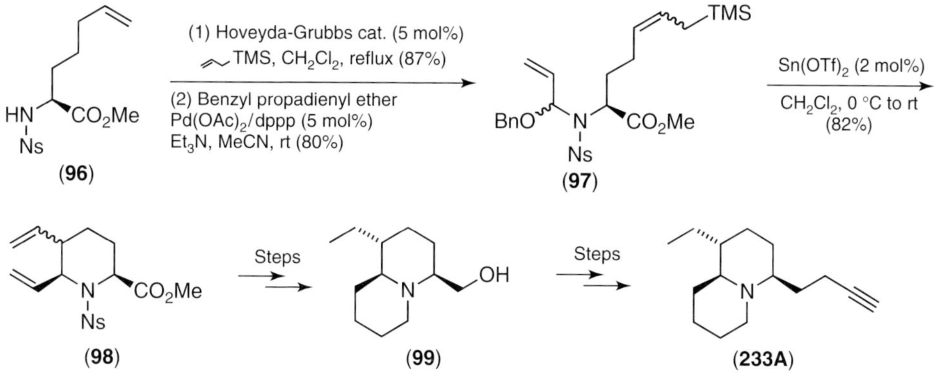

3.4 Monocyclic Piperidines and Carbocyclic Fused Systems | 115

Scheme 3.34

isomerization followed by cross-metathesis with allyltrimethylsilane to give the product, which was subjected to key amidopalladation with benzyl propadienyl ether to yield **97**. Sn(OTf)$_2$-catalyzed cyclization of **97** via acyliminium salt yielded the trisubstituted piperidine **98** as a 86:14 mixture of *trans*/*cis*-isomers. The major *trans*-isomer was converted to quinolizidine **99**, which is a key intermediate for the synthesis of poison-frog alkaloid **233A** (Scheme 3.34) [55].

The divergent synthesis of both 2,6-*cis*- and *trans*-3-*trans*-trisubstituted tetrahydropyridines from chiral organosilanes was reported by Panek. The first enantioselective total synthesis and determination of absolute stereochemistry of poison-frog alkaloid (−)-**217A** were achieved by this methodology via the quinolizidine **100**, as shown in Scheme 3.35. Treatment of preformed imine with TiCl$_4$ at low temperature yielded the desired product with good-to-excellent diastereoselectivity (8:1 to >30:1). Interestingly, the cis–trans selectivity depended upon the chirality of organosilane. Use of the 2,3-*anti*-organosilanes yielded the 2,6-*trans* products as the major isomer (9:1 to >30:1); on the other hand, use of the 2,3-*syn*-organosilanes yielded the 2,6-*cis* products (8:1 to >30:1) selectively [56].

Troin's group investigated very actively on the intramolecular Mannich-type cyclization reaction of chiral imine **101** to yield the 2,6-*cis*-disubstituted piperidine ring systems. This reaction was applied to the syntheses of highly substituted piperidines [57], pipecolic acid derivatives [58], and (±)-alkaloid **241D** (Scheme 3.36) [59].

Radical Cyclization The construction of chiral piperidines by means of radical cyclization reaction is a very interesting method, although very rarely used.

Snaith proposed a novel approach to 2,4-*trans*-disubstituted piperidines by the radical cyclization of 7-substituted 6-aza-8-bromooct-2-enoates **101** (Scheme 3.37) [60]. The use of tris(trimethylsilyl)silane instead of tributyltin hydride as the radical source enhanced the diastereoselectivity (>99:1). The trans selectivity is explained by the proposed transition state, as shown in Figure 3.2 involving a pseudo-A^{1,3} allylic strain.

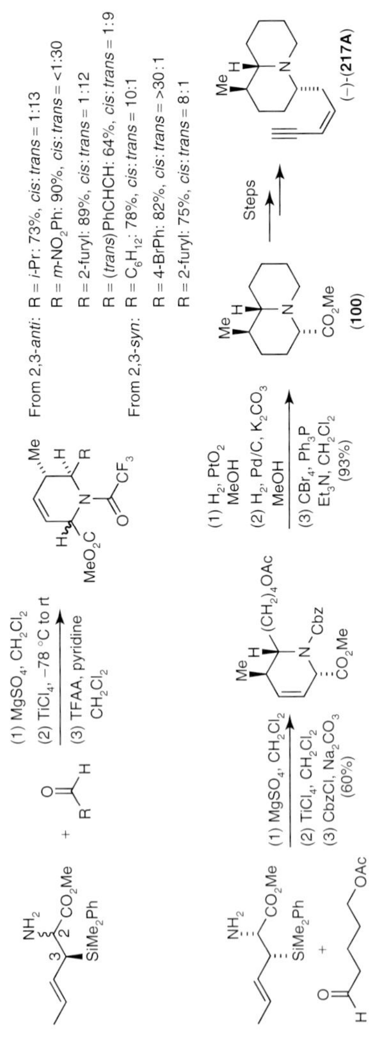

116 | 3 Asymmetric Synthesis of Six-Membered Ring Heterocycles

Scheme 3.35

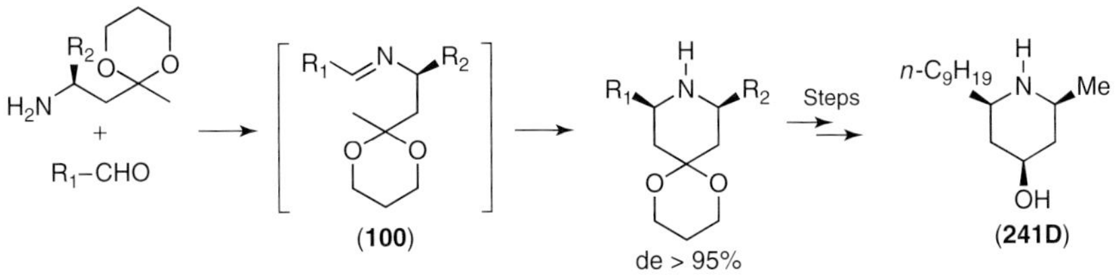

3.4 Monocyclic Piperidines and Carbocyclic Fused Systems

Scheme 3.36

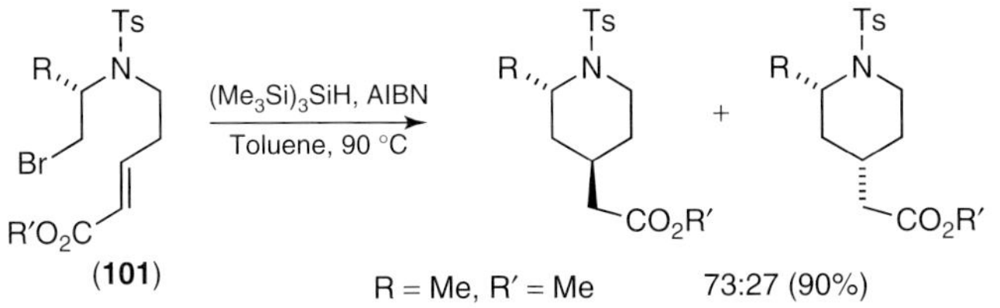

Scheme 3.37

R = Me, R′ = Me	73:27 (90%)
R = Bn, R′ = Me	92:8 (76%)
R = i-Pr, R′ = Me	97:3 (73%)
R = i-Pr, R′ = t-Bu	99:1 (75%)
R = sec-Bu, R′ = t-Bu	>99:1 (60%)

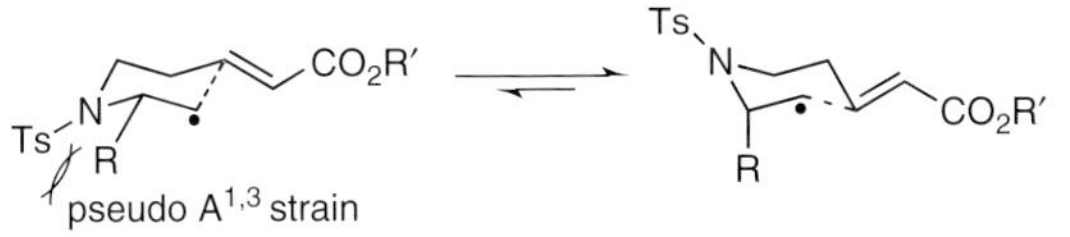

Fig. 3.2 Preferential conformer on the radical cyclization of **101**.

3.4.3
Cycloaddition Methods

3.4.3.1 [4 + 2] Azadiene Cycloaddition

The construction of chiral piperidines by using the aza[4 + 2] cycloaddition reaction as the key steps has been reported.

Hall's group achieved a new entry to palustrine alkaloids using the three-component sequential, aza[4 + 2] cycloaddition/allylboration/retro-sulfinyl-ene, reaction (Scheme 3.38). This approach was then applied to enantioselective synthesis of (−)-methyl dihydropalustramate. The key three-component sequential reaction smoothly proceeded to yield the product **102**, which was converted to (−)-methyl dihydropalustramate via oxazolizinone **103** [61].

3.4.3.2 [4 + 2] Acylnitroso Cycloaddition

The formation of bicyclic- or tricyclic-nitrogen heterocycles has been reported by using [4 + 2] or formal [3 + 3] cycloaddition reaction as the key step.

Kibayashi and Aoyagi studied the stereocontrolled intramolecular acylnitroso-Diels–Alder reaction to yield the bicyclic oxazino lactam. Oxidation of hydroxamic acid **104** with Pr_4NIO_4 in aqueous media (water–DMF = 50:1) yielded the

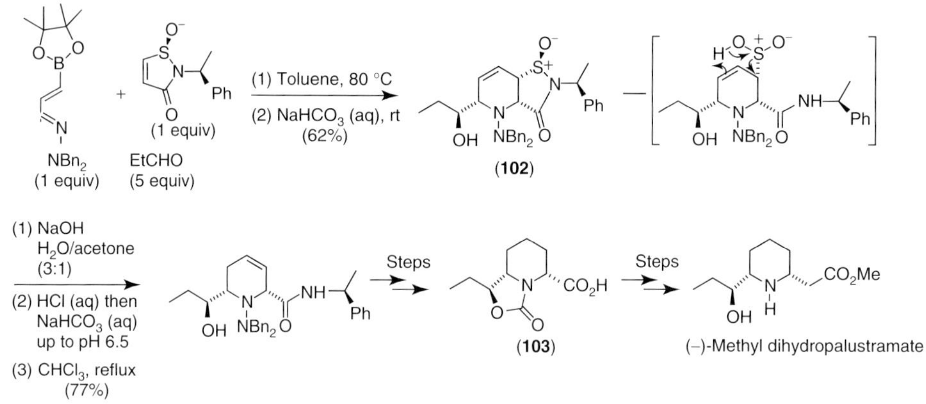

Scheme 3.38

cycloadduct **106** selectively (6.6:1) via acylnitroso compound **105**. Elaboration of **106** (stereoselective introduction of methyl substituent, construction of decahydroquinoline skeleton, and installation of the dienyl side chain) yielded the marine alkaloids (−)-lepadins A, B, and C [62, 63]. According to this method, the macrocyclic dilactones containing a 2,3,6-trisubstituted piperidine of (+)-azimine and (+)-carpaine have also been synthesized (Scheme 3.39) [64].

The Diels–Alder reaction of chiral 1,4-bis(arylsulfonyl)-1,2,3,4-tetrahydropyridine to yield the octahydroquinoline ring systems was achieved by Craig. Thus, the cycloaddition reaction of **107** with maleic anhydride or acrolein yielded the cycloadduct **108** or **109**, respectively. Formation of **108** or **109** indicated preferred endo approach of the dienophile toward the α-face of **107** (Scheme 3.40) [65].

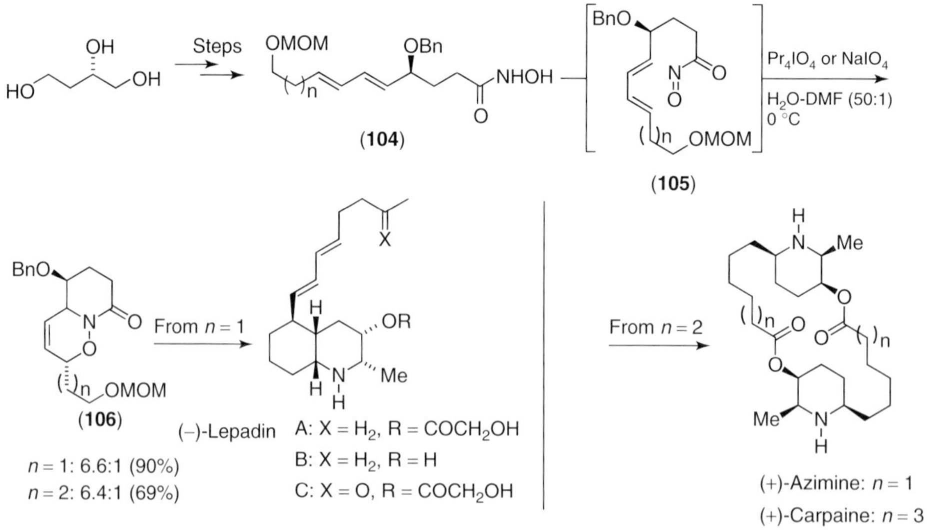

Scheme 3.39

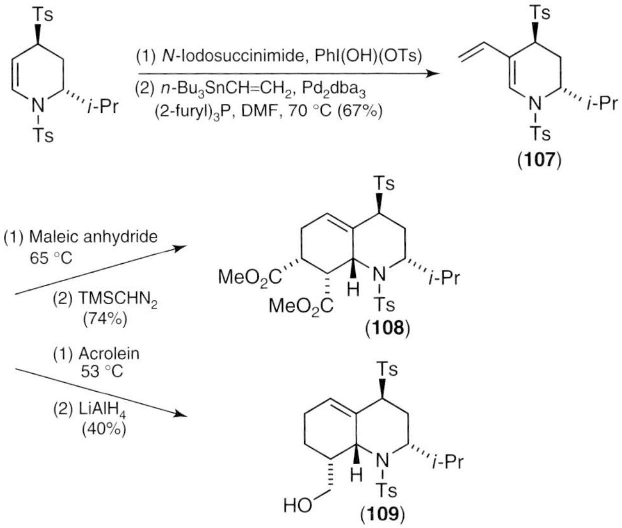

3.4 Monocyclic Piperidines and Carbocyclic Fused Systems

Scheme 3.40

3.4.3.3 [3 + 2] Cycloaddition

The [3 + 2] 1,3-dipolar cycloaddition reaction of nitrone is also an efficient methodology for the construction of chiral piperidine building blocks. Indeed, in this methodology, the cycloaddition allowed the formation of the C2–C3 bond of the piperidine ring along with the formation of the five-membered oxazolidine ring.

Poison-frog alkaloid (+)-allopumiliotoxin **323B•** has been synthesized by Holmes employing the intramolecular [3 + 2]-cycloaddition reaction of Z-N-alkenylnitrone to yield the precursor of 2,3,4-trisubstituted piperidine (Scheme 3.41). The key cycloaddition of **110** was conducted by heating at 70 °C in toluene to yield four isoxazolidine cycloadduct (32:5:8:8). The major isomer **111** was converted to 2,3,4-trisubstituted piperidine **112**, which was transformed into the ketone **113**. The stereoselective addition of Grignard reagent to **113** was performed to yield the desired diol **114** as the sole product. The selectivity on this addition reaction was explained by $A^{1,3}$-strain and Cieplak hypothesis (Figure 3.3). The diol **114** was converted to (+)-allopumiliotoxin **323B•** by the construction of the *exo*-olefin side chain [66].

3.4.4
Ring Transformation Methods

The construction of the piperidine ring can result from ring transformation and particularly from ring enlargement. The ring enlargement of five-membered ring systems such as pyrrolidine, lactone, or γ-lactam is also an effective method for the construction of six-membered nitrogen heterocycles. According to these

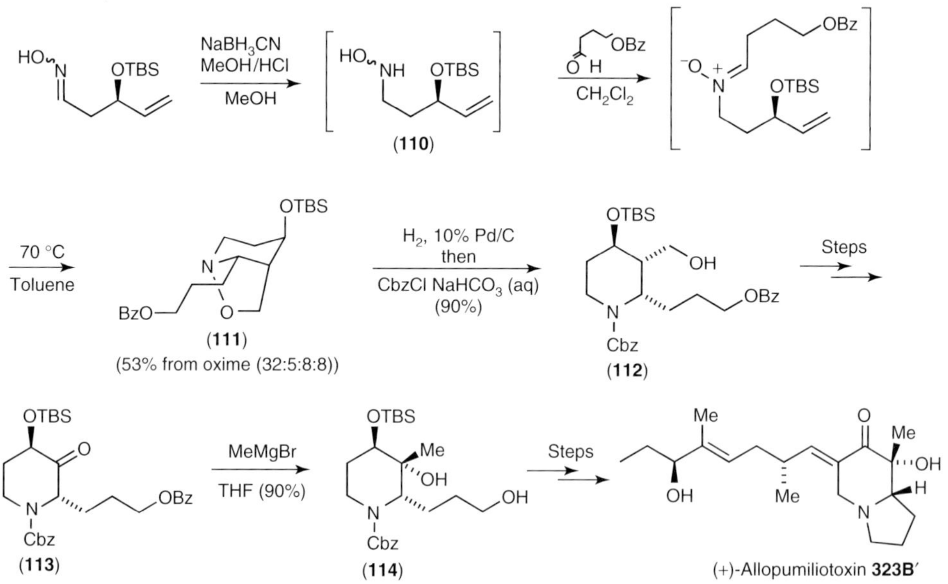

3 Asymmetric Synthesis of Six-Membered Ring Heterocycles

Scheme 3.41

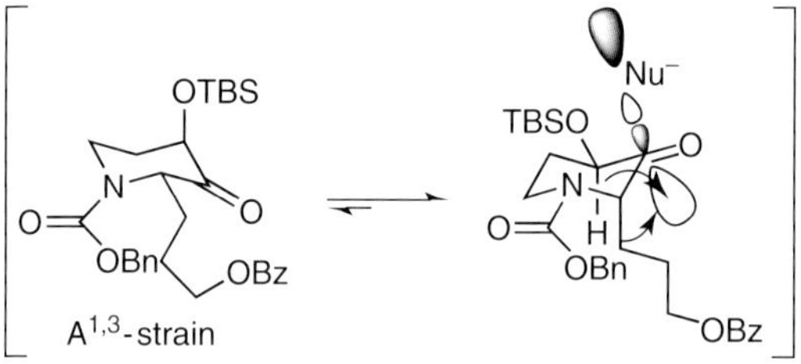

Fig. 3.3 Preferential conformer on Grignard-addition to **113**.

enlargement methodologies, functionalized piperidine or 2-piperidone chiral building blocks are elaborated in highly stereoselective manner.

3.4.4.1 Ring Enlargement of Pyrrolidines to Substituted Piperidines

A powerful method for the construction of piperidines is the ring transformation of prolinol derivatives. Cossy and Pardo described the ring expansion reaction of derivative **115**, derived from L-pyroglutamic acid, by treatment with MsCl followed by Et$_3$N to yield the 3,4,5-trisubstituted piperidine **116**, which was used for asymmetric formal synthesis of (−)-paroxetine (Scheme 3.42) [67].

The asymmetric synthesis of cis-2,3-disubstituted 3-piperidinol **120**, which is the useful intermediate for the synthesis of medicinally important neurokinin-1 receptor antagonists, has been reported by Lee [68] following a similar strategy. Thus, treatment of the prolinol derivative **118**, prepared by the Jacobsen's asymmetric epoxidation of olefin **117** as the key step, with MsCl followed by n-Bu$_4$NOAc

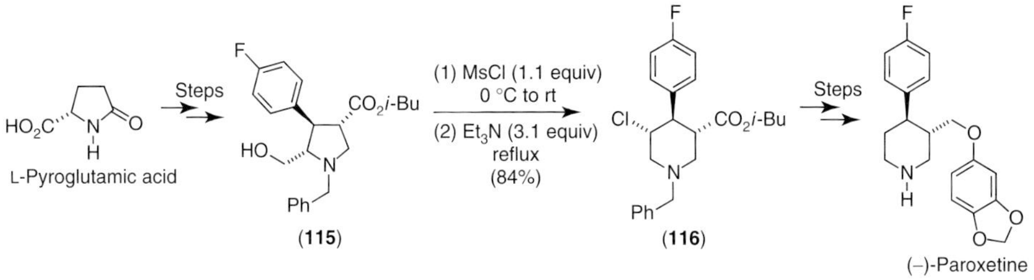

Scheme 3.42

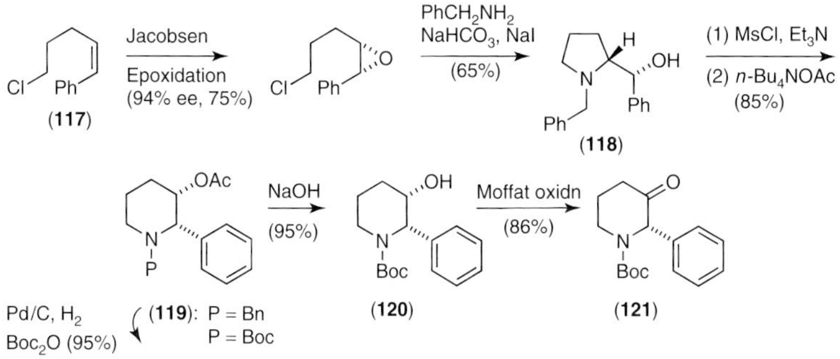

Scheme 3.43

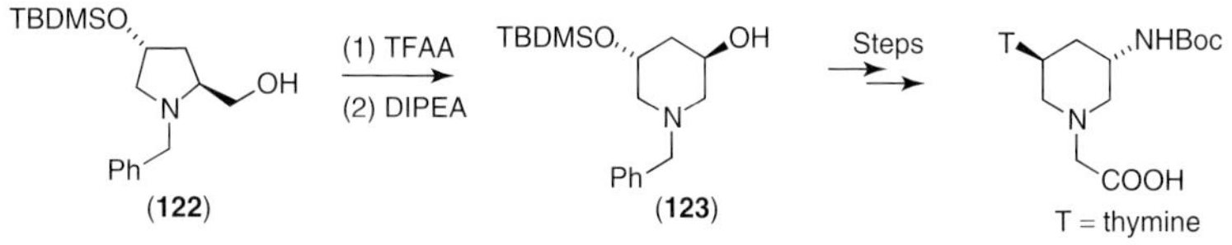

Scheme 3.44

yielded **119** with high enantiomeric purity. This piperidine was also converted to 3-piperidone **121** by Moffat oxidation with no racemization (Scheme 3.43).

In the synthesis of chiral piperidine peptide nucleic acid (PNA), the ring enlargement of prolinol **122** yielded the key intermediate *trans*-3,5-disubstituted piperidine **123** (Scheme 3.44) [69].

An original ring enlargement reaction of functionalized pyrrolidine **124** to piperidone **125** was proposed by Honda through a unique samarium-promoted C–N bond cleavage of **124** and simultaneous cyclization of the resulting amino ester to yield **125** (Scheme 3.45). The presence of pivalic acid as a proton source is essential in this transformation, and enantiospecific total synthesis of (−)-adalinine was also achieved. [70].

3.4.4.2 Ring Transformation of Lactones to 2-Piperidones

The lactones bearing a nitrogen functionality such as an amino or azide group on the γ-substituent are suitable substrates for this transformation.

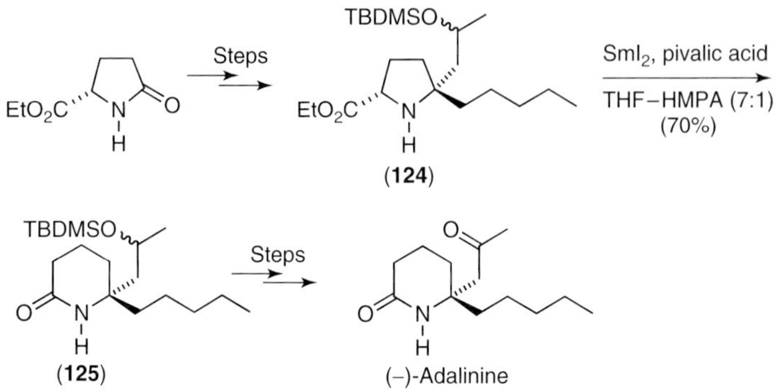

3 Asymmetric Synthesis of Six-Membered Ring Heterocycles

Scheme 3.45

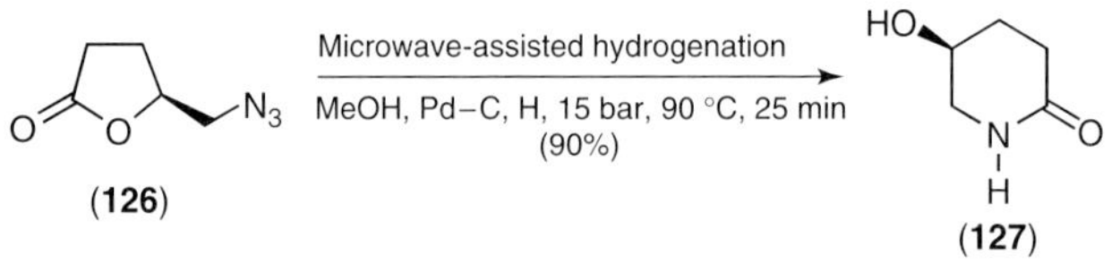

Scheme 3.46

Holzgrabe described the microwave-enhanced hydrogenation of azide-lactone **126** at the medium pressure to yield the 2-piperidone **127** in only 25 min with good yield (Scheme 3.46) [71].

3.4.4.3 Ring Enlargement of γ-Lactam to 2-Piperidones

Langlois described the synthesis of cis-4,5-disubstituted 2-piperidone **130** as a key intermediate for pseudodistomin C synthesis by ring expansion of aminomethyl-substituted γ-lactam **129** (Scheme 3.47). Thus, treatment of **129**, derived from (S)-pyroglutaminol via secondary alcohol **128**, in methanol at 65 °C for 24 h yielded transamidation product **130** with good yield, which was transformed into the marine alkaloid of pseudodistomin C [72].

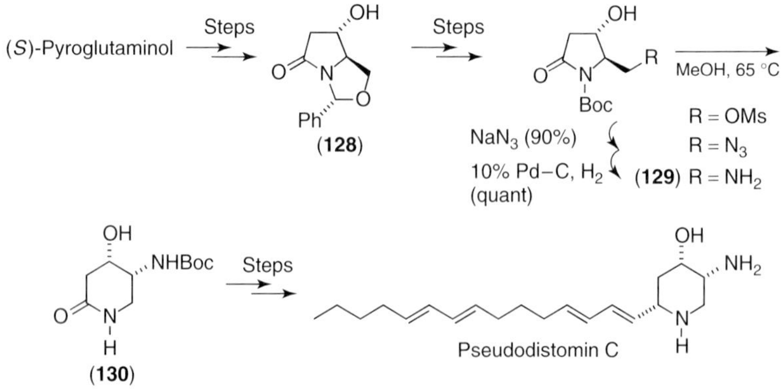

Scheme 3.47

3.4.5
Substitution of Already Formed Heterocycle

One strategy used in the asymmetric synthesis of piperidine was the preparation of a chiral and versatile building block that possesses the six-membered ring and can be further diversely substituted. This building block should be easily prepared with high yield and selectivity offering large possibilities of substitution. The substitutions should be stereocontrolled. The use of chiral auxiliaries is still the main strategy used, although some reactions using asymmetric catalysts were also reported.

3.4.5.1 Phenylglycinol-Derived Oxazolidine

Enantiomerically pure phenylglycinol is frequently used for the induction of the chiral center on piperidine ring system. Involved in an oxazolidine ring, it allowed various diastereomerically controlled reactions that were exploited by several research groups in the preparation of piperidine derivatives.

Husson reported the reductive decyanation of the 3-phenylhexahydro-5H-[1,3]oxazolo[3,2-a]pyridine-5-carbonitrile building block **86**, the key element of his elegant CN(R,S) strategy, with Raney nickel and its further application to other derivatives. This new piperidine scaffolds serve as stable 2-piperideine (enamine) equivalent in the rapid and efficient construction of 3-substituted piperidines **132** [73]. On the other hand, regioselective oxidative transformations of the N-(cyanomethyl)-oxazolidine system with potassium permanganate leading to enantiopure lactams have also been reported. Oxidation of **131** yielded the lactam **133**, which is a very versatile chiral building block for the synthesis of several alkaloids. This methodology can be applied to the conversion of **134** to chiral lactam **135** and piperidine **136**, as shown in Scheme 3.48 [74].

A unique approach to the synthesis of chiral nonracemic substituted [5,6]- and [6,6]-spiropiperidines was also described by the use of the CN(R, S) method

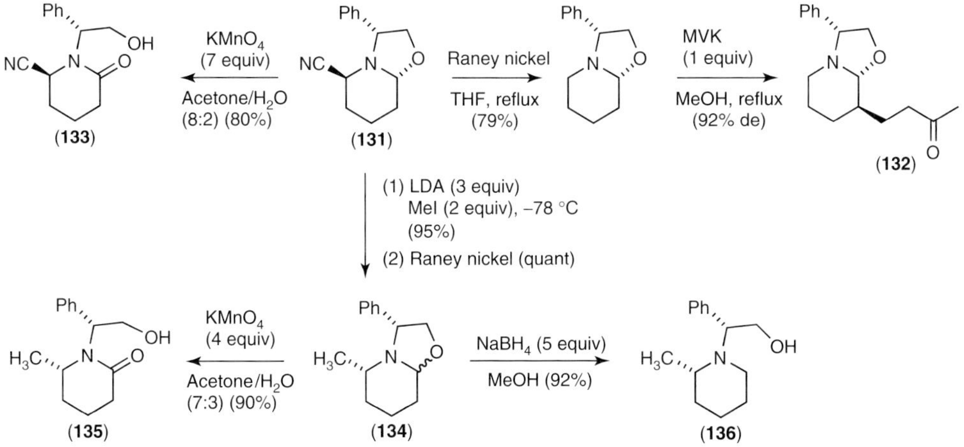

Scheme 3.48

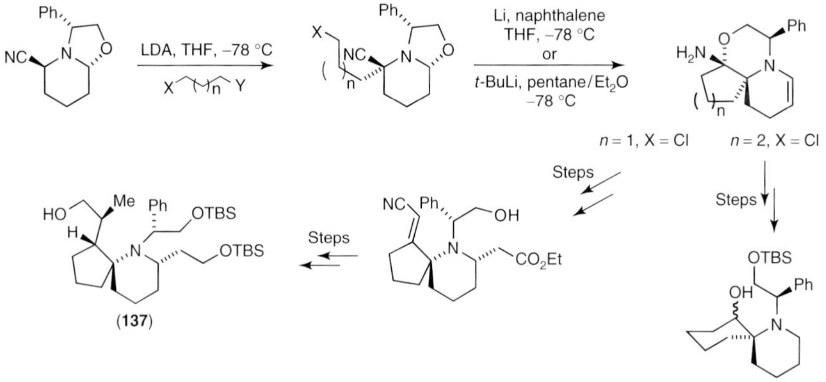

Scheme 3.49

(Scheme 3.49) [75]. Owing to the valuable application of this methodology to natural product synthesis, the advanced intermediate **137** in pinnaic acid series was achieved [76].

Lhommet developed the synthesis of chiral bicyclic β-enaminoester **138** by condensation of (S)-phenylglycinol on oxo alkynoate. The diastereoselective reduction of **138** was performed by hydrogenation of Pd(OH)$_2$ to yield the 2,6-*cis*-disubstituted piperidine **139** with an excellent enantioselectivity (>98%), which was a key intermediate previously reported in the total synthesis of (+)-calvine [77]. The construction of chiral bicyclic β-enaminoketone **140** by the same condensation procedure using the triketone instead of oxo alkynoate was also reported. The bicyclic β-enaminoketone **140** was converted to 2,3,6-trisubstituted piperidine **141** by hydrogenation of PtO$_2$ followed by hydrogenolysis of Pd(OH)$_2$ in highly stereoselective manner. The piperidine **141** was transformed into (−)-deoxocassine. On the other hand, the trisubstituted piperidine **142**, derived from **140**, was converted to (+)-isodeoxocassine via the epimerization on the 3-position (Scheme 3.50) [78].

Substitution reactions on the already formed heterocycle using the acyliminium ion strategy have been widely studied, and constitute an effective method for the construction of chiral piperidine ring.

Such strategy was the basis of Amat and Bosch enantioselective syntheses of piperidine alkaloids. Mono-substituted piperidine (R)-coniine, 2,6-*cis*-disubstituted piperidine alkaloid (2R,6S)-dihydropinidine, 2,6-*trans*-substituted piperidine alkaloids (2R, 6R)-lupetidine and (2R,6R)-solenopsin A, and indolizidine alkaloids (5R,8aR)-indolizidine **167B** and (3R,5S,8aS)-monomorine I were prepared starting from both enantiomers of a common Meyers' type lactam, (3R,8aS)-5-oxo-3-phenyl-2,3,6,7,8,8a-hexahydro-5H-oxazolo[3,2-a]pyridine. Addition of allyl moiety on **143** in the presence of TiCl$_4$ yielded the product in a diastereoselective manner. The selectivity observed here was rationalized by the nucleophilic attack upon the less hindered *Re* face on the iminium ion, whose conformation is restricted by the chelation, as shown in Figure 3.4. Thus, the mono-substituted piperidine alkaloid

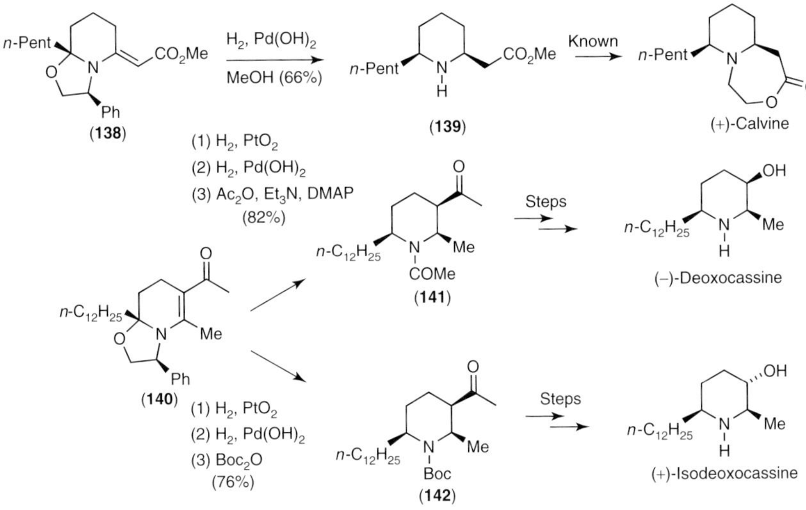

Scheme 3.50

of (−)-coniine was synthesized from **144** in several steps. Similarly, the addition of higher-order cyanocuprate to **143** in the presence of $BF_3 \cdot Et_2O$ yielded the adduct **145** in highly diastereoselective manner. Partial reduction of **145** with Red-Al yielded the masked iminium ion **146**, which was made to react with Grignard reagent to yield **147** as the only isolated product with good yield. This stereoselectivity was explained by stereoelectronic effects. Further reactions on **147** yielded 2,6-*cis*-disubstituted piperidine alkaloid of (−)-dihydropinidine. Likewise, the syntheses of (−)-indolizidine **167B** and (+)-monomorine I have also been achieved. On the other hand, the iminium reduction of **149** (R = Me or $C_{11}H_{23}$) derived from **148** with $NaBH_4$ yielded the 2,6-*trans*-disubstituted piperidines, which was converted to lupetidine and solenopsin A, respectively (Scheme 3.51) [79].

This methodology was extended to the synthesis of 3,5-*cis*- and *trans*-disubstituted, 2,5-disubstituted, and 2,3,5-trisubstituted enantiopure piperidines, and the indole alkaloids 20*R*- and 20*S*-dihydrocleavamine [80].

3.4.5.2 Asymmetric Michael Addition

Toyooka disclosed the stereodivergent construction of 2,3,5,6-tetrasubstituted piperidine ring system by sequential use of key Michael-type conjugate addition reaction of the enaminoesters **150** and **151** (Scheme 3.52). Thus, the first enantioselective synthesis of poison-frog alkaloid **223A** and the structural revision were achieved [81]. On the other hand, the first enantioselective synthesis of unique tricyclic poison-frog alkaloid ent-**205B** [82, 83] was also achieved. The observed remarkable stereoselectivity of key conjugate addition reaction can be rationalized by the stereoelectronic effect and is also consistent with Cieplak's hypothesis, as shown in Figures 3.5 and 3.6.

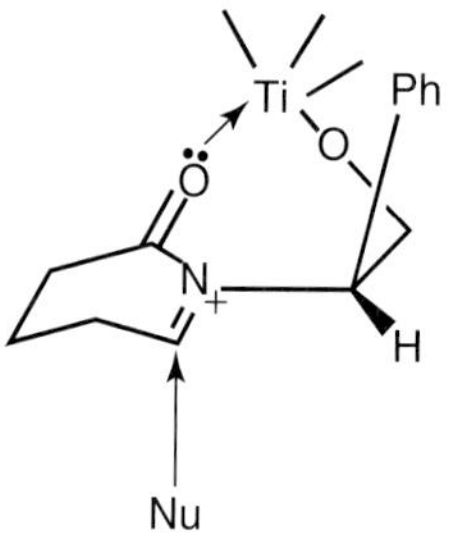

Fig. 3.4 Diastereoselective addition to **143** controlled by chelation.

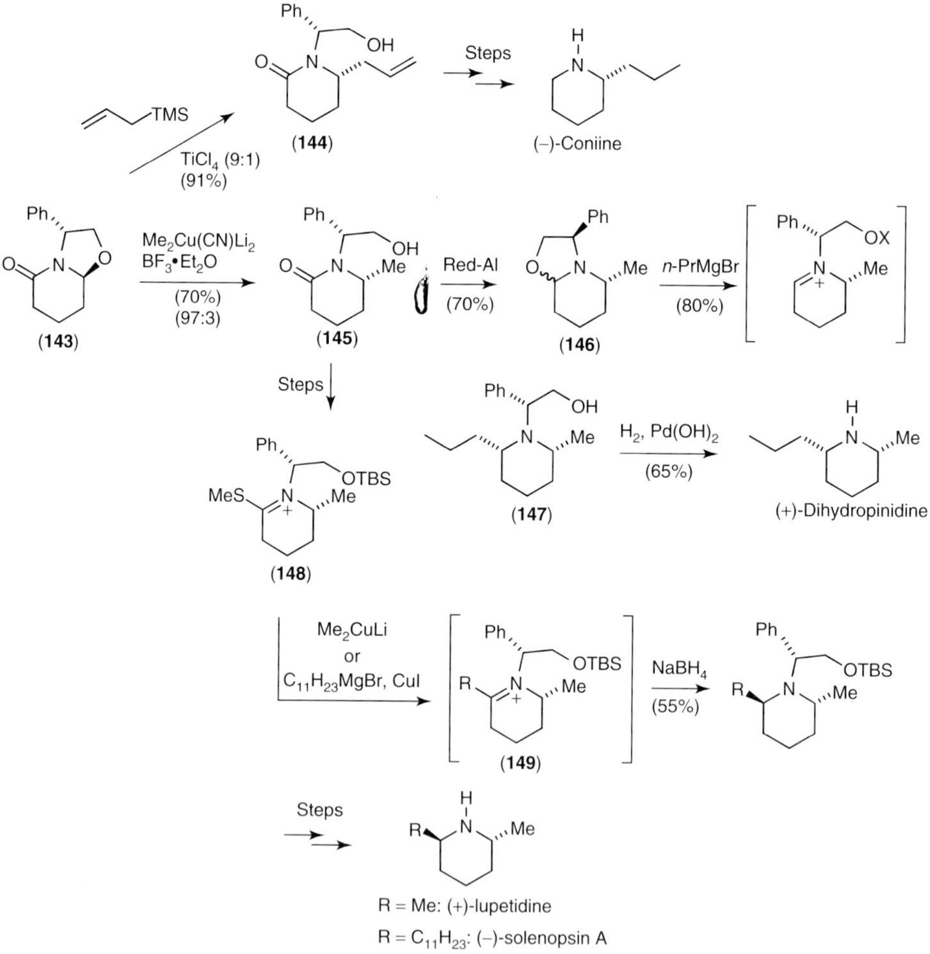

R = Me: (+)-lupetidine
R = C$_{11}$H$_{23}$: (−)-solenopsin A

Scheme 3.51

3.4 Monocyclic Piperidines and Carbocyclic Fused Systems

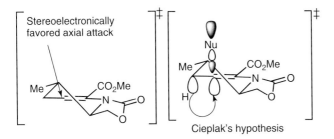

Scheme 3.52

Fig. 3.5 Stereochemical course of Michael-type addition to **151** (acyclic carbamate).

Fig. 3.6 Stereochemical course of Michael-type addition to **151** (cyclic carbamate).

3.4.5.3 Nitrone Cycloaddition

The 1,3-dipolar cycloaddition reaction of chiral six-membered ring of nitrone **153** was the key step in a convergent enantioselective synthesis of (+)-febrifugine.

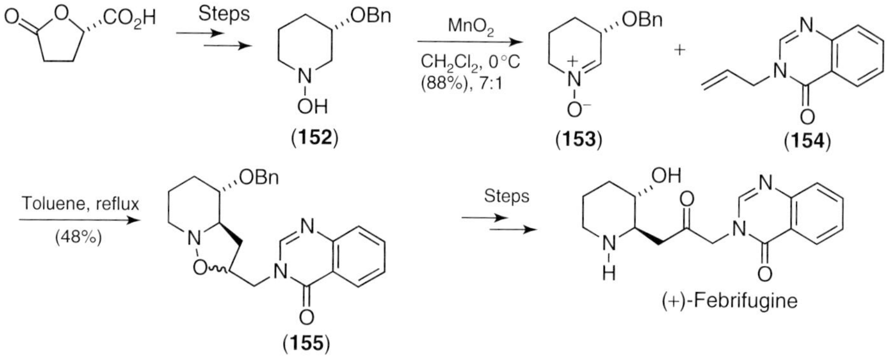

Scheme 3.53

Thus, the oxidation of piperidine **152**, which was derived from L-glutamic acid, with MnO_2 yielded a 7:1 mixture of separable regioisomeric nitrones. Key 1,3-dipolar cycloaddition reaction of **153** with quinazolone **154** yielded the adduct **155** as the major isomer. This isoxazolidine was converted to (+)-febrifugine in several steps, as shown in Scheme 3.53 [84].

3.4.5.4 Iminium Strategies

Yamazaki and Kibayashi described the construction of the quaternary center at the α-position of the piperidine ring using the acyliminium addition reaction on similar tricyclic N-acyl-N,O-acetal **156**. This method was applied to the asymmetric synthesis of (−)-adaline. Lewis acid promotes C–O bond cleavage to form the N-acyliminium ion, which preferably conforms with the hydrogen atom in the

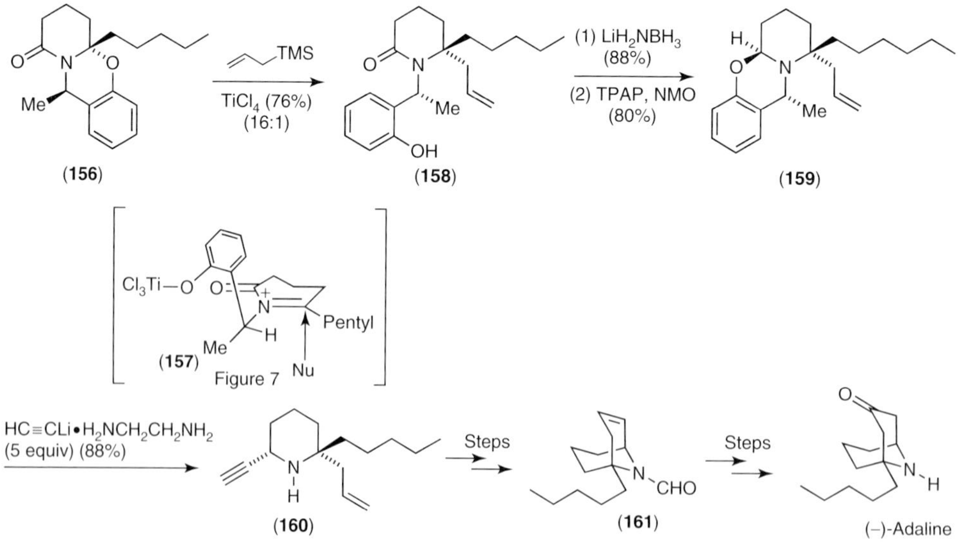

Scheme 3.54

inside position to minimize the 1,3-allylic strain. Subsequent nucleophilic addition to **157** is expected to occur from the less hindered bottom face leading to the *R* configuration in the piperidone to give **158** (Figure 3.7). Reduction of **158**, and then tetrapropyl-ammonium perrathenate (TPAP) oxidation of the resulting alcohol yielded **159**. Upon treatment of **159** with lithium acetylide ethylenediamine complex, the nucleophilic alkynylation proceeded with complete inversion of configuration at the reaction center with concomitant removal of the 1-(2-hydroxyphenyl)ethyl function via C–N bond cleav-to yield **160** as the sole diastereomer. Further elaborations on **160** via **161** yielded the (−)-adaline (Scheme 3.54) [85].

The asymmetric Morita–Baylis–Hillman type reaction of six-membered N,O-acetal with cyclic enones was reported by Aggarwal. In the presence of chiral sulfide **162** derived from camphor sulfonic acid and TMSOTf, high enantioselectivity was achieved (Scheme 3.55). In this reaction, the use of piperidine-based N,O-acetal yielded the adducts with improved enantioselectivity values compared to the pyrrolidine series.

When the acyclic enone methyl vinyl ketone (MVK) was employed, the enantioselectivity value of the corresponding adduct was very low [86].

Blaauw reported the total synthesis of (+)-epiquinamide by using a highly diastereoselective *N*-acyliminium ion allylation, and RCM to yield a quinolinone skeleton. The *N*-acyliminium allylation of **163** yielded the 2,3,6-trisubstituted piperidine **164**, which was converted to acrylamide **165**. The RCM reaction of **165** using Grubbs 2nd catalyst yielded the quinolinone **166**, which was transformed into (+)-epiquinamide in several steps, as shown in Scheme 3.56 [87].

In the following example, the synthesis of new six-membered multifunctional chiral building block **167** was obtained through the ring transformation of an optically active butenolide (as already illustrated in Section 3.4.4). Imide **167** is a very versatile chiral building block for the synthesis of natural alkaloids such as epiquinamide [88], homopumiliotoxin **223G** [88], deoxocassine [89], and biologically important molecules such as L-733,060 [90], L-733,061 [89], and CP-99,994 [90]. The key feature of Huang's approach is regio- and diastereoselective installation of the nucleophile on the 2-position of **167**. On treatment of piperidinol, derived from regioselective reduction of **167** with $NaBH_4$ and then with $BF_3 \cdot Et_2O$, the phenyl

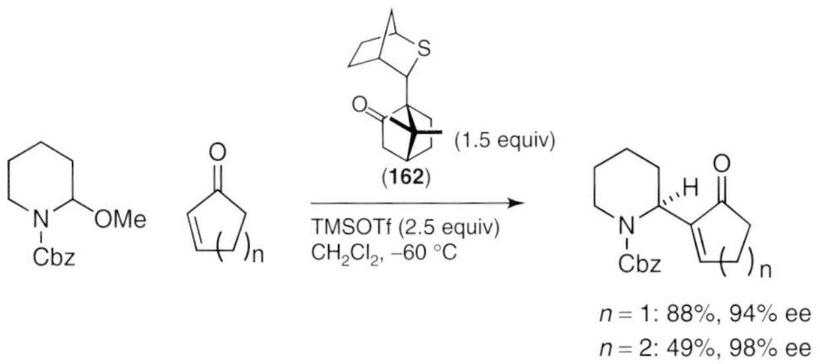

n = 1: 88%, 94% ee
n = 2: 49%, 98% ee

Scheme 3.55

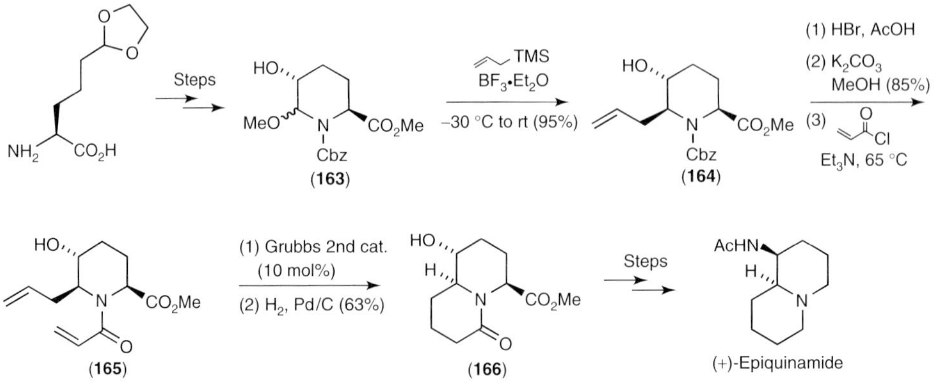

130 *3 Asymmetric Synthesis of Six-Membered Ring Heterocycles*

Scheme 3.56

migration from the *tert*-butyldiphenylsilyl (TBDPS) protective group smoothly proceeded to yield **168** in 80% isolated yield with high cis (>95%) selectivity. Reduction of lactam moiety and deprotection of *para*-methoxy-benzyl (PMB)-group followed by protection with (Boc)$_2$O yielded **169**. Further two-step treatment of **169** yielded (+)-L-733,060. On the other hand, the Boc-derivative **169** was transformed into the amino-derivative **171** via the reduction of the corresponding oxime **170**. Further elaboration yielded (+)-CP-99,994 (Scheme 3.57).

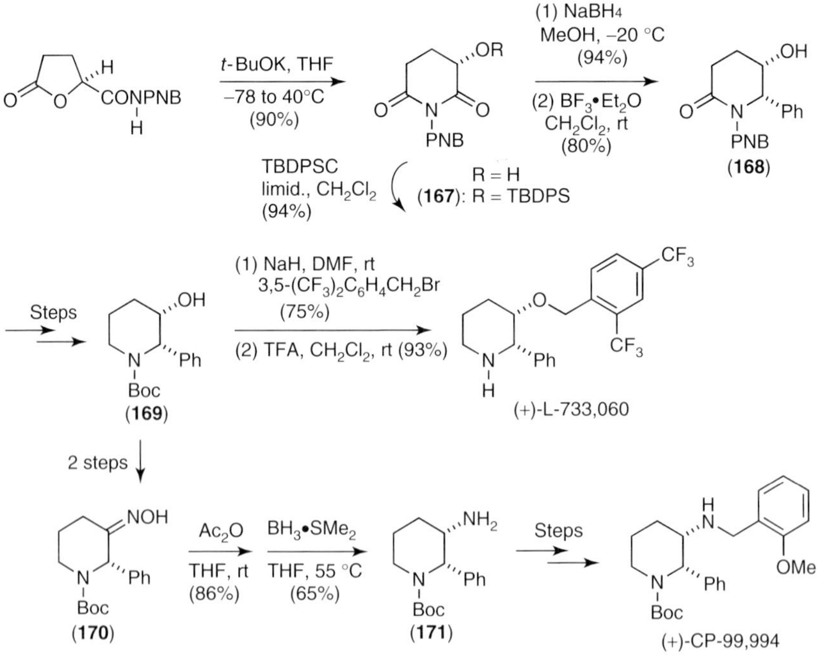

Scheme 3.57

3.4.5.5 Oxidative Methods

Oxidative methods, and particularly electrochemical process, proved to be valuable in the asymmetric synthesis of nitrogen compounds and piperidine derivatives. For example, the stereoselective synthesis of azasugars was obtained by Matsumura through the electrochemical oxidation of chiral piperidine derivatives. The enaminoester **100** was synthesized by anodic oxidation of **172** or **173** in 80% and 66% yield, respectively. After conversion of **174** to **175**, the latter was subjected again to electrochemical oxidation to yield **176**. The reduction of **176** with Et_3SiH yielded **177** as the major isomer. On the other hand, electrochemical oxidation of **178** or **179** yielded **180** in 62% and 51% yield, respectively, which was then converted to **181**. The electrochemical oxidation of **182** yielded **183**, whose α-acetoxy moiety was reduced with Et_3SiH to give **184** (Scheme 3.58) [91].

The sulfinyliminium salt instead of acyliminium ion is also effective for the enantioselective nucleophilic substitution on the six-membered ring system.

Royer reported the asymmetric synthesis of mono-substituted piperidine using the nucleophilic addition to the chiral N-sulfinyliminium salts. The first example

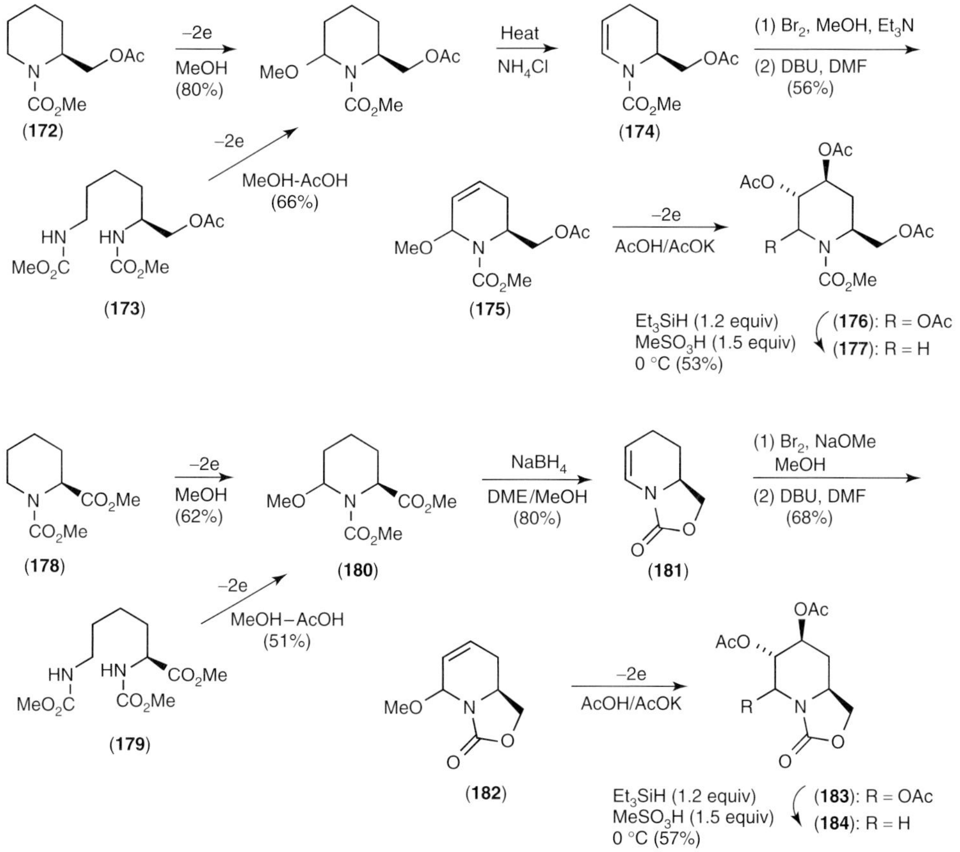

Scheme 3.58

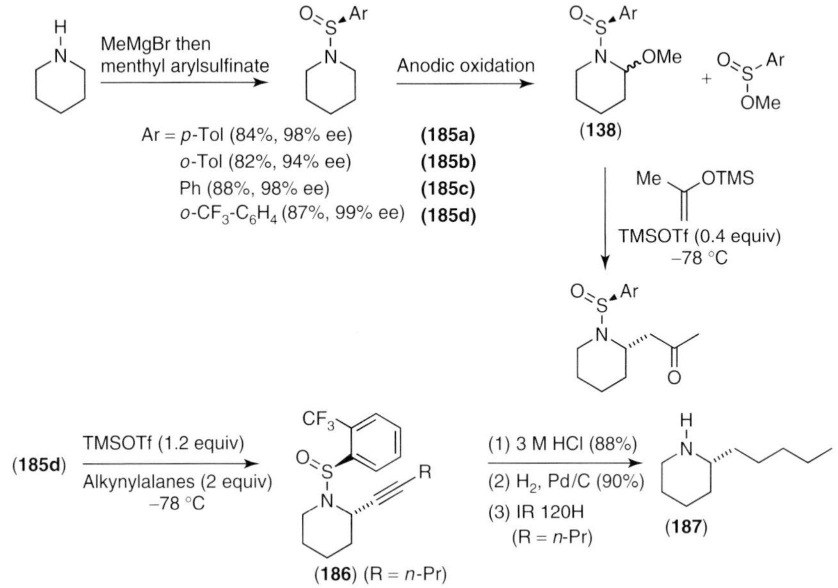

Scheme 3.59

Ar = p-Tol (84%, 98% ee) (185a)
o-Tol (82%, 94% ee) (185b)
Ph (88%, 98% ee) (185c)
o-CF$_3$-C$_6$H$_4$ (87%, 99% ee) (185d)

was the anodic oxidation of chiral sulfinylpiperidines, which was prepared from piperidine and menthyl arylsulfinates in high yields (82–88%) and enantioselectivity (94–99% ee). The anodic oxidation of **185** yielded α-methoxylated sulfinylamines along with the N–S bond cleavage products (~15%). Diastereoselective addition of trimethylsilylpropene to the corresponding iminium salts, derived from **185 d** with TMSOTf, proceeded to yield the adducts in acceptable diastereoselectivities (64–84% de). α-Alkynylation, as a further application of the chiral sulfinyliminium salt to the synthesis of mono-substituted piperidine, was also investigated. In this reaction, o-trifluoromethylsulfinylpiperidine showed best result, and the α-alkynylated products were obtained with high diastereoselectivities (92–97% de). The absolute configuration of the adduct was assigned by conversion of **186** to known **187** (Scheme 3.59) [92].

3.5
Fused Tri- or Bicyclic System with Bridgehead Nitrogen

Several syntheses of bicyclic systems with bridgehead nitrogen have already been described in the above sections as an illustration of synthetic methods used in the asymmetric synthesis of piperidines (see Sections 3.2, 3.3.4, 3.3.5, Nitrogen α–Carbanion, Iminium Addition, 3.3.2.3, 3.4.5.1, and 3.4.5.4). Some examples reported in the present section could also be considered as piperidines synthesis, though, in general, they are more specific methods directed toward the preparation of polycyclic systems.

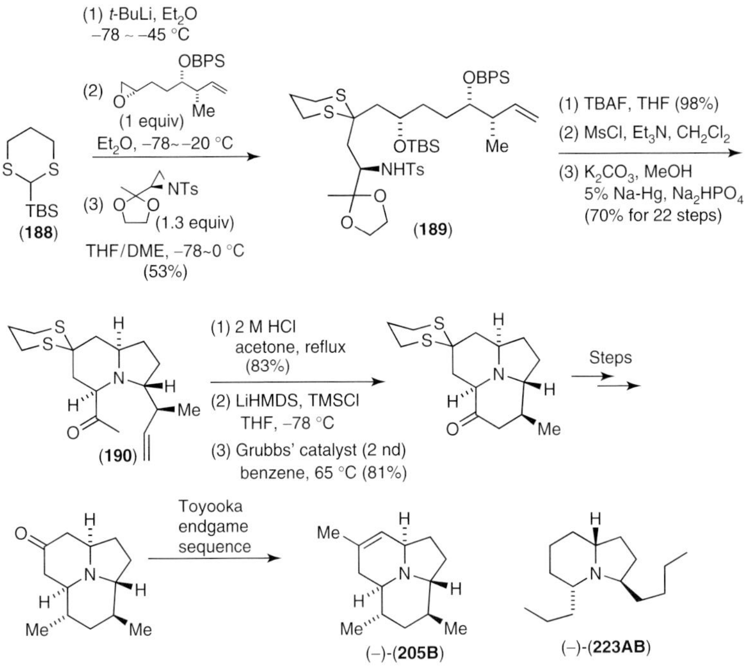

Scheme 3.60

A stereocontrolled total synthesis of natural enantiomer of poison-frog alkaloid (−)-**205B** has been achieved employing a one-pot construction of the embedded indolizidine skeleton as a key step. Thus, the dithiane three-component linchpin coupling reaction of **140** with aziridine and epoxide yielded the desired compound **141**. The one-pot construction of the indolizidine ring system was achieved by the removal of the silyl groups, bismesylation of the resulting diol, and treatment of the bismesylate with potassium carbonate in MeOH, followed by the addition of sodium amalgam to yield the indolizidine **142** (69% yield). This indolizidine was transformed into the natural enantiomer of (−)-**205B**. The indolizidine (−)-**223AB** was also synthesized by this convergent strategy (Scheme 3.60) [93, 94].

A highly diastereoselective quinolizidine ring closure reaction allowed Ma's group to achieve the total synthesis of marine alkaloids clavepictines A and B, and pictamine. Thus, the reaction of β-keto sulfone with chiral amine yielded the enamine sulfone **191**, which was then converted to alcohol **192** by iminium reduction and to precursor for the key intramolecular conjugate addition reaction. The key cyclization was performed by treatment of **193** with AlCl$_3$ followed by exposure of the liberated amine to aqueous NaHCO$_3$ to yield **195** via the most stable conformer **194**. After installation of the dienyl side chain by Julia coupling reaction, the total syntheses of clavepictines A and B, and pictamine were completed, as shown in Scheme 3.61 [95].

Angle's group reported the general synthesis of pyrroloquinolizidines using münchnone 1,3-dipolar cycloaddition as the key step. The strategy was applied to the

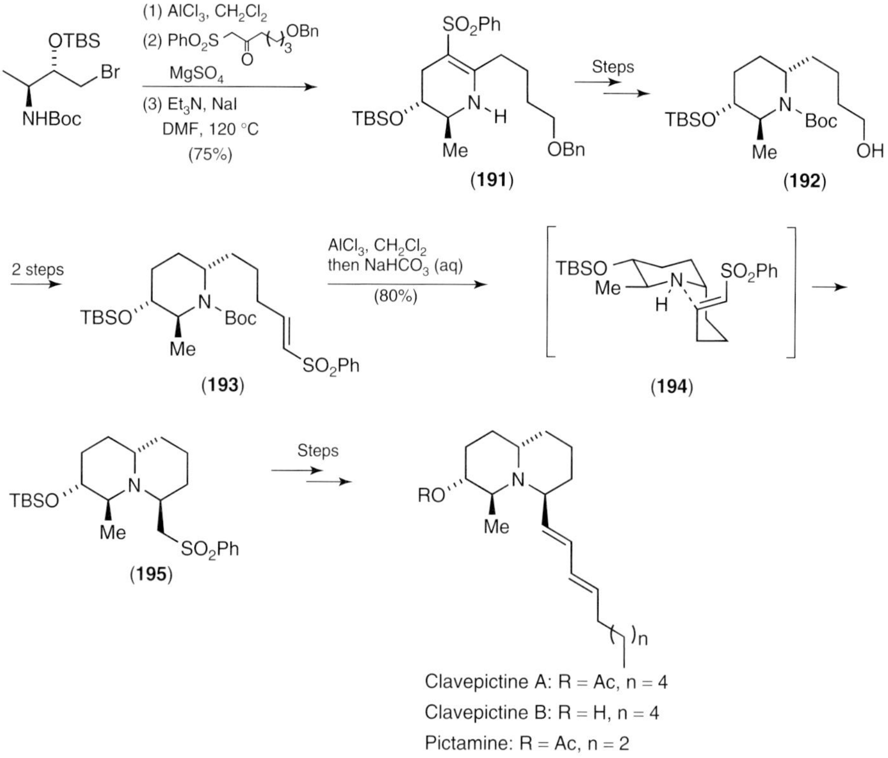

Scheme 3.61

synthesis of unnatural homolog of the ant-alkaloid mymicarin 215B. The chiral bicyclic lactam **197**, prepared from **196**, was subjected to 1,3-dipolar cycloaddition reaction with disubstituted alkyne **198** to yield the adduct **200**, which was then transformed into **201**, a homolog of mymicarin 215B. This is the first example of the

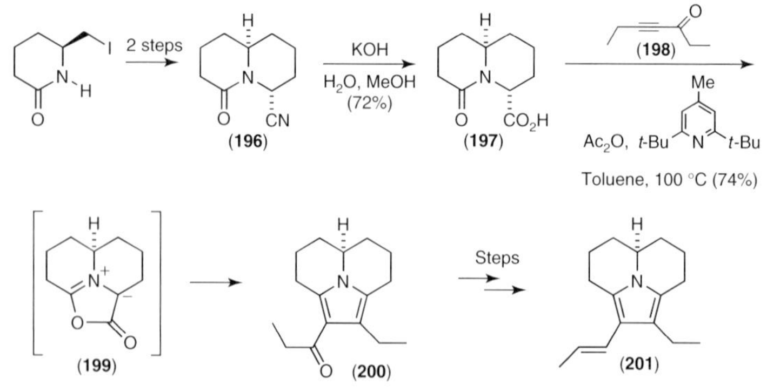

Scheme 3.62

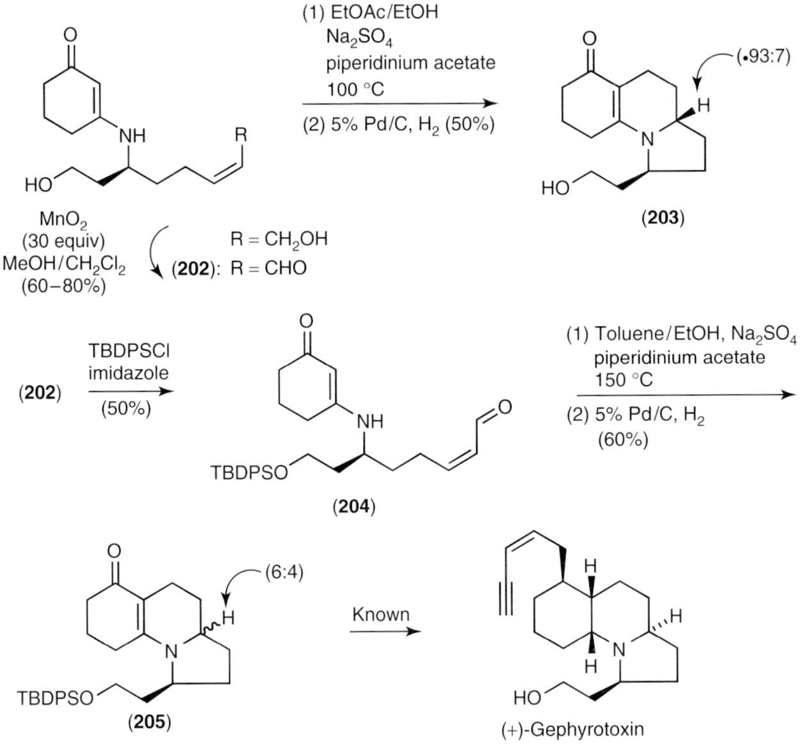

Scheme 3.63

1,3-dipolar cycloaddition using the tricyclic münchnone such as **199** (Scheme 3.62) [96].

A novel and highly stereoselective intramolecular formal [3 + 3] cycloaddition reaction of vinylogous amides tethered with α,β-unsaturated aldehydes allowed straightforward access to indolizidine skeleton. This methodology was extended to efficient formal synthesis of gephyrotoxin. Thus, the cycloaddition of **202** yielded the tricyclic compound **203** with high stereoselectivity. This ring system was key intermediate for the synthesis of gephyrotoxin; however, the stereochemistry of the newly formed position was opposite for gephyrotoxin synthesis. On the other hand, the cycloaddition of **204** yielded the desired cycloadduct **205**, which is the key intermediate of the Kishi's gephyrotoxin synthesis, although the stereoselectivity on the key cycloaddition reaction was low (Scheme 3.63) [97].

References

1. (a) Bailey, P. D., Millwood, P. A. and Smith, P. D. (1998) *Chem. Commun.*, 633–40; (b) Weinstraub, P. M., Sabol, J. S., Kane, J. M. and Borcherding, D. R. (2003) *Tetrahedron*, **59**, 2953–89; (c) Buffat, M. G. P. (2004) *Tetrahedron*, **60**, 1701–29;

(d) Laschat, S. and Dickner, T. (2000) *Synthesis*, 1781–813.
2 Comins, D. L. and Joseph, S. P. (1996) *Adv. Nitrogen Heterocycl.*, **2**, 251–94.
3 Charette, A. B., Grenon, M., Lemire, A., Pourashraf, M. and Martel, J. (2001) *J. Am. Chem. Soc.*, **123**, 11829–30.
4 Lemire, A. and Charette, A. B. (2005) *Org. Lett.*, **7**, 2747–50.
5 Ichikawa, E., Suzuki, M., Yabu, K., Albert, M., Kanai, M. and Shibasaki, M. (2004) *J. Am. Chem. Soc.*, **126**, 11808–9.
6 Sun, Z., Yu, S., Ding, Z. and Ma, D. (2007) *J. Am. Chem. Soc.*, **129**, 9300–1.
7 Yamada, S. and Morita, C. (2002) *J. Am. Chem. Soc.*, **124**, 8184–85.
8 (a) Kuethe, J. T. and Comins, D. L. (2000) *Org. Lett.*, **2**, 855–57; (b) Kuethe, J. T. and Comins, D. L. (2004) *J. Org. Chem.*, **69**, 5219–31.
9 Comins, D. L., Huang, S., McArdle, C. L. and Ingalls, C. L. (2001) *Org. Lett.*, **3**, 469–71.
10 Comins, D. L. and Williams, A. L. (2000) *Tetrahedron Lett.*, **41**, 2839–42.
11 Gotchev, D. B. and Comins, D. L. (2006) *J. Org. Chem.*, **71**, 9393–402.
12 Felpin, F.-X. and Lebreton, J. (2003) *Eur. J. Org. Chem.*, 3693–712.
13 Davis, F. A., Zhang, Y. and Li, D. (2007) *Tetrahedron Lett.*, **48**, 7838–40.
14 Felpin, F.-X., Boubekeur, K. and Lebreton, J. (2003) *Eur. J. Org. Chem.*, 4518–27.
15 Lee, H. K., Im, J. H. and Jung, S. H. (2007) *Tetrahedron*, **63**, 3321–27.
16 Holub, N., Neidhofer, J. and Blechert, S. (2005) *Org. Lett.*, **7**, 1227–29.
17 Rueping, M. and Antonchick, A. P. (2007) *Angew. Chem. Int. Ed.*, **46**, 4562–65.
18 Takahata, H., Suto, Y., Kato, E., Yoshimura, Y. and Ouchi, H. (2007) *Adv. Synth. Catal.*, **349**, 685–93.
19 Ouchi, H., Mihara, Y. and Takahata, H. (2005) *J. Org. Chem.*, **70**, 5207–14.
20 Montoro, R., Marquez, F., Llebaria, A. and Delgado, A. (2003) *Eur. J. Org. Chem.*, 217–23.
21 Davis, F. A., Xu, H. and Zhang, J. (2007) *J. Org. Chem.*, **72**, 2046–52.
22 Pu, X. and Ma, D. (2003) *J. Org. Chem.*, **68**, 4400–5.
23 Ma, N. and Ma, D. (2003) *Tetrahedron: Asymmetry*, **14**, 1403–6.
24 Ma, D. and Ma, N. (2003) *Tetrahedron Lett.*, **44**, 3963–65.
25 Davies, S. G. and Ichihara, O. (1991) *Tetrahedron: Asymmetry*, **2**, 183–86.
26 Garraffo, H. M., Jain, P., Spande, T. F. and Daly, J. W. (1997) *J. Nat. Prod.*, **60**, 2–5.
27 Yu, S., Zhu, W. and Ma, D. (2005) *J. Org. Chem.*, **70**, 7364–70.
28 Pu, X. and Ma, D. (2004) *Angew. Chem. Int. Ed.*, **43**, 4222–25.
29 Pu, X. and Ma, D. (2006) *J. Org. Chem.*, **71**, 6562–72.
30 Tanaka, K. and Katsumura, S. (2002) *J. Am. Chem. Soc.*, **124**, 9660–61.
31 Kobayashi, T., Nakashima, M., Hakogi, T., Tanaka, K. and Katsumura, S. (2006) *Org. Lett.*, **8**, 3809–12.
32 Kobayashi, T., Hasegawa, F., Tanaka, K. and Katsumura, S. (2006) *Org. Lett.*, **8**, 3813–16.
33 Wurz, R. P. and Fu, G. C. (2005) *J. Am. Chem. Soc.*, **127**, 12234–35.
34 Cossy, J., Willis, C., Bellosta, V. and BouzBouz, S. (2002) *J. Org. Chem.*, **67**, 1982–92.
35 Takahata, H., Ouchi, H., Ichinose, M. and Nemoto, H. (2002) *Org. Lett.*, **4**, 3459–62.
36 Concello'n, J. M., Riego, E., Rivero, I. A. and Ochoa, A. (2004) *J. Org. Chem.*, **69**, 6244–48.
37 Eskici, M. and Gallagher, T. (2000) *Synlett*, 1360–62.
38 Masuda, Y., Tashiro, T. and Mori, K. (2006) *Tetrahedron: Asymmetry*, **17**, 3380–85.
39 Kandula, S. R. V. and Kumar, P. (2006) *Tetrahedron*, **62**, 9942–48.
40 Singh, R. and Ghosh, S. K. (2002) *Tetrahedron Lett.*, **43**, 7711–15.
41 Lee, Y.-S., Shin, Y.-H., Kim, Y.-H., Lee, K.-Y., Oh, C.-Y., Pyun, S.-J., Park, H.-J., Jeong, J.-H.

and Ham, W.-H. (2003) *Tetrahedron: Asymmetry*, **14**, 87–93.
42. Enders, D. and Kirchhoff, J. H. (2000) *Synthesis*, 2099–105.
43. Yokoyama, H., Otaya, K., Kobayashi, H., Miyazawa, M., Yamaguchi, S. and Hirai, Y. (2000) *Org. Lett.*, **2**, 2427–29.
44. Yokoyama, H., Ejiri, H., Miyazawa, M., Yamaguchi, S. and Hirai, Y. (2007) *Tetrahedron: Asymmetry*, **18**, 852–56.
45. Hirai, Y., Watanabe, J., Nozaki, T., Yokoyama, H. and Yamaguchi, S. (1997) *J. Org. Chem.*, **62**, 776–77.
46. Makabe, H., Kong, L. K. and Hirota, M. (2003) *Org. Lett.*, **5**, 27–29.
47. Welter, C., Dahnz, A., Brunner, B., Streiff, S., Du1bon, P. and Helmchen, G. (2005) *Org. Lett.*, **7**, 1239–42.
48. LaLonde, R. L., Sherry, B. D., Kang, E. J. and Toste, F. D. (2007) *J. Am. Chem. Soc.*, **129**, 2452–53.
49. Abrunhosa-Thomas, I., Roy, O., Barra, M., Besset, T., Chalard, P. and Troin, Y. (2007) *Synlett*, 1613–15.
50. Kawabata, T., Kawakami, S. and Majumdar, S. (2003) *J. Am. Chem. Soc.*, **125**, 13012–13.
51. Davis, F. A. and Yang, B. (2005) *J. Am. Chem. Soc.*, **127**, 8398–407.
52. Davis, F. A. and Santhanaraman, M. (2006) *J. Org. Chem.*, **71**, 4222–26.
53. Davis, F. A., Chao, B. and Rao, A. (2001) *Org. Lett.*, **3**, 3169–71.
54. Monfray, J., Gelas-Mialhe, Y., Gramain, J.-C. and Remuson, R. (2005) *Tetrahedron: Asymmetry*, **16**, 1025–34.
55. Kinderman, S. S., de Gelder, R., van Maarseveen, J. H., Schoemaker, H. E., Hiemstra, H. and Rutjes, F. P. J. T. (2004) *J. Am. Chem. Soc.*, **126**, 4100–1.
56. Huang, H., Spande, T. F. and Panek, J. S. (2003) *J. Am. Chem. Soc.*, **125**, 626–27.
57. Glasson, S. R., Canet, J.-L. and Troin, Y. (2000) *Tetrahedron Lett.*, **41**, 9797–802.
58. Carbonnel, S., Fayet, C., Gelas, J. and Troin, Y. (2000) *Tetrahedron Lett.*, **41**, 8293–96.
59. Ciblat, S., Calinaud, P., Canet, J.-L. and Troin, Y. (2000) *J. Chem. Soc. Perkin Trans. 1*, 353–57.
60. Gandon, L. A., Russell, A. G., Güveli, T., Brodwolf, A. E., Kariuki, B. M., Spencer, N. and Snaith, J. S. (2006) *J. Org. Chem.*, **71**, 5198–207.
61. Tour, B. B. and Hall, D. G. (2004) *Angew. Chem. Int. Ed.*, **43**, 2001–4.
62. Ozawa, T., Aoyagi, S. and Kibayashi, C. (2000) *Org. Lett.*, **2**, 2955–58.
63. Ozawa, T., Aoyagi, S. and Kibayashi, C. (2001) *J. Org. Chem.*, **66**, 3338–47.
64. Sato, T., Aoyagi, S. and Kibayashi, C. (2003) *Org. Lett.*, **5**, 3839–42.
65. Adelbrecht, J.-C., Craig, D., Fleming, A. J. and Martin, F. M. (2005) *Synlett*, 2643–47.
66. Tan, C.-H. and Holmes, A. B. (2001) *Chem. Eur. J.*, **7**, 1845–54.
67. Cossy, J., Mirguet, O., Pardo, D. G. and Desmurs, J.-R. (2001) *Tetrahedron Lett.*, **42**, 5705–7.
68. Lee, J., Hoang, T., Lewis, S., Weissman, S. A., Askin, D., Volante, R. P. and Reider, P. J. (2001) *Tetrahedron Lett.*, **42**, 6223–25.
69. Lonkar, P. S. and Kumar, V. A. (2004) *Bioorg. Med. Chem. Lett.*, **14**, 2147–49.
70. Honda, T. and Kimura, M. (2000) *Org. Lett.*, **2**, 3925–27.
71. Heller, E., Lautenschläger, W. and Holzgrabe, U. (2005) *Tetrahedron Lett.*, **46**, 1247–49.
72. Langlois, N. (2002) *Org. Lett.*, **4**, 185–87.
73. Poupon, E., François, D., Kunesch, N. and Husson, H.-P. (2004) *J. Org. Chem.*, **69**, 3836–41.
74. François, D., Poupon, E., Kunesch, N. and Husson, H.-P. (2004) *Eur. J. Org. Chem.*, 4823–29.
75. Roulland, E., Cecchin, F. and Husson, H.-P. (2005) *J. Org. Chem.*, **70**, 4474–77.
76. Roulland, E., Chiaroni, A. and Husson, H.-P. (2005) *Tetrahedron Lett.*, **46**, 4065–68.
77. Calvet-Vitale, S., Vanucci-Bacqué, C., Fargeau-Bellassoued, M.-C. and Lhommet, G. (2005) *Tetrahedron*, **61**, 7774–82.

78 Noël, R., Vanucci-Bacqué, C., Fargeau-Bellassoued, M.-C. and Lhommet, G. (**2007**) *Eur. J. Org. Chem.*, 476–86.

79 Amat, M., Llor, N., Hidalgo, J., Escolano, C. and Bosch, J. (**2003**) *J. Org. Chem.*, **68**, 1919–28.

80 Amat, M., Lozano, O., Escolano, C., Molins, E. and Bosch, J. (**2007**) *J. Org. Chem.*, **72**, 4431–39.

81 Toyooka, N., Fukutome, A., Nemoto, H., Daly, J. W., Spande, T. F., Garraffo, H. M. and Kaneko, T. (**2002**) *Org. Lett.*, **4**, 1715–18.

82 Toyooka, N., Fukutome, A., Shinoda, H. and Nemoto, H. (**2004**) *Tetrahedron*, **60**, 6197–216.

83 Toyooka, N., Fukutome, A., Shinoda, H. and Nemoto, H. (**2003**) *Angew. Chem. Int. Ed.*, **42**, 3808–10.

84 Ashoorzadeh, A. and Caprio, V. (**2005**) *Synlett*, 346–48.

85 Itoh, T., Yamazaki, N. and Kibayashi, C. (**2002**) *Org. Lett.*, **4**, 2469–72.

86 Myers, E. L., de Vries, J. G. and Aggarwal, V. K. (**2007**) *Angew. Chem. Int. Ed.*, **46**, 1893–96.

87 Wijdeven, M. A., Botman, P. N. M., Wijtmans, R., Schoemaker, H. E., Rutjes, F. P. J. T. and Blaauw, R. H. (**2005**) *Org. Lett.*, **7**, 4005–7.

88 Huang, P.-Q., Guo, Z.-Q. and Ruan, Y.-P. (**2006**) *Org. Lett.*, **8**, 1435–38.

89 Liu, L.-X., Ruan, Y.-P., Guo, Z.-Q. and Huang, P.-Q. (**2004**) *J. Org. Chem.*, **69**, 6001–9.

90 Huang, P.-Q., Liu, L.-X., Wei, B.-G. and Ruan, Y.-P. (**2003**) *Org. Lett.*, **5**, 1927–29.

91 Furukubo, S., Moriyama, N., Onomura, O. and Matsumura, Y. (**2004**) *Tetrahedron Lett.*, **45**, 8177–81.

92 Turcaud, S., Sierecki, E., Martens, T. and Royer, J. (**2007**) *J. Org. Chem.*, **72**, 4882–85.

93 Smith, A. B. and Kim, D.-S. III (**2005**) *Org. Lett.*, **7**, 3247–50.

94 Smith, A. B. and Kim, D.-S. III (**2006**) *J. Org. Chem.*, **71**, 2547–57.

95 Yu, S., Pu, X., Cheng, T., Wang, R. and Ma, D. (**2006**) *Org. Lett.*, **8**, 3179–82.

96 Angle, S. R., Qian, X. L., Pletnev, A. A. and Chinn, J. (**2007**) *J. Org. Chem.*, **72**, 2015–20.

97 Wei, L.-L., Hsung, R. P., Sklenicka, H. M. and Gerasyuto, A. I. (**2001**) *Angew. Chem. Int. Ed.*, **40**, 1516–18.

4
Asymmetric Synthesis of Seven- and More-Membered Ring Heterocycles

Yves Troin and Marie-Eve Sinibaldi

Seven-membered ring nitrogen heterocycles (azepines) together with more-membered substrates, in particular azocines (eight-membered azacycles) and sometimes azonines (nine-membered azacycles), are important classes of compounds occurring in a range of natural and synthetic products with interesting biological activities leading to application in pharmaceutical research [1a,b,c]. However, the generation of these medium-sized ring molecules, in particular of the eight or more-membered ring compounds, constitutes always a challenge in organic synthesis.

We summarize herein the different asymmetric methods described for the elaboration of these skeletons.

4.1
Substituted Azepines

4.1.1
Generalities

The azepine (azacycloheptatriene system) exists in four tautomeric forms with the relative stability, 3H->1H->2H->4H-(Scheme 4.1).

Dihydro- and tetrahydroazepines are the most common azepine substrates forming the pharmacophores of many biologically active compounds [1d].

For example, simple N-substituted hexahydroazepines are interesting substrates which are recognized as antitussives, mydriatics, antispasmodics and oral hypoglycemics, and, azepan-2-one (ε-caprolactam) which has been industrially used in the production of perlon constitutes the cornerstone starting material in the preparation of more elaborated azepanes.

Moreover, variously substituted hexahydroazepines are compounds of particular interest. Thus, hexahydro-azepin-4-ol derivatives possess most often analgesic properties; for example, proheptazine is recognized to have twice the analgesic effect of morphine without addictive effects. Balanol, a bisubstituted hexahydro-azepine is an antineoplastic inhibitor of protein kinase C and plays an important role in cellular growth control, regulation and differentiation (Scheme 4.2) [1e].

Asymmetric Synthesis of Nitrogen Heterocycles. Edited by Jacques Royer
Copyright © 2009 WILEY-VCH Verlag GmbH & Co. KGaA, Weinheim
ISBN: 978-3-527-32036-3

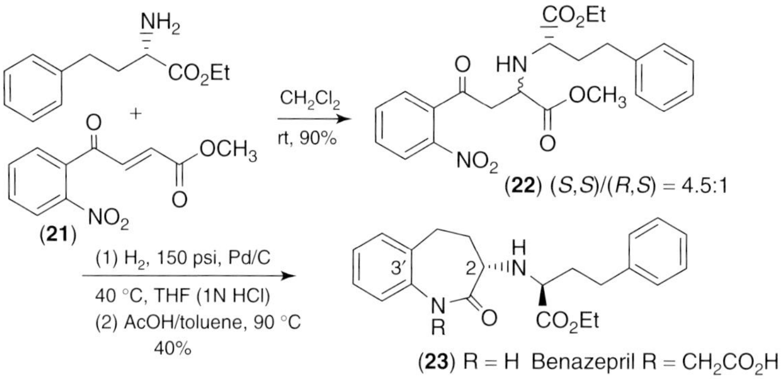

Scheme 4.12 Diastereoselective formal synthesis of benazepril.

4.1.2.2 Radical Cyclization

Numerous reports illustrated the construction of five- and six-membered heterocyclic rings by radical cyclization processes [11]. By contrast, less favorable cyclization to seven-membered ring has been less studied.

During their work devoted toward the total synthesis of (−)-balanol, Naito et al. [12a] described the radical cyclization of aldehyde **24a** with SmI_2 in the presence of Hexamethylphosphoric triamide (HMPA) leading preferentially to the *trans* seven-membered ring **25a** in 46% yield (Scheme 4.13). Selectivity has been explained by chelation between the $Sm(III)(HMPA)_n$ cation to the carbonyl and the oxime ether moieties. By the same way, Skrydstrup et al. [12b] synthesized the hexahydro-azepine **27** with high trans selectivity (10:1) by a Sm(II)iodide/HMPA induced pinacol-type cyclization of the corresponding acyclic hydrazone precursor **26**. Upon treatment with tributyltin hydride and 2,2′-azo*bis*(2-methylpropionitrile) (AIBN), oxime ether **24b** underwent stannyl radical addition–cyclization with high selectivity to the major *trans*-hexahydro-azepine fragment **25b** in an *exo-trig* reaction pathway (Scheme 4.13) [12c].

Finally, the enantiomerically pure hexahydro-azepine framework (3*R*, 4*R*)-**29a,b**, precursors of (−)-balanol, was obtained (i) after chemical optical resolution of the diastereomeric esters of the racemic alcohols **28** derived from **25** or (ii) by enzymatic optical resolution of **28** using the immobilized lipase *Pseudomonas* sp. (Scheme 4.14) [13].

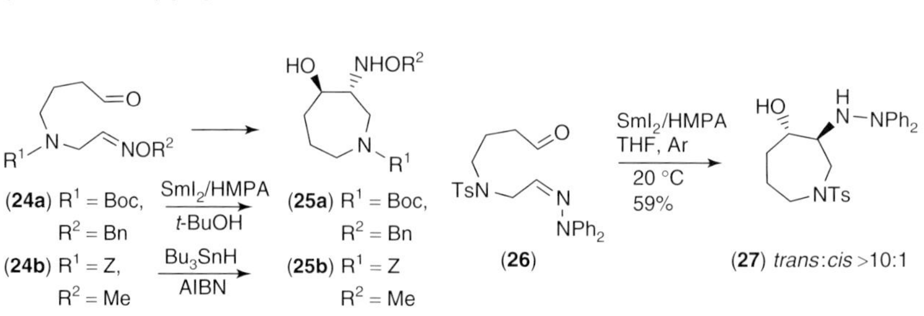

Scheme 4.13 Use of radical cyclization processes to azepine fragment of (−)-balanol.

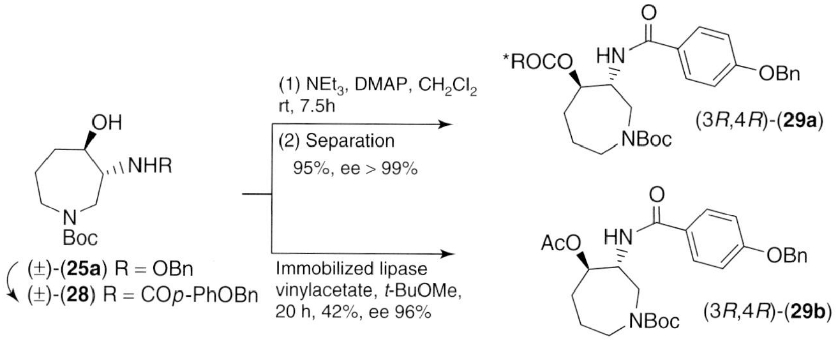

Scheme 4.14 Chemical or enzymatic optical resolution of ester (±)-**28**.

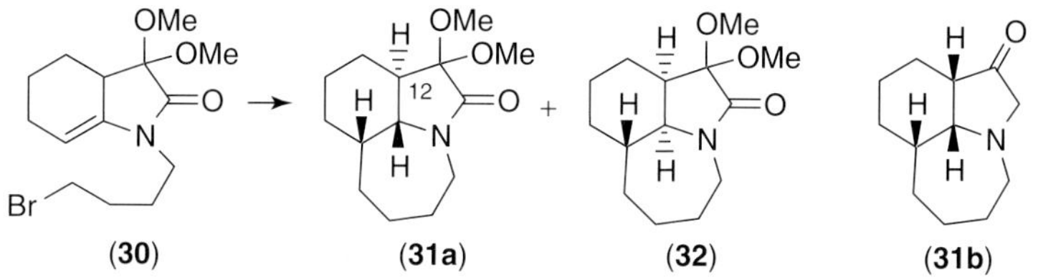

Scheme 4.15 Radical cyclization to azepino-indoline of *Stemona* alkaloids.

The azepino-indoline skeletons of the *Stemona* alkaloids have been efficiently prepared by radical cyclization [14] of hydroindolone **30** (Scheme 4.15). Thus, azepinoindolines **31a** and **32** were obtained in a 5:1 ratio and 85% overall yield using Ph$_3$SnH/AIBN in refluxing benzene under slow addition. The major formation of **31** arose from a 7-*endo-trig* radical cyclization process. Reduction of the lactam moiety followed by C-12 equilibration in acidic medium, permitted an access to **31b** owning another ring-fusion type of the *Stemona* compounds.

In 2000, Evans et al. [15] related the first examples of the intramolecular addition of acyl and alkyl radicals to oxauracil for the diastereoselective construction of *trans*-disubstituted-6,7-azabicycles **33** (Scheme 4.16).

Octahydrocyclopenta[b]azepines were obtained with excellent stereoselectivity by a 7-*endo* cyclization of various chiral vinylogous amides as reported by Cordes and Franke [16] in 2004. By this way, **34** was N-acylated with (S)-(−)-2-acetoxypropionyl chloride to give chiral enaminone **35** in 82% yield. Treatment of **35** with Bu$_3$SnH in the presence of the radical initiator 1,1′-azo-bis(cyclohexanecarbonitrile) (ACN)

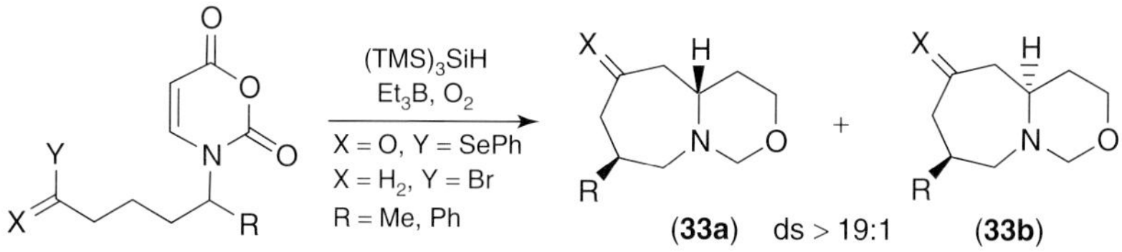

Scheme 4.16 *trans*-Disubstituted 6,7-azabicycles.

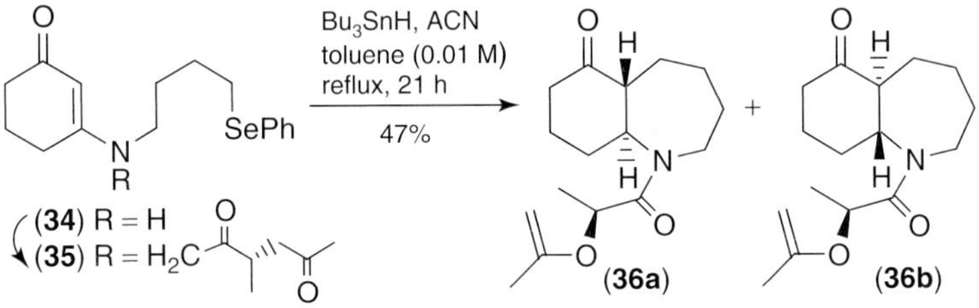

Scheme 4.17 Diastereoselective synthesis of octahydrocyclopenta[b]azepines.

afforded the enantiopure azabicycles **36a,b** via a 7-endo ring closure in 47% yield (2:1 ratio). Changing the chiral appendage on the nitrogen atom of **35** (use of (1S)-(−)-camphanic chloride) permitted the formation of a single trans-fused octahydrocyclopenta[b]azepine in 44% yield with the relative stereochemistry of **36a** (Scheme 4.17).

An original and efficient approach to chiral 1,3-perhydrobenzoxazines **39** precursors of 3-substituted hexahydroazepines has been described. Thus, compounds **38**, generated in two steps from **37**, were engaged in an intramolecular radical addition promoted by Bu_3SnH affording a nearly quantitative mixture of 7-endo **39a,b** and 6-exo **39c,d** cyclized compounds in a ratio depending upon the substituents at the α-position of the double bond (Scheme 4.18). When R^2 = Me and R^1 = H, the cyclization step delivered a mixture of three products **40a–c** in favor of the 7-endo cyclization (75, 12, and 13% yields, respectively). The major one **40a** was further transformed into the enantiopure chlorhydrate (S)-3-methylperhydroazepine **41** in 70% yield [17].

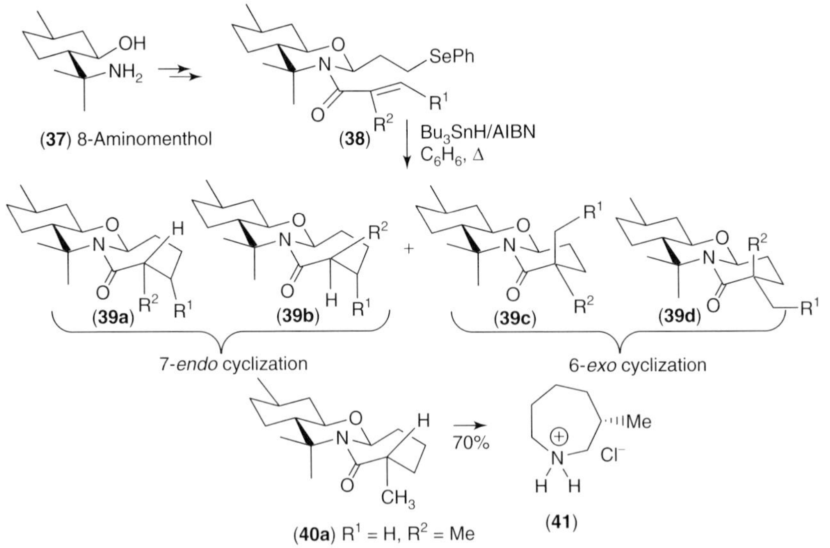

Scheme 4.18 Asymmetric approach to 3-substituted perhydroazepines.

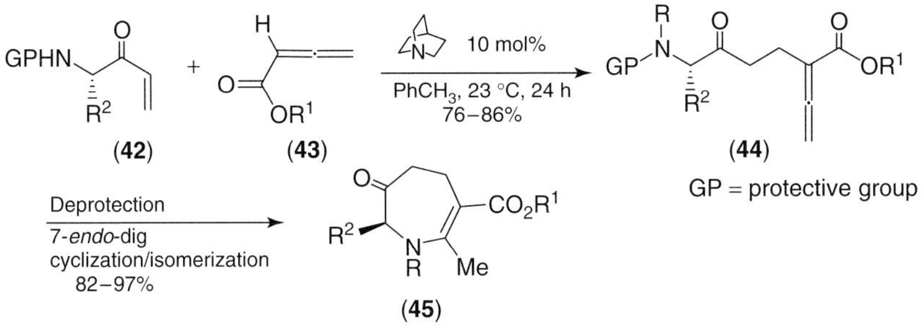

Scheme 4.19 Azepine from 7-*endo*-dig cyclization of allenoate esters.

Vinyl ketones **42** derived from N-protected amino acids were coupled to allenoate esters **43** to furnish **44** which, after amine deprotection followed by 7-*endo*-dig addition of nitrogen to the electron-deficient sp-carbon of the allene and isomerization, afforded the azepines **45** in excellent yields (Scheme 4.19) [18].

4.1.2.3 Intramolecular Cyclization

Pyne *et al.* [19] developed a convergent preparation of the tricyclic B,C,D-ring core structure of croomine, a natural product possessing a pyrrolo[1,2-*a*]azepine skeleton. The key precursor **48** was prepared by a tandem chiral vinyl epoxide aminolysis/ring-closing metathesis (RCM) sequence starting from amine **46** and epoxide **47** and was then engaged in an intramolecular S_N2 process to generate the 1H-pyrrolo[1,2-*a*]azepine **49** in a good yield of 68 % (Scheme 4.20).

Beak and Lee [20] reported an efficient asymmetric synthesis of 3,4,5- and 3,4,5,6-substituted hexahydroazepines **54** and **55** starting from enantio-enriched ene-carbamates **51** themselves obtained *via* a (−)-sparteine-mediated asymmetric lithiation of allylamines **50** (Scheme 4.21).

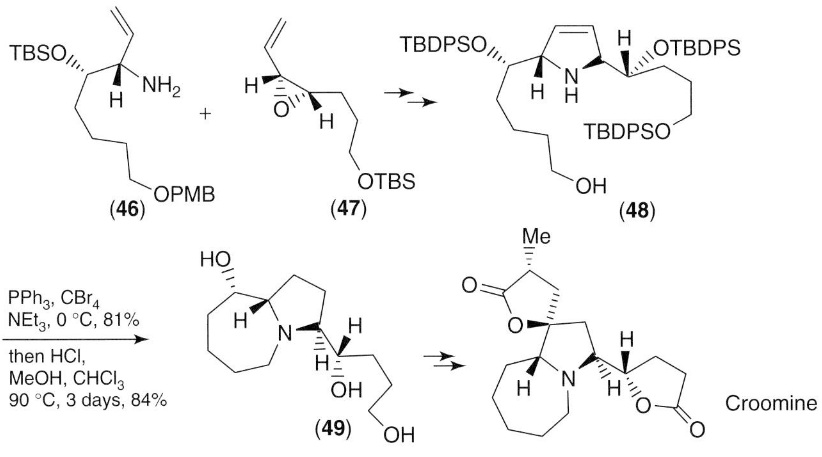

Scheme 4.20 Diastereoselective access to 1*H*-pyrrolo[1,2-*a*]azepine framework.

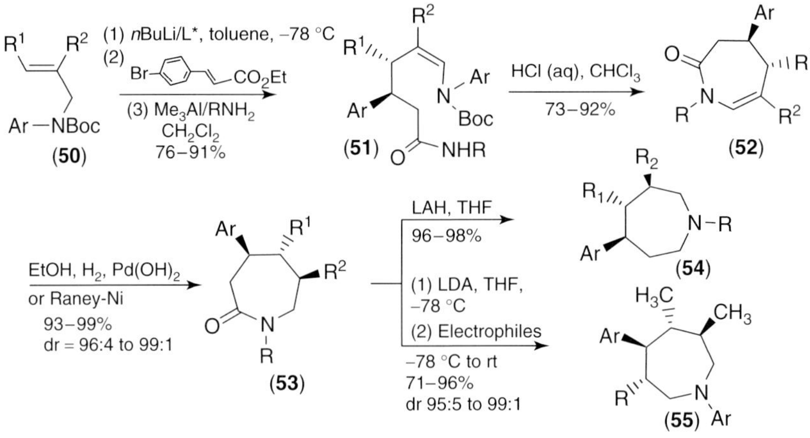

Scheme 4.21 Asymmetric synthesis of 4,5,6- and 3,4,5,6-substituted azepanes.

Acid hydrolysis of enamide **51** to aldehyde and subsequent cyclization led to the expected seven-membered ring system **52**. Hydrogenation of **52** afforded **53** possessing a *trans* stereochemistry at the C-6 position.

Reduction of lactam with LiAlH$_4$ gave efficiently trisubstituted compounds **54**, whereas substitution at the C-3 position of **53** (for example, R^1 = R^2 = CH$_3$, R = Ar) by selected electrophiles provided tetrasubstituted azepanes **55** in good yields and high diastereoselectivity (dr >95:5) (Scheme 4.21).

Optically active ω-bromocyanohydrin **56** easily synthesized through an enantioselective (R)-oxynitrilase-catalyzed reaction from its corresponding ω-bromoaldehyde was used as starting material to prepare azepan-3-ol **58** in a high enantiomeric excess. The crucial ring closure took place quantitatively using tBuOK on the (R)-protected bromoamino alcohol **57** with ee of 91% (Scheme 4.22) [21a].

Similar annelation process has been applied to the synthesis of the azepine core of (−)-galanthamine *via* the mesylate derivative of **59** but in modest yield (Scheme 4.23) [21b].

Williams *et al.* [22] reported in 2003 an enantioselective synthesis of (−)-stemonine in which, the aza-seven-membered ring is formed by a Staudinger-aza-Wittig cascade reaction of an aldehyde-azide **61** followed by NaBH$_4$ reduction. The chiral azide precursor **61** was itself easily prepared from the optically pure butyrolactone **60** through classical transformations.

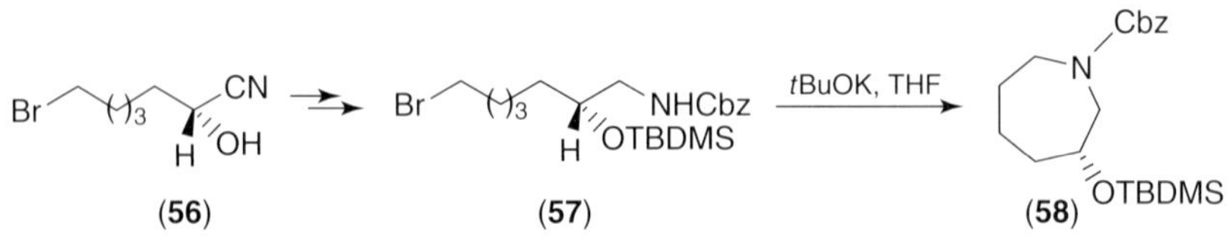

Scheme 4.22 Enantioselective synthesis of 3-hydroxyazepanes.

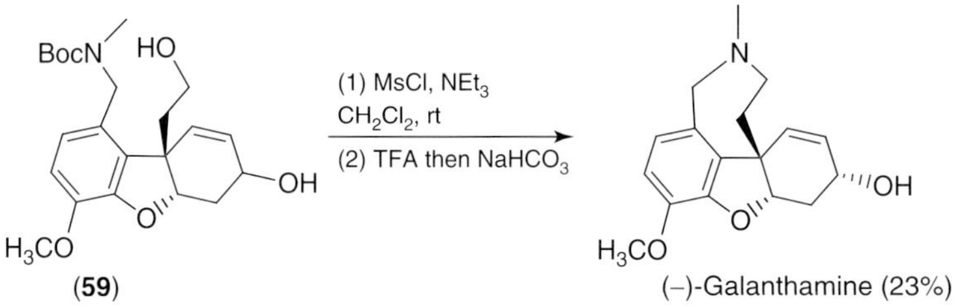

4.1 Substituted Azepines

Scheme 4.23 Synthesis of (−)-galanthamine.

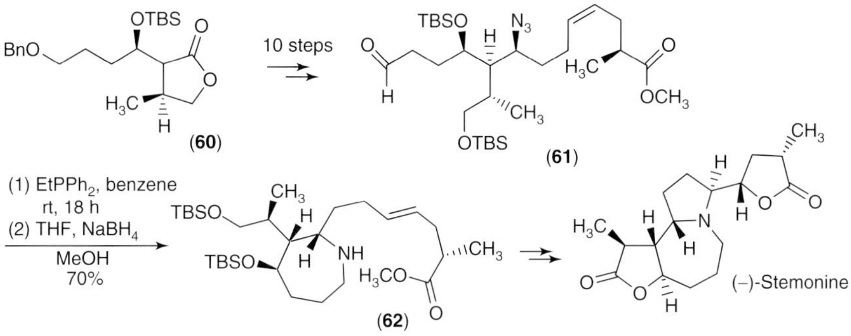

Scheme 4.24 Staudinger-aza-Wittig formation of the azepine ring.

Finally, iodine-induced cyclization of **62** led in a single step to the pyrrolidinobutyrolactone framework with good stereoselectivity which was then further transformed efficiently to the tricyclic fused (−)-stemonine (Scheme 4.24).

Reductive aminocyclization has been successfully adapted to the elaboration of seven-membered azacycles and we depict below some examples.

Palladium-catalyzed asymmetric allylic alkylation of halophenol **63** and α,β-unsaturated ester **64** led to intermediate **65** which was further transformed using a Heck cyclization process to the tricyclic compound **66**. Standard transformations afforded then compound **67** which underwent after Boc deprotection a final reductive intramolecular amination reaction to give (−)-3-deoxygalanthamine [23] in a good yield (Scheme 4.25).

Tri- and tetrahydroxyazepanes which were known as potent glycosidase inhibitors, were obtained in good yields from D-(−)-quinic acid. Oxidative cleavage of alcohols **68** or **69** afforded the corresponding dialdehydes which underwent a subsequent reductive aminocyclization using $BnNH_2$ and $NaBH(OAc)_3$. Hydrogenation gave azepanes **70** and **71**, which were in fact prepared in 10 steps with 17–34% yields from D-(−)-quinic acid (Scheme 4.26) [24].

Chiral nitrogen-containing seven-membered rings have been also efficiently prepared using 7-endo cyclization of olefins on N-acyliminium [25]. Thus, enamide **72** was obtained in three steps in a 1:1 ratio of two isomers from N-protected L-phenylalanine used as a chiral template. After separation by preparative high performance liquid chromatography (HPLC), isomer **72a** ($R^1 = CO_2Me$, $R^2 = H$) was efficiently cyclized by acidic treatment to the optically active azepine

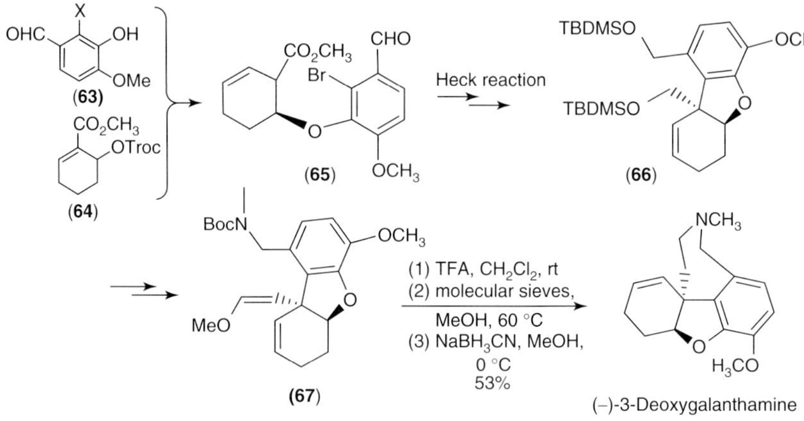

Scheme 4.25 Azepine ring from reductive intramolecular amination.

derivative **73** possessing the 5-7-6 tricyclic system of cephalotaxine alkaloid. The stereoselectivity of the reaction was excellent and was due to the equatorial conformation adopted by the phthalimide moiety in the acyliminium intermediate structure (Scheme 4.27) [26].

An asymmetric construction of the same 5-7-6 tricyclic core was reported through a Pictet–Spengler reaction [27]. Thus, conversion of the amido-alcohol **74** easily obtained from (S)-malic acid, to the lactams **75a,b** was realized in 70% yield with a good regioselectivity using triflic acid *via* an acyliminium cyclization process (Scheme 4.28).

The Pictet–Spengler reaction was also efficiently applied by Chida *et al.* [28] to the stereoselective synthesis of (+)-galanthamine starting from D-glucose which allowed the formation of the optically pure cyclohexene ring of the amide precursor **76**. Upon treatment with *para*-formaldehyde in the presence of trifluoroacetic acid (TFA), **76** generated the known seven-membered tetracyclic derivative **77**, precursor of the alkaloid in 67% yield (Scheme 4.29).

Using the same methodology, the azepine ring of (−)-lycoramine was elaborated by Malachowski *et al.* [29] in 2007 in the first enantioselective synthesis of this alkaloid.

Even if the Pummerer-cyclization applied to seven-membered ring systems is less efficient than for the six-membered analogs, benzazepine derivatives that are difficult to obtain by others methods have been synthesized using this reaction and a review discussed their preparation [30].

Others cyclization processes have been adapted to the edification of azepane framework.

For example, Hayes *et al.* [31a] created the azepine ring of the tetracyclic core of (−)-cephalotaxine [31b], using an intramolecular Heck cyclization. Thus, cyclization of **78**, prepared from *N*-Boc-L-proline ester in seven steps and 7% overall yield, mediated by Pd catalyst afforded the sole formation of the endocyclic olefin **79** in 62% yield (Scheme 4.30).

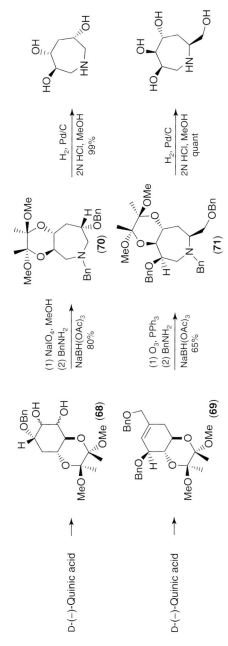

Scheme 4.26 Asymmetric synthesis of trihydroxy- and tetrahydroxyazepanes.

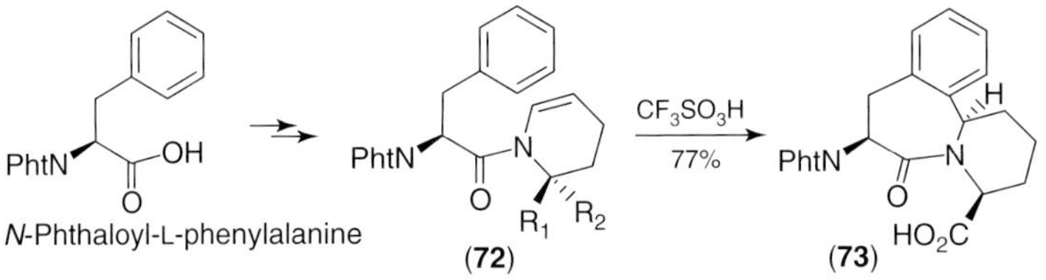

Scheme 4.27 Cyclization of olefins on N-acyliminium.

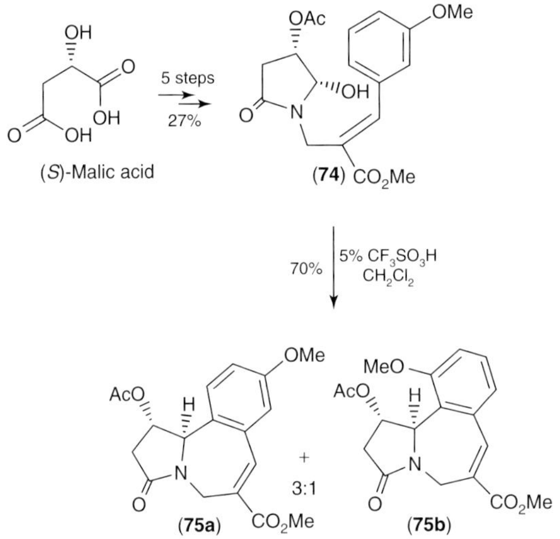

Scheme 4.28 Pictet–Spengler approach to the azepane ring.

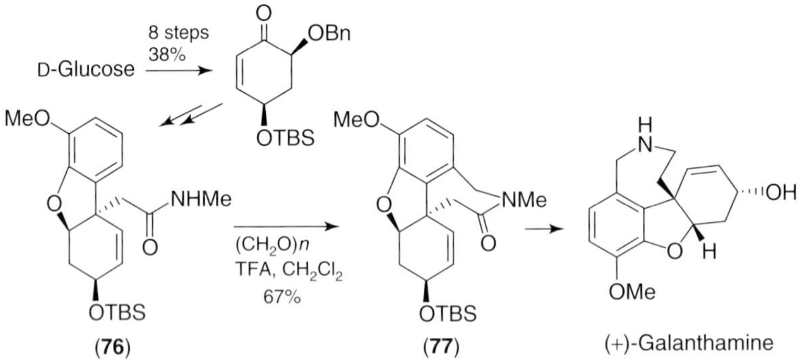

Scheme 4.29 Asymmetric synthesis of (+)-galanthamine via a Pictet–Spengler cyclization.

Azaspiro[4.4]nonane framework which constitutes the C/D ring moiety of cephalotaxine has received particular interest. Thus, compound **80**, prepared in an enantiomerically pure form using an original semipinacolic rearrangement of an α-hydroxyiminium ion by Royer et al. [32], was chosen as a pivotal compound in

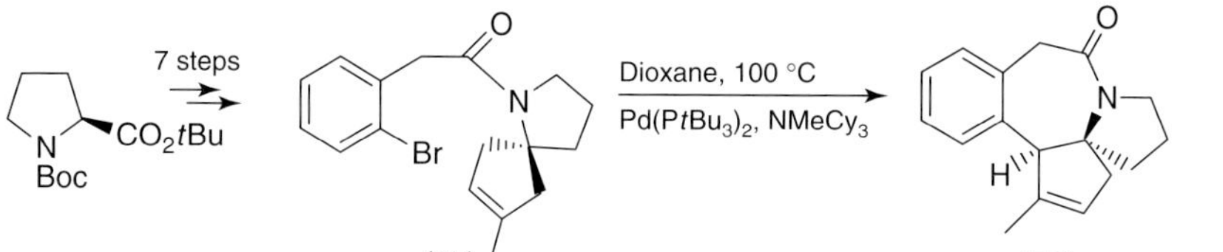

Scheme 4.30 Heck cyclization to seven-membered lactam.

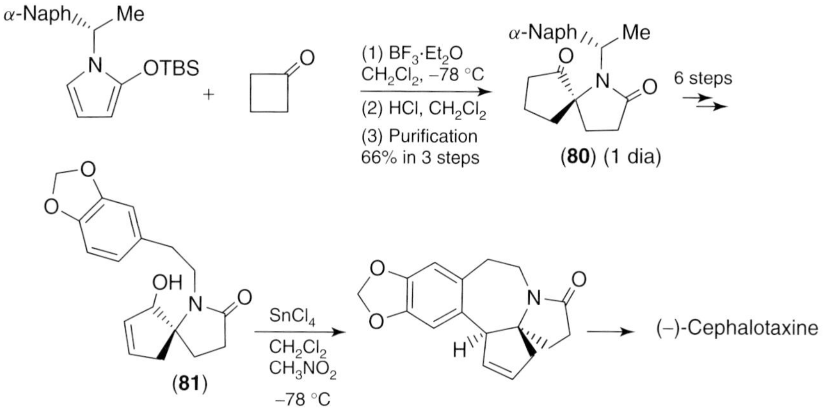

Scheme 4.31 Lewis acid cyclization to seven-membered azepane.

the preparation of (−)-cephalotaxine and analogs; the final azepine ring closure of 81 precursor was then performed using Lewis acid (Scheme 4.31).

4.1.2.4 Oxidative Phenol Coupling Reaction

Optically active (−)-galanthamine has been efficiently prepared from L-tyrosine methyl ester and 3,5-dibenzyloxy-4-methoxybenzaldehyde using as key step a phenolic oxidative coupling to create the seven-membered ring [33a].

An improved synthesis of the same alkaloid has been also depicted [33b,c] starting from the protected substrate **82**, issued from (R)-N-Boc-D-phenylalanine, and using, in this case, phenyliodine(III)bis-(trifluoroacetate) as the oxidant instead of Mn(acac)$_3$ for the oxidative coupling. In this case, the spirodienone **83** was obtained in 61% yield and was then further transformed into (−)-galanthamine. Formation of the last stereocenters of the final tetracycle is controlled by the chirality of the (R)-N-Boc-D-phenylalanine which, in fact, promotes a total asymmetric induction (Scheme 4.32).

4.1.2.5 The Ring Closure Metathesis

Because of increasing strain in the transition state and entropic influences, medium-sized rings, for example, 8–10 membered rings are the most difficult to prepare. Among all the cyclization reactions developed to obtain these

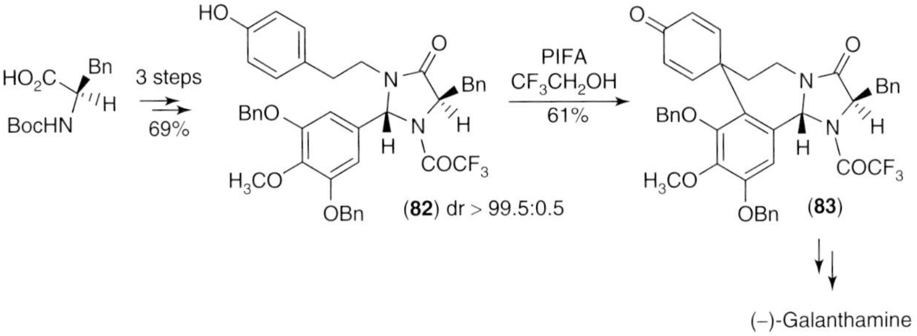

Scheme 4.32 Phenolic oxidative coupling to aza-seven-membered nucleus.

structures, the RCM provided an alternative efficient pathway to form heterocyclic systems, in particular nitrogen-containing compounds and even to prepare fused-nitrogen subunits present in the natural product area. Therefore, this method has been extensively developed these last years. In fact, numerous syntheses of the seven-membered ring of azepine-compounds using RCM [34] have also been investigated, all the approaches varying only on the strategy developed to prepare the diene and the enyne precursors **84** and **85** and on the type of catalysts used (Scheme 4.33). To facilitate the cyclization, several features were installed in the substrates **84** or **85** providing conformational constraint. These steric hindrances were generally obtained by using preexisting rings (like oxazolidine ring) or acyclic conformational constraints.

We introduce here, different routes developed for the RCM cyclization leading to the azepine precursors **84** and **85**.

Chiral 2-substituted azepines **88** were prepared from *bis*-olefinated oxazolidines **86** *via* the [5,3,0]-bicyclic lactam scaffold **87** [35]. These chiral templates were obtained by azeotropic condensation of 5-hexene-2-one with (R)-phenylglycinol followed by treatment with acryloyl chloride. An RCM ring closure of the major 2-(R)-N-acroyl oxazolidine **86** using 5 mol% of Grubbs ruthenium catalyst followed by reduction of the resulting double bond afforded the 5,7-bicyclic lactam **87b** in 98% yield as a single product (61% from (R)-phenylglycinol).

Stereoselective reduction followed by hydrogenolytic cleavage of the N-benzyl moiety gave finally the perhydroazepine (S)-**88b** in 80% yield. The same procedure applied to the *syn* epimer gave the (R)-**88a** in 85% yield (Scheme 4.34).

Chiral aldoxime ether (S)-**89** acted as chiral precursor of allylamine **91** in the RCM approach of 2-substituted tetrahydroazepines. Conversion of **89** to the hydroxylamine **90** followed by N-allylation using allylbromide in the presence of K_2CO_3 led to **91** which underwent an RCM reaction by heating in the presence of

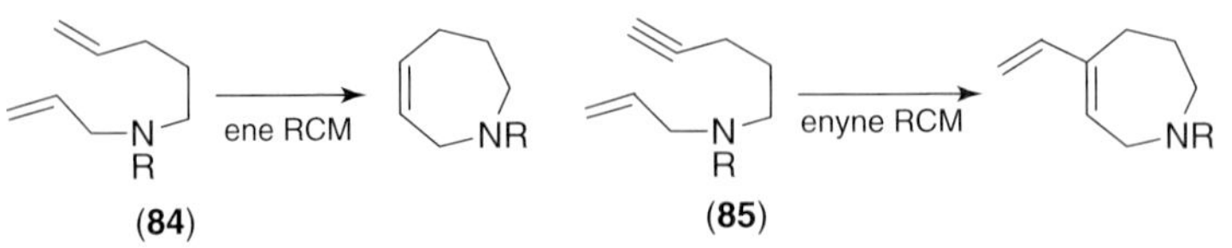

Scheme 4.33 RCM applied to azepine derivatives.

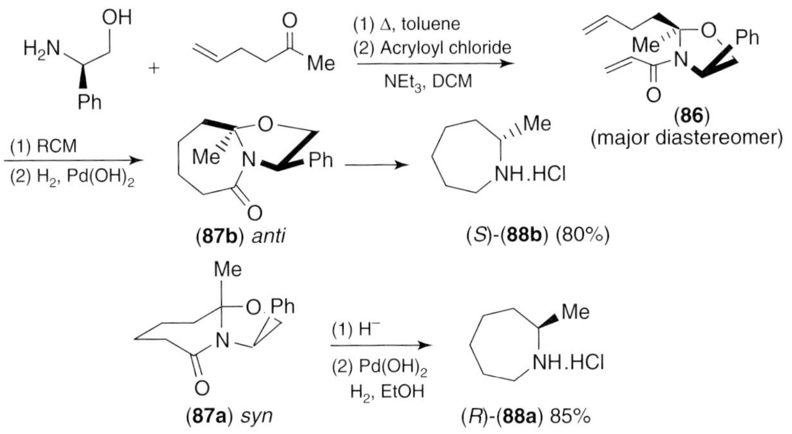

Scheme 4.34 Asymmetric synthesis of perhydroazepine starting from chiral oxazolidines.

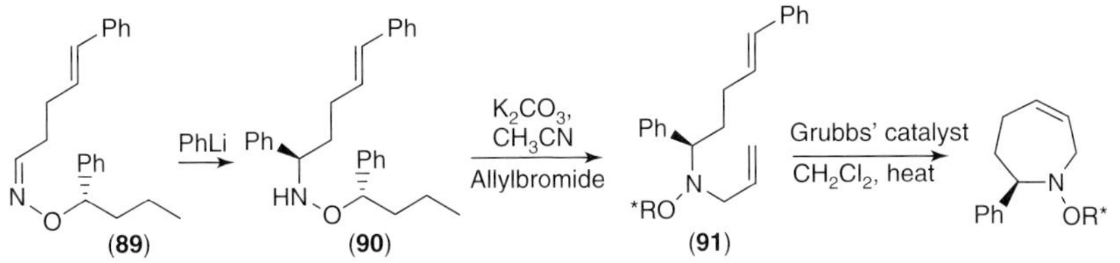

Scheme 4.35 Azepines from chiral aldoxime ether.

Grubbs catalyst giving the expected seven-membered heterocycle but in only 9% yield (Scheme 4.35) [36].

Fürstner and Thiel reported in 2000 [37] a short (eight steps) enantioselective route for the hexahydro-azepine moiety of (−)-balanol with an excellent overall yield. The approach needed the preparation of the diene **92** which was obtained through a four-step sequence starting with a Sharpless epoxidation of divinylcarbinol. RCM reaction of **92** was then conducted in the presence of the ruthenium indenylidene complex **93** giving the tetrahydroazepine **94** in 87% yield (Scheme 4.36).

The first total synthesis of the complex polycyclic *Stemona* alkaloid (−)-tuberostemonine, known for its properties against pulmonary tuberculosis and bronchitis, was reported by Wipf et al. [38] The synthesis of its [1,2-a]azepine core was on the basis of an RCM reaction of **96** to form the azepane skeleton. Pivotal intermediate **96** was formed by selective transformations of the building block **95**, while the latter was prepared from Cbz-L-tyrosine which served as a scaffold for the installation of nine of the 10 stereogenic carbons of tuberostemonine. Heating a solution of **96** in the presence of the ruthenium catalyst **97** led to the tricyclic precursor **98** of the alkaloid in 92% yield (Scheme 4.37).

Intramolecular Grubbs' olefin metathesis has also been used to obtain chiral fused 7-5- or 5-7-bicyclolactams in good yields [39a]. For example, the pyrrolo[1,2-a]azepine core of (−)-stemoamide was prepared either by an ene [39b]

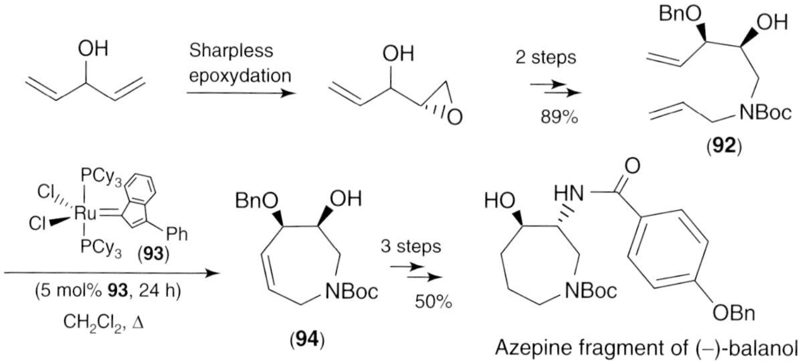

Scheme 4.36 Access to azepine fragment of (−)-balanol using chiral divinylcarbinol and RCM.

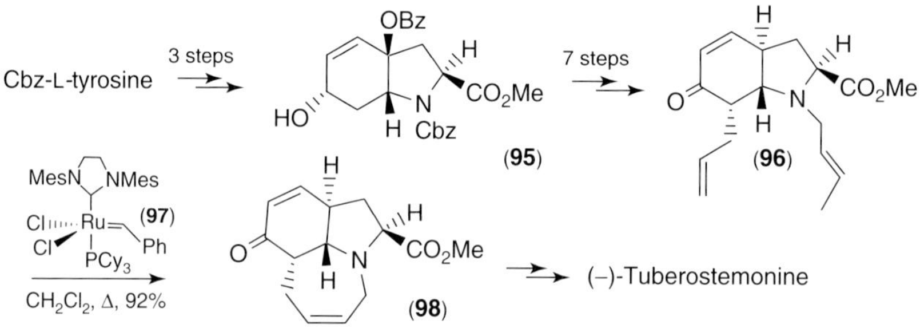

Scheme 4.37 Synthesis of [1,2-a]azepine core of (−)-tuberostemonine by RCM.

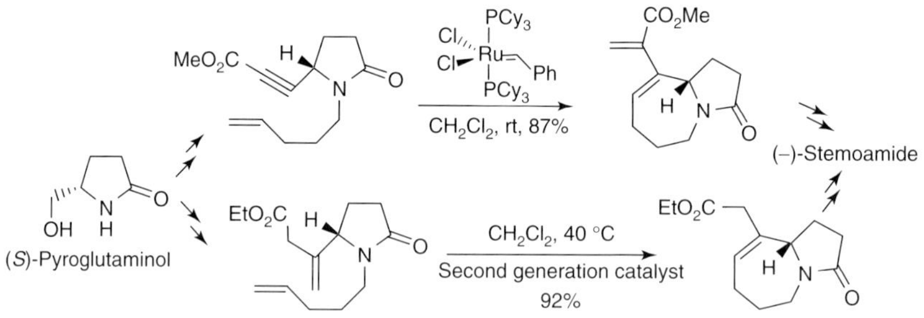

Scheme 4.38 Pyrrolo[1,2-a]azepine core of (−)-stemoamide.

or an enyne RCM [40a] reaction starting from protected (S)-pyroglutaminol, which introduced the chirality (Scheme 4.38).

Others successful applications of intramolecular enyne metathesis have been reported. Among them, we can mention the enantiospecific access to the 9-azabicyclo [4.2.1] skeleton core of (+)-anatoxin-a [40b] starting from D-methyl pyroglutamate and using, in this case, as the key enyne a *cis*-2,5-disubstituted pyrrolidine (Scheme 4.39).

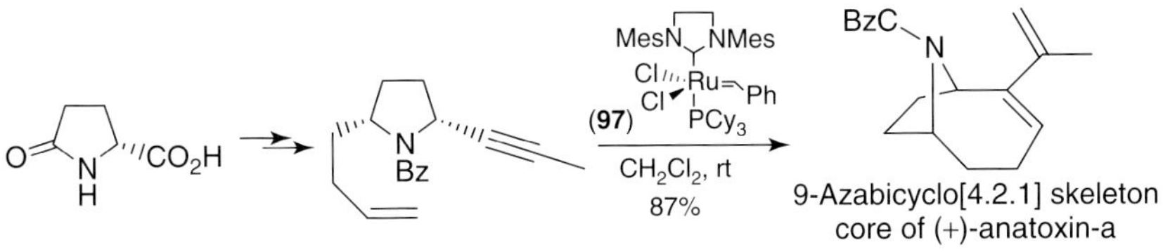

Scheme 4.39 Enantiospecific synthesis of (+)-anatoxine-a.

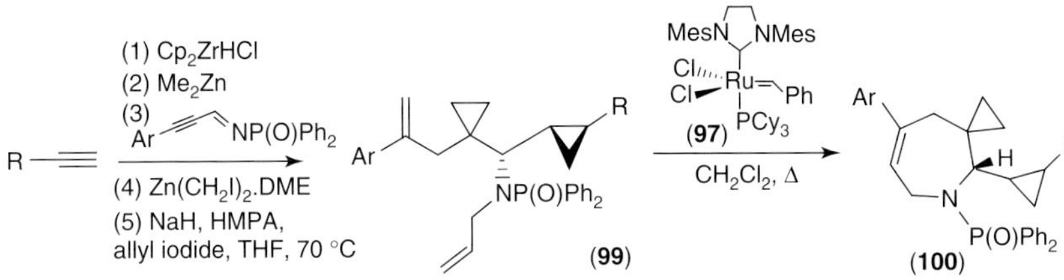

Scheme 4.40 Diastereoselective approach toward spiroazepines.

ω-Unsaturated dicyclopropylmethylamines obtained by multicomponent condensation of N-diphenylphosphinoylimines, alkynes, zirconocene hydrochloride, and diiodomethane constitutes interesting building block in the synthesis of spiro-azaheterocycles [41]. Thus, the N-alkylated C,C-dicyclopropylmethylamines **99** underwent a ring-closing reaction using second-generation Grubbs catalyst (10 mol%) leading to 1H-azepines **100** in 63–84% yield (Scheme 4.40).

Chiral alcohols **101** prepared in good diastereoselectivity (de > 90) from (−)-8-aminomenthol underwent RCM reaction with Grubbs' catalyst to afford 2-benzoylperhydro-1,3-benzoxazines **102**. The mol% of the catalyst used (4–15 mol%) together with the temperature of the reaction (20 °C to reflux) and its duration (20–240 h) depended of the nature of the substituent of the double bond [42]. The final transformation of **102** into the enantiopure 2,3,4,7-tetrahydro-1H-azepin-3-ols **103** was performed in three steps and 44–56% yield by (i) reductive ring opening of the N,O-acetal moiety with aluminum hydride, (ii) oxidation with pyridinium chloro-chromate (PCC) of the menthol, and (iii) basic cleavage (KOH, tetrahydrofuran (THF)/methanol) (Scheme 4.41).

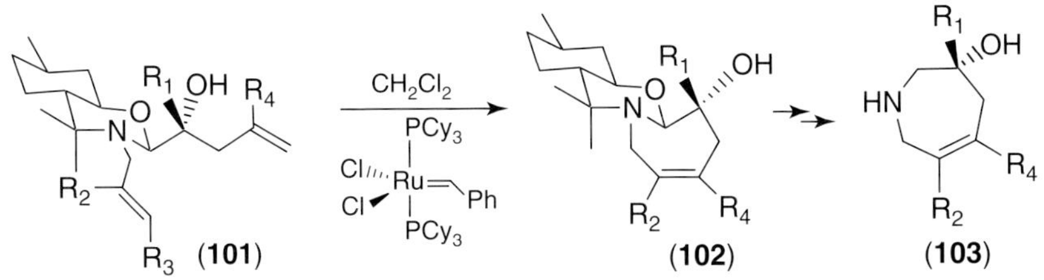

Scheme 4.41 Asymmetric synthesis of 2-benzoylperhydro-1,3-benzoxazines.

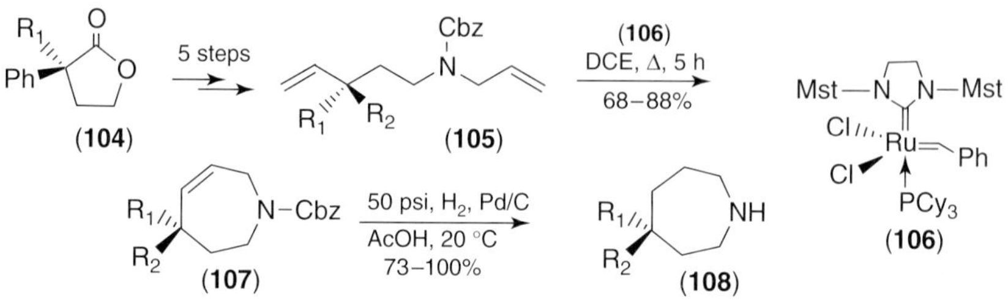

Scheme 4.42 Chiral 4-mono and 4,4'-disubstituted tetrahydroazepines.

A general strategy for the preparation of 4-mono and 4,4'-disubstituted tetrahydroazepines **108** was developed by Merschaert et al. [43] starting with chiral, commercially available or easy to prepare lactones **104**. The key aza derivatives **105** issued from **104** underwent then an RCM reaction using Grubbs' catalyst **106**, leading to the unstable seven-membered heterocycles **107** (68–88%) (Scheme 4.42).

Hydrogenolysis and concomitant reduction of the double bond furnished the tetrahydroazepines **108** in nearly quantitative yields (Scheme 4.42).

1,2,3,6-Tetrahydro-1H-azepin-3-ols are intermediates in the asymmetric synthesis of azasugars. A method for their stereoselective preparation was reported by Lee et al. [44a] starting from optically pure aziridine **109** and using, as the final ring closure pathway, an RCM reaction. Thus, the (2R)-aziridine **109**, generated by reaction between (R)-(+)-α-methylbenzylamine and ethyl 2,3-dibromopropionate, was transformed in 73% yield to the allyl ketone **110** using an Horner–Wadsworth–Emmons olefination. A stereoselective reduction of the ketone function led to the *erythro* alcohol (2R, 3S)-**111** in 96% yield using $NaBH_4/ZnCl_2$ in MeOH and to the *threo* (2R, 3R)-**112** exclusively by treatment with L-Selectride. The N-allylamine **113** was then obtained after regioselective ring opening of the chiral aziridine with a thiolate anion at the less sterically hindered C-1 position followed by alkylation with allylbromide. Subsequent RCM reaction of the resulting *bis*-olefin with second-generation Grubbs catalyst afforded enantiomerically pure (2S, 3S)- and (2S, 3R)-azepinols **114** and **115** from (2R, 3S)- and (2R, 3R)-alcohols **111** and **112**, respectively (Scheme 4.43).

Another tetrahydro-(1-H)-azepine-3-ol of similar structure has been also described using a similar methodology, the unique difference being in the synthesis of the chiral precursor which was here obtained through an Evans aldol reaction [44b].

Gmeiner et al. [45] developed a useful synthesis of (2S)-7-membered cyclic dipeptides mimics **119** using an RCM cyclization promoted by ruthenium catalyst of precursors **118**. The metathesis precursor **118** was readily available through a N,N'-dicyclohexylcarbodiimide (DCC)/HOBt-promoted peptide coupling of glycine derivative **116** with the secondary amine **117**, which was in turn prepared from glycine or alanine esters (Scheme 4.44).

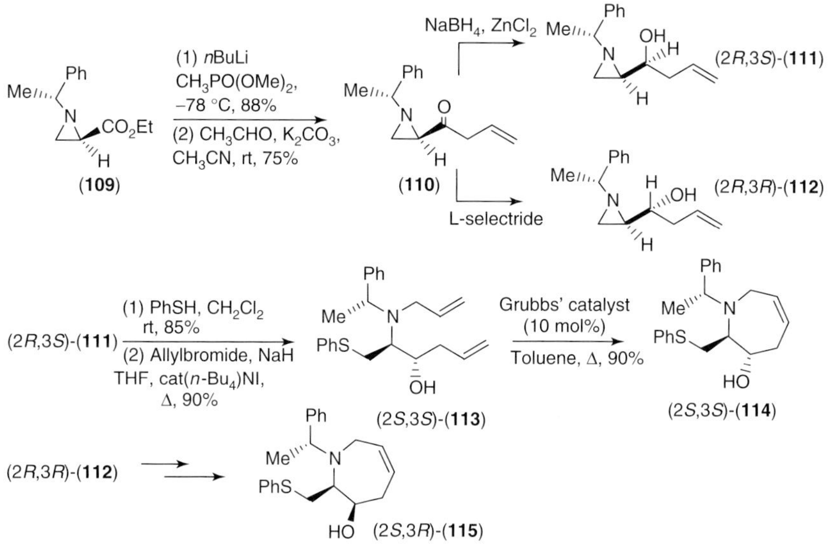

4.1 Substituted Azepines

Scheme 4.43 Chiral azepines from ring opening of aziridines followed by RCM reaction.

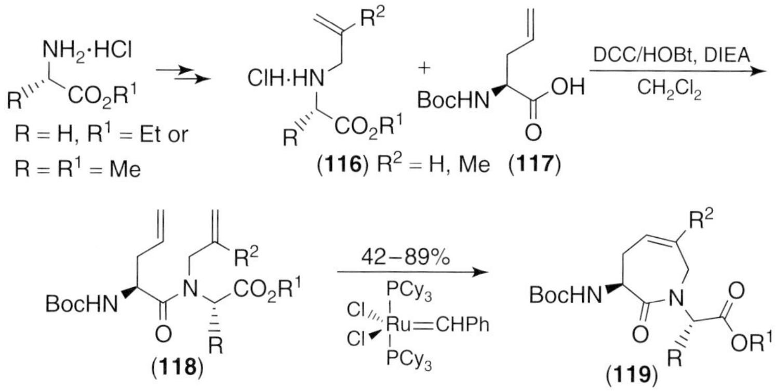

Scheme 4.44 Chiral 2-amino seven-membered lactams from aminoacids.

4.1.3
Cycloaddition Methods

Just a few cycloaddition methods were depicted to generate aza-seven-membered ring. The most commonly developed was the [4 + 3] cycloadditions which took place with moderate to high yields with complete regio and diastereoselectivity. Others attractive approaches exploited either 1,3-dipolar cycloaddition of nitrone or intramolecular Diels–Alder reactions (IMDA).

4.1.3.1 [5 + 2] Cycloaddition

A concise approach to the C-D-E part **121** of HHT, was reported in 2001 by Booker-Milburn et al. [46a] Intramolecular [5 + 2] photocycloaddition of substituted maleimide **120** led, after reopening of the four-membered ring, to azepine framework which was obtained as a mixture of two diastereoisomers (yield 75%, de 3.5:1) (Scheme 4.45).

The same cyclization process was applied also with success to the synthesis of the tetracyclic core of neotuberostemonine [46b] and of oxetanol-fused azepines [46c].

4.1.3.2 [4 + 3] Cycloaddition

Thermal cycloaddition reaction of alkenyl Fischer chiral carbene **122** derived from (−)-menthol with oxime **123** resulted in the formation of diastereomers **124** and **125** in 87% yield and a 7:3 ratio. Changing for (+)-menthol afforded the same diastereomers in 80% yield and a reverse 3:7 ratio. Crystallization of the major one followed by acid hydrolysis allowed the isolation of enantiomerically pure azepinones (Scheme 4.46) [47].

4.1.3.3 Nitrone Cycloaddition

Formation of azepane skeleton has been also envisaged through intramolecular 1,3-dipolar cycloaddition of nitrone to olefine, followed by N–O reductive cleavage processes.

An example of this method [48a] consists in the preparation of the pyrrolo[1,2-a]azepine framework **130** by N-allylcarbohydrate nitrone cycloaddition. The precursor **126** was classically obtained in six steps from the 1,2,5,6-di-O-isopropylidene-α-D-glucofurannose and was then transformed to the furanoside-fused pyrrolidine **127** as a 2:1 anomeric mixture in 73% yield using standard reactions. On treatment with N-methylhydroxylamine hydrochloride/K_2CO_3

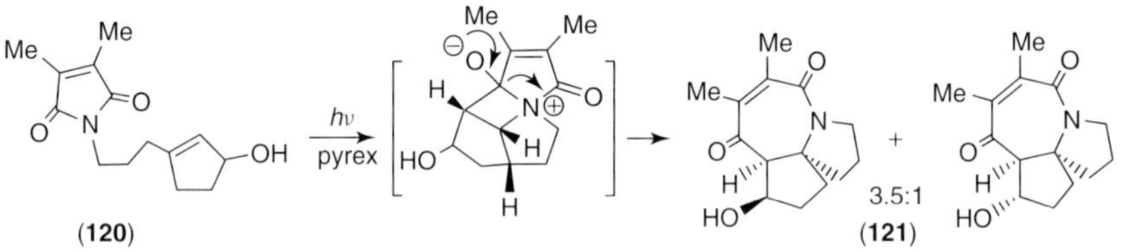

Scheme 4.45 Diastereoselective access to azepines derivatives by [5 + 2] photocycloaddition.

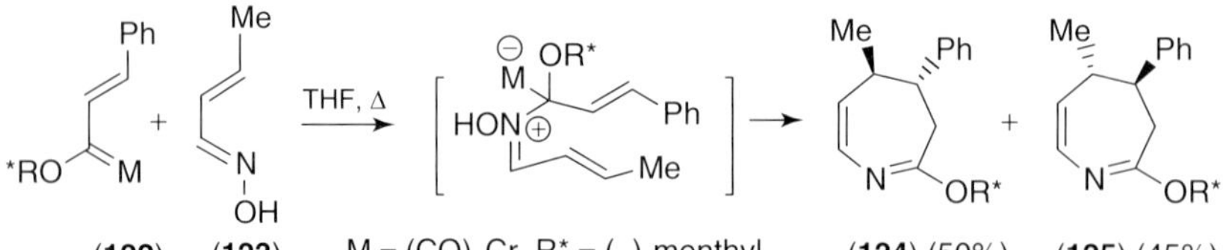

Scheme 4.46 Enantiomerically pure azepinone using [4 + 3] cycloaddition process.

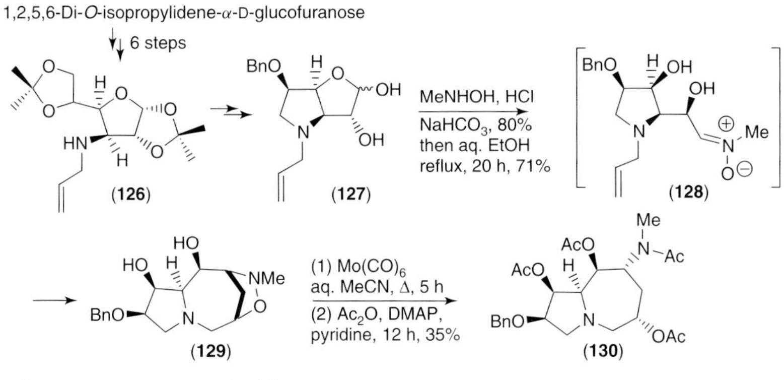

Scheme 4.47 Nitrone cycloaddition to azepine ring.

127 gave *exclusively* the bridged isoxazolidine **129** via the nitrone **128** in a 71% yield. Cleavage of this isoxazolidine ring was realized using Mo(CO)$_6$ in aqueous acetonitrile and led to the azabicyclic ring in 35% yield (Scheme 4.47).

Intramolecular nitrone cycloaddition allowed also an access to aminohydroxyazepines starting from amino acids as chiral pool [48b].

4.1.3.4 Intramolecular Diels–Alder Reactions (IMDA) – [4 + 2] Cycloaddition

The intramolecular version of the Diels–Alder reaction [49] has been applied with success to the elaboration of the seven-membered ring of the *Stemona* alkaloids.

Thus, Jacobi and Lee [50a] described an original concise synthesis of (−)-stemoamide using an IMDA-*retro*-Diels-Alder (DA) reaction of the propyne derivative **132** obtained in three steps from (*S*)-oxazole lactam **131** easily prepared on multigram scales with no racemization from (*S*)-L-pyroglutamic acid. Thermolysis of (*S*)-**132** in diethylbenzene at reflux gave the tricyclic azepine compound **134** owning the natural configurations and issued from rearrangement of tetracyclic derivative **133**. Subsequent selective reduction with NaBH$_4$/NiCl$_2$ afforded finally (−)-stemoamide (Scheme 4.48).

This cyclization mode has been applied with success to the synthesis of (−)-stenine, another member of the *Stemona* alkaloids [50b].

4.1.4
Ring Transformation Methods

Much attention has been focused upon the preparation of seven-membered rings incorporating one nitrogen atom using (i) rearrangement of cyclohexanones (Beckmann, Schmidt reactions or its intramolecular version, the Boyer reaction) or (ii) expansion of preexisting smaller rings. Sometimes, thermal and photochemical rearrangements are also combined with retro-aldol reaction.

We report in this section the asymmetric syntheses of azepines framework on the basis of all these methods [51], and, for better understanding, we present here

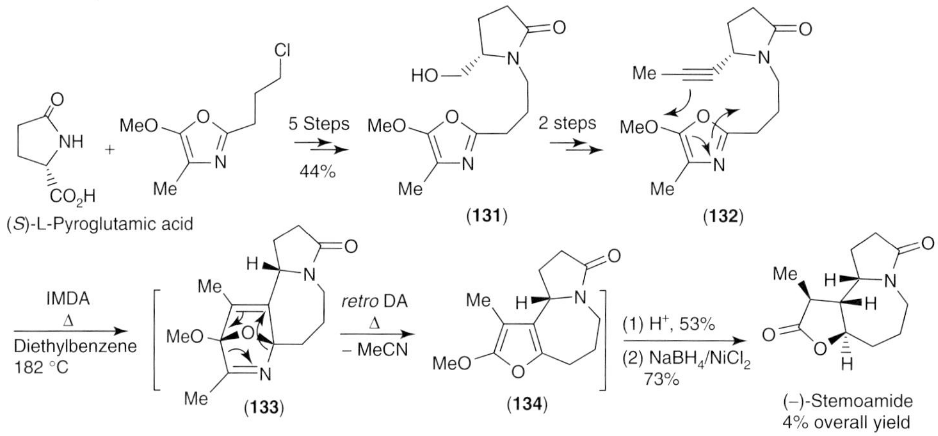

Scheme 4.48 Azepine ring from an IMDA reaction.

the literature results in two parts; the first one describing "classical reactions" of rearrangement and the second part introducing an access to seven-membered aza nucleus by enlargement of nitrogenous cycle of variable size.

4.1.4.1 Classical Methods

Aubé et al. [52a,b] converted ketone **135** to lactams **137a,b** in good yield and stereoselectivity (for example, when R^1 = Ph and R^2 = H, **137a/137b** = 7.2:1) via an "in situ-tethering" strategy using hydroxyalkyl azides **136** and a Lewis acid. The stereoselectivity may be explained by the equatorial attack of azide on the oxonium ion followed by migration of the pseudo-axial bond antiperiplanar to the leaving N_2 substituent (Scheme 4.49).

In a further study, the authors [52c] showed that the stereoselectivity of the reaction could be modulated by electronic and conformational effects. The ratio of isomers **137a,b** depended upon the nature (alkyl or aryl) of the R^2-substituents of the hydroxyl-azide **136** and even upon the electronic nature of the R^2-aryl group. Effective enhancement of selectivity was also detected using a linker containing a quaternary carbon like **138** (Scheme 4.50).

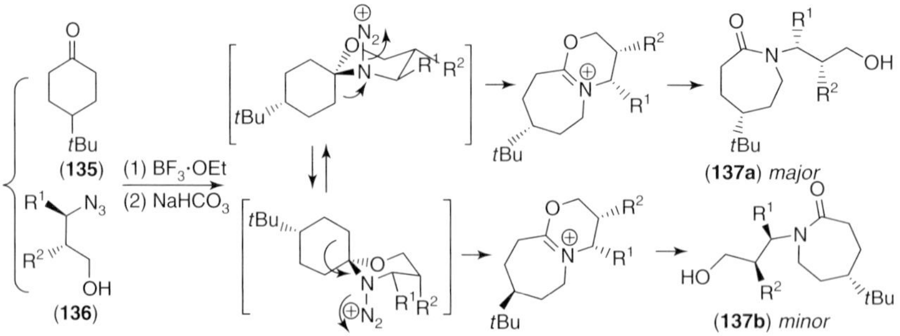

Scheme 4.49 Chiral caprolactam by asymmetric nitrogen insertion process.

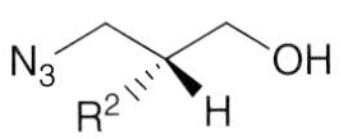

(136) R¹ = H

R² = Me **(137a)** : **(137b)** = 74 : 26
R² = i-Pr **(137a)** : **(137b)** = 88 : 12
R² = Ph **(137a)** : **(137b)** = 64 : 36
R² = 4-nitrophenyl **(137a)** : **(137b)** = 76 : 24
R² = 4-methoxyphenyl **(137a)** : **(137b)** = 47 : 53

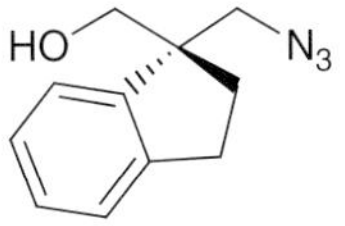

(138)
Yield of **(137)** = 85%,
(137a) : **(137b)** = 77 : 23

Scheme 4.50 Influence of the azide in the formation of the caprolactams.

This method allowed the stereoselective synthesis of substituted caprolactams by ring expansion of chiral ketones in a one-pot procedure with only one work-up providing diastereomers which were easy to separate. Moreover, the removal of the chiral group on the nitrogen could be accomplished efficiently and depended only upon the structure. A systematic study of this asymmetric Schmidt reaction has been published in 2003 by Aubé et al. [52d]

During their work devoted to the synthesis of a series of Gly–Gly-derived macrocycles containing a *cis*-aminocaproic acid (Aca) tether, Aubé et al. [52e] needed to prepare several linkers **141** which could be easily obtained by acid hydrolysis followed by esterification of chiral caprolactams. An example is depicted below: the caprolactam **140** resulted in excellent yield from light induced ring opening (photo-Beckmann rearrangement) of the chiral oxaziridines mixture **139**. The mixture of lactams thus obtained could be easily separated and the major one, **140**, afforded linker **141** with an excellent enantiomeric excess of 98% (Scheme 4.51).

Thermal regioselective Beckmann rearrangement (microwave 100 °C) [53a] of separated *p*-toluenesulfonyloximes **143** obtained from ketone **142** led to isomeric lactams **144** (Scheme 4.52).

An asymmetric Beckmann rearrangement/allylsilane-terminated cation cyclization cascade was used to elaborate the seven-membered ring of cephalotaxine framework (Scheme 4.53) [53b].

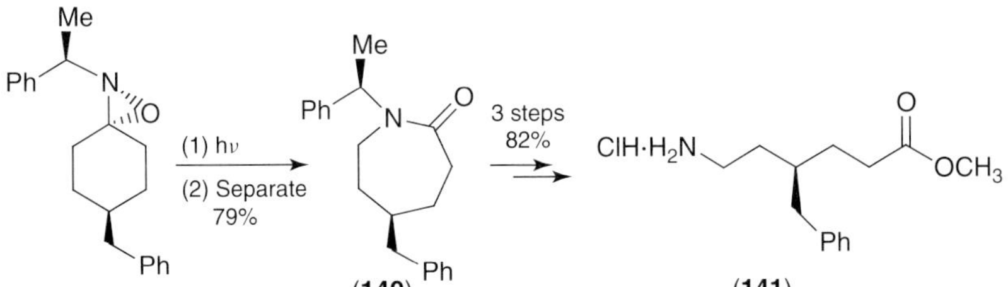

Scheme 4.51 Photolysis of oxazolidines to chiral caprolactams.

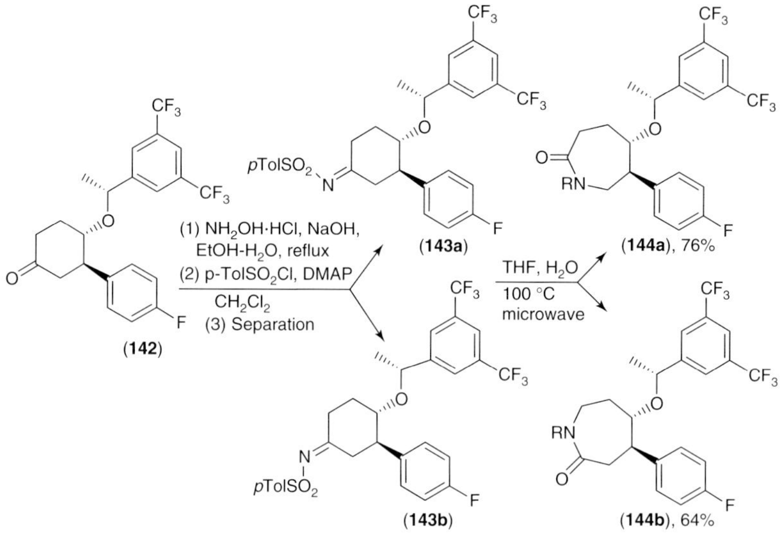

Scheme 4.52 Modified Schmidt rearrangement to caprolactams.

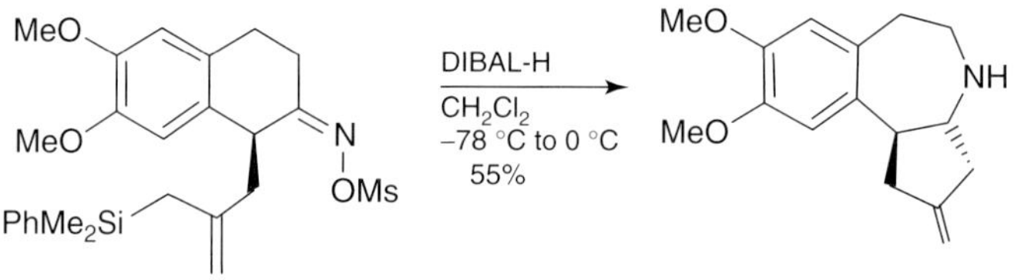

Scheme 4.53 Asymmetric Beckmann rearrangement.

4.1.4.2 Ring Expansion

The synthesis of azepines derivatives has been efficiently envisaged by ring expansion of preexisting smaller rings in particular piperidine, pyrrolidine, azetidinium salts, and pyranoside.

During their study toward the synthesis of balanol, Kato and Morie [54] developed the ring expansion of a *cis*-aziridinium cation **145** to form, together with the normal displacement piperidine derivative, the aza-seven-membered ring *via* an S_N2-type attack of the azide anion at the methyne carbon (Scheme 4.54).

Using the same approach starting from the symmetrical piperidine **146**, Brechbiel et al. [55] synthesized diastereomerically azepanes in about 90% yield by ring expansion using NaN_3 in Dimethyl-sulfoxide (DMSO) with good stereoselectivity; only the *cis* isomer was obtained *via* a backside attack at the methine carbon (Scheme 4.55).

It is well known that azidolactols underwent reductive ring enlargement in the presence of hydrogen to yield azepanes [56]. Applied to chiral lactol **147** and lactone **148**, obtained from the transformation of L-ascorbic acid, this reaction allowed the preparation of 3,4-disubstituted azepanes **149** and (5*S*, 6*S*)-caprolactam **150**, precursors of (+)- and (−)-balanol in excellent yields (Scheme 4.56) [57].

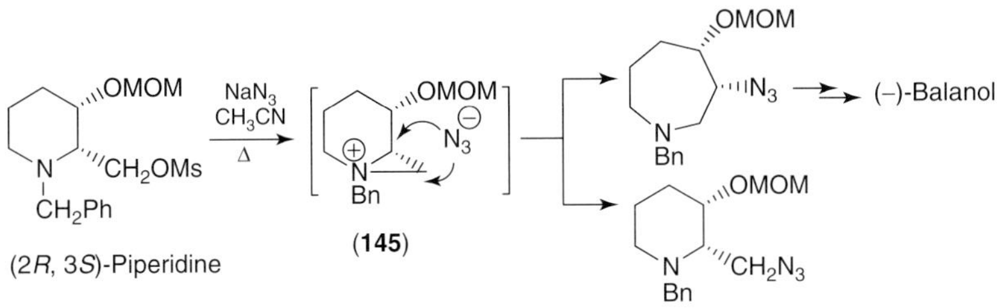

Scheme 4.54 Chiral azepine fragment of balanol *via* piperidine ring expansion.

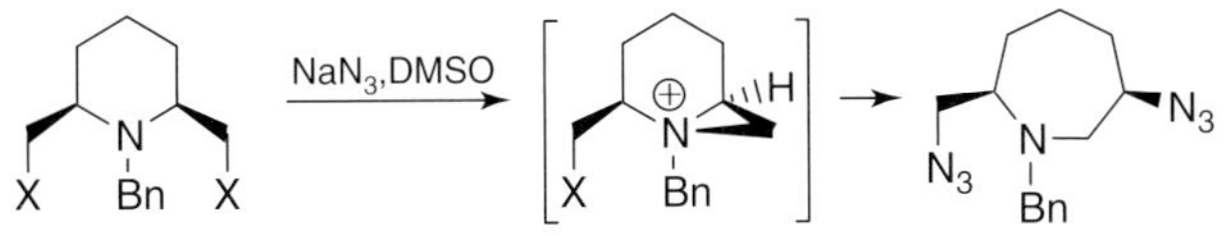

Scheme 4.55 Piperidine ring expansion to azepanes.

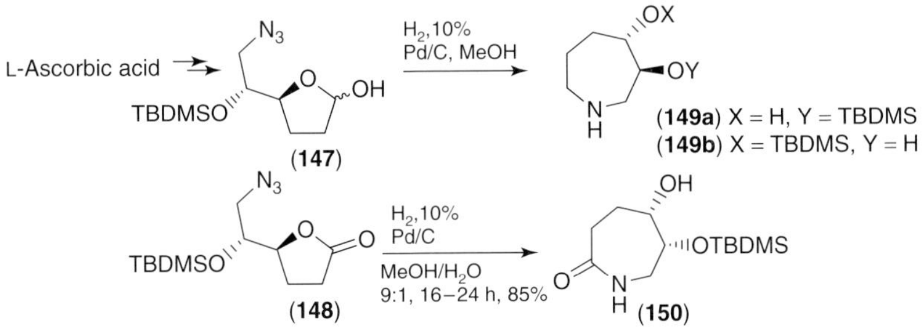

(146) X = Cl, Br

Scheme 4.56 Chiral lactones to substituted caprolactam.

Polyhydroxyazepanes (seven-membered iminocyclitols) [58], have been synthesized using an analogous method. For example, iminocyclitol **152** could be obtained starting from chiral azidolactol **151** derived from α-D-glucopyranoside, by a tandem Staudinger intramolecular aza-Wittig ring expansion with triphenyl phosphine in 87% yield (Scheme 4.57) [58c].

Chemo-enzymatic synthesis starting from benzylpyranoside produced polyhydroxyazepane **153** in only two steps using aqueous media and proceeding on a gram scale (Scheme 4.58) [58d].

In 2003 [59a], a simple and versatile method for the enantio- and diastereoselective synthesis of 2-substituted- and 2,7-disubstituted-3-aminoazepanes **156** and **157** was reported using a tandem ring enlargement/alkylation or reduction process starting from 2-cyano 6-oxazolopiperidine**154** (Scheme 4.59). Bicyclic aminal **155**, derived from **154** by standard reactions, could be then transformed into the *trans*-azepane **156** in yields between 80% and 96%.

This method permitted the stereoselective preparation of the indolyldiamine **156** (R^1 = N-methyl indole), carba-analog precursor of the tetracyclic azepane eudistomine, an antiviral and antitumoral compound (Scheme 4.58) [59b].

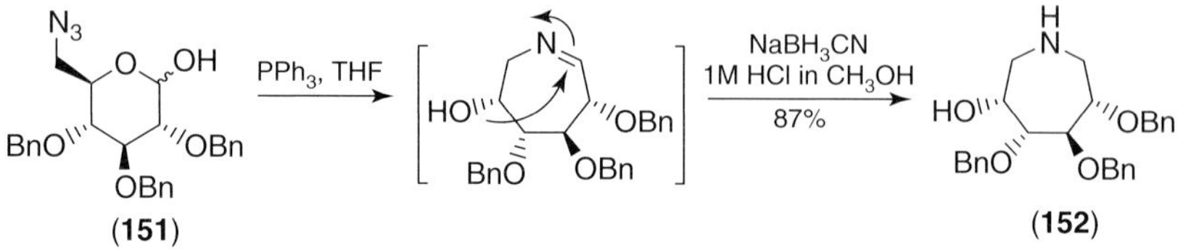

4 Asymmetric Synthesis of Seven- and More-Membered Ring Heterocycles

Scheme 4.57 Seven-membered iminosugar.

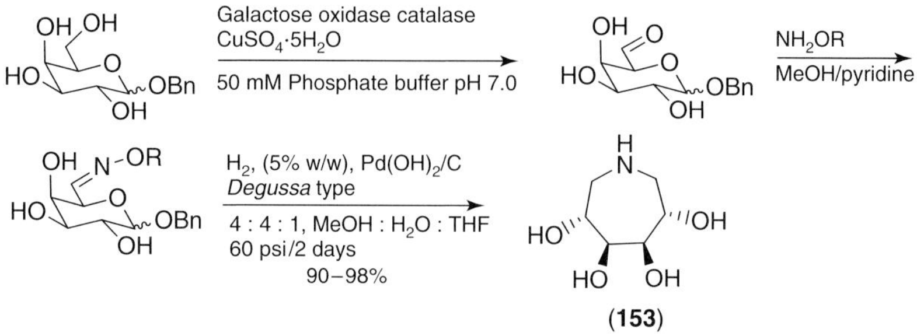

Scheme 4.58 Chemo-enzymatic synthesis of seven-membered iminocyclitol.

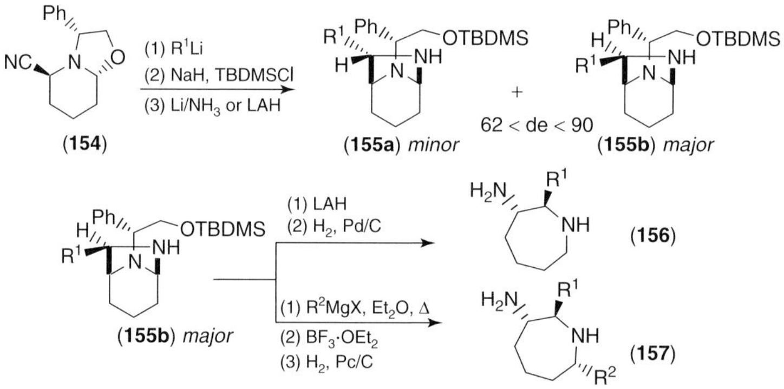

Scheme 4.59 Enantioselective synthesis of 2-substituted- or 2,7-disubstituted-3-aminoazepanes.

On the other hand, treatment of 154 with iso-butyl or benzyl Grignards, which act as nucleophilic reagents, led to 2,7-disubstituted-3-aminoazepanes 157 in good yields and excellent diastereoselectivity (de > 95%) (Scheme 4.59).

The chiral hexahydro-azepine 161 constitutes an interesting potential azepine derivative for the synthesis of more functionalized compounds. It could be rapidly prepared by ring opening of γ-aminoaldehyde 158 with (R)-tert-butylsulfinamide followed by reaction with diphenyl vinyl sulfonium 160. Derivative 161 existed as a 3:1 mixture of diastereomers easily separated by flash chromatography (Scheme 4.60). Further deprotection of the major aziridine has permitted, for example, the synthesis of the known precursor of (−)-balanol [60].

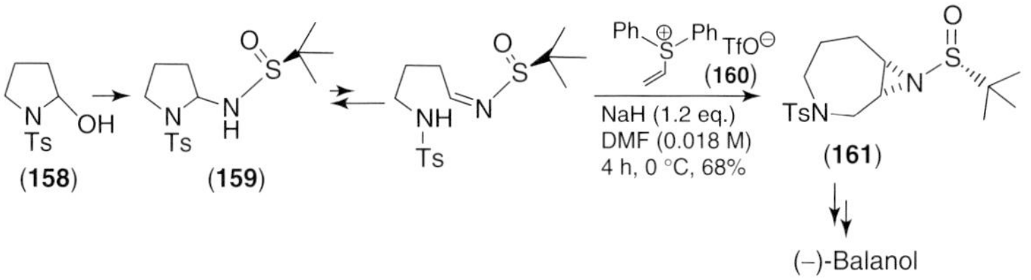

Scheme 4.60 Vinyl sulfonium salt in asymmetric synthesis of aziridine fused azepine.

The two-carbon ring expansion of heterocyclic compounds is an original method to elaborate azepine derivatives, which has been successfully used in the preparation of 3-oxo-pyrrolidine and azetidine substrates (*vide infra*).

Thus, a Michael Aldol Retro-Dieckmann (MARDi) sequence was used in converting 3-oxo-pyrrolidine **162** and acrolein in the presence of MeOH into a mixture of hydroxyazepane **163** (dr = 1.6:1) and tetrahydro-azepine **164** in 36% overall yield (Scheme 4.61) [61].

Ring expansion of enantiomerically pure 2-alkenylazetidinium trifluoromethanesulfonate salts **165** under basic condition permitted a clean access to enantiomerically pure substituted azepanes **166a** and **166b** with variable diastereoselectivity at the C-2 position through a [2,3] sigmatropic shift (Stevens rearrangement) of the ylide generated after selective deprotonation α to the ester group (Scheme 4.62) [62].

4.1.4.3 Substitution of Already Formed Heterocycles

Most of the syntheses of complex azepines derivatives associated with pharmaceutically important activities depicted in the literature have, as a corner stone, an already formed aza-seven-membered ring which has been formed from caprolactam derivatives [51b].

For example, asymmetric synthesis of fused seven-membered lactams **168** was reported employing an asymmetric hydrogenation of the caprolactam derivative **167** with [Rh(COD)-(+)-(2S, 5S)-Et-DUPHOS]OTf. This approach allowed

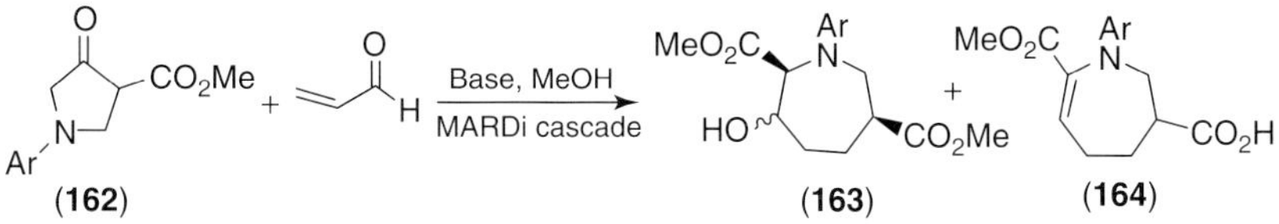

Scheme 4.61 Diastereoselective MARDi cascade to azepine ring.

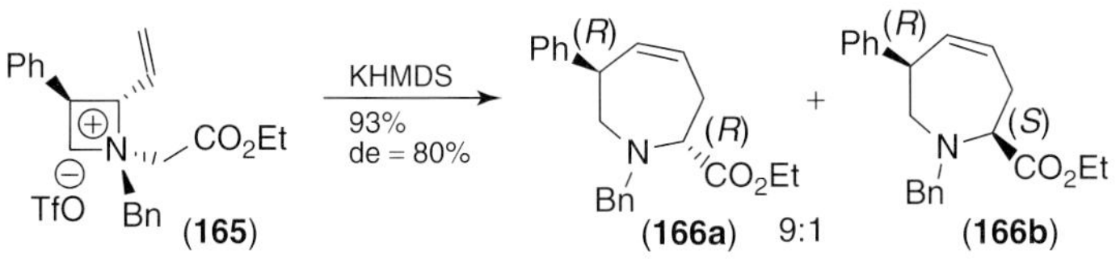

Scheme 4.62 Ring expansion of chiral 2-alkenylazetidinium trifluoromethanesulfonate salts.

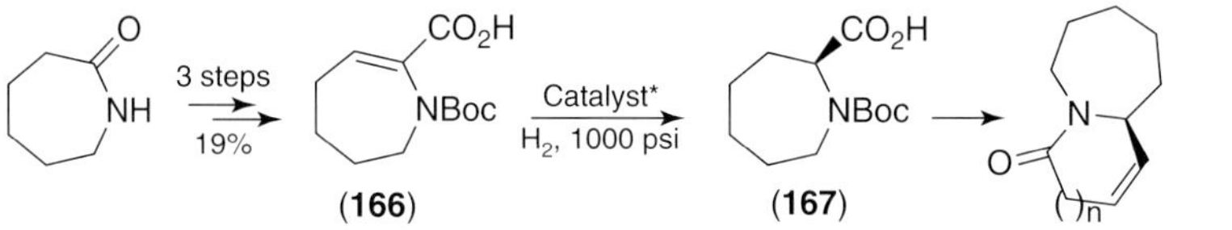

Scheme 4.63 Fused chiral seven-membered lactams from caprolactam.

the preparation of the (S)-N-protected-homopipecolic acid **168** in 98% yield and 90% ee. Acid **168** is then further transformed into the azabicyclic compounds **169a,b** in a four-step sequence using as a key reaction an RCM ring formation (Scheme 4.63) [63].

Construction of the azaspiro[5.6]dodec-9-ene **172**, a part of pinnatoxin A, a marine toxin of shellfish and dinoflagellate, has been efficiently elaborated by an asymmetric Diels–Alder reaction of the α-methylene caprolactam **170** with the diene **171** using chiral Lewis acid.

The precursor **170** was readily obtained in five steps and 54% overall yield from caprolactam itself (Scheme 4.64) The enantioselectivity observed in this synthesis arose from the chirality of the copper complex formed and the *exo* selectivity was due to the unfavorable steric interactions between the substituent in the diene and the chiral ligand of copper complex in the *endo*-transition state [64].

The ongoing interest in bengamides and analogs as clinical candidates for cancer chemotherapy has prompted the preparation of a lot of various original 2-N-substituted-aminocaprolactams by introducing the polyfunctional substituent either on the commercially available (2S)-aminocaprolactam, or, on the readily accessible chiral 4-hydroxy-2-aminocaprolactam [51b], [65a]. Thus, for example, synthesis of (+)-bengamide E relied on ring opening of substituted acetylenic β-lactone **174** by (S)-3-aminocaprolactam **173**. Subsequent stereoselective reduction of the acetylenic moiety led to (+)-bengamide E. A similar approach was also reported starting with γ-lactone **175** (Scheme 4.65) [65b,c,d,e].

The *meso* epoxide azepine compound **176** prepared by an RCM reaction of N-Teoc-N, N-*bis*homoallylamine constitutes a versatile substrate for the synthesis of hexahydro-disubstituted azepines. When treated with trimethylsilyl azide in the presence of the Jacobsen Cr(salen) catalyst epoxide **176** furnished, after cleavage of the silyl ether, the hydroxyazide **177** in 98% yield and 87% ee in two steps. Epoxide opening of **176** with *i*-butanol under the same reaction conditions led to

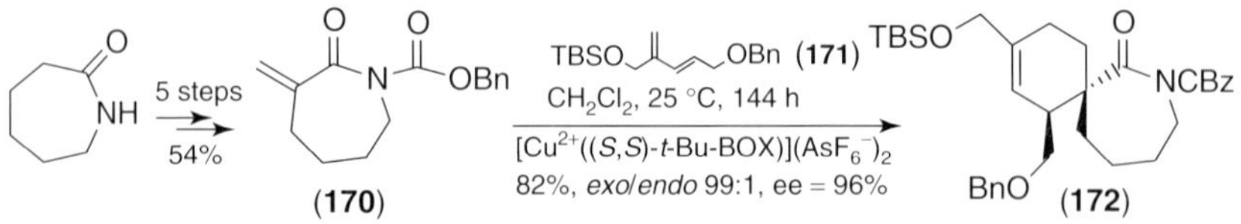

Scheme 4.64 α-Methylene caprolactam, precursor of the azaspirocyclic part of pinnatoxin A.

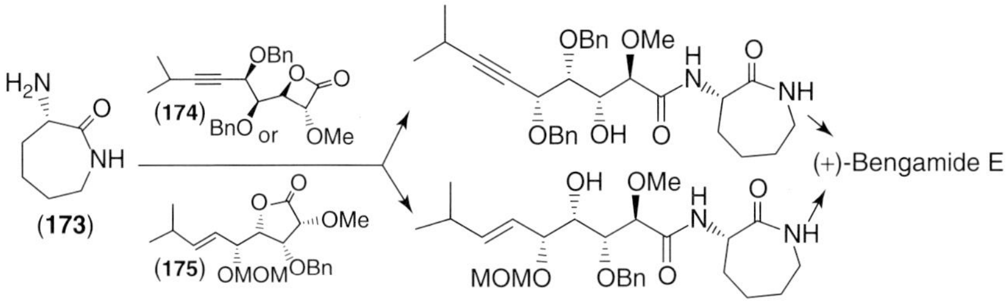

Scheme 4.65 Use of (S)-2-aminocaprolactam for the synthesis of azepine derivatives.

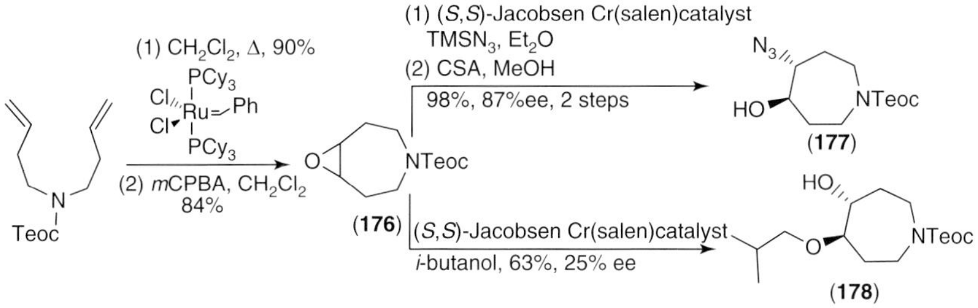

Scheme 4.66 Epoxide ring opening to chiral 4-hydroxy azepines.

the alcohol **178** in 63% yield but in this case with a modest enantiomeric excess (25%) (Scheme 4.66) [66].

4.2
Substituted Azocines

The commonly encountered azocines (π-equivalent heterocyclic analogues of cycloocta-tetraenes), fully or partly reduced, constitute a part of the skeleton of natural products with complex structures like apparicine, magallanesine, nakadomarin A and manzamine A, an indole alkaloid isolated from marine sponges with antitumor and antibiotic activities (Scheme 4.67) [67a].

Many of the synthetic azocine derivatives have found application as therapeutic agents because of their broad spectrum of properties [1d] together with the favorable flexibility of their ring which allow binding with species.

In particular azocin-2-ones have been used as sedatives, anticonvulsant, antihypertensive agents and the disubstituted ones find application as peptide analogues to mimic the type VI β-turn conformation of natural polypeptides [67b,c].

Fused-azocines like benzofuroazocines possess important activity in the central nervous system [67d] and the pyrrolido- and piperidino-azocines could serve as intermediates in alkaloids synthesis [67e].

The preparation of this aza-eight-membered ring, especially in highly functionalized form, using classical cyclization from an alicyclic chain by intramolecular

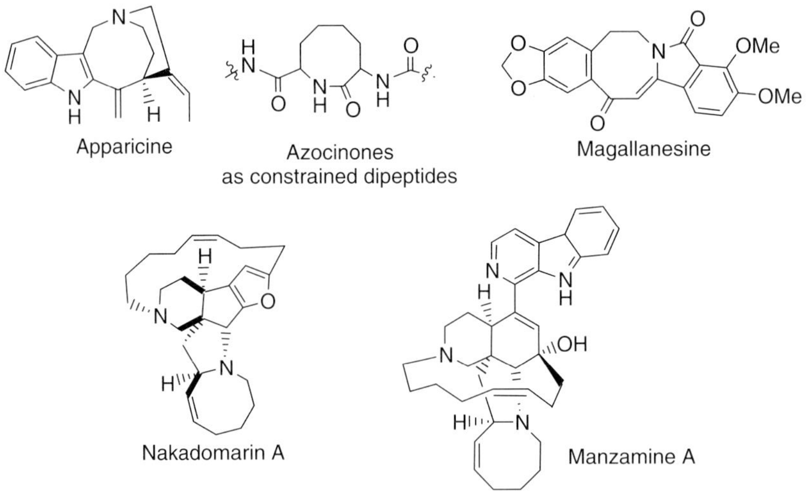

Scheme 4.67 Azocine compounds.

nucleophilic substitution remains very specific and is often slow and hampered by unfavorable enthalpies and entropies of the reaction [68].

Diverse other synthetic methodologies have been developed on the basis of intramolecular substitution, ring contraction or enlargement such as fragmentation of fused 5/5-ring system, [2 + 2] photochemical, or thermal cycloaddition reactions, and most recently by powerful RCM reaction that sometimes requires high dilution conditions for successful conversion.

4.2.1
Azocines from Intramolecular Nucleophilic Substitution

For example, azocine **181** has been efficiently synthezized in 76% yield by intramolecular basic substitution of the oxazolidinone iodide **180** easily obtained by exposure of the Sharpless epoxide **179** to sodium hydride and benzoylisocyanate [69] (Scheme 4.68).

Recently, fused indolo-azocines have been prepared by an intramolecular cyclization of alkyne derivatives of trytamine catalyzed by $AuCl_3$ in an 8-*endo*-dig process (Scheme 4.69) [70].

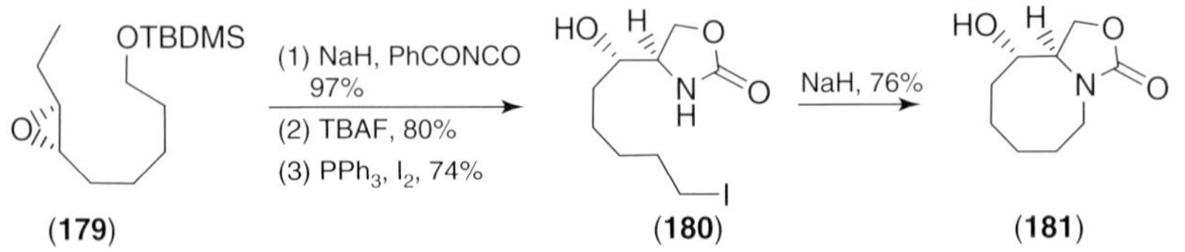

Scheme 4.68 Intramolecular cyclization to azocine.

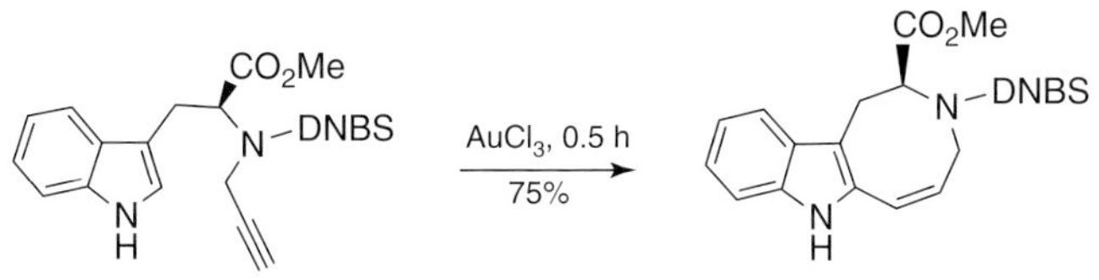

DNBS = dinitrobenzenesulfonyl group

Scheme 4.69 Cyclization to azocines catalyzed by gold.

Asymmetric synthesis of benzazocine framework **184**, an advanced intermediate in the synthesis of FR900482, has been achieved by Pd-catalyzed carbonylation-insertion-cyclization of an arylhydroxylamine with a tethered vinyl iodide **183**. Compound **183** was easily prepared from commercially available 3,5-dinitro-*p*-toluic acid in seven steps and 57% yield *via* the enantiomerically pure secondary alcohol **182** (ee > 99%). The best parameters for this reaction were a temperature ranging 65–80 °C, *N*, *N*-dimethylacetamide as solvent and $Pd(PPh_3)_2Cl_2$ as catalyst. By this way, eight-membered rings were obtained in 64–78% yield (Scheme 4.70) [71].

4.2.2
Ring Transformations Methods

Formation of substituted eight-membered lactams by intramolecular rearrangement of 5,6-dehydro-2-oxocanone **187** provided an access to either *trans*-3,8-disubstituted or 3-substituted eight-membered lactams **188** and **189**, from acids **185** or **186** in 89% and 85% yields, respectively (Scheme 4.71) [72].

Cope rearrangement of enantiomerically pure 2-azetidinone-tethered dienes **190**, prepared as single *cis*-diastereomers from cinnamylideneimine derived from *R*-(+)-*α*-methylbenzylamine, allowed the preparation of optically pure eight-membered lactams **191** in good yields (Scheme 4.72) [73].

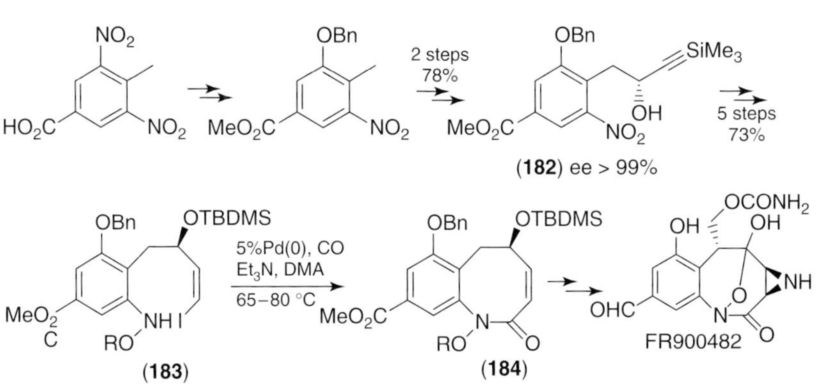

Scheme 4.70 Asymmetric approach to FR900482 benzazocine.

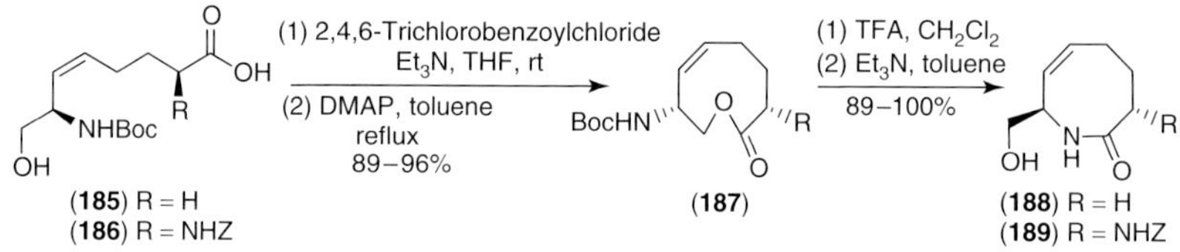

Scheme 4.71 2-Azocanones by lactone-to-lactam ring contraction of 2-oxonanones.

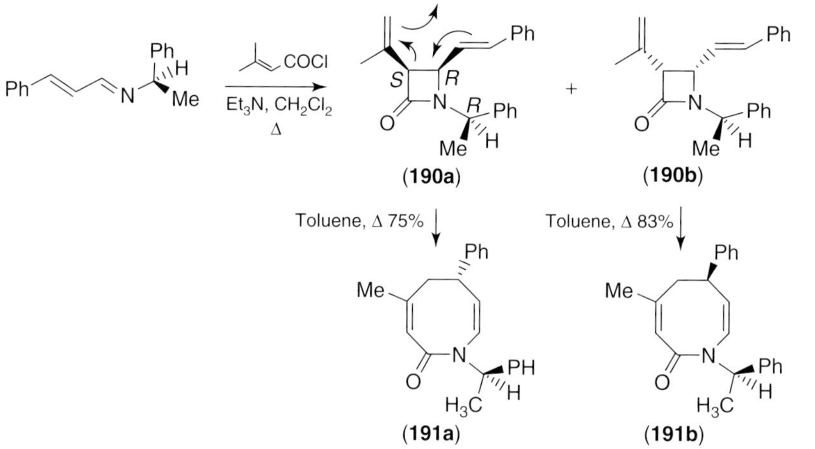

Scheme 4.72 Eight-membered lactams via [3,3] sigmatropic rearrangement of azetidinones.

4.2.3
Cycloaddition Approaches to Azocines

Thermal [2 + 2] cycloaddition reaction between 1,4-dihydropyridine **192** and diethylacetylenedicarboxylate in the presence of Al_2O_3, afforded a cyclobutene intermediate which underwent an electrocyclic ring opening resulting in the formation of the two diastereomers **193a** and **193b** in 30 and 44% yields, respectively (Scheme 4.73) [74].

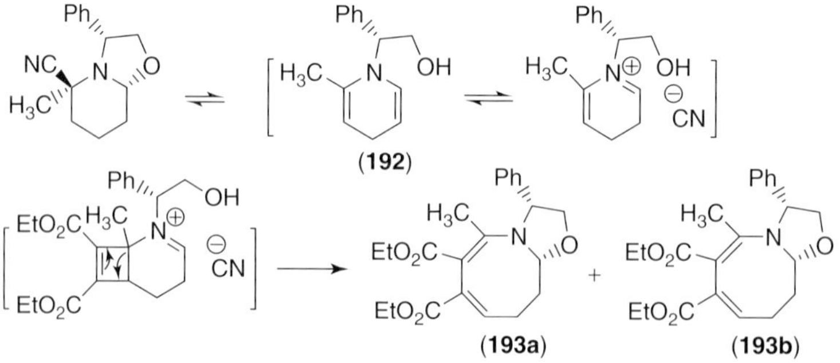

Scheme 4.73 Synthesis of tetrahydro-azocines by cycloaddition of latent 1,4-dihydropyridines.

4.2 Substituted Azocines

Another cycloaddition process allowing the stereoselective synthesis of 3,7-disubstituted azocin-2-one has been developed by Shea et al. [75].

Thus bridged bicyclic oxazinolactam **195b** was obtained as a single diastereomer in 85% yield (100% ds) by a type 2 intramolecular N-acylnitroso Diels–Alder reaction of the hydroxamic acid **194** ($R^2 = H$, $R^1 = Bn$) in the presence of n-Bu$_4$NIO$_4$. For example, bicycle[5.3.1]oxazinolactam **195a** could be transformed to the *trans*-azocinone **196** by alkylation followed by cleavage of the N–O bond, whereas oxazinolactam **195b** was converted to the *cis*-compound **197**. In the case of (R)-C-4 substituted hydroxamic acid **194** ($R^1 = H$, $R^2 = Bn$), Diels–Alder oxidation with n-Bu$_4$NIO$_4$ provided in 84% yield two diastereomers **195d** and **195e** in a 3.7:1 ratio in favor of the *syn* compound **195e** (Scheme 4.74).

4.2.4
Ring-Closing Metathesis

Synthesis of medium-ring cyclic amines has been reported by Schrock et al. [76a] using an asymmetric ring-closing metathesis (ARCM) reaction catalyzed by Mo catalysts in the absence of solvent. Thus, compound **198** in the presence of 5 mol% **199** delivered optically pure amine **200** (ee > 98%) in 93% yield (Scheme 4.75).

Enantioselective syntheses of benzaocine compounds (e.g. (+)-FR900482) have been completed with this method using Grubbs' catalyst [76b].

A library of constrained chiral eight-membered ring lactams has been prepared in high yields from two L-allylglycine residues **201** (Agy) employing an RCM reaction with first generation benzylidene ruthenium catalyst [77] (Scheme 4.76).

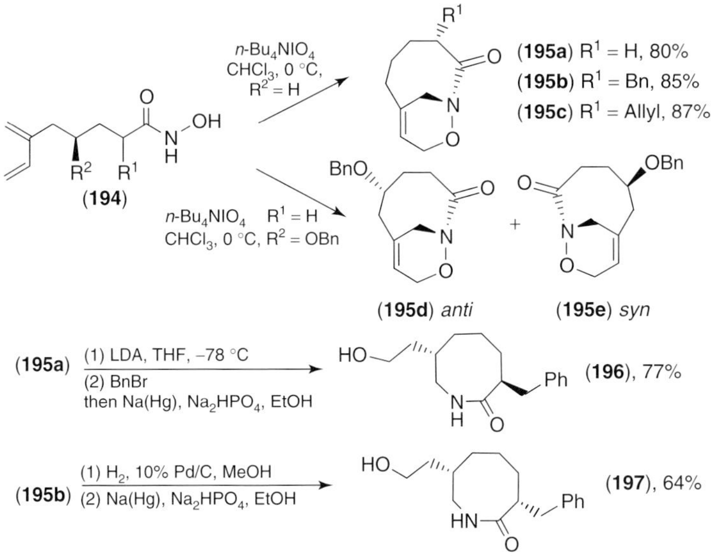

Scheme 4.74 Medium-ring lactams by diastereoselective intramolecular N-acylnitroso Diels–Alder reaction.

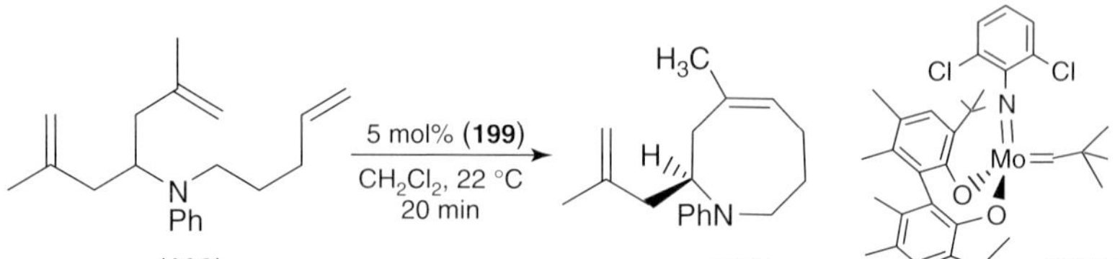

Scheme 4.75 Mo-Catalyzed asymmetric ring-closing metathesis to eight-membered amines.

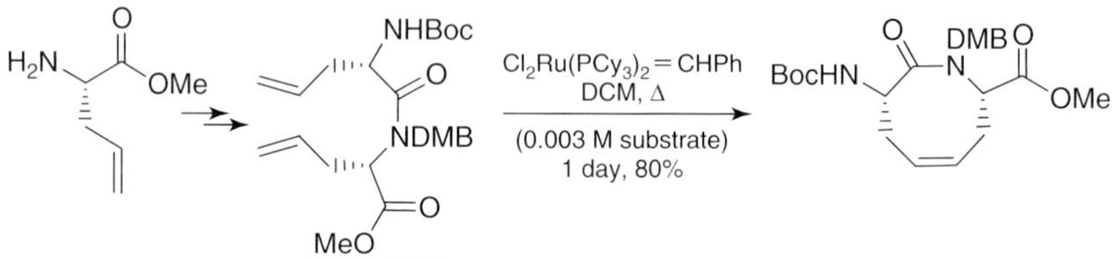

Scheme 4.76 Synthesis of eight-membered ring lactam by RCM reaction.

In view of the high potential of iminosugars for drug discovery and considering the increase of conformational flexibility that the eight-membered ring has brought, a synthesis of iminoalditols **203** has been developed from aminoheptenitols **202** using an RCM reaction.

The chirality at C-3, C-4 and C-5 came from the commercially available starting material, 2,3,4,6-tetra-O-benzyl-D-glucopyranose precursor of **202**. The epimer mixture at C-2 was easily separated by ion-exchange chromatography (Scheme 4.77) [78].

Enantiopure 1,2,3,4,5,8-hexahydroazocin-3-ols have been prepared by distereoselective addition (de > 95) of Grignard reagents to chiral perhydrobenzoxazines followed by an RCM reaction of the 1,9-azadienes thus obtained [42].

Moreover, the RCM reaction was adapted successfully to the synthesis of the eight-membered rings of more complex compounds, namely the alkaloids Manzamine A, Ircinal and Nakadomarin A in moderate to good yields [79].

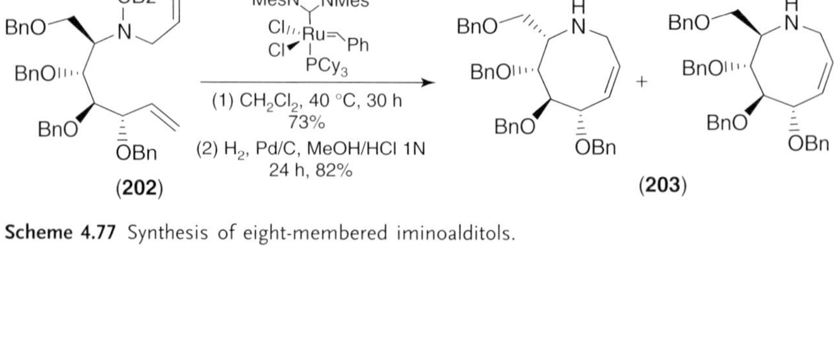

Scheme 4.77 Synthesis of eight-membered iminoalditols.

4.3
Substituted Large Nitrogen-Containing Rings

Among the large nitrogen-containing rings, the substituted nine-membered derivatives deserve particular attention. In fact, these nitrogen heterocycles, well known as azonine derivatives, are found as subunits in natural and pharmaceutically important molecules [1c,d], [51b] like rhazinilam, which interfere with tubulin polymerization, tuberostemonone, and cleavamine (Scheme 4.78).

Moreover, these medium-sized constrained ring systems, in particular the azocinones, serve as key intermediates in the synthesis of bicyclic amino compounds like indolizidines or quinolizidines [80] by selective transannular ring contractions.

The easiest way to generate a nine-membered nitrogen ring, for example lactam azocinones, consists of an intramolecular reaction of ω-amino esters. But, even using activating groups either at the carboxyl or amino functions the cyclization was often low yielding (Scheme 4.79) [81].

This process has been also used in indole series and allowed the synthesis of cleavamine [81c] and 5a-*homo*-vinblastine [81d].

Because of the interesting biological properties of rhazinilam, a lot of methods have been devoted to the synthesis of its skeleton.

Among them, palladium-catalyzed intramolecular coupling [82a] and carbon homologation to macrolactam *via* palladium-catalyzed carbonylation [82b] were the more efficient and afforded the nine-membered aza ring in good yields (Scheme 4.80).

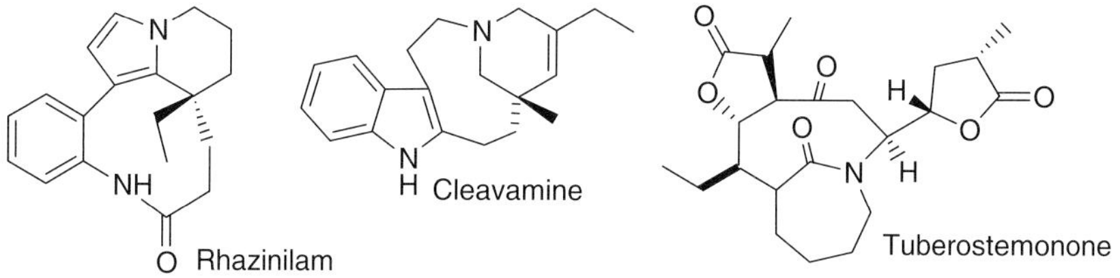

Scheme 4.78 Substituted Azonines.

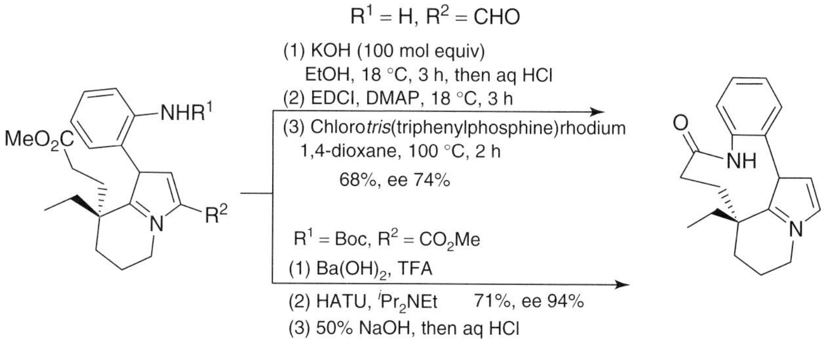

Scheme 4.79 Enantioselective synthesis of (−)-rhazinilam.

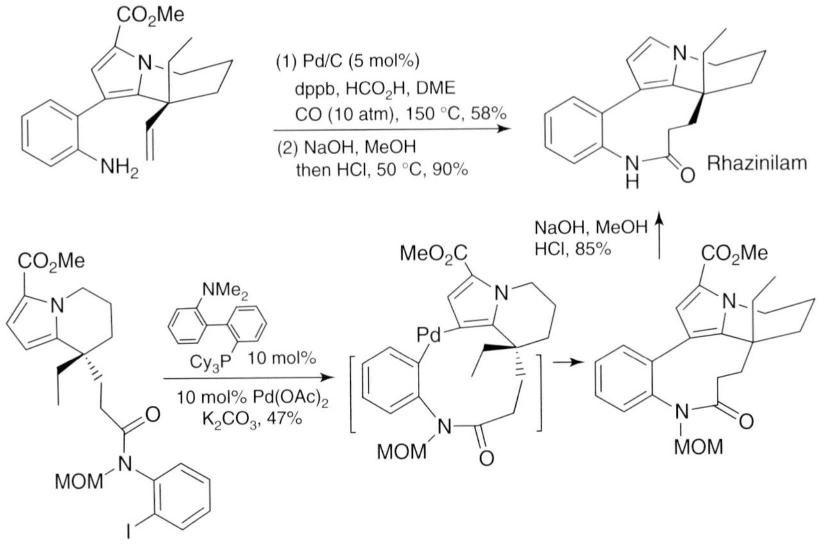

4 Asymmetric Synthesis of Seven- and More-Membered Ring Heterocycles

Scheme 4.80 Palladium-catalyzed coupling to rhazinilam skeleton.

Others methodologies have been developed to elaborate azocine derivatives in particular the ring enlargement of a smaller ring by N or C atoms insertion using fragmentation or rearrangement reactions. For example, oxidative cyclization of Cbz-tyrosine permitted the preparation of the hydroindole **204**, which when exposed to a mixture of iodobenzene diacetate and iodine, provided the azonane **205** as a single isomer in 72% yield *via* a radical-fragmentation-oxidation reaction (Scheme 4.81) [83].

Insertion of a nitrogen atom could also be realized by classical Beckmann rearrangement (Scheme 4.82) [84]

Applied to a more-membered ring, the photoinduced Beckmann ring expansion allowed, for example, the synthesis of the 10-membered framework of isohalichlorensin. Thus, photolysis at 254 nm in benzene of the chiral spirooxaziridine

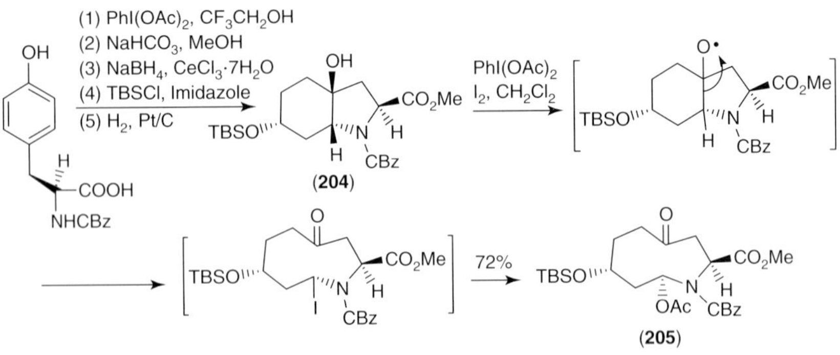

Scheme 4.81 Ring expansion of 4-hydroxyindoles.

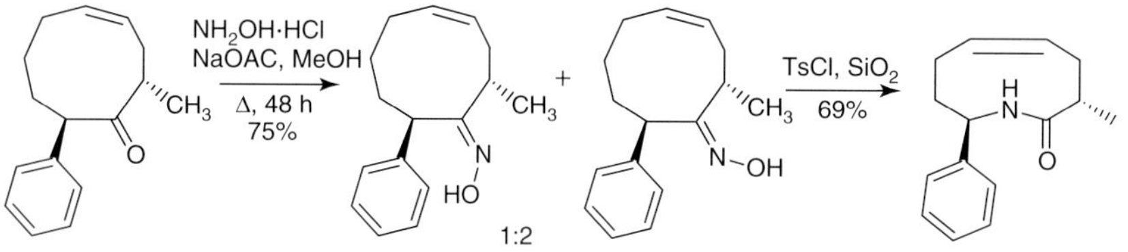

Scheme 4.82 Nine-membered ring lactam by Beckmann rearrangement.

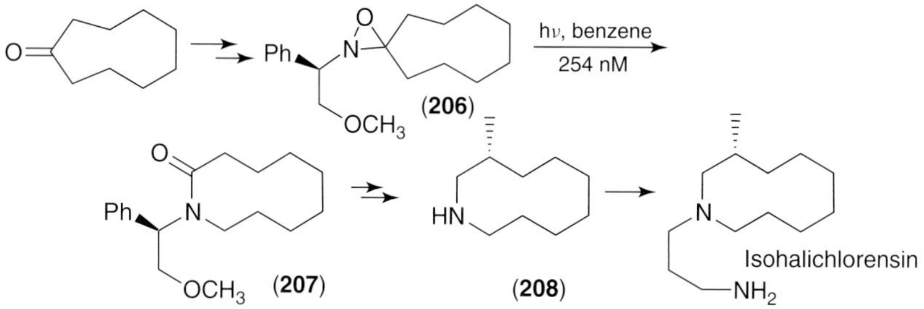

Scheme 4.83 Enantioselective access to 10-membered amine.

206 prepared from commercially available cyclononanone and (R)-(−)-1-amino-1-phenyl-2-methoxyethane, afforded the 10-membered lactam 207 in 57% yield, which could be further transformed in 54% yield to the (3R)-3-methylazacyclodecane 208, precursor of the natural alkaloid isohalichlorensin (Scheme 4.83) [85].

Strained azocinones bearing an E-double bond within their skeletons are of particular interest because they constitute the key intermediates for the synthesis of diverse natural products. An original access to these compounds has been developed by zwitterionic aza-Claisen rearrangement of vinyl pyrrolidines with carboxylic acid fluorides [86]. This reaction led to enantiomerically pure azocinones with (i) exclusive generation of the E-double bond in the ring (ii) yields ranging 35–92% depending on the acid used (better yields were observed with phenylacetyl and chloroacetyl fluoride), and (iii) 3,4-syn/anti diastereoselectivity [86a] strongly influenced by the nature of the substituent (R^2) in the starting vinyl pyrrolidine.

For example, unsubstituted vinyl pyrrolidine 210a obtained from 209a led to a mixture 211a with almost no distereoselectivity. In contrast, when the bulky *tert*-butylsilyloxy (TBSO) group was used, 1,4-chirality transfer was nearly complete. Thus, pyrrolidine 210b, easily prepared from *trans*-4-hydroxy-L-(−)-proline 209b, reacted with various carboxylic acids (R^3 = Cl, F) leading, after a Lewis acid activation of the resulting acyclic precursor followed by a base induced deprotonation of the acylimmonium salt, to a zwitterion which could exist in two transition states (TS), a chairlike and a boatlike TS. The final rearrangement proceeded then diastereoselectively *via* a preferred boatlike transition state affording predominantly (*anti/syn*: 10/1) the 3,4-*syn*-disubstituted (E)-azocinone 211b with a *trans* relative configuration of the stereogenic centers C-3 and C-8 (Scheme 4.84) [86b].

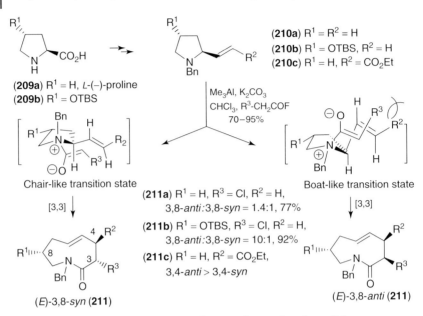

Scheme 4.84 Aza-Claisen rearrangement of TBSO-substituted vinyl pyrrolidines.

In contrario, when the vinylic group of the pyrrolidine was substituted (**210c**) a change in the diastereoselectivity was observed and the 3,4-*anti* substituted azoninone **211c** arising from the preference for a chairlike transition state was majority formed as a result of the repulsive interactions between the substituent and the ester group detected in the boatlike transition state [86c].

Enantioselective aza-Claisen rearrangement has been applied to the synthesis of 10-membered lactam starting from *trans*-2,3-disubstituted acyl piperidine **212**. The diastereoselectivity of this reaction induced by an amide enolate is due to (i) the favorable chair–chair like TS and (ii) the formation of the preferred amide (*Z*)-enolate. The lactam **213**, formed in 74% yield as the sole product, was further converted into the aglycone part of the macrolactam antibiotic fluvirucinine A_1 in 10 steps and 16% yield by standard transformations (Scheme 4.85) [87a].

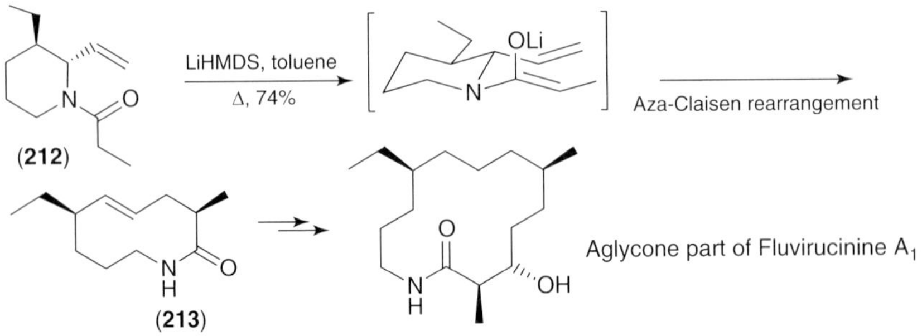

Scheme 4.85 Asymmetric aza-Claisen rearrangement.

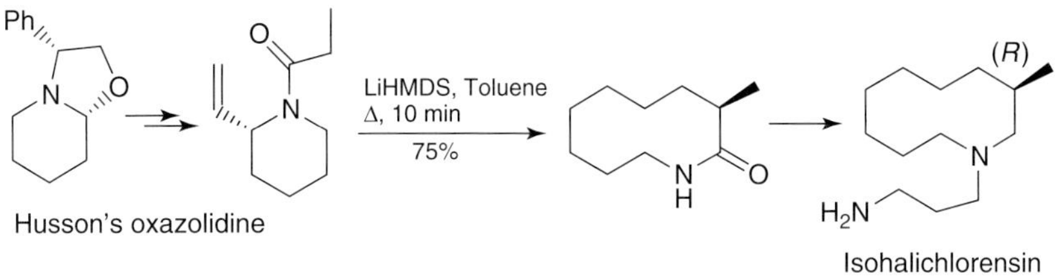

Scheme 4.86 Ring expansion to large-membered amine.

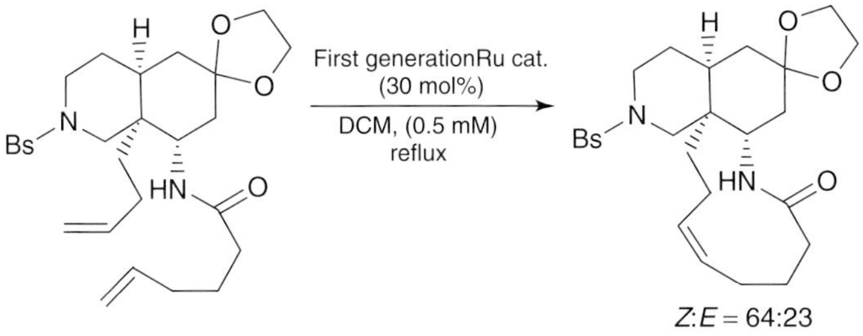

Scheme 4.87 Formation of the C ring of Manzamine B by RCM reaction.

The same method, starting from (R)-N-propionyl-2-vinyl piperidine obtained from Husson's oxazolopiperidine, allowed also the synthesis of the 10-membered diamine isohalichlorensin [87b] (Scheme 4.86).

More recently, the applicability of the RCM reaction in the synthesis of medium-sized rings [34], which permits high yields and tolerates many functional groups, has been reported for the synthesis of 10-membered [88] and larger azacycles. For example, the 11 aza ring of manzamine B and the 15-membered azacycle of (+)-nakadamorin A were elaborated by this method (Scheme 4.87).

References

1 (a) Smalley, R. K. (**1984**) in *Comprehensive Heterocyclic Chemistry*, (ed W. Lwowski), Pergamon press, Bristol, Vol. **7**, pp. 545–46; (b) Devon, T. K. and Scott, A. I. (**1972**) *Handbook of Naturally Occuring Compounds*, Academic Press, New York and London, Vol. **II**; (c) Meigh, J.-P. K. (**2004**) *Science of Synthesis*, (ed S. M. Weinreb), Thieme, Stuttgart, New York, Vol. **17**, pp. 829–30; (d) Evans, P. A. and Holmes, A. B. (**1991**) *Tetrahedron*, **47**, 9131–66 and references cited therein; (e) O'Hagan, O. (**1997**) *Nat. Prod. Rep.*, 637–52; (f) Nishizuka, Y. (**1992**) *Science*, **258**, 607–14; (g) Hansen Jr, D. W., Peterson, K. B., Trivedi, M., Kramer, S. W., Webber, R. K., Tjoeng, F. S., Moore, W. M., Jerome, G. M., Kornmeier, C. M., Manning, P. T., Connor, J. R., Misko, T. P., Currie, M. G. and Pitzele, B. S. (**1998**) *J. Med. Chem.*, **41**, 1361–66; (h) Thale, Z., Kinder, F. R., Bair, K. W., Bontempo, J., Czuchta, A. M., Versace, R. W., Phillips, P. E., Sanders, M. L., Wattanasin, S. and Crews, P. (**2001**) *J. Org. Chem.*, **66**, 1733–41; (i) Walz, A. J. and Miller, M. J. Org. Lett., **2002**, **4**, 2047–50; (j) Miller, W. H., Alberts, D. P., Bhatnagar, P. K., Bondinell, W. E., Callahan, J. F., Calvo, R. R., Cousins,

R. D., Erhard, K. F., Heerding, D. A., Keenan, R. M., Kwon, C., Manley, P. J., Newlander, K. A., Ross, S. T., Samanen, J. M., Uzinskas, I. N., Venslavsky, J. W., Yuan, C. C.-K., Haltiwanger, R. C., Gowen, M., Hwang, S.-M., James, I. E., Lark, M. W., Rieman, D. J., Stroup, G. B., Azzarano, L. M., Salyers, K. L., Smith, B. R., Ward, K. W., Johanson, K. O. and Huffman, W. F. (2000) *J. Med. Chem.*, **43**, 22–26; (k) Smith, B. M., Smith, J. M., Tsai, J. H., Schultz, J. A., Gilson, C. A., Estrada, S. A., Chen, R. R., Park, D. M., Prieto, E. B., Gallardo, C. S., Sengupta, D., Thomsen, W. J., Saldana, H. R., Whelan, K. T., Menzaghi, F., Webb, R. R. and Beeley, N. R. A. (2005) *Bioorg. Med. Chem. Lett.*, **15**, 1467–70; (l) Boeglin, D., Bonnet, D. and Hibert, M. (2007) *J. Comb. Chem.*, **9**, 487–500; (m) Kricka, L. J. and Ledwith, A. (1974) *Chem. Rev.*, **74**, 101–23; (n) Zhang, A., Neumeyer, J. L. and Baldessarini, R. J. (2007) *Chem. Rev.*, **107**, 274–302; (o) Cho, H., Murakami, K., Nakanishi, H., Fujisawa, A., Isoshima, H., Niwa, M., Hayakawa, K., Hase, Y., Uchida, I., Watanabe, H., Wakitani, K. and Aisaka, K. (2004) *J. Med. Chem.*, **47**, 101–9; (p) Pilli, R. A., da Conceição Ferreira de Oliveira, M. (2000) *Nat. Prod. Rep.*, **17**, 117–27; (q) Brem, B., Seger, C., Pacher, T., Hofer, O., Vajrodaya, S. and Greger, H. (2002) *J. Agric. Food Chem.*, **50**, 6383–88; (r) Marco-Contelles, J., do Carmo Carreiras, M., Rodriguez, C., Villarroya, M. and García, A. G. (2006) *Chem. Rev.*, **106**, 116–33; (s) Zhu, D.-C., Zittoun, R. and Marie, J.-P. (1995) *Bull. Cancer*, **82**, 987–95; (t) Hitt, E. (2002) *Lancet Oncol.*, **3**, 259.

2 Humphrey, J. M. and Chamberlin, A. R. (1997) *Chem. Rev.*, **97**, 2243–66.

3 Singh, J., Kronenthal, D. R., Schwinden, M., Godfrey, J. D., Fox, R., Vawter, E. J., Zhang, B., Kissick, T. P., Patel, B., Mneimne, O., Humora, M., Papaioannou, C. G., Szymanski, W., Wong, M. K. Y., Chen, C. K., Heikes, J. E., DiMarco, J. D., Qiu, J., Deshpande, R. P., Gougoutas, J. Z. and Mueller, R. H. (2003) *Org. Lett.*, **5**, 3155–58.

4 Masse, C. E., Morgan, A. J. and Panek, J. S. (2000) *Org. Lett.*, **2**, 2571–73.

5 Boeckman Jr, R. K., Clark, T. J. and Shook, B. C. (2002) *Org. Lett.*, **4**, 2109–12.

6 Edwards, D. J., Pritchard, R. G. and Wallace, T. W. (2003) *Tetrahedron Lett.*, **44**, 4665–68.

7 Meyers, A. I. and Brengel, G. P. (1997) *Chem. Commun.*, 1–8.

8 Penhoat, M., Levacher, V. and Dupas, G. (2003) *J. Org. Chem.*, **68**, 9517–20.

9 (a) Gosselin, F. and Lubell, W. D. (2000) *J. Org. Chem.*, **65**, 2163–71; (b) Alibés, R., Blanco, P., Casas, E., Closa, M., De March, P., Figueredo, M., Font, J., Sanfeliu, E. and Álvarez-Larena, A. (2005) *J. Org. Chem.*, **70**, 3157–67; (c) Angiolini, M., Araneo, S., Belvisi, L., Cesarotti, E., Checchia, A., Crippa, L., Manzoni, L. and Scolastico, C. (2000) *Eur. J. Org. Chem.*, **14**, 2571–81.

10 Yu, L.-T., Huang, J.-L., Chang, C.-Y. and Yang, T.-K. (2006) *Molecules*, **11**, 641–48.

11 Yet, L. (2000) *Chem. Rev.*, **100**, 2963–3007.

12 (a) Miyabe, H., Torieda, M., Kiguchi, T. and Naito, T. (1997) *Synlett*, **5**, 580–82; (b) Riber, D., Hazell, R. and Skrydstrup, T. (2000) *J. Org. Chem.*, **65**, 5382–90; (c) Naito, T., Nakagawa, K., Nakamura, T., Kasei, A., Ninomiya, I. and Kigushi, T. (1999) *J. Org. Chem.*, **64**, 2003–9.

13 Miyabe, H., Torieda, M., Inoue, K., Tajiri, K., Kiguchi, T. and Naito, T. (1998) *J. Org. Chem.*, **63**, 4397–407.

14 Rigby, J. H., Laurent, S., Cavezza, A. and Heeg, M. J. (1998) *J. Org. Chem.*, **63**, 5587–91.

15 Evans, P. A., Manangan, T. and Rheingold, A. L. (2000) *J. Am. Chem. Soc.*, **122**, 11009–10.

16 Cordes, M. and Franke, D. (**2004**) *Synlett*, **11**, 1917–20.
17 Andrés, C., Duque-Soladana, J.-P., Iglesias, J.-M. and Pedrosa, R. (**1999**) *Tetrahedron Lett.*, **40**, 2421–24.
18 Evans, C. A., Cowen, B. J. and Miller, S. J. (**2005**) *Tetrahedron*, **61**, 6309–14.
19 (a) Pyne, S. G., Davis, A. S., Gates, N. J., Hartley, J. P., Lindsay, K. B., Machan, T. and Tang, M. (**2004**) *Synlett*, **15**, 2670–80; (b) Lindsay, K. B. and Pyne, S. G. (**2004**) *Synlett*, **15**, 779–82.
20 Joong, S. and Beak, P. (**2006**) *J. Am. Chem. Soc.*, **128**, 2178–79.
21 (a) Monterde, M. I., Nazabadioko, S., Rebolledo, F., Brieva, R. and Gotor, V. (**1999**) *Tetrahedron: Asymmetry*, **10**, 3449–55; (b) Satcharoen, V., McLean, N. J., Kemp, S. C., Camp, N. P. and Brown, R. C. D. (**2007**) *Org. Lett.*, **9**, 1867–69.
22 Williams, D. R., Shamim, K., Reddy, J. P., Amato, G. S. and Shaw, S. M. (**2003**) *Org. Lett.*, **5**, 3361–64.
23 Trost, B. M., Tang, W. and Toste, F. D. (**2005**) *J. Am. Chem. Soc.*, **127**, 14785–803.
24 (a) Shih, T.-L., Yang, R.-Y., Li, S.-T., Chiang, C.-F. and Lin, C.-H. (**2007**) *J. Org. Chem.*, **72**, 4258–61; (b) Painter, G. F. and Falshaw, A. (**2000**) *J. Chem. Soc., Perkin Trans. 1*, **7**, 1157–59.
25 Royer, J., Bonin, M. and Micouin, L. (**2004**) *Chem. Rev.*, **104**, 2311–52.
26 Flynn, G. A., Giroux, E. L. and Dage, R. C. (**1987**) *J. Am. Chem. Soc.*, **109**, 7914–15.
27 Marson, C. M., Pink, J. H., Hall, D., Hursthouse, M. B., Malik, A. and Smith, C. (**2003**) *J. Org. Chem.*, **68**, 792–98.
28 (a) Tanimoto, H., Kato, T. and Chida, N. (**2007**) *Tetrahedron Lett.*, **48**, 6267–70; (b) Malachowski, W. P., Paul, T. and Phounsavath, S. (**2007**) *J. Org. Chem.*, **72**, 6792–96.
29 Malachowski, W. P., Paul, T. and Phounsavath, S. (**2007**) *J. Org. Chem.*, **72**, 6792–96.
30 Bur, S. K. and Padwa, A. (**2004**) *Chem. Rev.*, **104**, 2401–32.
31 (a) Worden, S. M., Mapitse, R. and Hayes, C. J. (**2002**) *Tetrahedron Lett.*, **43**, 6011–14; (b) Liu, Q., Ferreira, E. M. and Stoltz, B. M. (**2007**) *J. Org. Chem.*, **72**, 7352–58.
32 Planas, L., Perard-Vires, J., Royer, J. (**2004**) *J. Org. Chem.*, **69**, 3087–92.
33 (a) Shimizu, K., Tomioka, K., Yamada, S.-I. and Koga, K. (**1978**) *Chem. Pharm. Bull.*, **26**, 3765–71; (b) Node, M., Kodama, S., Hamashima, Y., Baba, T., Hamamichi, N. and Nishide, K. (**2000**) *Angew. Chem. Int. Ed.*, **40**, 3060–62; (c) Kodama, S., Hamashima, Y., Nishide, K. and Node, M. (**2004**) *Angew. Chem. Int. Ed.*, **43**, 2659–61.
34 Deiters, A. and Martin, S. F. (**2004**) *Chem. Rev.*, **104**, 2199–238.
35 Meyers, A. I., Downing, S. V. and Weiser, M. J. (**2001**) *J. Org. Chem.*, **66**, 1413–19.
36 Hunt, J. C. A., Laurent, P. and Moody, C. J. (**2002**) *J. Chem. Soc., Perkin Trans. 1*, **21**, 2378–89.
37 Fürster, A. and Thiel, O. R. (**2000**) *J. Org. Chem.*, **65**, 1738–42.
38 Wipf, P., Rector, S. R. and Takahashi, H. (**2002**) *J. Am. Chem. Soc.*, **124**, 14848–49.
39 (a) Hanessian, S., Sailes, H., Munro, A. and Therrien, E. (**2003**) *J. Org. Chem.*, **68**, 7219–33; (b) Torssell, S., Wanngren, E. and Somfai, P. (**2007**) *J. Org. Chem.*, **72**, 4246–49.
40 (a) Kinoshita, A. and Mori, M. (**1996**) *J. Org. Chem.*, **61**, 8356–57; (b) Brenneman, J. B. and Martin, S. F. (**2004**) *Org. Lett.*, **6**, 1329–31.
41 Wipf, P., Stephenson, C. R. J. and Walczak, M. A. A. (**2004**) *Org. Lett.*, **6**, 3009–12.
42 Pedrosa, R., Andrés, C., Gutiérrez-Loriente, A. and Nieto, J. (**2005**) *Eur. J. Org. Chem.*, 2449–58.
43 Delhaye, L., Merschaert, A., Diker, K. and Houpis, I. N. (**2006**) *Synthesis*, **9**, 1437–42.
44 (a) Lee, H. K., Im, J. H. and Jung, S. H. (**2007**) *Tetrahedron*, **16**, 3321–27; (b) Lee Trout, R. E. and Marquis, R. W. (**2005**) *Tetrahedron Lett.*, **46**, 2799–801.

45 Hoffmann, T., Waibel, R. and Gmeiner, P. (2003) *J. Org. Chem.*, **68**, 62–69.

46 (a) Booker-Milburn, K. I., Dudin, L. F., Anson, C. E. and Guile, S. D. (2001) *Org. Lett.*, **3**, 3005–8; (b) Booker-Milburn, K. I., Hirst, P., Charmant, J. P. H. and Taylor, L. H. J. (2003) *Angew. Chem. Int. Ed.*, **42**, 1642–44; (c) Booker-Milburn, K. I., Baker, J. R. and Bruce, I. (2004) *Org. Lett.*, **6**, 1481–84.

47 Barluenga, J., Tomás, M., Ballesteros, A., Santamaría, J., Carbajo, R. J., López-Ortiz, F., García-Granda, S. and Pertierra, P. (1996) *Chem. Eur. J.*, **2**, 88–97.

48 (a) Nath, M., Mukhopadhyay, R. and Bhattacharjya, A. (2006) *Org. Lett.*, **8**, 317–20; (b) Liu, Y., Maden, A. and Murray, W. V. (2002) *Tetrahedron*, **58**, 3159–70.

49 Takao, K.-I., Munakata, R. and Tadano, K.-I. (2005) *Chem. Rev.*, **105**, 4779–807.

50 (a) Jacobi, P. A. and Lee, K. (2000) *J. Am. Chem. Soc.*, **122**, 4295–303; (b) Morimoto, Y., Iwahashi, M., Kinoshita, T. and Nishida, K. (2001) *Chem. Eur. J.*, **7**, 4107–16.

51 (a) Kantorowski, E. J. and Kurth, M. J. (2000) *Tetrahedron*, **56**, 4317–53; (b) Nubbemeyer, U. (2001) *Topics in Current Chemistry*, Springer-Verlag, Berlin Heidelberg, Vol. 216, pp. 125–96.

52 (a) Aubé, J. (1997) *Chem. Soc. Rev.*, **26**, 269–77; (b) Gracias, V., Milligan, G. L. and Aubé, J. (1995) *J. Am. Chem. Soc.*, **117**, 8047–48; (c) Katz, C. E. and Aubé, J. (2003) *J. Am. Chem. Soc.*, **125**, 13948–49; (d) Sahasrabudhe, K., Gracias, V., Furness, K., Smith, B. T., Katz, C. E., Reddy, D. S. and Aubé, J. (2003) *J. Am. Chem. Soc.*, **125**, 7914–22; (e) MacDonald, M., Vander Velde, D. and Aubé, J. (2001) *J. Org. Chem.*, **66**, 2636–42.

53 (a) Elliott, J. M., Carlson, E. J., Chicchi, G. G., Dirat, O., Dominguez, M., Gerhard, U., Jelley, R., Jones, A. B., Kurtz, M. M., Tsao, K. I. and Wheeldon, A. (2006) *Bioorg. Med. Chem. Lett.*, **16**, 2929–32; (b) Schinzer, D., Abel, U. and Jones, P. G. (1997) *Synlett*, **5**, 632–34.

54 Morie, T. and Kato, S. (1998) *Heterocycles*, **48**, 427–31.

55 Chong, H.-S., Ganguly, B., Broker, G. A., Rogers, R. D. and Brechbiel, M. W. (2002) *J. Chem. Soc., Perkin Trans. 1*, **18**, 2080–86.

56 Paulsen, H. and Todt, K. (1967) *Chem. Ber.*, **100**, 512–20.

57 Herdeis, C., Mohareb, R. M., Neder, R. B., Schwabenländer, F. and Telser, J. (1999) *Tetrahedron: Asymmetry*, **10**, 4521–37.

58 (a) Le Merrer, Y., Poitout, L., Depezay, J.-C., Dosbaa, I., Geoffroy, S. and Foglietti, M.-J. (1997) *Bioorg. Med. Chem.*, **5**, 519–33; (b) Morís-Varas, F., Qian, X.-H. and Wong, C.-H. (1996) *J. Am. Chem. Soc.*, **118**, 7647–52; (c) Li, H., Zhang, Y., Vogel, P., Sinaÿ, P. and Blériot, Y. (2007) *Chem. Commun.*, 183–85; (d) Andreana, P. R., Sanders, T., Janczuk, A., Warrick, J. I. and Wang, P. G. (2002) *Tetrahedron Lett.*, **43**, 6525–28; (e) Liu, T., Zhang, Y. and Blériot, Y. (2007) *Synlett*, **6**, 905–8.

59 (a) Cutri, S., Bonin, M., Micouin, L., Husson, H.-P. and Chiaroni, A. (2003) *J. Org. Chem.*, **68**, 2645–51; (b) Cutri, S., Diez, A., Bonin, M., Micouin, L. and Husson, H.-P. (2005) *Org. Lett.*, **7**, 1911–13.

60 Unthank, M. G., Hussain, N. and Aggarwal, V. K. (2006) *Angew. Chem. Int. Ed.*, **45**, 7066–69.

61 Coquerel, Y., Bensa, D., Doutheau, A. and Rodriguez, J. (2006) *Org. Lett.*, **8**, 4819–22.

62 Couty, F., Durrat, F., Evano, G. and Marrot, J. (2006) *Eur. J. Org. Chem.*, 4214–23.

63 Lim, S. H., Ma, S. and Beak, P. (2001) *J. Org. Chem.*, **66**, 9056–62.

64 Ishihara, J., Horie, M., Shimada, Y., Tojo, S. and Murai, A. (2002) *Synlett*, **3**, 403–6.

65 (a) Xu, D. D., Waykole, L., Calienni, J. V., Ciszewski, L., Lee, G. T., Liu, W., Szewczyk, J., Vargas, K., Prasad, K., Repiš, O. and Blacklock, T. J.

(2003) *Org. Process Res. Dev.*, **7**, 856–65; (b) Marshall, J. A. and Luke, G. P. (**1993**) *J. Org. Chem.*, **58**, 6229–34; (c) Mukai, C., Kataoka, O. and Hanaoka, M. (**1994**) *Tetrahedron Lett.*, **35**, 6899–902; (d) Mukai, C., Moharram, S. M., Kataoka, O. and Hanaoka, M. (**1995**) *J. Chem. Soc., Perkin Trans. 1*, **22**, 2849–54; (e) Sarabia, F. and Sánchez-Ruiz, A. (**2005**) *J. Org. Chem.*, **70**, 9514–20.

66 Smith III, A. B., Cho, Y. S., Zawacki, L. E., Hirschmann, R. and Pettit, G. R. (**2001**) *Org. Lett.*, **3**, 4063–66.

67 (a) Alvarez, M. and Joule, A. (**2001**) *Alkaloids, Chemistry and Biology*, Academic Press, New York, Vol. **57**, pp. 235–72; (b) Thorsett, E. D., Harris, E. E., Aster, S. D., Peterson, E. R., Snyder, J. P., Springer, J. P., Hirshfield, J., Tristram, E. W., Patchett, A. A., Ulm, E. H. and Vassil, T. C. (**1986**) *J. Med. Chem.*, **29**, 251–60; (c) Derrer, S., Davies, J. E. and Holmes, A. B. (**2000**) *J. Chem. Soc., Perkin Trans. 1*, **17**, 2943–56; (d) Tadic, D., Linders, J. T. M., Flippen-Anderson, J. L., Jacobson, A. E. and Rice, K. C. (**2003**) *Tetrahedron*, **59**, 4303–614; (e) Vskressensky, L. G., Borisova, T. N., Kulikova, L. N., Varlamov, A. V., Catto, M., Altomare, C. and Carotti, A. (**2004**) *Eur. J. Org. Chem.*, **14**, 3128–35.

68 (a) Torisawa, Y., Motohashi, Y., Ma, J., Hino, T. and Nakagawa, M. (**1995**) *Tetrahedron Lett.*, **36**, 5579–80; (b) Winkler, J. D., Stelmach, J. E. and Axten, J. (**1996**) *Tetrahedron Lett.*, **37**, 4317–18; (c) Uchida, H., Nishida, A. and Nakagawa, M. (**1999**) *Tetrahedron Lett.*, **40**, 113–16; (d) Paleo, M. R., Aurrecoechea, N., Jung, K.-Y. and Rapoport, H. (**2003**) *J. Org. Chem.*, **68**, 130–38.

69 Winkler, J. D., Stelmach, J. E. and Axten, J. (**1996**) *Tetrahedron Lett.*, **37**, 4317–18.

70 Ferrer, C., Amijs, C. H. M. and Echavarren, A. M. (**2007**) *Chem. Eur. J.*, **13**, 1358–73.

71 (a) Baran, P. S. and Corey, E. J. (**2002**) *J. Am. Chem. Soc.*, **124**, 7904–5; (b) Trost, B. M. and Ameriks, M. K. (**2004**) *Org. Lett.*, **6**, 1745–48.

72 Derrer, S., Feeder, N., Teat, S. J., Davies, J. E. and Holmes, A. B. (**1998**) *Tetrahedron Lett.*, **39**, 9309–12.

73 Alcaide, B., Rodríguez-Ranera, C. and Rodríguez-Vicente, A. (**2001**) *Tetrahedron Lett.*, **42**, 3081–83.

74 Lallemand, M.-C., Chiadmi, M., Tomas, A., Kunesch, N. and Husson, H.-P. (**1995**) *Tetrahedron Lett.*, **36**, 2053–56.

75 (a) Chow, C. P., Shea, K. J. and Sparks, S. M. (**2002**) *Org. Lett.*, **4**, 2637–40; (b) Sparks, S. M., Chow, C. P., Zhu, L. and Shea, K. J. (**2004**) *J. Org. Chem.*, **69**, 3025–35.

76 (a) Dolman, S. J., Sattely, E. S., Hoveyda, A. H. and Schrock, R. R. (**2002**) *J. Am. Chem. Soc.*, **124**, 6991–97; (b) Fellows, I. M., Kaelin Jr, D. E. and Martin, S. F. (**2000**) *J. Am. Chem. Soc.*, **122**, 10781–87.

77 (a) Creighton, C. J. and Reitz, A. B. (**2001**) *Org. Lett.*, **3**, 893–95; (b) Creighton, C. J., Leo, G. C., Du, Y. and Reitz, A. B. (**2004**) *Bioorg. Med. Chem.*, **12**, 4375–85.

78 Godin, G., Garnier, E., Compain, P., Martin, O. R., Ikeda, K. and Asano, N. (**2004**) *Tetrahedron Lett.*, **45**, 579–81.

79 (a) Ono, K., Nakagawa, M. and Nishida, A. (**2004**) *Angew. Chem. Int. Ed.*, **43**, 2020–23; (b) Humphrey, J. H., Liao, Y., Ali, A., Rein, T., Wong, Y.-L., Chen, H.-J., Courtney, A. K. and Martin, S. F. (**2002**) *J. Am. Chem. Soc.*, **124**, 8584–92; (c) Nagata, T., Nakagawa, M. and Nishida, A. (**2003**) *J. Am. Chem. Soc.*, **125**, 7484–85; (d) Ahrendt, K. A. and Williams, R. M. (**2004**) *Org. Lett.*, **6**, 4539–41; (d) Martin, S. F., Humphrey, J. M., Ali, A. and Hillier, M. C. (**1999**) *J. Am. Chem. Soc.*, **121**, 866–67.

80 (a) Diederich, M. and Nubbemeyer, U. (**1996**) *Chem. Eur. J.*, **2**, 894–900; (b) Sudau, A., Münch, W., Bats,

J.-W. and Nubbemeyer, U. (**2002**) *Eur. J. Org. Chem.*, **19**, 3304–14.

81 (a) Banwell, M. G., Beck, D. A. S. and Willis, A. C. (**2006**) *Arkivoc*, **iii**, 163–74; (b) Liu, Z., Wasmuth, A. S. and Nelson, S. G. (**2006**) *J. Am. Chem. Soc.*, **128**, 10352–53; (c) Amat, M., Escolano, C., Lozano, O., Llor, N. and Bosch, J. (**2003**) *Org. Lett.*, **5**, 3139–42; (d) Kuehne, M. E., Cowen, S. D., Xu, F. and Borman, L. S. (**2001**) *J. Org. Chem.*, **66**, 5303–16.

82 (a) Bowie Jr, A. L., Hughes, C. H. and Trauner, D. (**2005**) *Org. Lett.*, **7**, 5207–9; (b) Johnson, J. A., Li, N. and Sames, D. (**2002**) *J. Am. Chem. Soc.*, **124**, 6900–3.

83 Wipf, P. and Li, W. (**1999**) *J. Org. Chem.*, **64**, 4576–77.

84 Olson, G. L., Voss, M. E., Hill, D. E., Kahn, M., Madison, V. S., and Cook, C. M. (**1990**) *J. Am. Chem. Soc.*, **112**, 323–33.

85 Usuki, Y., Hirakawa, H., Goto, K. and Iio, H. (**2001**) *Tetrahedron: Asymmetry*, **12**, 3293–96.

86 (a) Nubbemeyer, U. (**1995**) *J. Org. Chem.*, **60**, 3773–80; (b) Sudau, A., Münch, W. and Nubbemeyer, U. (**2000**) *J. Org. Chem.*, **65**, 1710–20; (c) Laabs, S., Schermann, A., Sudau, A., Diederich, M., Kierig, C. and Nubbemeyer, U. (**1999**) *Synlett*, **1**, 25–28.

87 (a) Suh, Y.-G., Kim, S.-A., Jung, J.-K., Shin, D.-Y., Min, K.-H., Koo, B.-A. and Kim, H.-S. (**1999**) *Angew. Chem. Int. Ed.*, **38**, 3545–47; (b) Zheng, J.-F., Jin, L.-R. and Huang, P.-Q. (**2004**) *Org. Lett.*, **6**, 1139–42. (c) Heinrich, M. R., Steglich, W., Banwell, M. G. and Kashman, Y. (**2003**) *Tetrahedron*, **59**, 9239–47.

88 Matsumura, T., Akiba, M., Arai, S., Nakagawa, M. and Nishida, A. (**2007**) *Tetrahedron Lett.*, **48**, 1265–68.

Part Two
Asymmetric Synthesis of Nitrogen Heterocycles With More Than One Heteroatom

5
Asymmetric Synthesis of Three- and Four-Membered Ring Heterocycles with More Than One Heteroatom

Steve Lanners and Gilles Hanquet

5.1
Introduction

Three- and four-membered N-containing heterocycles are invested with a special allure that is derived from their apparent simplicity and spartan architecture. Thus, these systems are multifaceted, and the literature continues to provide evidence of their diversity, both in terms of preparative routes and subsequent transformations. Among them, those containing two or more heteroatoms have, for the most part, only been synthesized in the second half of the twentieth century, even though some of them exhibit a synthetically useful balance between stability and reactivity. They are often used as versatile and selective reagents and in some cases, as synthetic intermediates or for their biological properties. While syntheses of such racemic small heterocycles have already been reviewed [1], only few papers were devoted to their stereoselective preparation.

This chapter highlights the stereoselective preparation of three- and four-membered N-heterocycles, both containing two or more heteroatoms, from the recent literature. We will generally focus on stereoselective transformations that lead to enantiomerically enriched N-heterocycles.

5.2
Three-Membered N-Heterocycles with Two Heteroatoms

Three-membered N-heterocycles with two heteroatoms were discovered only after 1950. A great number of them have been synthesized since then, using generally simple procedures familiar to synthetic chemists decades before. The saturated C–N–N rings (diaziridines) and their dehydrogenation products (diazirines) play the most important role in this field together with the C–N–O rings (oxaziridines). The latter are the most widely represented family in enantiomerically pure form as a result of their important applications in asymmetric synthesis [2]. Three-membered

Asymmetric Synthesis of Nitrogen Heterocycles. Edited by Jacques Royer
Copyright © 2009 WILEY-VCH Verlag GmbH & Co. KGaA, Weinheim
ISBN: 978-3-527-32036-3

N-heterocycles containing sulfur or phosphorus atoms are less common, because of their fewer applications and their difficult syntheses.

The stereochemistry of these small rings has received considerable attention mainly because of the chirality of the nitrogen atom and the generally appreciable barrier to its inversion (Equation 5.1) [3].

$$\begin{array}{c}\text{(structure)}\end{array} \rightleftharpoons \begin{array}{c}\text{(structure)}\end{array} \tag{5.1}$$

Enantioselective preparations of these small rings are mainly based on substrate-controlled diastereoselective ring construction starting from enantiomerically pure precursors. In many cases, this strategy is mainly based on nitrogen-heteroatom ring closure as depicted in Equation 5.2 [1e].

$$\begin{array}{c}\text{(structure)}\end{array} \longrightarrow \begin{array}{c}\text{(structure)}\end{array} \tag{5.2}$$

Some examples of photoisomerization of open chain 1,3 dipoles such as nitrones, azomethinimines and linear diazo compounds have also been employed but are less common.

5.2.1
Diaziridines

In general, the configuration at the nitrogen atom of trialkylamines ($NR^1R^2R^3$) is invertible at room temperature, and they are impossible to isolate in optically active form. However, inversion of the N-atom in a three-membered ring system becomes remarkably slow [4]. The inversion barriers (108–113 kJ mol^{-1}) have been determined by Mannschreck et al. in 1969 by means of NMR studies [5]. Generally, monocyclic diaziridines in solution and in the solid or gaseous state exist as 1,2-trans-isomers because the 1,2-cis-form is destabilized by n–n interactions of the nitrogen lone pairs and nonbonded interactions of N-substituents [6].

The first optically active diaziridines were reported in 1974 [7], and their absolute configurations were established by X-ray diffraction analysis. Chiroptical properties [8] and configurational stability were studied by racemization kinetics [9]. In 1979, Mannschreck et al. applied chiral chromatography to the optical resolution of diaziridines and also measured the inversion barriers by racemization kinetics [10]. In all known examples, the inversion barriers of diaziridines did not exceed 117 kJ mol^{-1} and are related to the substitution pattern of the three-membered ring [2, 11].

Most enantiomerically pure diaziridines described in the literature result from optical resolution of a racemic mixture using chiral chromatography or crystallization [12]. Syntheses of racemic diaziridines almost always proceed by N–N bond formation, using Schmitz's methodology, which consists of mixing an aldehyde or a ketone with an amine and an electrophilic amination agent [13]. The latter can be a chloramine, hydroxylamine-O-sulfonic acid (HOSA), or an O-sulfonylated hydroxylamine. As enantioselective electrophilic amination reagents remain unusual, the most interesting route to chiral nonracemic diaziridines is based on diastereoselective preparations starting from enantiomerically pure amines or ketones.

5.2.1.1 Substrate-Controlled Diastereoselective Diaziridination Using Chiral Enantiomerically Pure Amines

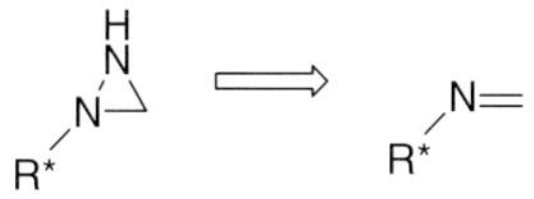

Ito et al. have studied the enantioselective N-atom transfer from chiral diaziridine **1** to α,β-unsaturated amides leading to chiral aziridines with good diastereo- and enantioselection (ee 96%, *trans* maj.). Diaziridine **1** was prepared (under Schmitz's conditions) [13b] with (1*R*)-1-phenylethylamine from cyclohexanecarboxaldehyde as a mixture of two diastereomers **1/1'** (7:3) separable after column chromatography [14]. The configurations of **1** and **1'** were assigned by analogy to the diastereoselectivity reported for the preparation of **2**, bearing two methyl substituents on the endocyclic carbon, and its diastereoisomer **2'** [15]. While **2** gradually isomerized to its diastereoisomer at room temperature, **1** was found to be stable and could be used as enantioselective nitrogen-atom transfer reagent (Scheme 5.1). The presence of the *cis*-methyl group in **2** probably facilitates the isomerization.

Chiral nonracemic diaziridines have also been prepared from imines bearing chiral substituents on nitrogen using an [(arenesulfonyl)oxy]carbamate as aminating agent (Scheme 5.2) [16].

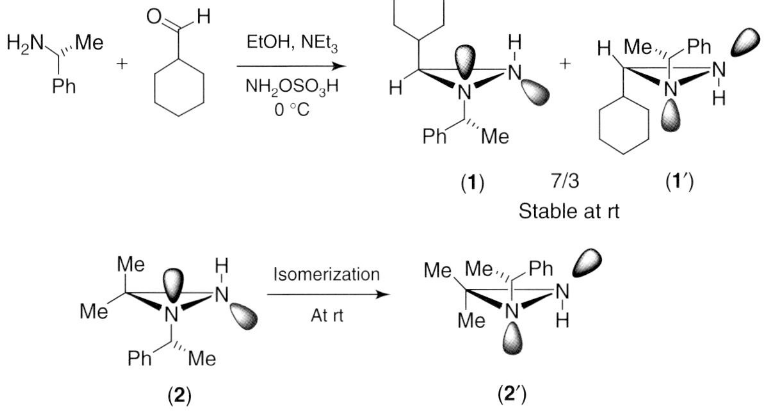

Scheme 5.1

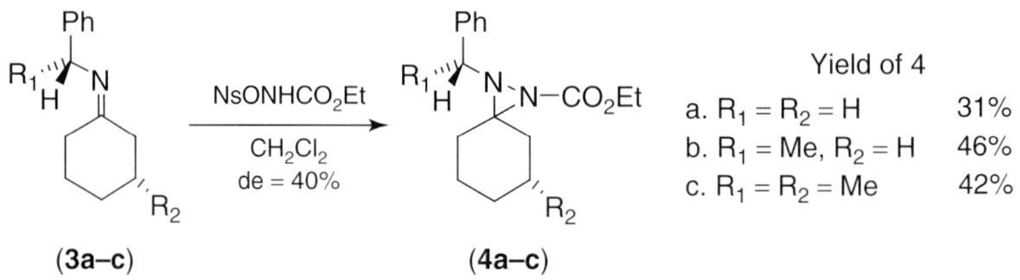

Scheme 5.2

	Yield of 4
a. $R_1 = R_2 = H$	31%
b. $R_1 = Me$, $R_2 = H$	46%
c. $R_1 = R_2 = Me$	42%

The yields were modest but the diastereoisomeric mixtures of **4b** and **4c** are easily separated by flash chromatography and spirodiaziridines **4b** and **4c** were obtained as pure diastereoisomers.

5.2.1.2 Substrate-Controlled Diastereoselective Diaziridination Using Chiral Enantiomerically Pure Ketones

Vasella et al. have prepared glycosylidene diaziridines **5** as precursors of the corresponding diazirines [17]. The starting ketone was converted to (glycosylidene)amino sulfonate **6** [18] and finally treated with ammonia under pressure (Scheme 5.3). These diaziridines are mixtures of *trans*-configurated diastereoisomers. The main (S,S)-configured isomer **5S** is stabilized by a weak intramolecular H-bond from *pseudoaxial* N—H to RO-C(2) [19].

This study suggests that the addition of the amine to lactone oxime sulfonate is kinetically controlled.

Vasella et al. showed later that the Schmitz conditions [13] could be applied directly on the corresponding ketone (validone) if a persilylated pyranoside nucleus was used [20]. The trimethylsilyl protecting group plays a crucial role for the formation of the diaziridine increasing the yields from 28% (for deprotected validone) [21] to 50%.

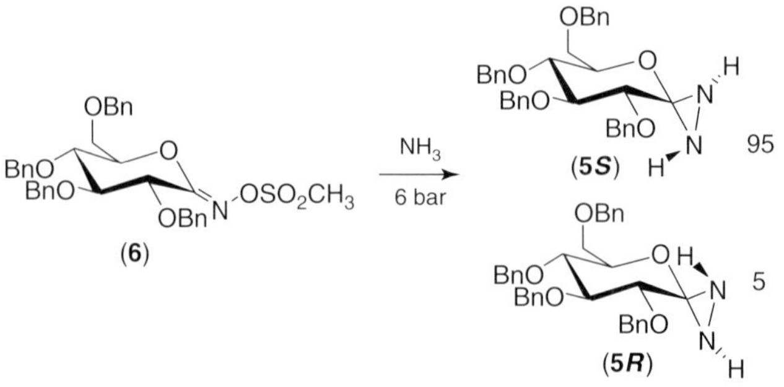

Scheme 5.3

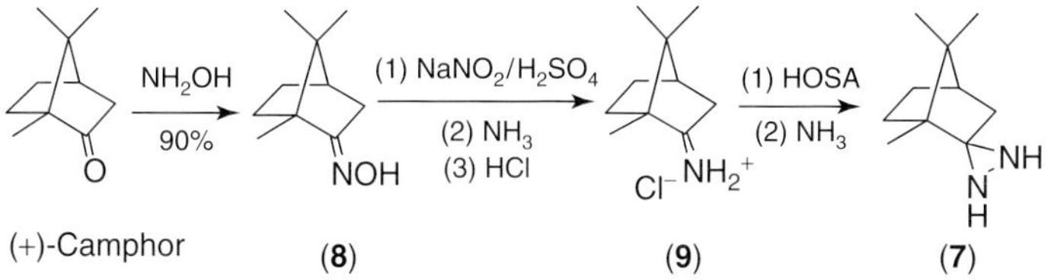

5.2 Three-Membered N-Heterocycles with Two Heteroatoms

Scheme 5.4

2-Hydrazicamphane (**7**), which is also a precursor of the corresponding diazirine, was synthesized in enantiomerically pure form starting from (+)-camphor (Scheme 5.4) [22]. (+)-Camphor was converted to the corresponding oxime **8**, which after a nitrosation/rearrangement sequence followed by aminolysis afforded the imine hydrochloride **9** in good yields. Treatment of imine **9** with HOSA and NH_3 delivered enantiomerically pure 2-hydrazicamphane **7**.

Schmitz's conditions [13b] ave been applied to 14-hydroxydihydromorphinones to prepare the corresponding diaziridines with interesting opioid activity [23].

5.2.2
Diazirines

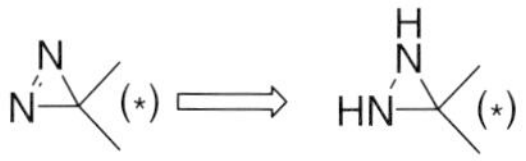

The recent emergence of diazirines as popular carbene precursors can be attributed to their relative stabilities with regard to acids, bases, and heat when compared with other sources [24]. The chiroptical properties of enantiomerically pure diazirines are of particular interest because the diazirine chromophore is very well suited for circular dichroism (CD) studies [25]. This was applied recently in studies dealing with the induced CD as a probe for structural analysis of supramolecular complexes [26]. Furthermore, the fact that diazirines are easily photoactivated by long-wave UV light makes them suitable for biochemical applications [27].

As diazirines are the dehydrogenation products of diaziridines, their synthesis is simple when the corresponding diaziridines are available. The preparation of 2-azicamphane **10**, which represents one of the very few examples of chiral enantiomerically pure diazirines, is performed by oxidation of 2-hydrazicamphane **7** (Scheme 5.5) [28].

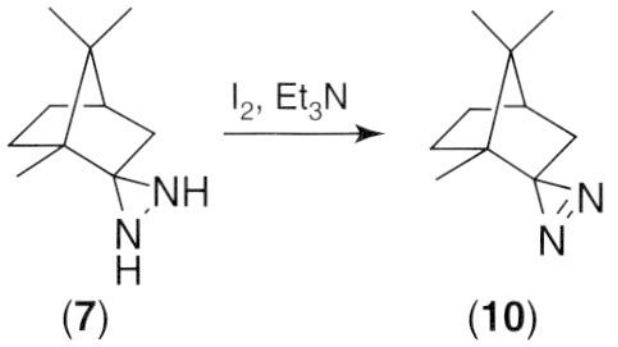

Scheme 5.5

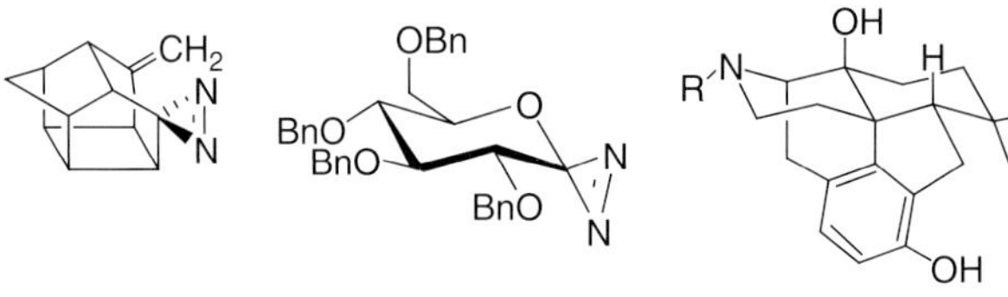

Fig. 5.1 Structures of diazirines **11**, **12** and **13**.

The chiral nonracemic diazirines **11** [29], **12** [17] and **13** [22] depicted in Figure 5.1 were also prepared by oxidation of the corresponding diaziridines.

5.2.3
Oxaziridines

The oxaziridines, three-membered rings containing nitrogen, oxygen, and carbon, have been of mechanistic and structural interest since their introduction in a classic paper by Emmons in 1957 [30]. This paper not only describes the preparation and structural proof of a new heterocycle but also contains the basis for most oxaziridine chemistry known so far, including rearrangement chemistry and decomposition of oxaziridines by low-valent metal salts [31]. During the following decade, oxaziridine chemistry remained confidential and this heterocycle became of interest for stereochemists in the 1970s because of the chirality of the nitrogen atom and the appreciable barrier to its inversion. This barrier to inversion was determined to be $105-130\,\text{kJ}\,\text{mol}^{-1}$ [32]. It turns out that the combination of placing a nitrogen in a three-membered ring (which destabilizes the 120° angle of the sp^2-like transition state involved in pyramidal inversion) with attaching the electron-withdrawing oxygen atom (which opposes the increased s-orbital character of nitrogen during inversion) makes simple oxaziridines the most interesting class of compounds containing a bona fide nitrogen stereogenic center [33]. Oxaziridines have also been shown to epimerize photochemically through a nitrone intermediate [34].

While the inversion barrier is considerable in N-alkyl oxaziridines [35] and N-halogenated oxaziridines [36], it is smaller when the N-substituent is capable of π-conjugation (or hyperconjugation). N-aryl as well as N-acyl oxaziridines both have inversion barriers around $90\,\text{kJ}\,\text{mol}^{-1}$ [37, 38]. N-sulfonyl and N-phosphinoyl oxaziridines also exhibit a lower barrier to inversion as a result of hyperconjugation present in the system [39].

Synthetic applications of chiral oxaziridines concern photochemical rearrangement reactions affording chiral amides or lactams with high levels of enantio- or regioelectivity [40], and asymmetric heteroatom transfer reactions. Oxaziridines can be used as both electrophilic aminating and oxygenating agents in their reactions with a wide variety of nucleophiles. In spite of this dual reactivity, the predominance of one process over another can be affected by varying the substitution pattern on nitrogen. In general, oxaziridines with small groups on nitrogen (H, Me) [41, 42] or aryl, acyl, carboxyamido [43], alkoxycarbonyl [44] groups act as aminating agents, whereas those with bulky [45] or strongly electron-withdrawing groups on nitrogen such as sulfonyl, sulfamyl, or phosphinoyl [46] groups preferentially transfer the oxygen atom [47]. Perfluoroalkyl oxaziridines [48] and oxaziridinium salts [49, 50] are highly reactive electrophilic oxygen-atom transfer reagents. The epoxidation of double bonds mediated by oxaziridinium salts are generally performed *in situ* via an organocatalytic process using the corresponding chiral iminium salts in substoichiometric amounts and Oxone as oxygen source [31, 51].

Enantioselective oxygen-atom transfer reactions onto nucleophiles, such as olefins, enolates or sulfides [52] represent the essential part of the applications of chiral oxaziridines in asymmetric synthesis. Chiral oxaziridines can be produced via oxidation of chiral nonracemic imines, oxidation of achiral imines with a chiral nonracemic peracid, via separation of diastereoisomeric oxaziridines produced from achiral peracid oxidation using chromatography [53] or enzymatic resolution [54]. A few resolutions of racemic oxaziridines using complexation or inclusion with chiral host complexes have also been reported [55]. Several nonoxidative methods have been also developed but are less general: photocyclization of nitrones carrying a chiral substituent and photocyclization of achiral nitrones in an optically active environment. Finally, derivatization at the nitrogen atom of chiral nonracemic stable N–H oxaziridines can therefore provide a useful alternative method to peracidic oxidation of chiral nonracemic imines, for the preparation of N-functionalized oxaziridines.

The electrophilic amination of ketones [13b] and the double 1,4 conjugate addition of hydroxamic acids to propiolates [56] to give racemic oxaziridines have been described but no asymmetric versions of these methods have been developed so far.

5.2.3.1 Chiral Peracidic Oxidation of Achiral Imines

The first preparation of an oxaziridine, and still the most widespread method today, was the oxidation of an imine with a peracid. Rapidly, attempts to obtain enantiomerically enriched oxaziridines using chiral nonracemic peracids have been performed. The most commonly used oxidizing agent is (S)-(+)-peroxycamphoric acid (monoperoxycamphoric acid (MPCA) 14) [57]. In most instances, the degree of asymmetric induction afforded by MPCA is rather low [35, 58]. Nevertheless, the use of crystalline MPCA at −78 °C in a carefully selected solvent system can give optical purities of up to 60% for oxaziridine 15 [59]. (Scheme 5.6)

Modest enantioselectivities (5–33% optical yield) have been observed when mixing different imines with *meta*-chloroperbenzoic acid (*m*-CPBA) in the presence of optically active carbinols [60]. The better selectivities were observed with chiral acyclic or aromatic trifluoromethyl-carbinols and are correlated with the reaction temperature and the relative amount of chiral solvent [61]. As one has to consider

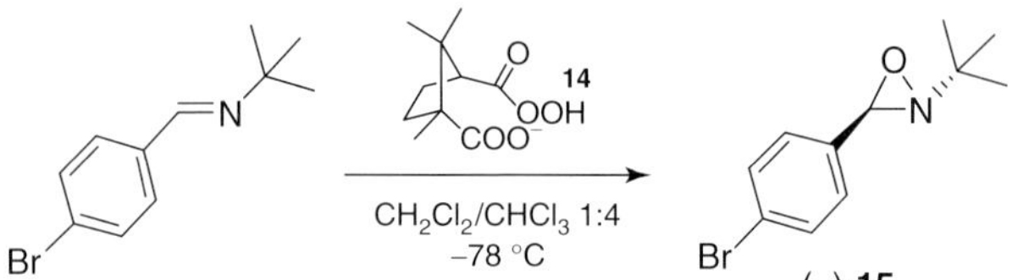

Scheme 5.6

that in the presence of chiral alcohols the peracid can be hydrogen-bonded to the solvent to produce a chiral peracid form, the authors proposed a chiral carbinol-imine solvation via hydrogen bonding between the imine nitrogen and the chiral protic solvent.

5.2.3.2 Achiral Peracidic Oxidation of Chiral Nonracemic Imines

Diastereoselection of substrate-controlled peracidic oxidations of chiral nonracemic imines is generally excellent. The mechanism of oxidation of imines has been studied and found to proceed through a two-step sequence, through cleavage of π bonding followed by the elimination of one molecule of carboxylic acid, rather than a concerted oxygen transfer [62–64]. This method is the most widely used for preparation of chiral oxaziridines of high optical purities, starting from enantiomerically pure imines. In the same manner as that in diaziridines, the stereogenic center of the imine can come either from the ketone or the amine part. In the latter case, α–methylbenzylamine is widely used because of its low cost and ease of removal under dissolving metal conditions. This methodology will be illustrated by the most representative examples for each kind of N-substituted oxaziridines.

N-Unsubstituted Oxaziridines

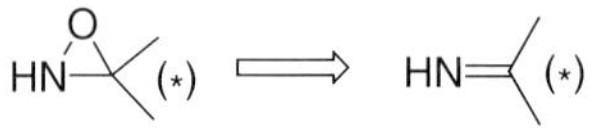

Although the first N–H oxaziridine was reported in the 1960s [65], because of their general instability, only a few N–H oxaziridines have been prepared and utilized for their ability to transfer an amino group to various nucleophiles [41, 66]. The preparation of the first stable enantiomerically pure chiral N–H oxaziridines has been accomplished only very recently by Page et al. [67]. These N–H oxaziridines **17** and **18**, derived from (1R)-(+)-camphor and (1R)-(+)-fenchone were remarkably stable and used as asymmetric electrophilic nitrogen-atom transfer reagents on various ester- and nitrile-containing carbon nucleophiles [24]. Classical treatment of ketones with HOSA or a precursor of a chloramine to prepare N–H oxaziridines [13] were totally ineffective on camphor or fenchone. Thus, Page et al. decided to oxidize the primary (N–H) imines **19** and **20**, which were prepared from the corresponding oximes **21** and **22** via the nitrimines **23** and **24**, with m-CPBA (Scheme 5.7). The N–H oxaziridines **17** and **18** were obtained with good overall

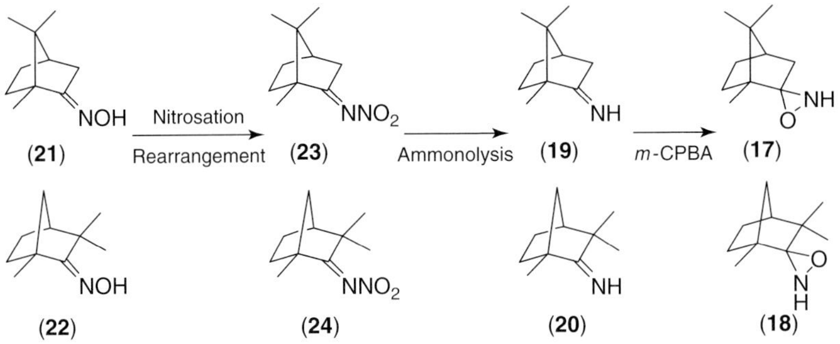

5.2 Three-Membered N-Heterocycles with Two Heteroatoms

Scheme 5.7

yields in enantiomerically pure form as diastereoisomeric mixture (6/4) at the pyramidal nitrogen atom. The diastereoisomers observed arise from the two configurations at the nitrogen atom resulting from *endo* attack of the oxidant on the imine **19** and *exo* attack of the oxidant on the imine **20**.

N-Alkyl Oxaziridines

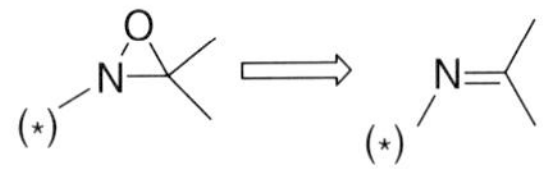

The pioneering work concerning diastereoselective peracidic oxidation of imines has been performed on Schiff bases [68] bearing a chiral substituent on nitrogen, namely, (S) or (R)-α-phenylethylamine. In the case of acyclic imines, the configuration of the C–N double bond influences the stereochemical outcome of the reaction. Using imines such as **26** derived from symmetrical ketones, a mixture of N-epimers **25a,b** is obtained (Scheme 5.8) [69].

When aldimines or imines derived from unsymmetrical ketones are employed, four oxaziridines are obtained [70]. A pro-S attack of the peracid is proposed to account for the observed absolute configurations of oxaziridines **28** (Scheme 5.9).

A stereospecific oxidation of chiral *trans*-aldimines leading to the corresponding (E)-oxaziridines has been described when a nitrile–hydrogen peroxide system was used as oxygen source instead of *m*-CPBA [71].

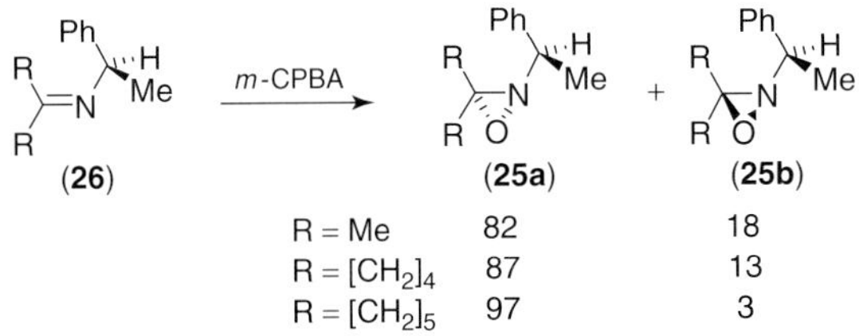

R	(25a)	(25b)
R = Me	82	18
R = [CH$_2$]$_4$	87	13
R = [CH$_2$]$_5$	97	3

Scheme 5.8

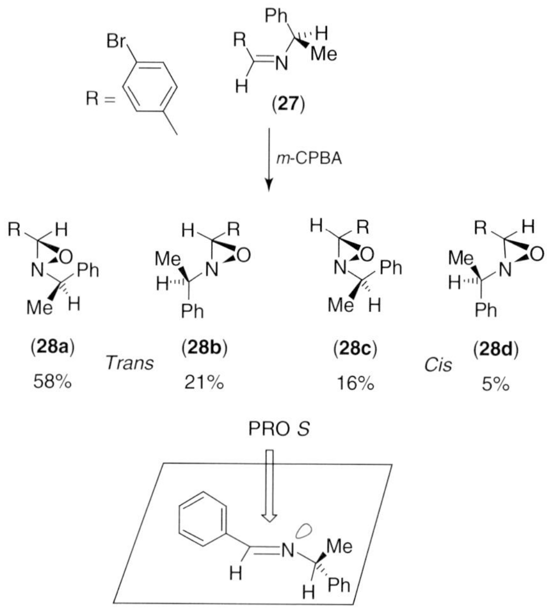

Scheme 5.9

Chiral nonracemic spirocyclic oxaziridines are important compounds in stereoselective synthesis as a result of their possible stereospecific photorearrangement leading to chiral lactams [72]. Stereospecific cyclic expansion occurs generally in which the group *syn* to the chiral N-substituent of the oxaziridine undergoes preferential migration to nitrogen upon photolysis. By use of chiral nitrogen substituents (typically α-methylbenzyl) on symmetrical cyclic imines, it is possible to control the absolute stereochemistry of the oxaziridine nitrogen atom [73]. Oxaziridines derived from six-membered cyclic imines tend to form via equatorial attack [74] and are also subject to control by a chiral group on nitrogen substituent (Scheme 5.10) [22].

Condensation of a 4-substituted cyclohexanone generates a mixture of diastereomeric imines **29a–b**. Addition of the peracid to either imine can afford products of equatorial attack or axial attack, with the former being favored. Finally, N–O bond closure occurs such that the newly generated nitrogen stereogenic center emerges preferentially with the *unlike* [75] relative configuration when α-methylbenzylamine was used for imine formation. The selectivity toward the equatorial/unlike product increases as the bulkiness of the peracid and of the substituent of the cyclic imine become more important [76]. Imines prepared from cyclohexanones containing bulky substituents, presumably have a superior ability to prejudice the six-membered ring into one chairlike conformation over the other.

If a stereogenic center is located on the cyclohexane skeleton such as in the case of (R)-(+)-3-methylcyclohexanone **31**, the overall stereochemical control results from a matched effect between this stereogenic center and the benzylic stereocenter on the nitrogen [77]. (Scheme 5.11)

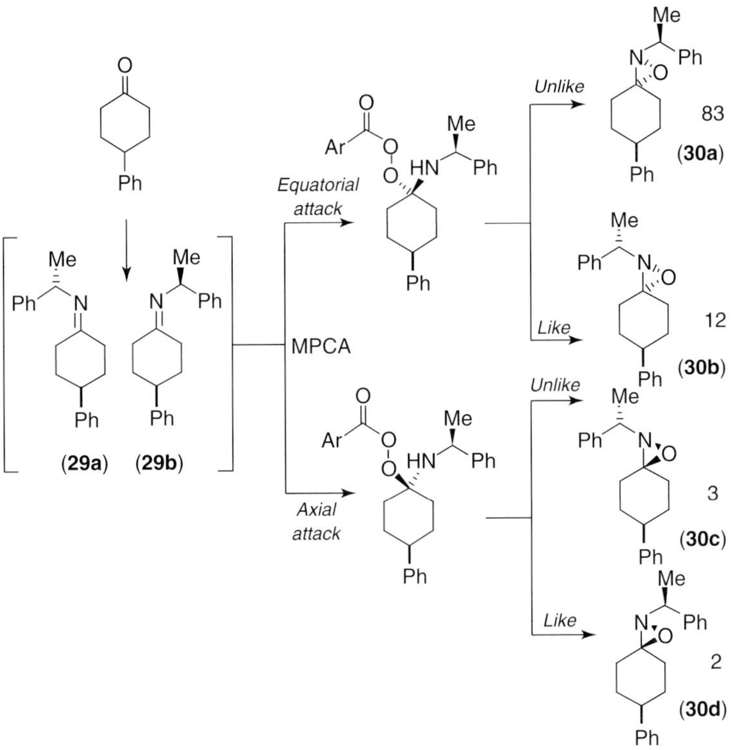

5.2 Three-Membered N-Heterocycles with Two Heteroatoms | 199

Scheme 5.10

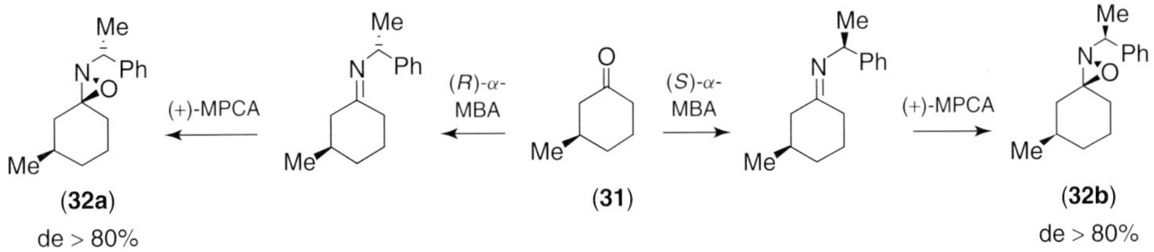

Scheme 5.11

Condensation of (R)-α-methylbenzylamine and (S)-α-methylbenzylamine on (R)-(+)-3-methylcyclohexanone **31** followed by peracidic oxidation using (+)-MPCA delivers the oxaziridines **32a** and **32b** with an excellent diastereoselectivity. In each case, attack of the oxidizing agent takes place from an equatorial direction, leading to the *cis* relationship between the C-3 methyl group and the oxaziridine oxygen. In addition, oxidation affords products in which the benzylic stereogenic center and the oxaziridine nitrogen have *unlike* [75] relative stereochemistry [78]. (+)-MPCA is used only for its steric hindrance as any given stereocenter of this peracid is too far away from the forming bonds to have an effect on the face selectivity of the oxygen addition. Interestingly, photorearreangment of oxaziridines **32a–b** affords the corresponding regioisomeric lactams stereospecifically.

The peracid oxidation of imines derived from 2-alkyl-substituted cyclohexanones gives more complex mixtures because of partial epimerization at C-2 position during imine formation and oxidation of the enamine tautomer [79]. The stereoselection of the reaction drops dramatically leading to a number of very minor oxaziridines accompanying one or two major isomers in the reaction mixture [72].

The effect of polar substituents at the C-2 position on the stereoselectivity of oxaziridine formation [80] has also been recently studied on the same substrates [81]. In contrast, the general equatorial attack by *m*-CPBA observed onto 4- and 3-alkylsubstituted cyclohexylimines or 2-methoxy-substituted cyclohexylimines can be changed into axial attack for 2-hydroxylated cyclohexylimines in favorable circumstances. A *syn* hydroxyl-directed approach of the peracid in the same manner as *syn*-directed epoxidation of allylic alcohols is proposed to take into account this unusual stereoselectivity.

N-Alkyl Oxaziridines as Precursors of *N*-Quaternarized Oxaziridines

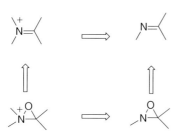

N-alkyl oxaziridines are sluggish for epoxidation of alkenes and even for sulfoxidations but quaternarization of their nitrogen atom significantly enhances the oxygen transfer reactivity [49, 50]. N-Alkylation leads to oxaziridinium salts that are among the most efficient agents for oxygen transfer onto nucleophilic substrates. It has also been demonstrated that oxygen-atom transfer from *N*-alkyl oxaziridines could be promoted by Brønsted or Lewis acids [82, 83].

Since the first report of steroidal oxaziridinium salt **33** by Lusinchi and coworkers in 1976 [84] and the establishment that such N-quaternarized oxaziridines are powerful electrophilic reagents for oxyfunctionalization of organic substrates [49, 50], their potential as asymmetric epoxidating agents has been intensively pursued (Figure 5.2) [85].

Chiral nonracemic oxaziridinium salts are generally generated *in situ* by peracidic oxidation of the corresponding enantiomerically pure iminium salts which are used for epoxidations in substoichiometric amounts in a catalytic cycle [50, 51, 85].

Only few optically active oxaziridinium salts have been isolated so far and most of them result from N-methylation of the corresponding enantiomerically pure oxaziridines. This alkylation reaction does not affect the configuration of the three-membered ring. The peracidic oxidation of iminium salts also leads to optically active oxaziridinium salts, but generally their isolation via this pathway is

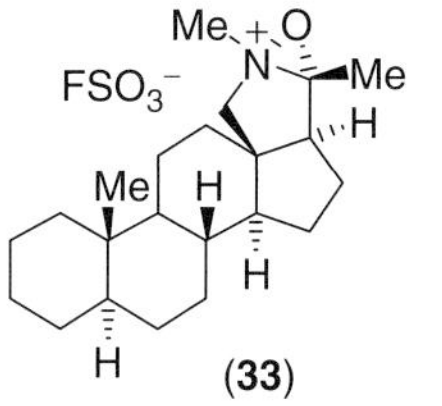

5.2 Three-Membered N-Heterocycles with Two Heteroatoms | 201

Fig. 5.2 Structure of the first isolated oxaziridinium salt.

more difficult. We only focus on the preparation of oxaziridinium salts that have been isolated or clearly detected in solution.

Steroid-Based N-alkyl Oxaziridines Enantiomerically pure steroid-based oxaziridines **34** [86] and **37** [87] have been obtained by a stereoselective peracid oxidation of the corresponding imines **36** and **38**. A total selectivity of the peracid attack for the α face of **36** and the β face of **38** was observed (Scheme 5.12). Peracidic oxidation of

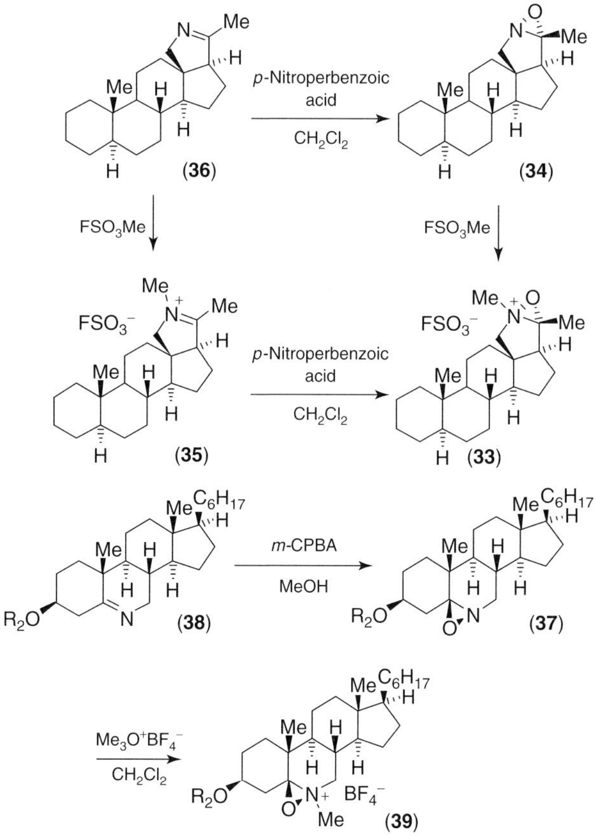

Scheme 5.12

iminium salt **35** afforded oxaziridinium salt **33** with the same complete α-selectivity [84]. Oxaziridine **34** can also be prepared by treatment of the corresponding nitrone by *p*-tolylsulfonyl chloride in basic medium [88]. Oxaziridinium salts **33** and **39** result from methylation of the corresponding enantiomerically pure oxaziridines **34** and **37**.

Dihydroisoquinoline-based N-alkyl oxaziridines In 1993, the enantiomerically pure oxaziridinium salt **40** was isolated by crystallization and fully characterized including X-ray diffraction [89, 90]. It was conveniently prepared from benzaldehyde and (1*S*, 2*R*)-(+)-norephedrine via enantiomerically pure dihydroisoquinoline **41** following two possible stereoselective pathways already described for oxaziridinium **33** (Scheme 5.13).

While the peracidic oxidation of the iminium **42** was fully stereoselective at room temperature, that of the parent imine **41** had to be performed at low temperature ($-45\,°C$) [82b, 91]. At room temperature, the peracidic oxidation of the dihydroisoquinoline **41** was not stereoselective and even not chemoselective as the corresponding nitrone was detected as side product. The *cis* relative configuration observed between the oxygen of the oxaziridine ring and the methyl substituent of oxaziridine **43** and oxaziridinium **40** can result from minimization of torsional strain during the formation of the three-membered ring [92].

Catalytic asymmetric epoxidations mediated by the dihydroisoquinolinium salt **44** derived from a chiral amine delivered good level of enantioselectivity for various olefins [93]. Nonaqueous conditions developed by Page *et al.* have allowed NMR spectroscopy to be carried out on the reaction mixture and show the stereoselective formation of the corresponding oxaziridinium salt **45** (Scheme 5.14) [94].

Binaphtyl-based N-alkyl Oxaziridines Recently, enantiomerically pure oxaziridines **46** and **47** have been prepared as acid-promoted asymmetric sulfoxidation reagents [95]. They resulted from a stereospecific oxidation of the corresponding enantiomerically pure imine by *m*-CPBA in methanol, leading to single enantiomers (Figure 5.3). The formation of the S_C, R_N isomer can be interpreted as a result of a steric

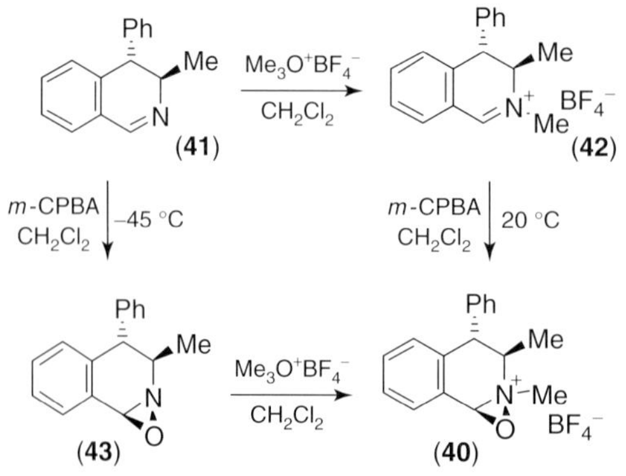

Scheme 5.13

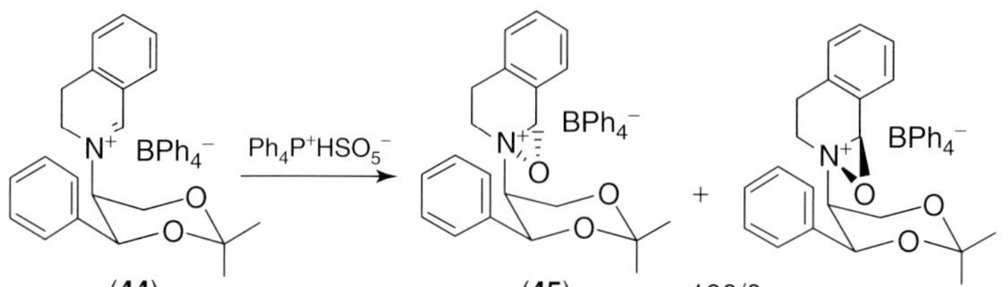

Scheme 5.14

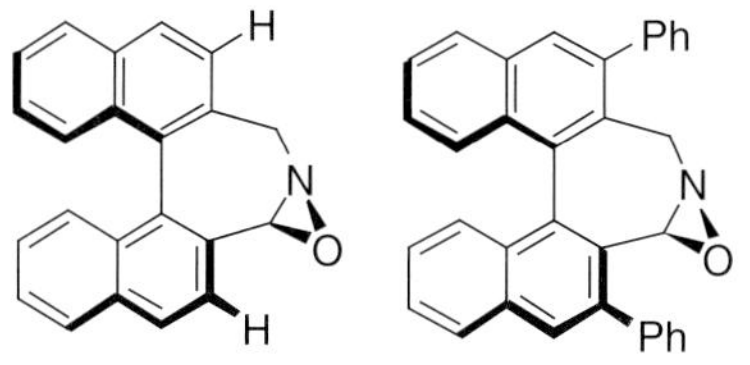

Fig. 5.3 Structures of oxaziridines **46** and **47**.

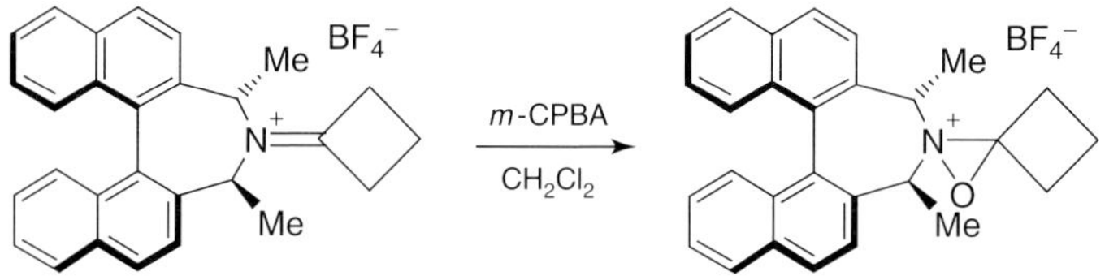

Scheme 5.15

repulsion between the arene hydrogen and an oxygen of the perester of the minor intermediate [92].

Binaphtyl-based iminium salt **48** with C_2-symmetry has been transformed by peracidic oxidation into the enantiomerically pure oxaziridinium salt **49** that has been clearly characterized in solution by NMR spectroscopy (Scheme 5.15). A dichloromethane solution of the oxaziridinium salt **49** was used at low temperature to epoxidize various olefins with enantioselectivities up to 75% [96].

N-Acyl and N-Alkoxycarbonyl Oxaziridines

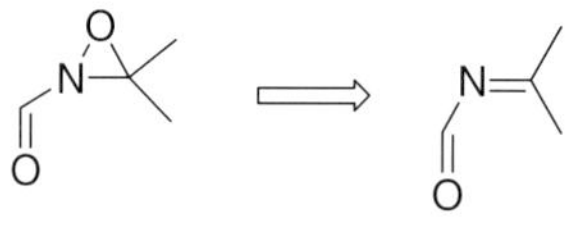

N-acyl and mostly N-alkoxycarbonyl oxaziridines have found extensive use as a source of electrophilic nitrogen [43, 97, 98]. An asymmetric variant of this methodology would be useful as the development of efficient chiral nonracemic

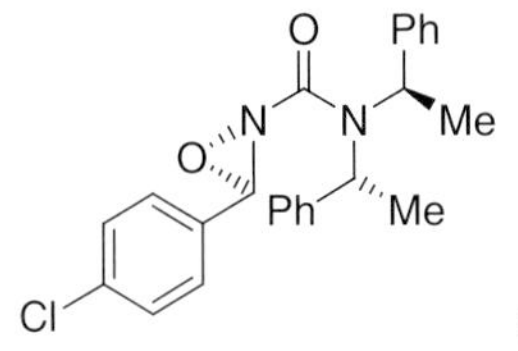

Fig. 5.4 Examples of chiral N-substituted oxaziridines.

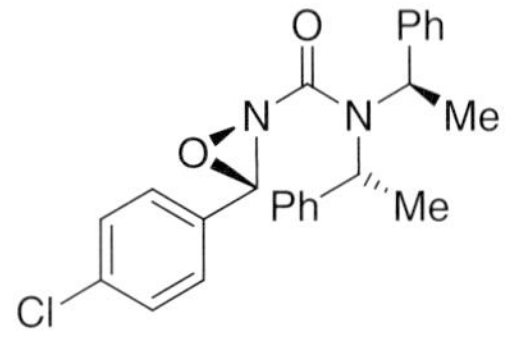

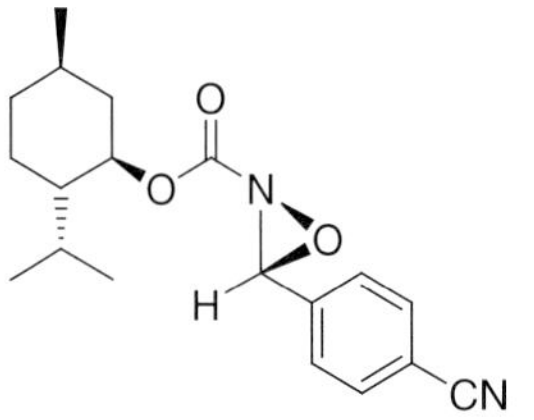

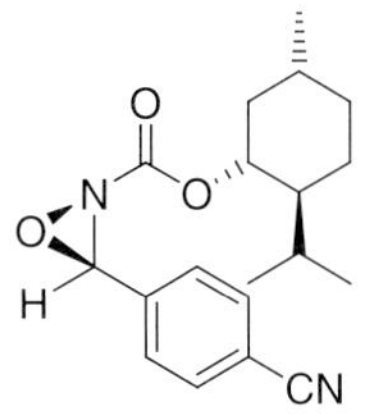

reagents for asymmetric transfer of electrophilic nitrogen to organic substrates is an important contemporary goal.

Only two chiral N-protected oxaziridines **50** [99] and **51** [100] have been prepared so far, both by the same group. The stereogenic center of the starting imines is located on the amine part for all of them (Figure 5.4).

N-Carboxamido oxaziridines **50** were prepared by biphasic basic m-CPBA oxidation of the corresponding N-carboxamidoimines as a mixture of two diastereoisomers which are *trans* isomers with opposite configuration at the ring carbon. They have been used as the first asymmetric sulfimidation reagents [82] of thioethers and aziridination reagents of olefins [101] giving interesting enantiomeric excesses.

N-Alkoxycarbonyl oxaziridines **51** were prepared using the m-CPBA/BuLi system [43] as oxidant. The compounds were single diastereoisomers at the ring carbon disclosing a high diastereoselectivity with respect to facial attack on the imine carbon. Asymmetric lithium enolate amination was performed with oxaziridines **51** with modest enantiomeric excesses [83].

N-Sulfonyl Oxaziridines

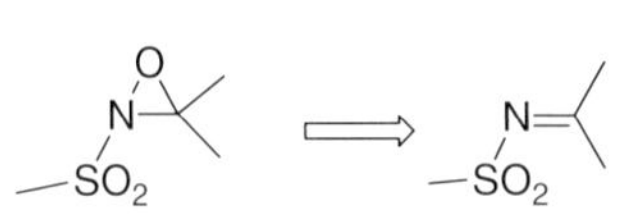

The development of reagents for the reagent-controlled asymmetric oxygen-atom transfer onto prochiral nucleophiles with high enantioselectivity is an important synthetic goal. To date, N-sulfonyl oxaziridines represent the most versatile and general active oxygen compounds able to oxidize many prochiral nucleophiles with high enantioselection [31, 47].

5.2 Three-Membered N-Heterocycles with Two Heteroatoms

N-Sulfonyl oxaziridines **52** were the first examples of oxaziridines to have a substituent other than carbon or hydrogen attached to nitrogen. These stable oxaziridines are readily prepared by biphasic-buffered oxidation of chiral sulfonimines **53** with m-CPBA or potassium peroxymonosulfate [102]. The latter resulted from condensation of enantiomerically pure sulfonamides with the diethylacetal of an aromatic aldehyde. Since oxidation of sulfonimines give only (E)-2-sulfonyl oxaziridines [103], just two oxaziridine diastereoisomers having the S,S and R,R configurations at the three-membered ring are obtained upon oxidation of **53**. The best ratio of **52** (S,S/R,R) obtained reached 65/35 and was improved by fractional crystallizations to give optical yields (30–100%) depending on the nature of the aromatic substituent (Scheme 5.16) [104].

In order to increase the efficiencies of chiral 2-sulfonyl oxaziridines **52**, Davis et al. decided to vary the groups attached to the oxaziridine N and C atoms in a systematic manner. They found that chiral 2-sulfamyl oxaziridines **54** could be prepared by oxidation of the corresponding chiral 2-sulfamylimines, in enantiomerically pure form after separation of diastereoisomers by chromatography [105]. These oxaziridines were particularly efficient as enantioselective epoxidation reagents (ee up to 80%) when the aromatic substituent were perfluorinated (Figure 5.5).

3-Substituted 1,2-benzisothiazole-1,1-dioxide **55a-b** has been also developed as new N-sulfonyl oxaziridine [106]. It was obtained by biphasic oxidation of the corresponding imines in a 85/15 mixture of diastereoisomers. Several crystallizations from ethanol gave pure (+)-**55a** and (+)-**55b**, which were less effective than N-sulfamyl oxaziridines **54** in their asymmetric oxidations (Figure 5.6).

The camphor-derived oxaziridines **56–62** are of particular interest for two reasons: first, they are derived from materials readily available from the chiral pool with a rigid structure, thus reducing the problem of separation of isomers during the synthesis as observed for oxaziridines **52**, **54**, and **55**. Secondly, these oxaziridines

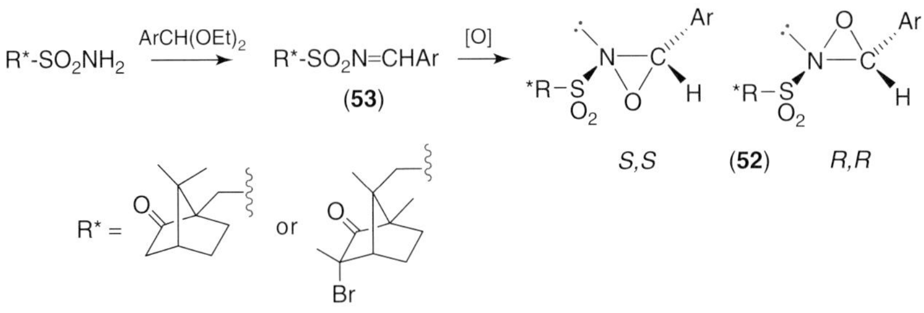

Scheme 5.16

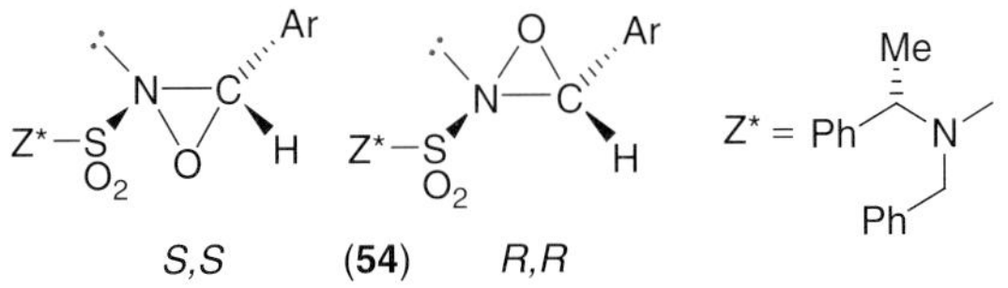

Fig. 5.5 Structure of N-sulfamyloxaziridines **54**.

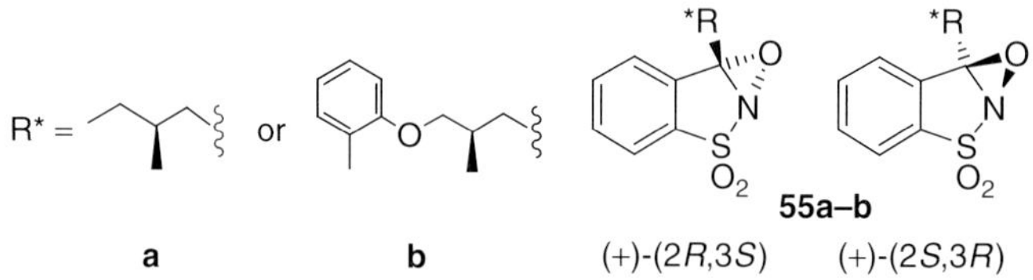

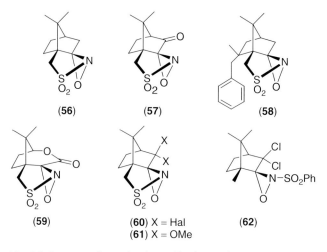

Fig. 5.6 Structure of N-sulfonyloxaziridines **55**.

Fig. 5.7 Structure of camphoryl N-sulfonyloxaziridines.

are among the most efficient reagents, specially **60** and **62**, for enantioselective sulfoxidations of thioethers and enantioselective α-hydroxylation of carbonyl compounds. Not only is the product stereochemistry predictable, but the enantiomeric excesses often exceed 90% (Figure 5.7) [31, 47], [107, 108].

An important advantage of the camphorylsulfonyl oxaziridines is that upon oxidation of the corresponding imines, they are obtained as single isomers because the exo face of the C–N double bond is blocked by the methyl group of the methylene bridge.

Some *exo*-camphorylsulfonyl oxaziridines have been also described. For example, *exo*-oxaziridne **63** has been prepared by m-CPBA oxidation of camphor imines **64** [109]. This unexpected result is probably due to the conformation of the imine which prevents attack of the peracid from the *endo* direction (Scheme 5.17).

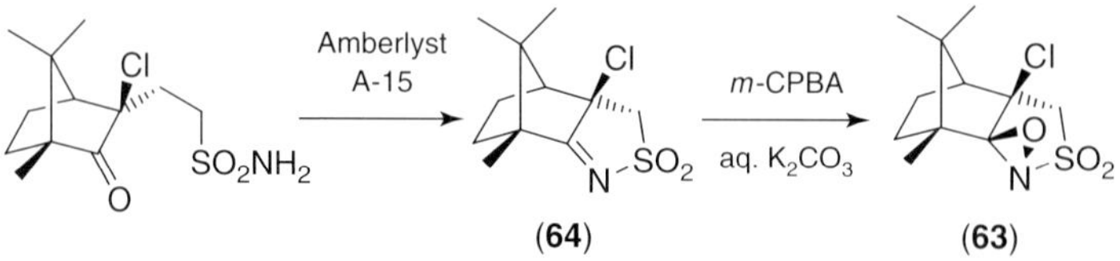

Scheme 5.17

N-Phosphinoyl Oxaziridines

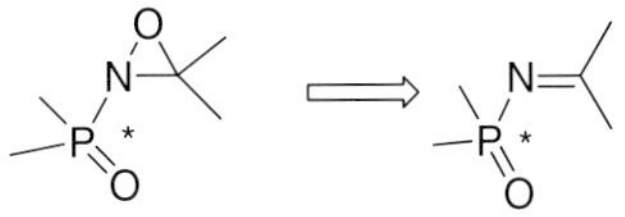

N-Phosphinoyl oxaziridines, like their better established N-sulfonyl analogs are generally prepared by peracidic oxidation of the corresponding imines [110]. Optically active N-phosphinoyl oxaziridines **65** and **66** containing a chiral phosphorus center have been prepared and evaluated as enantioselective oxygen-atom transfer reagents onto thioethers (Scheme 5.18) [111].

Oxidation of N-Phosphinoylimine **67** (75% ee) prepared from scalemic R-P-mesityl-P-phenylphosphinic amide (75% ee) with anhydrous m-CPBA/potassium fluoride complex [110] afforded a mixture of diastereoisomeric oxaziridines **65** and **66** (2.6:1) with total retention of the configuration at the phosphorus atom (75% ee).

5.2.3.3 Photocyclization of Nitrones

Photocyclization of Achiral Nitrones in Optically Active Environment Optically active oxaziridines can be obtained with generally modest enantioselectivity (optical purity 35%) by photoisomerization of achiral aldo- and keto-nitrones in the presence of a chiral solvent [112]. The best selectivities were obtained with (+)-(S) or (−)-(R)-2,2,2-trifluoro-1-phenylethanol [113]. More interesting is the irradiation of complexes of nitrones **68** and **69** in a crystalline inclusion with optically active (−)-1,6-di(o-chlorophenyl)-1,6-diphenylhexa-2,4-diyne-1,6-diol that affords the corresponding oxaziridines with high enantioselectivity (>95% ee) (Figure 5.8) [114].

Photocyclization of Chiral Nitrones Photorearrangment of chiral nitrones to optically active oxaziridines has been found to occur in achiral solvents with optimum diastereoisomeric excess of 20% [113]. Because of low stereoselectivities observed, this method remained confidential.

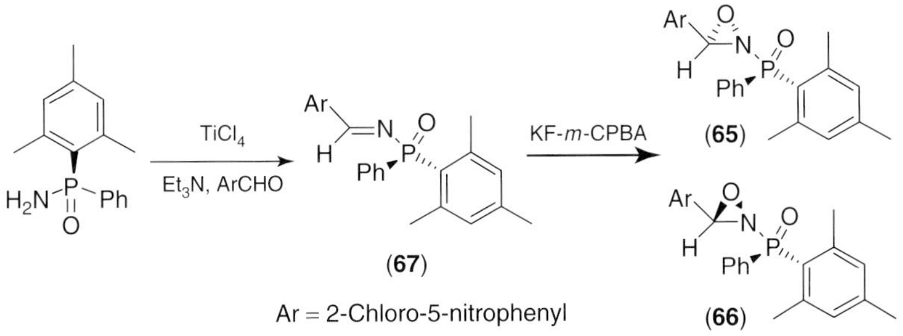

Scheme 5.18

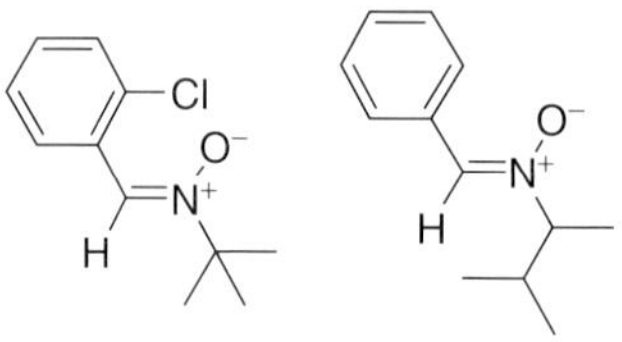

(68)   (69)

Fig. 5.8 Structure of nitrones **68** and **69**.

5.3
Four-Membered N-Heterocycles with Two Heteroatoms

Compared to the three-membered heterocycles mentioned above, their four-membered homologs give rise to a larger variety of structures as for each combination of heteroelements there are two possible constitutional isomers (i.e. the 1,2- and 1,3-di-X-cyclobutanes). In addition, a wide variety of synthetic methods, centered mostly around [2+2]-cycloadditions and similar processes have been developed for their synthesis, and have been comprehensively reviewed [1]. As previously, we only cover recent methods that stereoselectively access enantiomerically enriched heterocycles.

Despite their structural diversity, the four-membered N-heterocycles containing two (or more) heteroatoms have received much less attention than their smaller-ring congeners as far as their use as reagents or biologically active substrates is concerned. There are, however, a few notable exceptions. There is, for example, a fundamental interest in the electronic structure of some of the unsaturated four-membered ring structures in order to elucidate to what extent the presence of the heteroatoms renders these systems aromatic or antiaromatic [115, 116]. On the other hand, 1,2-diazetidine-3-ones, 1,3-diazetidine-2-ones (both often referred to as aza-β-lactams), 1,2-thiazetidine-1,1-dioxides (β-sultams) and azaphosphetidines (β-phospholactams) have received more attention from the synthetic community as analogs of β-lactams and thus potential antibiotics [117]. Not surprisingly, more general strategies for the synthesis of the latter structures have been sought, but their asymmetric synthesis remains underdeveloped.

5.3.1
Diazetidines

Diazetidines, the four-membered N-heterocycles, containing a second nitrogen atom comprise two isomeric types of structures, namely, 1,2-diazetidines **70** and 1,3-diazetidines **71** (Figure 5.9).

(70)   (71)

Fig. 5.9 General structures of diazetidine isomers.

Despite the fact that 1,2-diazetidines have been described as early as 1931 [118] and that the most general method for the construction of the more prominent aza-β-lactams, namely, the reaction of azostilbenes with ketenes, dates back to 1941 [119], reports on chiral nonracemic diazetidines are scarce. Even the more recent and general strategies for the construction of these β-lactam analogs have not been developed to deliver enantiomerically enriched products [120].

The first report concerning the synthesis of a scalemic 1,2-diazetidine-3-one was made in the context of the synthesis of dipeptides of L-proline with unnatural amino acids [121]. In an Ugi-type multicomponent reaction, N-amino-L-proline was reacted with an aldehyde and cyclohexylisonitrile to afford, presumably via the seven-membered intermediates 72, the enantiomerically enriched diazetidines 73 in modest optical purities ranging from 22 to 35%. The latter were converted to the corresponding dipeptides by reductive cleavage of the N–N bond (Scheme 5.19).

More recently, a diastereoselective addition of hydrazine to the uridine-derived phenylselenones 74 was described as part of the synthesis effort toward nucleoside analogs [122]. 1,2-Diazetidines 75 were obtained in good yield as single diastereoisomers (Scheme 5.20).

A simple procedure for the synthesis of chiral nonracemic 3-substituted 1,2-diazetidines has been published recently by Ma et al. [123]. The authors describe an approach based on the proline-catalyzed α-hydrazination of aldehydes followed by aldehyde reduction and cyclization under Mitsunobu conditions or via the corresponding mesylate. Although most of their study was conducted with racemic proline, as a proof of principle 3-phenylpropanal 76 was reacted in the presence of (R)-proline and reduced *in situ* to afford 2-hydrazinoalcohol 77, which was then cyclized to the diazetidine 78 in a one-pot procedure using methanesulfonyl

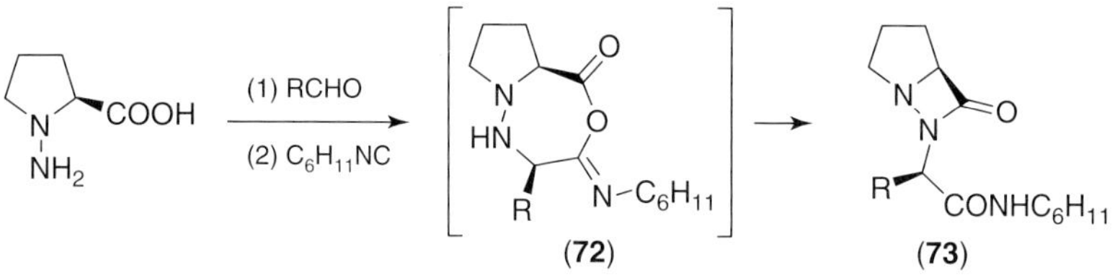

Scheme 5.19

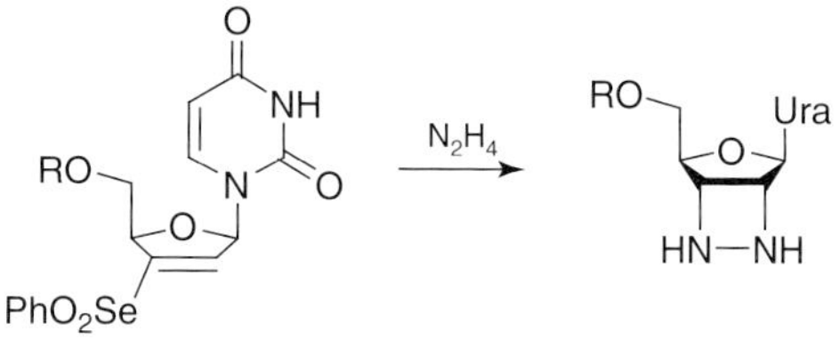

74a: R = MMTr
74b: R = H

75a: R = MMTr, 60%
75b: R = H, 62%

Scheme 5.20

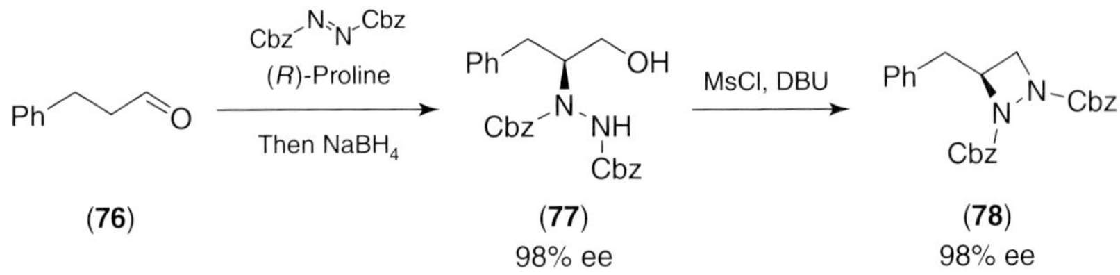

210 | 5 Asymmetric Synthesis of Three- and Four-Membered Ring Heterocycles

Scheme 5.21

chloride and DBU (Scheme 5.21). Both **77** and **78** were obtained with an excellent enantiomeric excess of 98%.

While these few methods are available for the synthesis of 1,2-diazetidines, their 1,3-isomers are even less studied. In particular, to our knowledge, there are no general methods described for the stereoselective synthesis of enantiomerically enriched 1,3-diazetidines. There are two reports on the synthesis of amino acid-derived diazetidinones by the reactions of tryptophan derivatives with alkyl isocyanates [124]; however, their structural assignments have later been questioned [125].

5.3.2
Oxazetidines

Oxazetidines are the four-membered heterocycles containing one nitrogen and one oxygen atom. As for their diazetidine counterparts, two possible constitutional isomers, i.e. the 1,2- **79** and 1,3-diazetidines **80**, are possible (Figure 5.10).

While the chemistry of these compounds has been comprehensively reviewed on several occasions [1], there are only scattered reports on their asymmetric synthesis – a characteristic they share with their diazetidine analogs (see Section 5.3.1). One might argue that the limited availability of stereoselective synthetic methods is due to the poor stability of many oxazetidines. However, the scarcity of target molecules containing the oxazetidine motif can also be regarded as a reason for the apparent lack of interest from the synthetic community.

Symptomatically, the isolation and first structural assignment of the halipeptins [126] – cyclic depsipeptides that were thought to contain a 1,2-oxazetidine moiety – immediately triggered studies toward the synthesis of this unusual fragment [127]. However, these studies only confirmed the structural revision that had been published in the meantime and supported the hypothesis that halipeptin contains a thiazole ring instead [128], an assumption that was later ascertained by total synthesis [129].

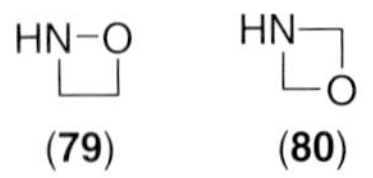

Fig. 5.10 General structures of oxazetidine isomers.

5.3 Four-Membered N-Heterocycles with Two Heteroatoms

Recently, two methods for the synthesis of enantiomerically enriched 1,2-oxazetidines have been reported by Florio et al. Their interest in these structures arose when, during the course of studies aimed toward the synthesis of α,β-unsaturated oxazolines, they discovered that 1,2-oxazetidines were formed during the reaction of α-lithiated oxazolines with nitrones [130]. Careful optimization of the reaction conditions allowed the highly diastereoselective synthesis of 1,2-oxazetidines **81a,b** from readily accessible α-chloro-oxazoline **82** via lithiation and addition to nitrones **83a,b** (Scheme 5.22) [131]. The reaction was found to be limited to the use of aromatic nitrones and enantiomerically pure oxazetidines **84a,c,d** and *ent-***84a,c** could be obtained from the valinol-derived α-chloro-oxazolines **85** and *ent-***85**.

The same authors recently reported a similar synthesis of enantiomerically enriched oxazetidines on the basis of the diastereoselective addition of lithiated aryloxiranes to nitrones [132]. Having established the excellent diastereoselection of this process, they demonstrated its applicability to the synthesis of optically enriched substrates. Indeed, optically active oxiranyllithium reagents **86** and *ent-***86** add to nitrones **83c** and **87a,b** to afford hydroxylamines **88a–c**, which upon treatment with NaOH in *i*-PrOH undergo an intramolecular epoxide opening to form oxazetidines **89a–c** with er of 98:2 (Scheme 5.23).

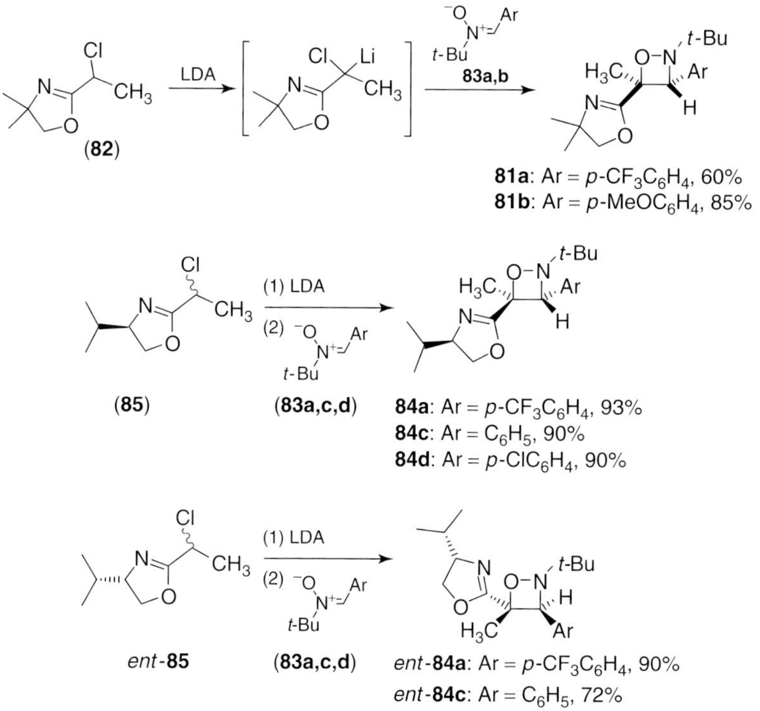

Scheme 5.22

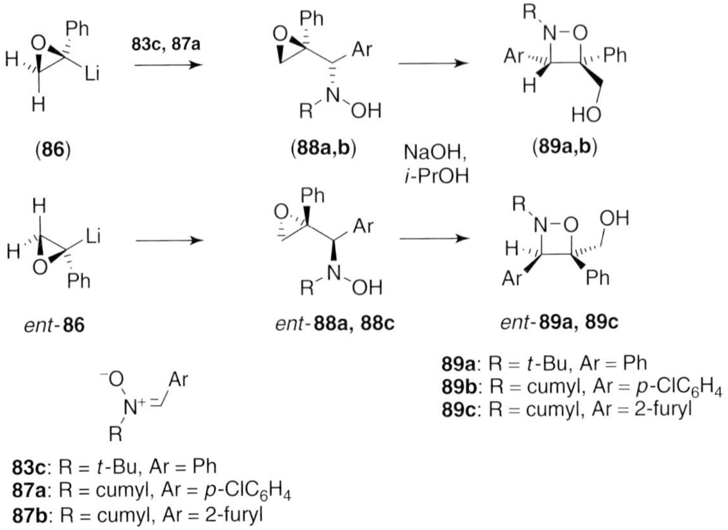
83c, 87a
(86)
(88a,b)
NaOH,
i-PrOH
(89a,b)
ent-86
ent-88a, 88c
ent-89a, 89c
89a: R = t-Bu, Ar = Ph
89b: R = cumyl, Ar = p-ClC6H4
89c: R = cumyl, Ar = 2-furyl
83c: R = t-Bu, Ar = Ph
87a: R = cumyl, Ar = p-ClC6H4
87b: R = cumyl, Ar = 2-furyl

Scheme 5.23

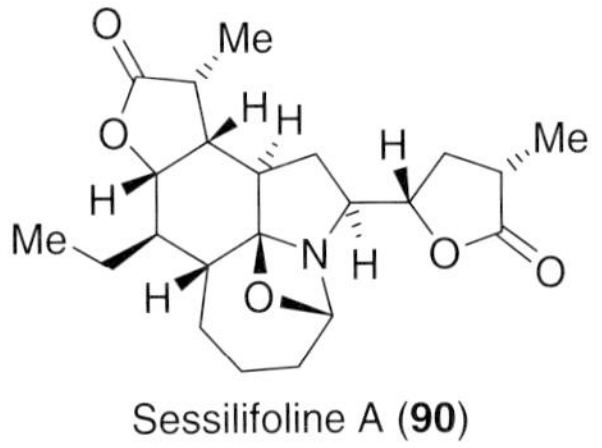
Sessilifoline A (90)

Fig. 5.11 Structure of sessilifoline A.

As was the case for 1,3-diazetidines, there are, to our knowledge, no methods described for the asymmetric synthesis of 1,3-oxazetidines. Again, it has to be emphasized that the strained bis-aminal functionality of these heterocycles makes them particularly reactive and unstable. However, the natural product sessilifoline A (**90**, Figure 5.11) was recently assigned a structure containing a 1,3-oxazetidine [133]. Hopefully, interesting natural product targets such as this one will rekindle the interest of the synthetic community in these rare heterocycles.

5.3.2.1 Thiazetidines

Thiazetidines are the four-membered heterocycles containing one nitrogen and one sulfur atom. Again, two types of constitutional isomers are possible: the 1,2- **91** and 1,3-thiazetidines **92** (Figure 5.12). In addition, the sulfur atom may be oxidized, and thus give rise to thiazetidine-1-oxides **93** and thiazetidine-1,1-dioxides **94, 95**.

The chemistry of this family of heterocycles has been comprehensively reviewed [1]. Again, our focus will be on the stereoselective synthesis of enantiomerically enriched products, and in this area there are, to our knowledge no reports concerning compound types **91, 92, 93** and **94**. As suggested for the oxazetidines

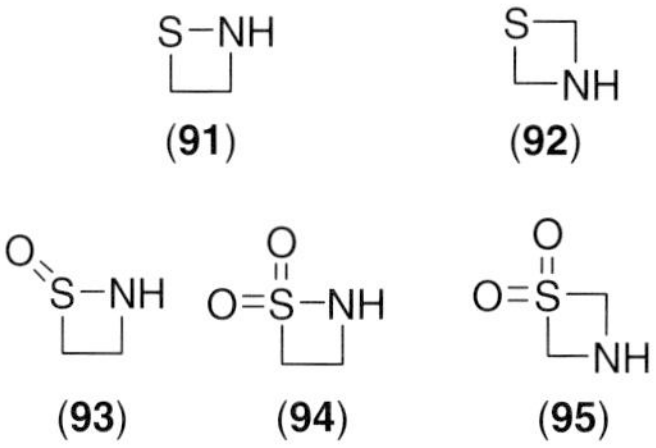

5.3 Four-Membered N-Heterocycles with Two Heteroatoms

Fig. 5.12 General structures of thiazetidines.

(*vide supra*), one might argue that it is the lack of attractive target structures that has prevented a more vivid interest in these structurally intriguing compounds. On the other hand, 1,2-thiazetidine-1,1-dioxides (β-sultams) of general structure **94** have received far more attention. We have mentioned the fundamental interest in the latter as more reactive analogs of β-lactam antibiotics [117]. This has been the main motivation of synthetic chemists to successfully develop approaches for their stereoselective preparation. In the following, we will distinguish between chiral β-sultams bearing stereogenic elements within or outside the four-membered ring and subdivide the synthetic methods accordingly.

A simple and general synthesis of enantiomerically pure chiral β-sultams has been reported recently by Page *et al.* [134]. They report the reaction of (chlorosulfonyl)dimethylacetyl chloride **96** with a series of amino acids to afford the corresponding enantiomerically enriched β-sultams **97** (Scheme 5.24).

Otto *et al.* have recently described the synthesis of β-sultams bearing stereogenic centers on and outside the four-membered ring. While their first approach used only racemic β-sultam **98** that was derivatized on nitrogen by the use of chiral acid chlorides and on the carboxylate end with amino acids [135], they later developed an asymmetric approach to the same types of compounds (Scheme 5.25) [136]. Starting from orthogonally protected aspartic acid **99**, chemoselective reduction of the carboxylate group afforded amino alcohol **100**. The latter was transformed into bromide **101** and nucleophilic displacement produced sulfide **102**. The hydrochloride **103** was obtained after Boc deprotection, and was cyclized to (*S*)-**98** via a one-pot oxidative chlorination-base induced ring-closure sequence.

Several other examples of cyclization strategies starting from amino acids have been reported [137]. In particular, β-sultam **104**, obtained from (*R*)-cysteine, has been used to prepare oxazaborolidine catalyst **105** *in situ* by reaction with borane (Scheme 5.26) [138]. The latter was able to promote the enantioselective reduction

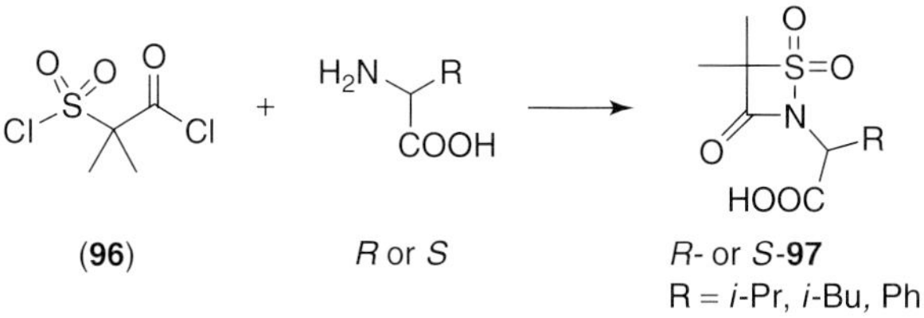

Scheme 5.24

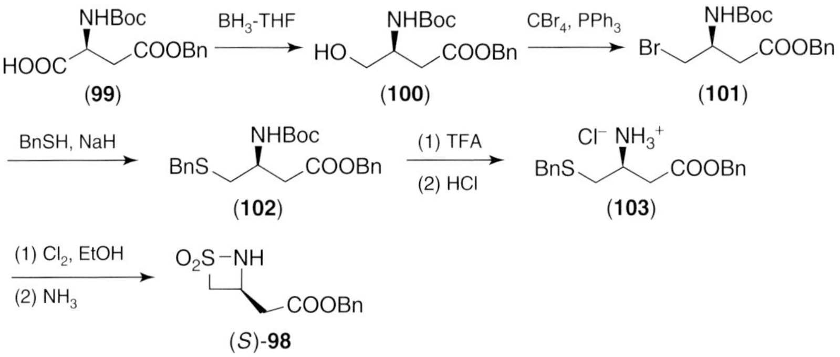

Scheme 5.25

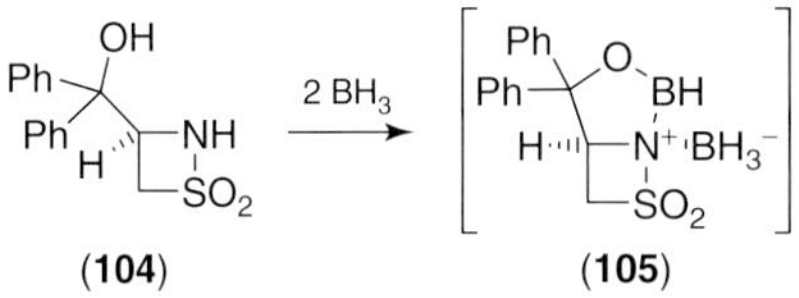

Scheme 5.26

of acetophenone with a good enantiomeric excess (81%), even though the reaction was run at an unusually high temperature (66 °C).

The same ring closure of open-chain β-amino sulfonates featured in the previous examples has also been put to profit by the Enders group. Their efforts in this area were initially seen as an extension of the conjugate addition methodology of chiral amines to α,β-unsaturated sulfonates that had been developed in order to prepare taurine derivatives [139]. The best results were obtained with the RAMBO chiral hydrazine **106**, which was used in a three-step sequence involving Lewis acid-catalyzed conjugate addition to sulfonates **107**, followed by N−N bond cleavage and Cbz protection to afford β-amino sulfonates **108** with enantiomeric excesses in excess of 96% (Scheme 5.27) [140]. Subsequent sulfonate hydrolysis and chlorination afforded sulfonyl chlorides **109**. Cbz removal and base-induced cyclization afforded the desired β-sultams **110** that displayed no erosion of enantiomeric purity.

In a second approach by the Enders group, 3,4-disubstituted β-sultams were targeted, and the general strategy modified in order to include a diastereoselective alkylation of SAMP-hydrazone **111** [141]. The latter was transformed into β-amino sulfide **112** by a known sequence. Removal of the benzyl group afforded **113**, which was oxidized and chlorinated to obtain **114** (Scheme 5.28). The same deprotection and cyclization conditions used previously finished the synthesis of the cis-3,4-disubstituted β-sultams **115**.

An interesting synthesis of β-sultams, on the basis of a ring opening-ring closing strategy starting from isoxazolidines **116**, was recently described by Caddick et al. [142]. Although this strategy has been applied successfully to the diastereoselective

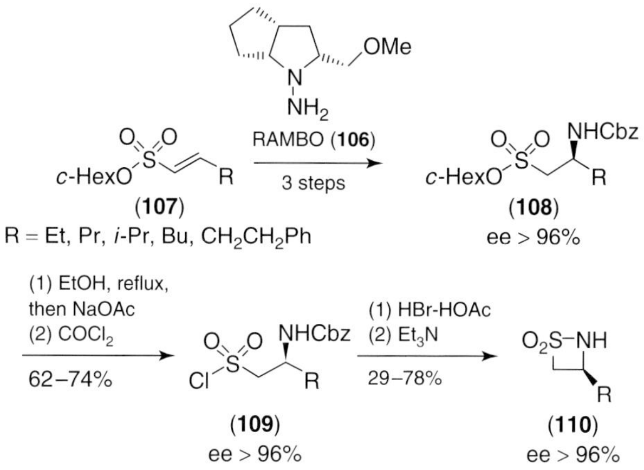

Scheme 5.27

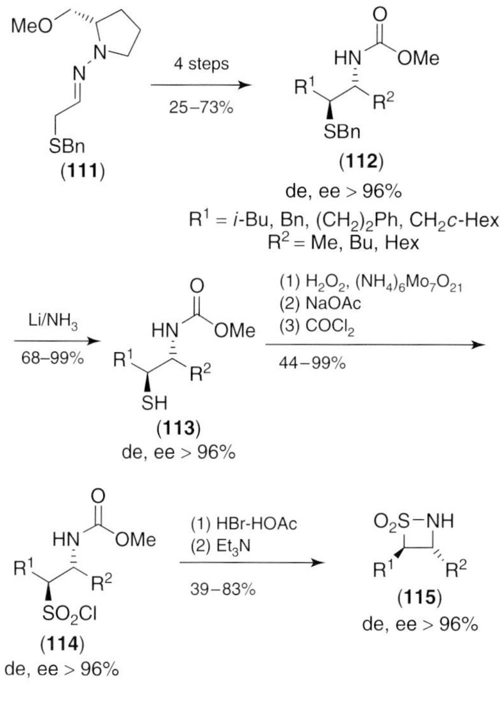

Scheme 5.28

synthesis of the sultams **117** bearing a large variety of aryl substituents, no asymmetric version is available yet (Scheme 5.29).

All the previously described methods proceed by cyclization involving S–N bond formation. There is only one example of a stereoselective synthesis of β-sultams by a C–N bond formation. Del Buttero et al. used chiral (tricarbonyl)chromium arene

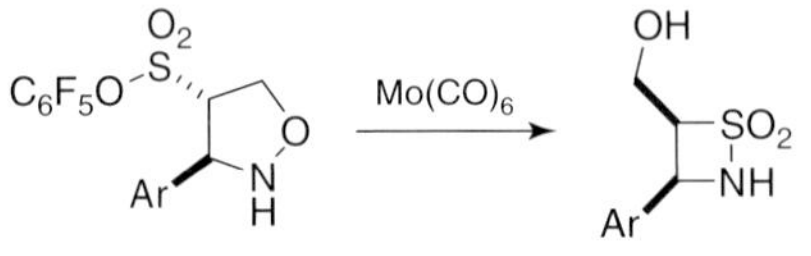

Scheme 5.29

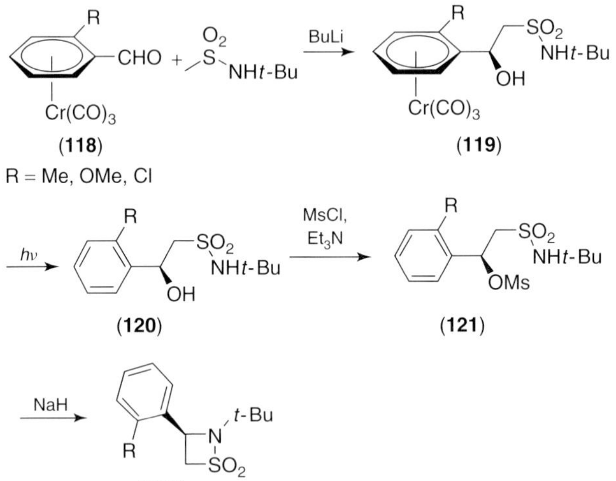

Scheme 5.30

complexes **118** as starting point for their approach (Scheme 5.30) [143]. Diastereoselective addition of the dianion of *tert*-butyl methanesulfonamide afforded alcohols **119**, which were deprotected under photochemical conditions and the resulting alcohols **120** transformed into their corresponding mesylates **121**. Deprotonation of the sulfonamide nitrogen by NaH effected the intramolecular S_N2 ring closure to afford β-sultams **122**.

Another appealing strategy for the formation of β-sultams is the [2 + 2] cycloaddition of sulfenes with imines. Despite the fact that this type of methodology had been known for several years [144], and that attempts at diastereoselective versions had been reported in solution [145] and on solid phase [146] with moderate success, there was no general asymmetric version available until the recent report by the Peters group [147]. A variety of sulfonyl chlorides **123** were used as sulfene precursors and reacted with electron-poor imines **124** in the presence of quinine as chiral catalyst. *cis*-β-Sultams **125** were obtained in good yields and diastereo- and enantioselectivities (Scheme 5.31).

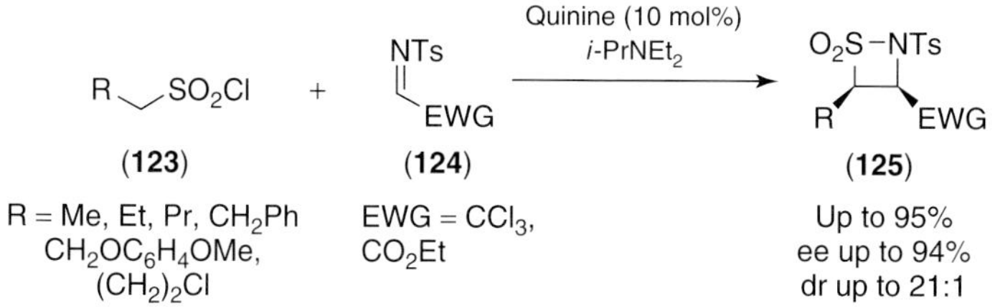

Scheme 5.31

R = Me, Et, Pr, CH$_2$Ph, CH$_2$OC$_6$H$_4$OMe, (CH$_2$)$_2$Cl

EWG = CCl$_3$, CO$_2$Et

Up to 95%
ee up to 94%
dr up to 21:1

5.4
Conclusions

As illustrated above, three- and four-membered N-heterocycles with two heteroatoms are prepared by a wide variety of synthetic methods. There are several methods to synthesize certain ring systems in enantiomerically enriched form and, in many cases, as a result of its high inversion barrier, their nitrogen atom is stereogenic. Whereas the three-membered rings, and mainly enantiomerically pure oxaziridines, have been used as chiral substrates or chiral reagents in a variety of unique stereoselective transformations, their four-membered congeners, such as β-sultams, have been mainly prepared for their potential biological applications. Despite these diverse applications and the efforts outlined above, the enantioselective synthesis of these heterocycles remains underdeveloped. Especially in the area of the hetero-azetidines, many subclasses lack preparative methods of general scope, and certain constitutional isomers have never been synthesized asymmetrically. Undoubtedly, because of their growing interest, the chemistry of three- and four-membered N-heterocycles will provide synthetic challenges for some time to come.

References

1 (a) Maitland, D. J. (**1975**) *Three-membered Rings. Saturated Heterocycl. Chem.*, **3**, 1; (b) Livingstone, R. (**1973**) Three- and four-membered heterocyclic rings, in *Rodd's Chemistry of Carbon Compounds* (2nd edn.) IV(Pt.A), 1; (c) Schmitz, E. (**1997**) Three-membered rings with two heteroatoms, in *Rodd's Chemistry of Carbon Compounds* (2nd edn. 2nd sup.), IV(Pt.A), 91; (d) Hewson, A. T. (**1997**) Four-membered rings containing two or three heteroatoms, in *Rodd's Chemistry of Carbon Compounds* (2nd edn. 2nd sup.) IV(Pt.A), 235; (e) Schmitz, E. (**1984**) *Comprehensive Heterocyclic Chemistry* (eds A. R. Katritzky and C. W. Rees), Pergamon, Vol. **7**, p. 195 (Three-membered rings); (f) Timberlake, J. W. and Elder, S. S. (**1984**) *Comprehensive Heterocyclic Chemistry* (eds A. R. Katritzky and C. W. Rees), Pergamon, Vol. **7**, p. 449 (Four-membered rings); (g) Mason, T. J. (**1985**) Three-membered ring systems, in *Heterocyclic Chemistry*, Lanchester Polytechic Coventry, UK, Vol. **4**, p. 1; (h) See also in (**1991–1992**) *Methoden der Organischen Chemie*, (Houben-Weyl), Vol. **E16**, Georg Thieme Verlag, Stuttgart, p. 1031, 1271.

2 (a) Kostyanovsky, R. G., Murugan, R. and Sutharchanadevi, M. (**1996**)

Diaziridines and diazirines, in *Comprehensive Heterocyclic Chemistry* (ed A. Padwa), Elsevier, **1A**, p. 347; (b) Davis, F. A., Reddy, R. T. and Thimma, R. (**1996**) *Oxaziridines and oxazirines* in Comprehensive Heterocyclic Chemistry (ed A. Padwa), Elsevier, Vol. **1A**, p. 365.

3 Jennings, W. B. and Boyd, D. R. (**1992**) *Cyclic Organonitrogen Stereodynamics*, (eds J. B. Lambert and Y. Takeuchi), Wiley-VCH Verlag GmbH, Chapt. 5, New York, Weinheim, Cambridge, p. 105.

4 Mannschreck, A., Radeglia, R., Grundemann, E. and Ohme, R. (**1967**) *Chem. Ber.*, **100**, 1778.

5 Mannschreck, A. and Seitz, W. (**1969**) *Angew. Chem.*, **8**, 212.

6 Kostyanovsky, R. G., Shustov, G. V., Starovoitov, V. V. and Chervin, I. I. (**1998**) *Mendeleev Commun.*, **3**, 113.

7 (a) Kostyanovsky, R. G., Polyakov, A. E. and Markov, V. I. (**1974**) *Izv. Akad. Nauk. SSSR Ser. Khim.*, 1671; (b) Kostyanovsky, R. G., Polyakov, A. E. and Shustov, G. V. (**1976**) *Tetrahedron Lett.*, **17**, 2059.

8 (a) Shustov, G. V., Kadorkina, G. K., Varlamov, S. V., Kachanov, A. V., Kostyanovsky, R. G. and Rauk, A. (**1992**) *J. Am. Chem. Soc.*, **114**, 1616; (b) Kostyanovsky, R. G., Korneev, V. A., Chervin, I. I., Voznesensky, V. N., Puzanov, Y. V. and Rademacher, P. (**1996**) *Mendeleev Commun.*, 106.

9 (a) Dyachenko, O. A., Atovmyan, L. O., Aldoshin, S. N. and Polyakov, A. E. (**1976**) *J. Chem. Soc. Chem Commun.*, 50; (b) Shustov, G. V., Kadorkina, G. K., Kostyanovsky, R. G. and Rauk, A. (**1988**) *J. Am. Chem. Soc.*, **110**, 1719.

10 (a) Häkli, H., Mintas, A. and Mannschreck, A. (**1979**) *Chem. Ber.*, **112**, 2028–32; (b) Mintas, A., Mannschreck, A. and Klasinc, L. (**1981**) *Tetrahedron*, **37**, 667.

11 Trapp, A., Schurig, V. and Kostyanovsky, R. G. (**2004**) *Chem. Eur. J.*, **10**, 951.

12 Kotyanovsky, R. G., Shutov, G. V. and Zaichenko, N. L. (**1982**) *Tetrahedron*, **38**, 949.

13 (a) Church, R. F. R., Kende, A. S. and Weiss, M. J. (**1965**) *J. Am. Chem. Soc.*, **87**, 2665; (b) Schmitz, E. and Ohme, R. (**1961**) *Chem. Ber.*, **94**, 2166.

14 Ishibara, H., Hori, K., Sugihara, H., Ito, Y. N. and Katsuki, T. (**2002**) *Helv. Chim. Acta*, **85**, 4272.

15 Shustov, G. V., Polyak, F. D., Nosava, V. S., Lieina, G. V., Nikiforovitch, R. G. and Kostyanovsky, R. G. (**1988**) *Khim. Getrotsikl. Soedin.*, **11**, 1461.

16 Fioravanti, S., Olivierei, L., Pellacani, L. and Tardella, A. T. (**1998**) *Tetrahedron Lett.*, **39**, 6391.

17 Briner, K. and Vasella, A. (**1989**) *Helv. Chim. Acta*, **72**, 1371.

18 Beer, D. and Vasella, A. (**1985**) *Helv. Chim. Acta*, **68**, 2254.

19 Bernet, B., Mangholz, S. E., Briner, K. and Vasella, A. (**2003**) *Helv. Chim. Acta*, **86**, 1488.

20 Kapferer, P., Birault, V., Poisson, J. F. and Vasella, A. (**2003**) *Helv. Chim. Acta*, **86**, 2211.

21 Kurz, G., Lehmann, J. and Thieme, R. (**1985**) *Carbohydr. Res.*, **136**, 125.

22 Kupfer, R., Rosenberg, M. G. and Brinker, U. H. (**1996**) *Tetrahedron Lett.*, **37**, 6647.

23 Ko, R. J., Gupte, S. M. and Nelson, W. L. (**1984**) *J. Med. Chem.*, **27**, 1727.

24 Moss, R. A. (**2006**) *Acc. Chem. Res.*, **39**, 267.

25 Hutov, G. V., Varlamov, S. V., Rauk, A. and Kostyanovsky, R. G. (**1990**) *J. Am. Chem. Soc.*, **112**, 3403.

26 (a) Krois, D. and Brinker, U. H. (**1998**) *J. Am. Chem. Soc.*, **120**, 11627; (b) Bobek, M. M., Krois, D. and Brinker, U. H. (**2000**) *Org. Lett.*, **2**, 1999.

27 (a) Morita, C., Hashimoto, K., Okuno, T. and Shirahama, H. (**2000**) *Heterocycles*, **52**, 1163; (b) Husain, S. S., Forman, S. A., Kloczewiak, M. A., Addona, G., Olsen, R. W. Pratt, M. B., Cohen, J. B. and Miller, K. W. (**1999**) *J. Med. Chem.*, **10**, 169; (c) Grassi, D., Lipuner, W., Aebi, M.,

Brunner, J. and Vasella, A. (**1997**) *J. Am. Chem. Soc.*, **119**, 10992.
28. Krois, D. and Brinker, U. H. (**2001**) *Synthesis*, **3**, 379.
29. Majerski, Z., Veljkovic, J. and Kaselj, M. (**1988**) *J. Org. Chem.*, **53**, 2662.
30. Emmons, W. D. (**1957**) *J. Am. Chem. Soc.*, **79**, 5739.
31. Davis, F. A. and Sheppard, A. C. (**1989**) *Tetrahedron*, **45**, 5703.
32. Bjorgo, J. and Boyd, D. R. (**1973**) *J. Chem. Soc., Perkin Trans. 2*, 1575.
33. Forni, A., Garuti, G., Moretti, I., Torre, G., Andreetti, G. D., Bocelli, G. and Sagarabotto, P. (**1978**) *J. Chem. Soc., Perkin Trans. 2*, 401.
34. Bjorgo, J., Boyd, D. R., Campbell, R. M. and Neill, D. C. (**1976**) *J. Chem. Soc., Chem. Commun.*, 162.
35. Boyd, D. R. (**1968**) *Tetrahedron Lett.*, **9**, 4561.
36. Shustov, G. V., Varlamov, S. V., Chervin, A. E., Aliev, R. G., Kosyanovky, R. G., Kim, D. and Rauk, A. (**1989**) *J. Am. Chem. Soc.*, **111**, 4210.
37. Ono, H., Splitter, J. S. and Calvin, M. (**1973**) *Tetrahedron Lett.*, **42**, 4107.
38. Jennings, W. B., Watson, S. and Boyd, D. R. (**1992**) *J. Chem. Soc., Chem. Commun.*, 1078.
39. Jennings, W. B., Watson, S. and Tolley, M. S. (**1987**) *J. Am. Chem. Soc.*, **109**, 8099.
40. Aubé, J. (**1997**) *Chem. Soc. Rev.*, **26**, 269.
41. Schmitz, E. and Andreae, S. (**1991**) *Synthesis*, 327.
42. Page, P. C. B., Limousin, C. and Murell, V. L. (**2002**) *J. Org. Chem.*, **67**, 7787.
43. Armstrong, A., Edmonds, I. D. and Swarbrick, M. A. (**2003**) *Tetrahedron Lett.*, **44**, 5335.
44. Vidal, J., Damestoy, S., Guy, L., Hannachi, J. C., Aubry, A. and Collet, A. (**1997**) *Chem. Eur. J.*, **32**, 1691.
45. Hata, Y. and Watanabe, M. (**1981**) *J. Org. Chem.*, **46**, 610.
46. Jennings, W. B., Kochanewicz, M. J., Lovely, J. C. and Boyd, D. R. (**1994**) *J. Chem. Soc., Chem. Commun.*, 2569.
47. Davis, F. A. and Chen, B. C. (**1992**) *Chem. Rev.*, **92**, 919.
48. Petrov, A. V. and Resnati, G. (**1996**) *Chem. Rev.*, **96**, 1809.
49. (a) Hanquet, G., Lusinchi, X. and Millet, P. (**1993**) *Tetrahedron*, **49**, 423; (b) Hanquet, G. and Lusinchi, X. (**1994**) *Tetrahedron Lett.*, **34**, 5299; (c) Hanquet, G. and Lusinchi, X. (**1994**) *Tetrahedron*, **50**, 12185; (d) Hanquet, G. and Lusinchi, X. (**1997**) *Tetrahedron*, **53**, 13727.
50. Adam, W., Saha-Möller, C. R. and Ganeshpure, P. A. (**2001**) *Chem. Rev.*, **101**, 3499.
51. Hanquet, G., Lusinchi, X. and Millet, P. (**1991**) *C. R. Acad. Sci. Paris*, **313**(SII), 625.
52. Davis, F. A. (**2006**) *J. Org. Chem.*, **24**, 8993.
53. Widmer, J. and Keller-Schierlein, W. (**1974**) *Helv. Chim. Acta*, **57**, 657.
54. Bucciarelli, M., Forni, A., Moretti, I. and Prati, F. (**1988**) *J. Chem. Soc., Chem. Commun.*, 1614.
55. Toda, F. and Ochi, M. (**1996**) *Enantiomer*, **1**, 85.
56. Zong, K., Shin, S. I. and Ryu, E. K. (**1998**) *Tetrahedron Lett.*, **39**, 6227.
57. (a) Morisson, J. D., Mosher, H. S. (**1971**) in *Asymmetric Organic Reactions*, American Chemical Society, Washington, DC, p. 336; (b) Montanari, F., Moretti, I. and Torre, G. (**1973**) *Gazz. Chem. Ital.*, **103**, 681; (c) Belzecki, C. and Motowicz, D. (**1975**) *J. Chem. Soc., Chem. Commun.*, 244.
58. (a) Montanari, F., Moretti, J. and Torre, G. (**1968**) *J. Chem. Soc., Chem. Commun.*, 1694; (b) Montanari, F., Moretti, J. and Torre, G. (**1969**) *J. Chem. Soc., Chem. Commun.*, 1086.
59. Pirkle, W. H. and Rinaldi, P. L. (**1977**) *J. Org. Chem.*, **12**, 2080.
60. Forni, A., Moretti, I. and Torre, G. (**1977**) *J. Chem. Soc., Chem. Commun.*, 731.
61. Bucciarelli, M., Forni, A., Moretti, I. and Torre, G. (**1980**) *J. Chem. Soc., Perkin Trans. I*, 2152.
62. Ogata, Y. and Sawaki, Y. (**1973**) *J. Am. Chem. Soc.*, **95**, 4687.

63 Azman, A., Koller, J. and Plesnicar, B. (1979) *J. Am. Chem. Soc.*, **101**, 1107.

64 Wang, Y., Chakalamannil, S. and Aubé, J. (2000) *J. Org. Chem.*, **65**, 5120.

65 (a) Schmitz, E., Ohme, R. and Murawski, D. (1961) *Angew. Chem.*, **73**, 708; (b) Schmitz, E. and Ohme, R. (1964) *Chem. Ber.*, **97**, 2521.

66 Choong, I. C. and Ellman, J. A. (1999) *J. Org. Chem.*, **64**, 6528.

67 Page, P. C. B., Limousin, C., Murell, V. L., Laffan, D. D. P., Bethell, D., Slawin, A. M. Z. and Smith, T. A. D. (2000) *J. Org. Chem.*, **65**, 4204.

68 Roelofsen, D. P. and van Bekkum, H. (1973) *Rec. Trav. Chim. Pays-Bas*, **91**, 605.

69 Belzecki, C. and Mostowicz, D. (1975) *J. Chem. Soc., Chem. Commun.*, 244.

70 (a) Belzecki, C. and Mostowicz, D. (1975) *J. Org. Chem.*, **40**, 3879; (b) Belzecki, C. and Mostowicz, D. (1977) *J. Org. Chem.*, **42**, 3917.

71 Kraïem, J., Kacem, Y., Khiari, J. and Hassine, B. B. (2001) *Synth. Commun.*, **31**, 263.

72 Aubé, J., Hammond, M., Gherardini, E. and Takusagawa, F. (1991) *J. Org. Chem.*, **56**, 499.

73 Usuki, Y., Wang, Y. and Aubé, J. (1995) *J. Org. Chem.*, **60**, 8028.

74 Oliveros, E., Rivière, M. and Lattes, A. (1976) *Org. Magn. Res.*, **8**, 601.

75 Seebach, D. and Prelog, V. (1982) *Angew. Chem. Int. Ed.*, **21**, 654.

76 Aubé, J., Wang, Y., Hammond, M., Tanol, M., Takusagawa, F. and van Velde, D. (1990) *J. Am. Chem. Soc.*, **112**, 4879.

77 Aubé, J., Hammond, M., Gherardini, E. and Takuagawa, F. (1991) *J. Org. Chem.*, **56**, 499; (b) Correction (1991) *J. Org. Chem.*, **56**, 4086.

78 Bucciarelli, M., Forni, A., Moretti, L. and Torre, G. (1977) *J. Chem. Soc., Perkin Trans. 2*, 1339.

79 For examples of the oxidation of imines containing adjacent alkyl groups, see: ref [69], ref [72], (a) Oliveros, E., Lattes, A. and Riviere, M. (1979) *Nouv. J. Chim.*, **3**, 739; (b) Olivieros, E., Lattes, A. and Riviere, M. (1980) *J. Heterocycl. Chem.*, **17**, 107.

80 For examples of the oxidation of imines containing adjacent heteroatom-containing groups, see: (a) Felluga, F., Nitti, P., Itacco, G. and Valentin, E. (1992) *J. Chem. Res. Synop.*, 86; (b) Czarnocki, Z. (1992) *J. Chem. Res. Synop.*, 334; (c) Wolfe, M. S., Dutta, D. and Aubé, J. (1997) *J. Org. Chem.*, **62**, 654.

81 Wang, Y., Chackalamannil, S. and Aubé, J. (2000) *J. Org. Chem.*, **65**, 5120.

82 (a) Hanquet, G., Lusinchi, X. and Milliet, P. (1988) *Tetrahedron Lett.*, **29**, 2817; (b) Bohé, L., Lusinchi, M. and Lusinchi, X. (1999) *Tetrahedron*, **55**, 155.

83 Schoumacker, S., Hamelin, O., Téti, S., Pécaut, J. M. and Fontecave, M. (2005) *J. Org. Chem*, **70**, 301.

84 (a) Milliet, P., Picot, A., Lusinchi, X. (1976) *Tetrahedron Lett.*, **17**, 1573; (b) Picot, A., Milliet, P. and Lusinchi, X. (1976) *Tetrahedron Lett.*, **17**, 1577; (c) Milliet, P., Picot, A. and Lusinchi, X. (1981) *Tetrahedron*, **37**, 4201.

85 Wong, O. A. and Shi, Y. (2008) *Chem. Rev.*, **108**, 3598.

86 Dadoun, H., Alazard, J. P. and Lusinchi, X. (1981) *Tetrahedron*, **37**, 1525.

87 Del Rio, R. E., Wang, B. and Bohé, L. (2007) *Org. Lett.*, **12**, 2265.

88 Alazard, J. P., Khemis, B. and Lusinchi, X. (1975) *Tetrahedron*, **31**, 1427.

89 Hanquet, G., Lusinchi, X. and Bohé, L. (1993) *Tetrahedron Lett.*, **45**, 7271.

90 Hanquet, G., Lusinchi, M., Chiaroni, A. and Riche, C. (1995) *Acta Cryst.*, **C51**, 2047.

91 Bohé, L., Lusinchi, M. and Lusinchi, X. (1999) *Tetrahedron*, **55**, 141.

92 Washington, I. and Houk, K. N. (2000) *J. Am. Chem. Soc.*, **122**, 2948.

93 (a) Page, P. C. B., Buckley, B. R., Barros, D., Ardakani, A. and Marples, B. A. (2004) *J. Org. Chem.*, **69**, 3595; (b) Page, P. C. B., Buckley, B. R., Heaney, H. and Blacker, A. J. (2005) *Org. Lett.*, **7**, 375.

94. Page, P. C. B., Barros, D., Buckley, B. R. and Marples, B. A. (**2005**) *Tetrahedron: Asymmetry*, **16**, 3488.
95. Akhatou, A., Cheboub, K., Rahini, M., Ghosez, L. and Hanquet, G. (**2007**) *Tetrahedron*, **63**, 6232.
96. Akhatou, A., Hanquet, G. and Ghosez, L. to be published.
97. Schmitz, E., Fechner-Simon, H. and Schramm, S. (**1969**) *Liebigs Ann. Chem.*, **725**, 1.
98. (a) Vidal, J., Guy, L., Sterin, S. and Collet, A. (**1993**) *J. Org. Chem.*, **58**, 4791; (b) Vidal, J., Hannachi, J. C., Hourdin, G., Mulatier, J. C. and Collet, A. (**1998**) *Tetrahedron Lett.*, **39**, 8845.
99. Armstrong, A., Edmonds, I. D. and Swarbrick, M. E. (**2003**) *Tetrahedron Lett.*, **44**, 5335.
100. Armstrong, A., Atkin, M. A. and Swallow, S. (**2001**) *Tetrahedron: Asymmetry*, **12**, 535.
101. Armstrong, A., Edmonds, I. E. and Swarbrick, M. E. (**2005**) *Tetrahedron Lett.*, **46**, 2207.
102. Davis, F. A., Stringer, O. D., Jenkins, R. H., Awad, S. B., Watson, W. H. and Galloy, J. (**1982**) *J. Am. Chem.Soc.*, **104**, 5412.
103. Davis, F. A. Jr, Lamendola, J., Nadir, U., Kluger, E. W., Sedergran, T. C., Panunto, T. W., Billmer, R., Jenkins, R., Turci, I. E., Watson, W. H., Chen, J. S. and Kimura, M. (**1980**) *J. Am. Chem. Soc.*, **102**, 2000.
104. Davis, F. A., McCauley, J. and Harakal, L. E. (**1984**) *J. Org. Chem.*, **49**, 1465.
105. Davis, F. A., Mc Cauley, J. P., Chattopadhyay, S., Harakal, M. E., Towson, J. C., Watson, H. W. and Tavanaiepour, I. (**1987**) *J. Am. Chem. Soc.*, **109**, 3370.
106. Davis, F. A., Reddy, T., Mc Cauley, J. P., Przeslawski, R. M. and Harakal, M. E. (**1991**) *J. Org. Chem.*, **56**, 809.
107. Verfürth, U. and Herrmann, R. (**1990**) *J. Chem. Soc., Perkin Trans. 1*, 2919.
108. Davis, F. A., Weismiller, M. C., Murphy, C. K., Reddy, R. T. and Chen, B. C. (**1992**) *J. Org. Chem.*, **57**, 7274.
109. Davis, F. A., Reddy, R. E., Kasu, P. V. N., Portonovo, P. S. and Carrol, P. J. (**1997**) *J. Org. Chem.*, **62**, 3625.
110. Jennings, W. B., Malone, J. F., Mc Guckin, M. R., Rutherford, M., Saket, M. and Boyd, D. R. (**1988**) *J. Chem. Soc., Perkin Trans. 2*, 1145.
111. Jennings, W. B., Kochanewycz, M. J., Lovely, C. K. and Boyd, D. R. (**1994**) *J. Chem. Soc., Chem. Commun.*, 2569.
112. Boyd, D. R. and Neill, D. C. (**1977**) *J. Chem. Soc., Chem. Commun.*, 51.
113. Boyd, D. R., Cambell, R. M., Coulter, B., Grimshaw, J., Neill, D. C. and Jennings, W. B. (**1985**) *J. Chem. Soc., Perkin Trans. 1*, 849.
114. Toda, F. and Tanaka, K. (**1987**) *Chem. Lett.*, **16**, 2283.
115. Breton, G. W. and Martin, K. L. (**2002**) *J. Org. Chem.*, **67**, 6699–704.
116. Mucsi, Z., Kötvélyesi, T., Viskolcz, B., Csizmadia, I. G., Novák, T. and Keglevich, G. (**2007**) *Eur. J. Org. Chem.*, 1759–67.
117. Page, M. I. and Laws, A. P. (**2000**) *Tetrahedron*, **56**, 5632–38.
118. Hoogeveen, A. P. J. (**1931**) *Rec. Trav. Chim. Pays-Bas*, **50**, 669–78.
119. Cook, A. H. and Jones, D. G. (**1941**) *J. Chem. Soc.*, 184–87.
120. (a) Lawton, G., Moody, C. J. and Pearson, C. J. (**1987**) *J. Chem. Soc., Perkin Trans. 1*, 877–84; (b) Lawton, G., Moody, C. J. and Pearson, C. J. (**1987**) *J. Chem. Soc., Perkin Trans. 1*, 885–97; (c) Lawton, G., Moody, C. J., Pearson, C. J. and Williams, D. J. (**1987**) *J. Chem. Soc., Perkin Trans. 1*, 899–902; (d) Taylor, E. C., Davies, H. M. L. and Hinkle, J. S. (**1986**) *J. Org. Chem*, **51**, 1530–36 and references cited therein.
121. Achiwa, K. and Yamada, S. (**1974**) *Tetrahedron Lett.*, **20**, 1799–802.
122. Tong, W., Wu, J.-C., Sandström, A. and Chattopadhyaya, J. (**1990**) *Tetrahedron*, **46**, 3037–60.
123. Miao, W., Xu, W., Zhang, Z., Ma, R., Chen, S.-H. and Li, G. (**2006**) *Tetrahedron Lett.*, **47**, 6835–37.
124. (a) Brana, M. F., Garrido, M., Lopez Rodriguez, M. L. and Morcillo, M. J. (**1987**) *Heterocycles*, **26**, 95–100; (b) Brana, M. F., Garrido, M.,

Hernando, J. L., Lopez Rodriguez, M. L. and Morcillo, M. J. (1987) *J. Heterocycl. Chem.*, **24**, 1725–27.

125 Claesson, A. (1988) *Heterocycles*, **27**, 2087–90.

126 Randazzo, A., Bifulco, G., Giannini, C., Bucci, M., Debitus, C., Cirino, G. and Gomez-Paloma, L. (2001) *J. Am. Chem. Soc.*, **123**, 10870–76.

127 Snider, B. B. and Duvall, J. R. (2003) *Tetrahedron Lett.*, **44**, 3067–70.

128 Della Monica, C., Randazzo, A., Bifulco, G., Cimino, P., Aquino, M., Izzo, I., De Riccardis, F. and Gomez-Paloma, L. (2002) *Tetrahedron Lett.*, **43**, 5707–10.

129 (a) Yu, S., Pan, X., Lin, X. and Ma, D. (2005) *Angew. Chem. Int. Ed.*, **44**, 135–38; (b) Nicolaou, K. C., Kim, D. W., Schlawe, D., Lizos, D. E., de Noronha, R. and Longbottom, D. A. (2005) *Angew. Chem. Int. Ed.*, **44**, 4925–29.

130 (a) Capriati, V., Degennaro, L., Florio, S. and Luisi, R. (2001) *Tetrahedron Lett.*, **42**, 9183–86; (b) Capriati, V., Degennaro, L., Florio, S. and Luisi, R. (2002) *Eur. J. Org. Chem.*, 2961–69.

131 Luisi, R., Capriati, V., Florio, S. and Piccolo, E. (2003) *J. Org. Chem.*, **68**, 10187–90.

132 Capriati, V., Florio, S., Luisi, R., Salomone, A. and Corrado, C. (2006) *Org. Lett.*, **8**, 3923–26.

133 Qian, J. and Zhan, Z.-J. (2007) *Helv. Chim. Acta*, **90**, 326–31.

134 Tsang, W.-Y., Ahmed, N., Hemming, K. and Page, M. I. (2007) *Org. Biomol. Chem.*, **5**, 3993–4000.

135 Röhrich, T., Abu Thaher, B. and Otto, H.-H. (2004) *Monatsh. Chem.*, **135**, 55–68.

136 Röhrich, T., Abu Thaher, B., Manicone, N. and Otto, H.-H. (2004) *Monatsh. Chem.*, **135**, 979–99.

137 See for example: (a) Mielniczak, G. and Łopusiński, A. (1999) *Heteroatom. Chem.*, **10**, 61–67; (b) Meinzer, A., Breckel, A., Abu Thaher, B., Manicone, N. and Otto, H.-H. (2004) *Helv. Chim. Acta*, **87**, 90–105.

138 Trentmann, W., Mehler, T. and Martens, J. (1997) *Tetrahedron: Asymmetry*, **8**, 2033–43.

139 Enders, D. and Wallert, S. (2002) *Synlett*, 304–6.

140 (a) Enders, D. and Wallert, S. (2002) *Tetrahedron Lett.*, **43**, 5109–111; (b) Enders, D., Wallert, S. and Runsink, J. (2003) *Synthesis*, 1856–68.

141 Enders, D., Moll, A., Schaadt, A., Runsink, J. and Raabe, G. (2003) *Eur. J. Org. Chem.*, 3923.

142 De, A. K., Lewis, K., Mok, J., Tocher, D. A., Wilden, J. D. and Caddick, S. (2006) *Org. Lett.*, **8**, 5513–18.

143 Baldoli, C., Del Buttero, P., Perdicchia, D. and Pilati, T. (1999) *Tetrahedron*, **55**, 14089–96.

144 Szymonifke, M. J. and Heck, J. V. (1989) *Tetrahedron Lett.*, **30**, 2869–72.

145 Iwama, T., Kataoka, T., Muraoka, O. and Tanabe, G. (1998) *J. Org. Chem.*, **63**, 8355–60.

146 Gordeev, M. F., Gordon, E. M. and Patel, D. W. (1997) *J. Org. Chem.*, **62**, 8177–81.

147 Zajac, M. and Peters, R. (2007) *Org. Lett.*, **9**, 2007–10.

6
Asymmetric Synthesis of Five-Membered Ring Heterocycles with More Than One Heteroatom

Catherine Kadouri-Puchot and Claude Agami

Asymmetric syntheses of chiral five-membered heterocycles containing two heteroatoms, one of which being nitrogen, have been the subject of several studies. Such compounds indeed are of manifold significance. First, they are intrinsically interesting owing to their various biological properties: anti-inflammatory ability (Δ^2-pyrazolines), antibacterial activity (oxazolidinones and pyrazolidinones), ligands for dopamine receptors and platelet glycoprotein antagonists (isoxazolines), and anti-inflammatory drugs (thiazolines) can be quoted as very fragmentary examples.

In the second place, these heterocycles are extraordinary and useful reagents for the synthesis of a great variety of optically active molecules and most of them are the basis of new methodologies that are now widely used. In this context, many distinguished chemists have linked their names to processes that embody exceptional versatility in building elaborate optically active molecules often obtained in an enantiopure form.

Last but not least, many members of this family (especially, oxazolines and isoxazolines) are used as chiral ligands in metal-catalyzed enantioselective transformations; this property might appear as the most promising one and research in that direction is fast expanding.

Owing to these outstanding features, as can be expected, many reviews pertinent to this area are available and they will be cited in due course below. Therefore, the present review will focus on the more recent outcomes, and the most widely known results of the last 15 years will be described.

6.1
Five-Membered Heterocycles with N and O Atoms

6.1.1
Oxazolidines

Chiral oxazolidines may be viewed as masked forms of aldehydes, and as such are widely employed as chiral auxiliaries in asymmetric synthesis. Since the

Asymmetric Synthesis of Nitrogen Heterocycles. Edited by Jacques Royer
Copyright © 2009 WILEY-VCH Verlag GmbH & Co. KGaA, Weinheim
ISBN: 978-3-527-32036-3

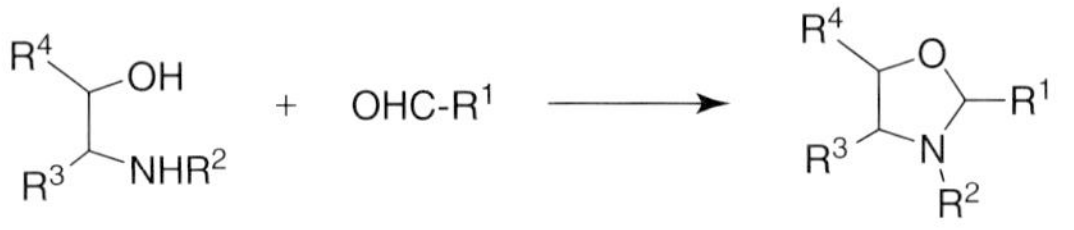

Scheme 6.1

stereoselective construction of this type of five-membered heterocycle is especially straightforward, this property has allowed numerous applications [1].

Basically, oxazolidines are easily formed by an acid-catalyzed reaction between an aldehyde (or an acetal) and a suitable 1,2-amino alcohol derived from an α-amino acid or belonging to the ephedrine family (Scheme 6.1).

This transformation involves the prior formation of an iminium ion, which undergoes an intramolecular nucleophilic addition from the hydroxy group. This cyclization, which is the main pathway leading to such heterocycles [2], is nevertheless a disfavored 5-endo-trig process [3], and is unmanageable in the case of an unsubstituted amino group ($R^2 = H$ in Scheme 6.1) reacting with aromatic aldehydes. Actually, in that case, the condensation (Scheme 6.2 R = Ar) leads to a mixture of the oxazolidine **1** (as an epimeric mixture) and its tautomeric hydroxy-imine **2**, the ratio between them being dependent on many experimental features [4]; however, oxazolidine is always produced as the minor tautomer in this case. When the aldehyde is not an aromatic one, the imine tautomer **2** (R ≠ Ar) appears as a trace product but the created oxazolidine still appears as two epimers in nearly equal amounts. On the other hand, the fact that the cyclization goes to completion with all N-substituted amino alcohols (Scheme 6.1) can be viewed as a consequence of the classical Thorpe–Ingold effect.

As stated above, the case of oxazolidine formation from N-substituted amino alcohol does not show any of these inconveniencies: yields and stereoselectivities are excellent. However, the formation of N-alkyloxazolidines has to be treated separately from that of N-tosyl and N-Boc oxazolidines.

6.1.1.1 N-Alkyloxazolidines

Considerable confusion about the structure of these heterocycles prevailed in the early stages, since an X-ray analysis published in 1971 [5] specified that condensation of ephedrine and p-bromobenzaldehyde led to a 2,4-*trans*-oxazolidine configuration. Many subsequent works [6, 7] have contributed to correct this assertion; it is now unambiguously established that, in all cases, the major oxazolidine isomer shows a relative 2,4-*cis*-configuration (the single crystal studied by Neelakantan [5] should not be representative of the dissolved material). Actually this cis configuration is

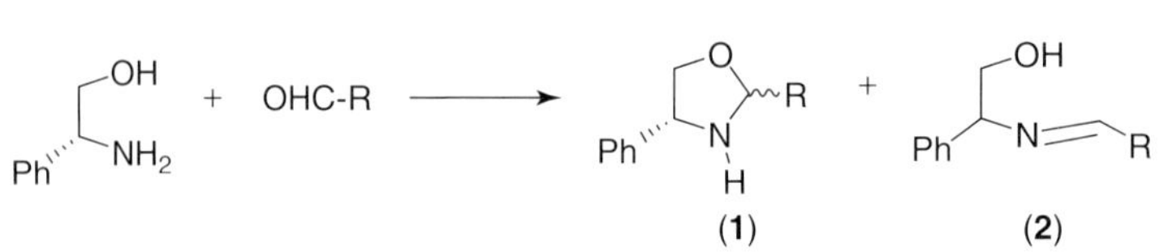

Scheme 6.2

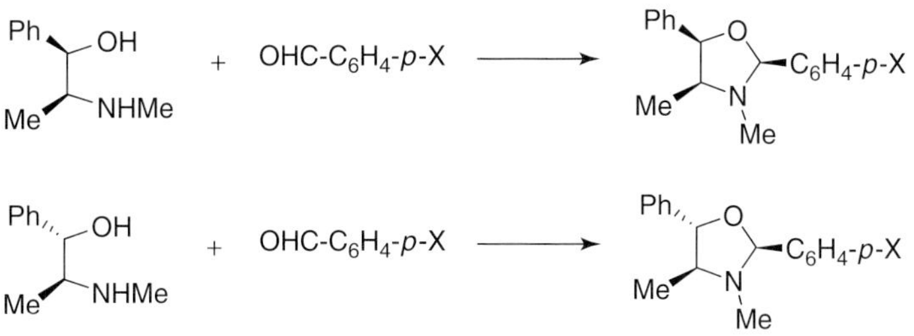

6.1 Five-Membered Heterocycles with N and O Atoms | 225

Scheme 6.3

the result of a thermodynamic control and the less stable trans structure may, in some cases, appear solely as a transient species depending on the reaction medium [8]. Scheme 6.3 depicts two representative examples of oxazolidine formations from (−)-ephedrine [7] and from (+)-pseudoephedrine [8], respectively.

The fact that the 2,4-*cis* configuration is more stable that the trans one can be ascribed to the favorable trans/trans array between substituents on the C-2, N-3 and C-4 centers. This also explains why both C-2 epimers are equally stable when N-unsubstituted amino alcohols are involved.

The most used chiral amino alcohols are ephedrine [7, 9], pseudoephedrine [4], valinol [10], phenylglycinol [11], and their N-substituted derivatives. Recently [12], amino alcohols prepared from L-methionine and S-methyl-L-cysteine were employed for the synthesis of oxazolidines which were later used in the actively explored field [13] of asymmetric palladium-catalyzed allylations.

Oxazolidines are also valuable tools in the most challenging field of organofluorine compounds. A large array of amino alcohols was reacted with trifluoromethyl aldehyde, which afforded oxazolidines **3** in excellent yields but as an epimeric mixture (Scheme 6.4); this last fact is unimportant since these oxazolidines were transformed into their corresponding imine counterparts **4**, which exist as unique stereoisomers [14]. A highly diastereoselective organometallic addition to these imines affords the interesting chiral amino alcohols **5**.

Analogous trifluoromethyl-substituted oxazolidines also served as chiral auxiliaries for the alkylation of amides and for the synthesis of various β-amino acids, β-amino alcohols, and γ-amino alcohols via Reformatsky- and Mannich-type reactions [15]. Recently, it has been reported [16] that the oxazolidine formed from ephedrine and salicylaldehyde can be used as a catalytic ligand in the addition of diethylzinc to a variety of aldehydes.

An original methodology arose when a dialdehyde (namely glutaraldehyde) instead of an aldehyde was treated with phenylglycinol ($R^1 = C_6H_5$, $R^2 = H$) or

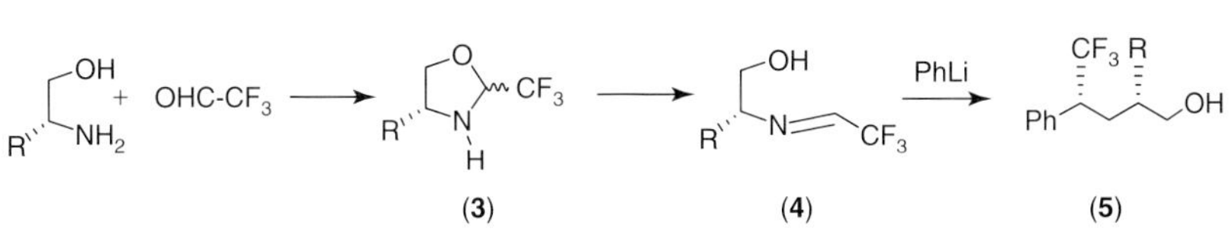

Scheme 6.4

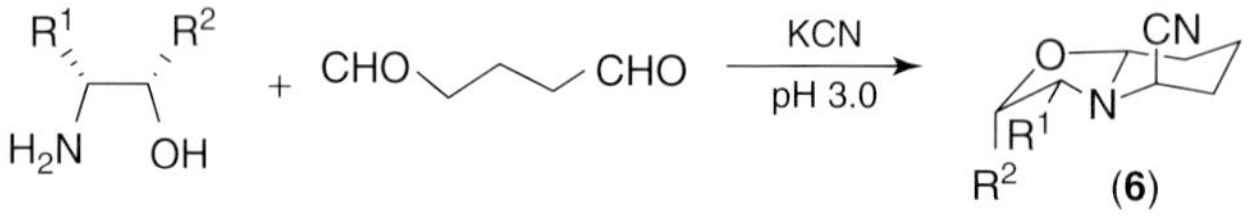

Scheme 6.5

norephedrine ($R^1 = CH_3$, $R^2 = C_6H_5$) as the amino alcohol counterpart, in the presence of potassium cyanide (Scheme 6.5); this led to the bicyclic oxazolidine **6** as a single isomer in which the classical 2,4-*cis* structure is still present (in that case, however, the favored trans junction of the bicyclic system should be taken into account to explain the stereoselectivity) [17a]. On the other hand, an anomeric effect explains the axial position of the cyano group, and this geometry is a crucial element for the numerous syntheses that were developed from this compound as starting material. This methodology, which permits the synthesis of many natural and unnatural derivatives containing either the piperidine or the pyrrolidine moieties, was dubbed the CN(R, S) method by its authors, Husson et al. [18].

Some years after the preceding reports, Katritzky et al. described a similar condensation that also involves dialdehydes: glutaraldehyde [19a] and succinaldehyde [19b] which were made to react with phenylglycinol in the presence of benzotriazole (instead of potassium cyanide). Whereas the condensation product **7** formed from glutaraldehyde was obtained as a mixture of epimers and regioisomers [19c], this was not the case when succinaldehyde was the starting material; in this last case, the pyrrolidine homolog **8** was obtained as the single, more stable trans isomer (Scheme 6.6).

Another occurrence of this 2,4-*cis* structure in a bicyclic oxazolidine was found in the oxazolopiperidine **10**, which results from the highly stereoselective intramolecular cyclization (Scheme 6.7) of the pyridinium salt **9** [20].

The cyclization depicted in Scheme 6.8 shows a cyclization that is related to the preceding one: oxazolidine **12** was synthesized from β-hydroxy tertiary amine N-oxides **11** [21]. Here again, the product shows the usual 2,4-*cis* geometry.

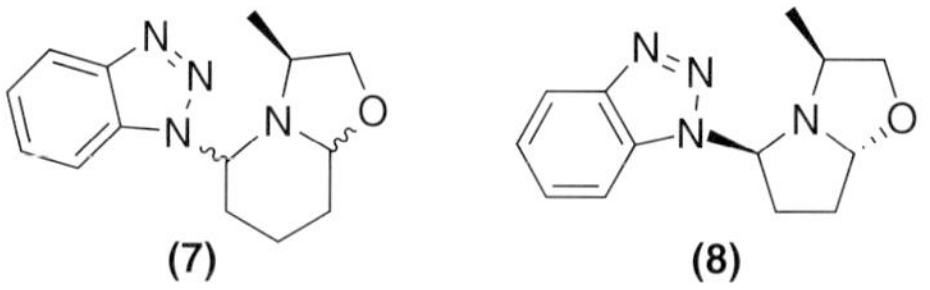

Scheme 6.6

Scheme 6.7

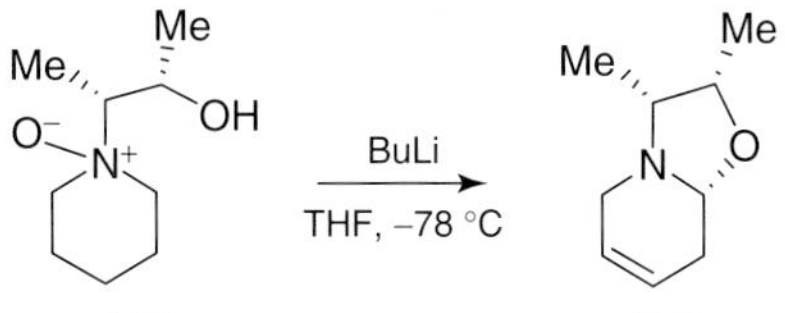

6.1 Five-Membered Heterocycles with N and O Atoms

Scheme 6.8

A moderate diastereoselectivity, favoring now the 2,4-*trans* geometry in compound **13**, was reported during the iodocyclization of 1,4-dihydropyridines **14** [22]. It should be noted that an iminium ion does not come into play (as in the preceding reactions): cyclization occurs on an iodonium ion and the reaction product may not be thermodynamically controlled (Scheme 6.9).

Intermediate iminium ions are also involved during the partial reduction of δ-lactams whose intramolecular cyclization (Scheme 6.10) leads oxazolopiperines. Thus, compound **16** was obtained from the readily available bicyclic lactam **15** and eventually cyclohexenone **17** was produced in >99% ee [23].

δ-Lactam **18** can be converted into oxazolopiperidine **19** through addition of a Grignard reagent (Scheme 6.11). The cis configuration observed in the product was attributed to an $A^{1,2}$ allylic strain. [24]

Other chiral bicyclic compounds analogous to the preceding ones were recently synthesized by an original method: 2-pyrrolidino-1-ethanol was oxidatively cyclized to the corresponding oxazolopyrrolidines by treatment with Me_3NO [25].

This reaction is catalyzed by an iron carbonyl complex, which led to a 2,4-*trans* geometry of the oxazolidine moiety since, owing to the preferred cis ring junction of this [3.3.0] system, this structure is the more stable. Though the mechanism of this process is not fully elucidated, an iminium ion seems to be a likely intermediate (Scheme 6.12).

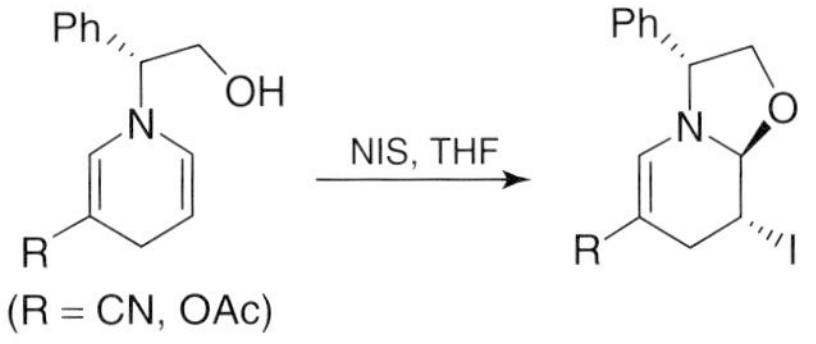

Scheme 6.9

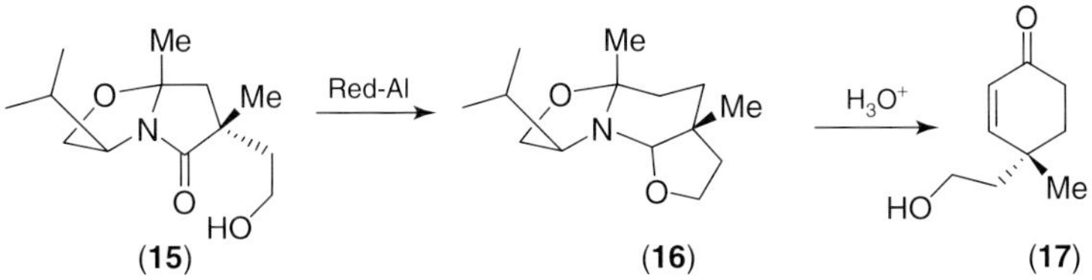

Scheme 6.10

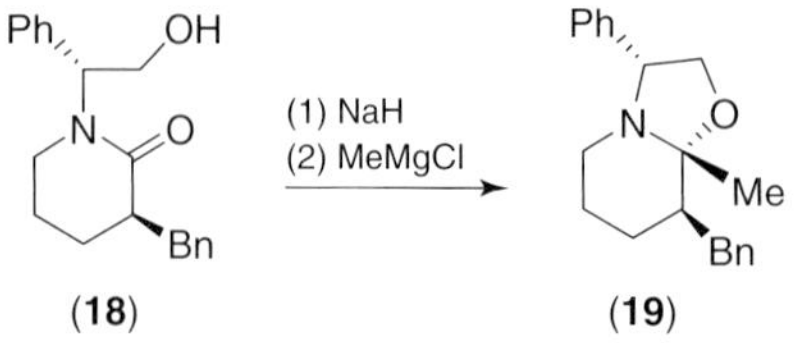

Scheme 6.11

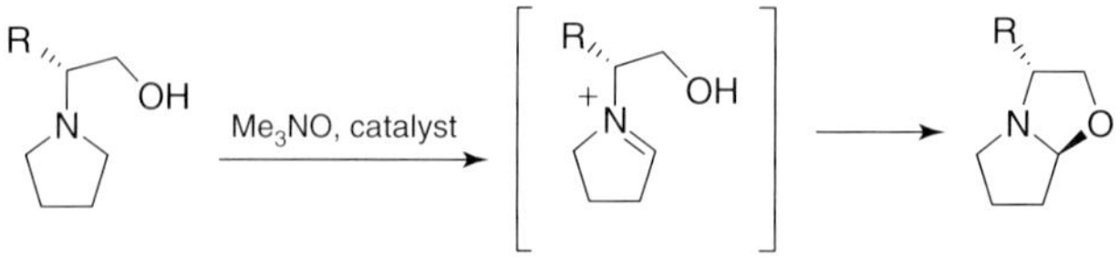

Scheme 6.12

6.1.1.2 N-Tosyl and N-Boc Oxazolidines

Scolastico et al. [26] made use of ketals, instead of aldehydes, as substrates for the synthesis of oxazolidines **20**, which present a very interesting distinctive feature: the heterocycle nitrogen is substituted by various electron-withdrawing groups (EWGs; Cbz, Ts, CO_2-t-Bu, CO_2Me, etc.). However, the tosyl N-substituent proved to be the most productive since it allows the formation of the 2,4-cis epimer in a totally stereoselective mode (Scheme 6.13). A structural study [27] showed that, when substituted by this group, the nitrogen is no longer planar (as with the other groups) and, as described above, results in the formation of this thermodynamically favored geometry. A great deal of work has been published that reports the various applications of such oxazolidines as chiral inductors in asymmetric synthesis [1]; most applications concern stereoselective transformations on an unsaturated moiety attached on the C-2 center of the oxazolidine (see an example in Scheme 6.13).

Subsequently, the same authors [28], concurrently with Hoppe's group [29], reported a new stereoselective method of producing stereoisomerically pure N-tosyl oxazolidines (Scheme 6.14) via the formation of an intermediate oxocarbenium **21**, reacting with a nucleophile such as an enoxysilane or an allylstanane.

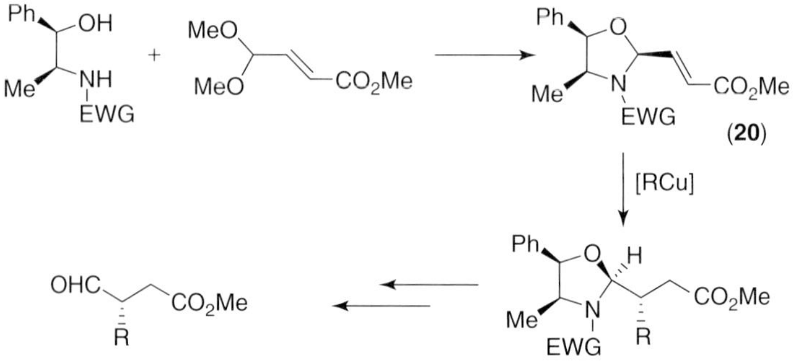

Scheme 6.13

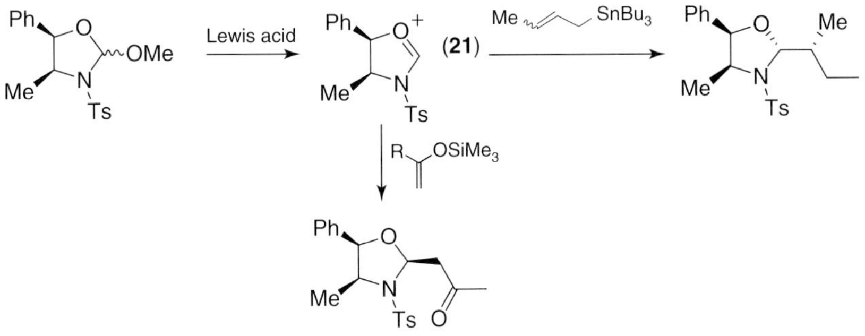

Scheme 6.14

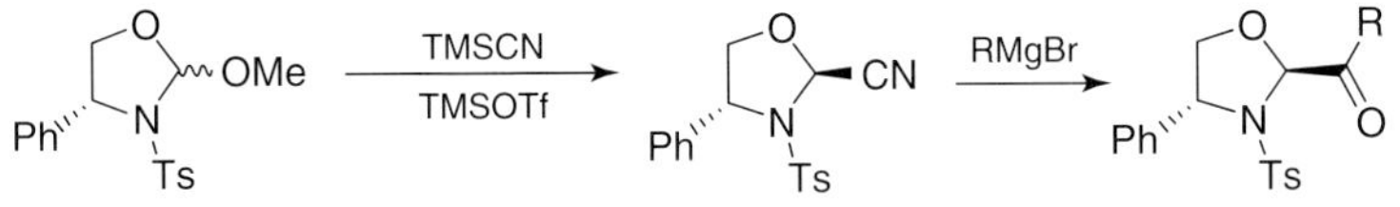

Scheme 6.15

These various methodologies led the way to the synthesis of some of the most useful oxazolidine derivatives: that is, the N-tosyl and N-Boc-2-acyloxazolidines. Actually, Hoppe [30] published the synthesis (Scheme 6.15) of the N-tosyl derivative via the addition the Lewis acid–catalyzed addition of a cyanide ion to an oxocarbenium ion analogous to **21** (Scheme 6.14), and shortly thereafter Colombo [31] described a synthesis of N-Boc-2-acyloxazolidine by an organometallic method (Scheme 6.16). Both methods afford very conveniently 2-acyloxazolidines **22** or **23**.

The use of such 2-acyloxazolidines as chiral substrates for many asymmetric transformations was extensively studied by Couty et al. [1]. These authors described a new method that affords such synthons (Scheme 6.17) via a condensation between ethylglyoxylate and phenylglycinol or norephedrine [32]. The produced oxazolidines **24**, once saponified, were transformed into their corresponding Weinreb amides **25**, which were, in turn, treated with Grignard reagents in order to produce the required 2-acyloxazolidines **26** in a totally stereoselective way.

Such a methodology allows the formation of a great variety of substrates, which undergo many asymmetric transformatons. One example [33] of such a process is given in Scheme 6.18.

Before closing this paragraph, it is worth mentioning that chiral oxazolidines produced from ketones, instead of aldehydes, have been scarcely described. Besides

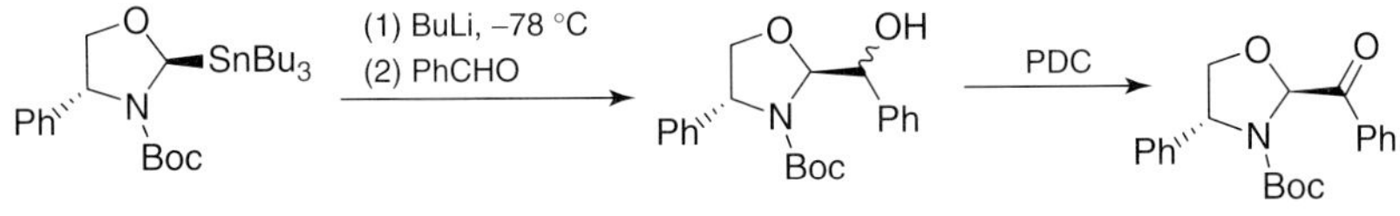

Scheme 6.16

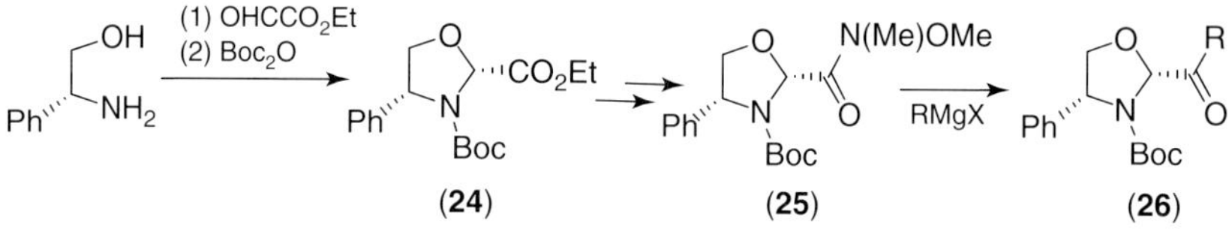

Scheme 6.17

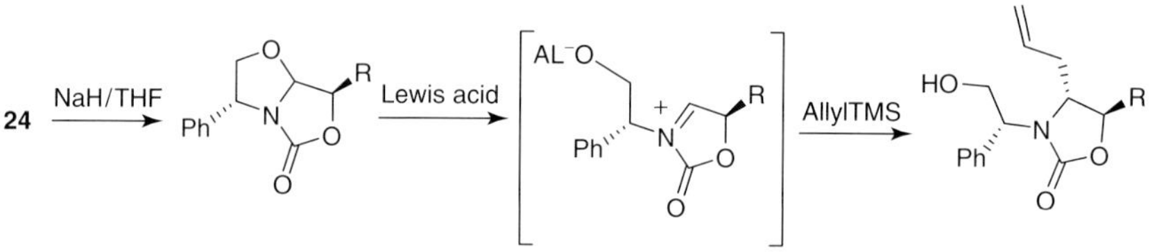

Scheme 6.18

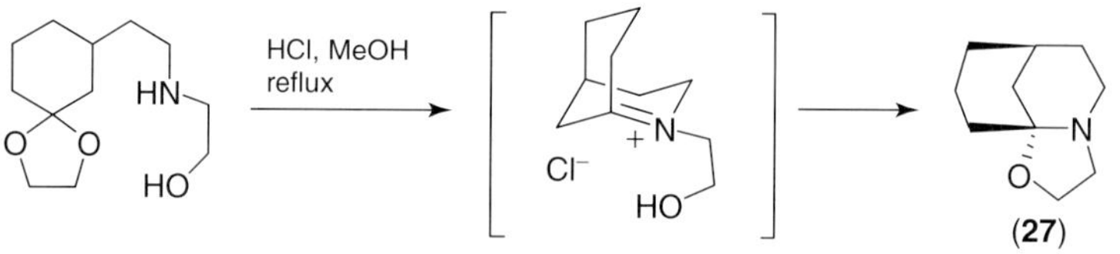

Scheme 6.19

the well-known fact that ketones are much less reactive than aldehydes as regards this type of ketalization, it may be emphasized that in such oxazolidine derivatives the essential control of the geometry of the quaternary C-2 center would be very enigmatic. Clearly, there is no such a problem when this center is not a stereogenic one as in the special case of the oxazolidine derived from acetone, which was recently synthesized by using fluoroalkanesulfonyl azides [34a].

On the other hand, the structural constraints inherent to the tricyclic nature of oxazolidine **27** explain why it was formed as a single diastereoisomer in the reaction depicted in Scheme 6.19 [34b,c]. Interestingly, this cyclization involves the addition on an anti-Bredt iminium ion.

6.1.2
Oxazolines (4,5-dihydrooxazoles)

In the field of asymmetric synthesis, oxazolines hold a preeminent position, since they come into play as chiral auxiliaries as well as ligands in catalytic systems. Asymmetric carbon–carbon bond formation is of great significance for the synthesis of enantiopure compounds; in this respect, Meyers' methodology, which makes use of such moieties in order to create definite stereogenic centers in carboxylic derivatives, is considered as one of the most fruitful methods in this area [35]. This process was extended with equal success to aromatic substitutions and aryl couplings [36, 37]. On the other hand, oxazoline and, especially, bisoxazolines are recognized as very fruitful ligands for asymmetric catalysis and are concerned with

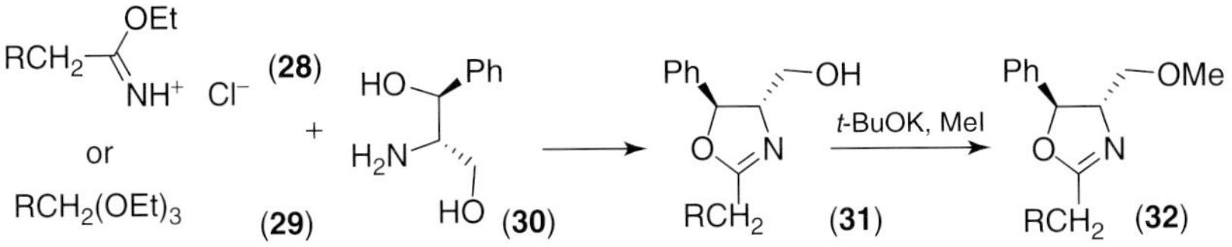

6.1 Five-Membered Heterocycles with N and O Atoms | 231

Scheme 6.20

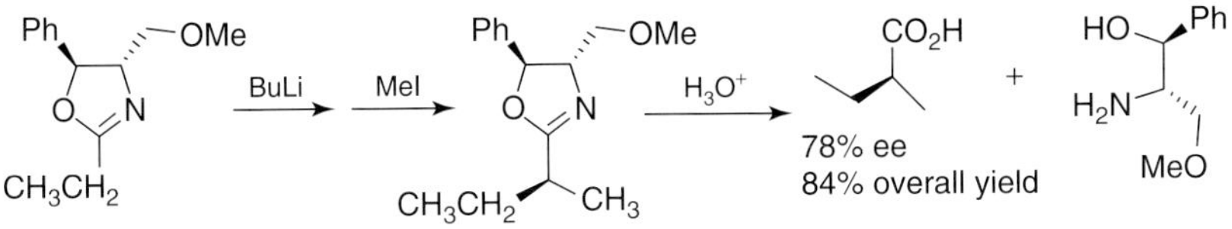

Scheme 6.21

a very impressive array of important reactions, such as for example, aldol reactions, homo- and hetero-Diels–Alder reactions, cyclopropanations, allylic substitutions, and so on [38]. The popularity of these compounds is mainly due to their ease of preparation, which rests on a relatively few but very efficient methods.

As regards the preparation of these versatile heterocycles, it should be underscored that they are masked forms of carboxylic acids; however, carboxylic acids themselves are sometimes inadequate to synthesize oxazolines owing to the weakness of their electrophilic reactivity. Meyers [39] showed that it would be worthwhile to convert them into much more reactive derivatives: that is, imidates of nitriles **28** or orthoesters **29**. These substrates react very smoothly with various enantiopure amino alcohols, and the oxazolines **31** are obtained, maintaining the configuration of the initial stereocenters (Scheme 6.20). The first amino alcohol that was used in this way was (+)-(1S,2S)-1-phenyl-2-amino-1,3-propanediol **30** (this enantiomer is commercially available since it was a basic material during the industrial synthesis of chloramphenicol). The CH_2OMe polar appendage in reagent **32** was thought to play a crucial role during the asymmetric transformations that were operated on the carboxylic moiety. Indeed, it can act as a ligand, which, jointly with a nitrogen atom, chelates a metal cation (in the most cases Li^+), thereby forcing the attacking nucleophile to enter from a specific side of the molecule.

Once the subsequent asymmetric transformation is realized, the amino alcohol is easily recovered from the acidic hydrolysis of the product (see Scheme 6.21 for an example).

Since then, many other amino alcohols have been employed in order to synthesize such chiral oxazolines [40]. Among them, the usual reduced derivatives of α-amino acids and derivatives of the ephedrine family frequently reported as substrates (Scheme 6.22) are phenylalaninol **33**, valinol **34**, phenylglycinol **35**, ephedrine **36**, pseudoephedrine **37**, and so on. Special mention may be made to a camphor-derived amino alcohol **38**, which leads to oxazolines whose carboxylic moiety was alkylated with excellent diastereoselectivity without the need for any polar appendage [41].

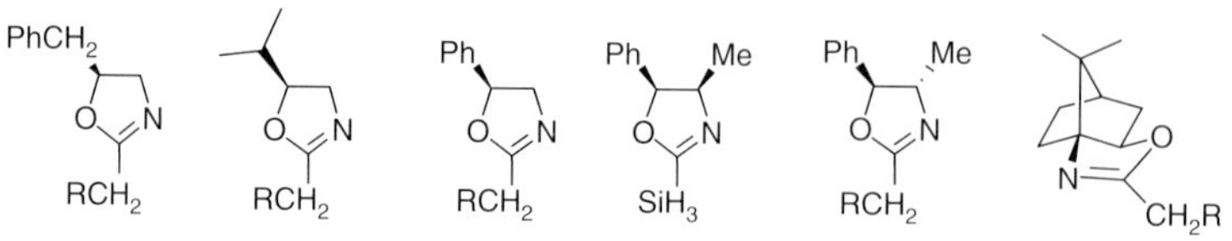

Scheme 6.22

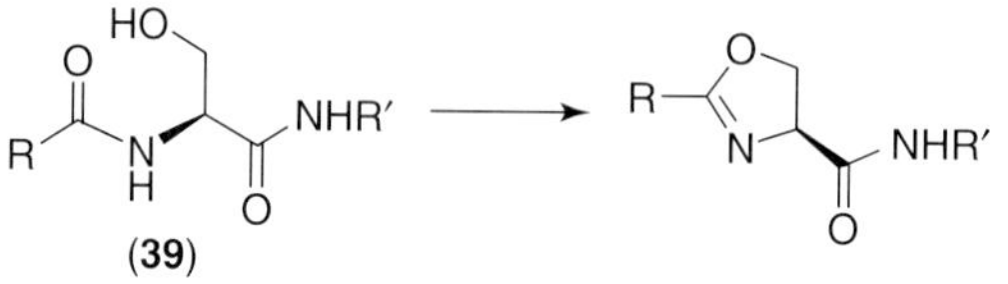

Structures (33)–(38): oxazoline derivatives with various substituents (PhCH₂, iPr, Ph, Ph/Me, Ph/Me, camphor-derived).

Scheme 6.23

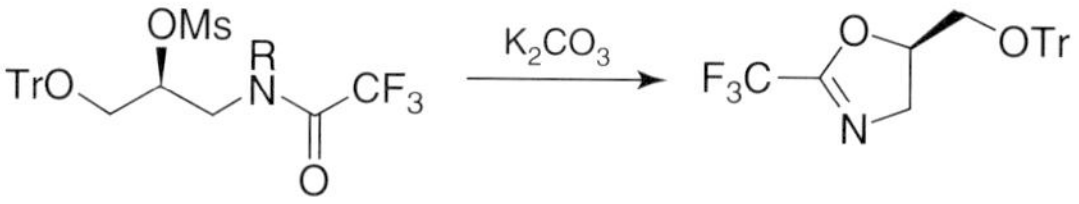

N-acyl-β-hydroxy-α-amino acid amide (39) cyclization to oxazoline.

Scheme 6.24

Cyclization of TrO–CH(R)–CH(OMs)–NH–C(O)CF₃ with K$_2$CO$_3$ to give 2-CF$_3$-5-OTr oxazoline.

Actually, Meyers has observed that such extra ligands may not always be necessary in order to achieve highly stereoselective synthesis [37].

Apart from their outstanding involvement in asymmetric synthesis, enantiopure oxazolines are found as elements of miscellaneous natural products and biologically active molecules [42]. Therefore, it is understandable that a great deal of work has been devoted to their preparation.

The most expeditious way of synthesizing 1,2-oxazolines is the reaction of 1,2-amino alcohols (belonging to or simply derived from the chiral pool) with acid derivatives as presented by Meyers (see above). Intramolecular substitution has also contributed to such syntheses. Thus, various dehydrating catalysts have been reported as being effective in the dehydrative cyclizations of N-acyl-β-hydroxy-α-amino acid amides **39** (Scheme 6.23), and a catalytic procedure using molybdenum oxide has recently been reported for the corresponding reaction applied to N-acylserine and N-acylthreonine derivatives [43].

Another type of a related intramolecular cyclization involves the amide group as the nucleophile (Scheme 6.24). In this case, the hydroxy moiety has to be converted into mesylate and the ring closure occurs stereospecifically with inversion [44]. This procedure has the advantage of allowing the formation of oxazolines substituted only on the C-5 center; these compounds are otherwise difficult to obtain owing to the lack of accessibility of the corresponding amino alcohols.

To the best of our knowledge, there is only one work reporting a genuine enantioselective synthesis of enantiopure oxazolines (all the other cases actually are mostly diastereoselective processes). Komatsu et al. [45] described the transformation of styrene (Scheme 6.25) derivatives into optically active oxazolines when they are reacted with acyl chlorides in the presence of the salen complex **40**.

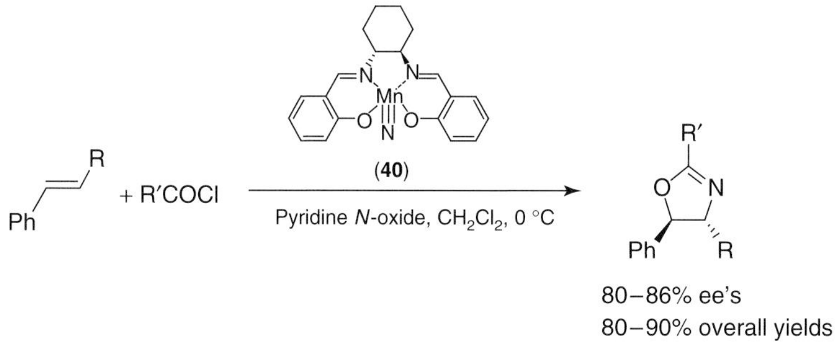

6.1 Five-Membered Heterocycles with N and O Atoms | 233

Scheme 6.25

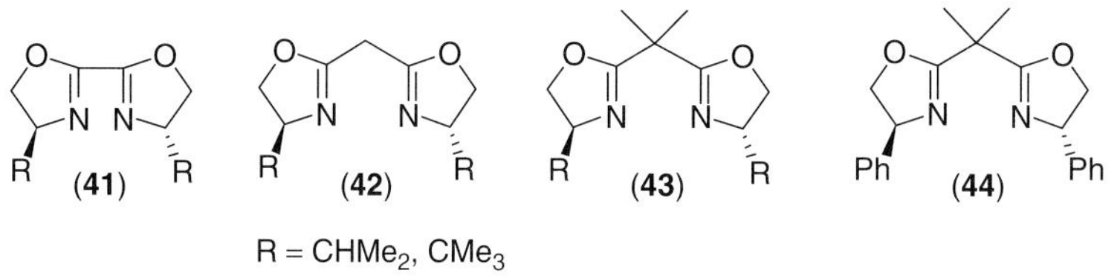

80–86% ee's
80–90% overall yields

Scheme 6.26

R = CHMe$_2$, CMe$_3$

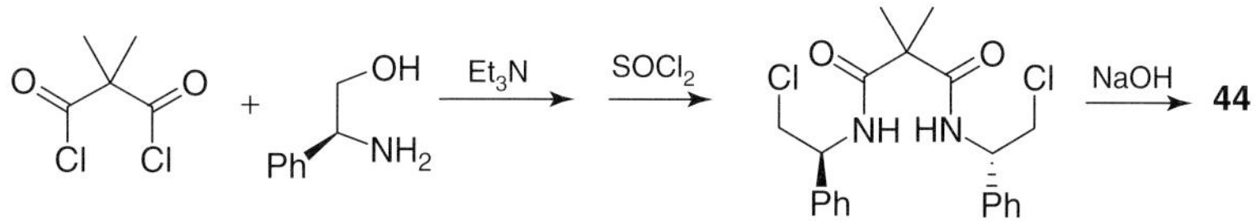

Scheme 6.27

The first report of chiral oxazolines as ligands for catalytic asymmetric transformations goes back to 1986 [46]. Since then, much works has been devoted to the preparation and application of these compounds, and, as mentioned before, a number of reviews have covered this topic [38]. However, the seminal studies by both Evans [47a] and Corey [47b] (Scheme 6.26) should not be overlooked because they introduced the exceptionally fruitful C$_2$-symmetric bisoxazoline ligands **41–43** [47a] and **44** [47b] in this field (Box ligands).

The procedure that was used to prepare these first Box ligands is univocal: the spacer comes from a symmetrically disubstituted malonic acid derivative, and the cyclization from the chiral aminoalcohol moiety is somewhat trivial (Scheme 6.27). This procedure still is the most used one, and many variants have been presented, in particular by Denmark [48] and Hanessian [49]: sulfonates as leaving group, various dehydrating agents, malononitrile instead of malonyl chloride, and so on. All these interesting variants have been extensively described in Desimoni and Jorgensen's review [38d].

The second general procedure can be viewed as an extension of the preceding one; it is based on the substitution of the methylene hydrogens of the malonyl moiety and has led to numerous other Box ligands (see Scheme 6.28 for a representative example [50] of this method).

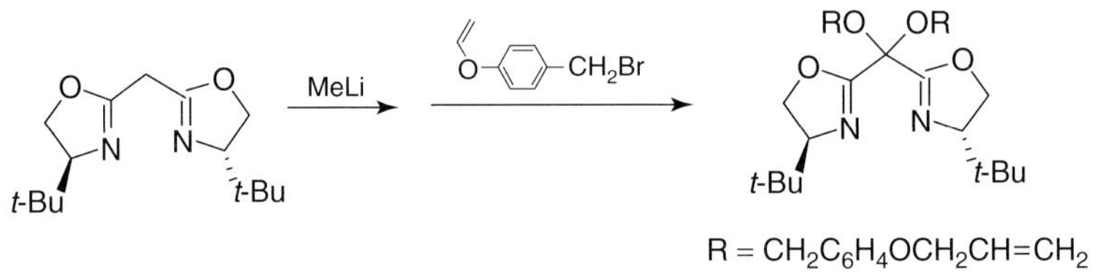

Scheme 6.28

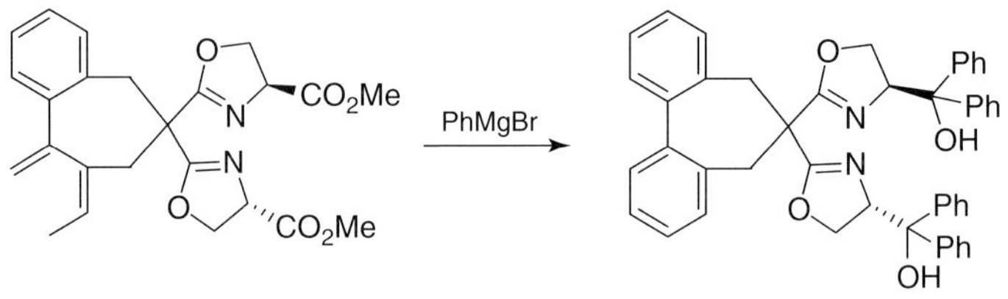

Scheme 6.29

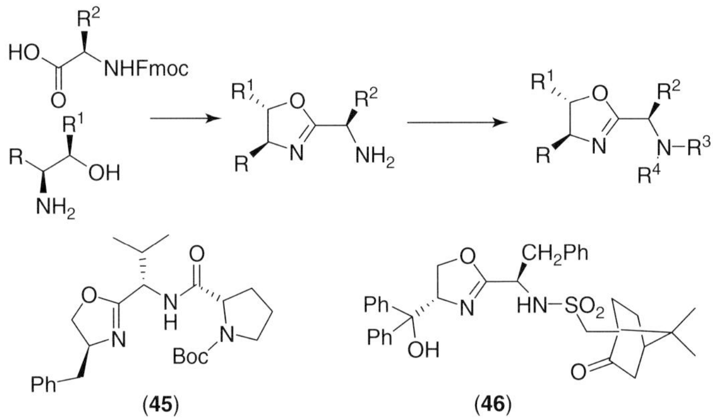

Scheme 6.30

Another procedure consists in the manipulation of groups already present in the ligand or in the spacer as depicted in Scheme 6.29 [51].

Recently, nonsymmetric Box-containing ligands appeared to be very promising in the asymmetric catalysis of various reactions. Thus, Sigman exploited a modular approach to ligands (for example **45** and **46**) composed of both oxazoline and another basic moiety (Scheme 6.30) acting as templates during the catalysis of enantioselective Diels–Alder reaction [52a], addition of allylic halides to aldehydes [52b] as well as addition of allyl bromide to ketones [52c].

Very recently, another set of 16 nonsymmetrical Box ligands were tested in the chromium-catalyzed Nozaki–Hiyama–Kishi allylation of benzaldehyde [53]. Compounds **47** and **48** are the most promising in this respect (Scheme 6.31).

Now, we cannot overemphasize the remarkable creativity that is still displayed in this field, and a large number of various ligands have been prepared and tested in

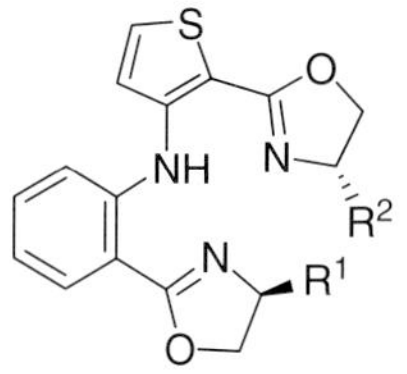

47: R^1 = t-Bu, R^2 = Bn
48: R^1 = Bn, R^2 = t-Bu

Scheme 6.31

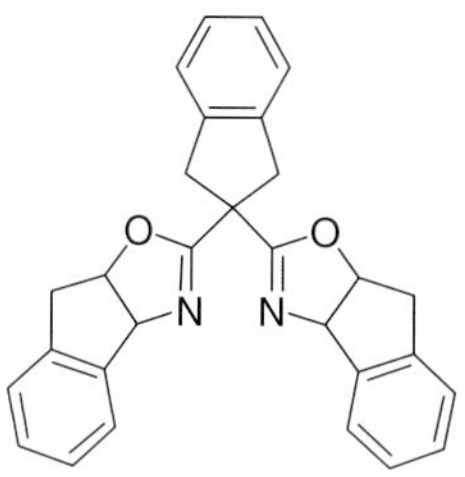

Scheme 6.32

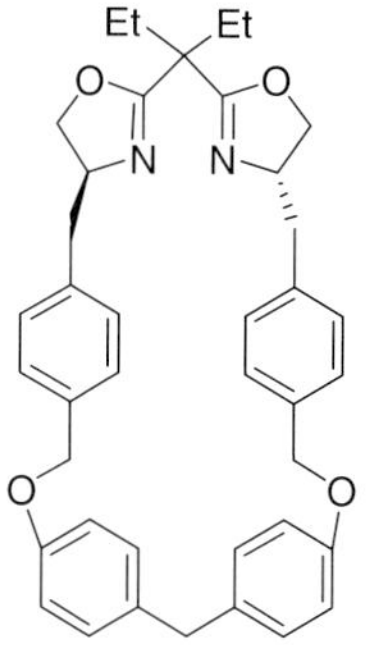

a variety of metal-catalyzed enantioselective reactions (see for instance compounds **49** [54] and **50** [55] in Scheme 6.32).

Extensive research is still ongoing in this very fruitful field, and recently the use of such ligands supported on poly(ethylene glycol) has been reported [56]; enantioselective catalysis can thus be envisioned in aqueous solution.

6.1.3
Oxazolidinones

6.1.3.1 Oxazolidin-2-ones

Oxazolidin-2-ones constitute an important class of oxazolidine derivatives. Since the first report in 1981 [57], chiral oxazolidin-2-ones (namely, Evans auxiliaries) have been widely used as chiral auxiliaries in numerous asymmetric organic reactions [58, 59], such as aldol condensations [60], Diels–Alder reactions [61], 1,3-dipolar cycloadditions [62], and radical reactions [63]. Moreover, since the 1980s, oxazolidinones represent an exciting new class of synthetic antibacterial agents [64]. Given the high synthetic utility of these heterocycles, asymmetric synthesis of oxazolidinones has attracted much attention. As the oxazolidines described above, one of the important chiral sources for the preparation of enantiopure 1,3-oxazolidin-2-ones is β-amino alcohols [65]. Reactions of amino alcohols with carbonyl derivatives such as phosgene [66], trichloromethylchloroformate (diphosgene) [67, 68], triphosgene [69], carbonyldiimidazole [70], and diethylcarbonate [71] have led to oxazolidinones in which the configurations of the stereocenters of the starting amino alcohols are retained. The construction of oxazolidinones from N-carbamate derivatives of

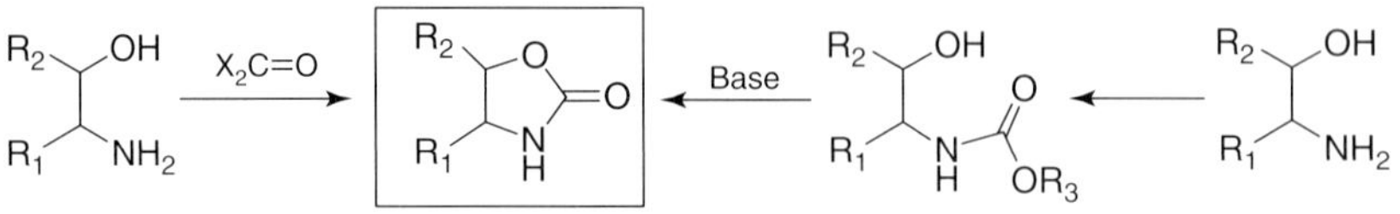

Scheme 6.33

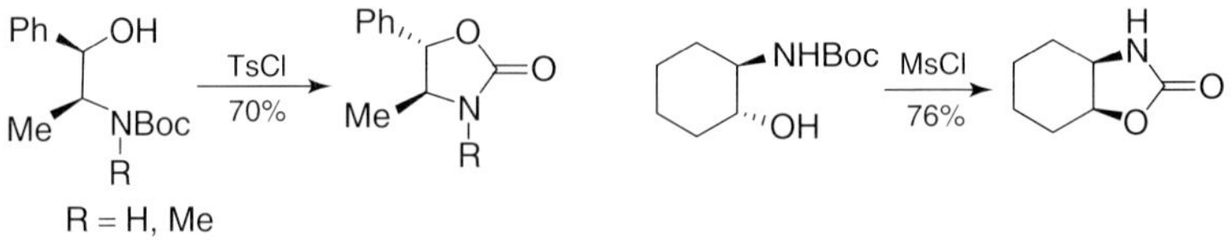

Scheme 6.34

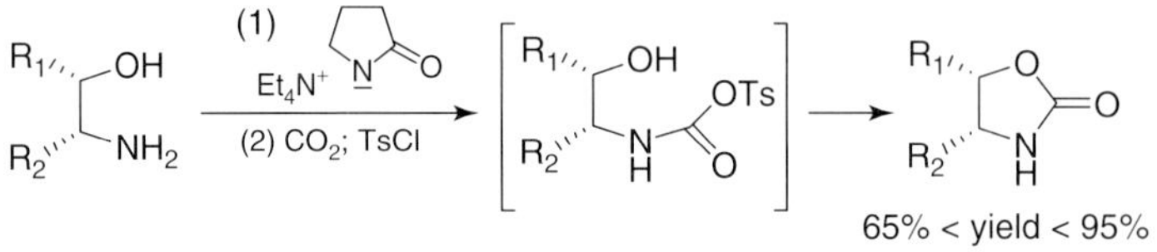

Scheme 6.35

1,2-aminoalcohols occurred also without changes in the configuration of the carbon bearing the hydroxyl group, when the carbamate was treated by a base, as shown in Scheme 6.33 [72, 73].

In contrast, an inversion of configuration, as depicted in Scheme 6.34, was observed via an S_N2 pathway from N-Boc derivatives of 1,2-amino alcohols, when the hydroxyl group was transformed into a good leaving group (reaction with TsCl [74], MsCl [75]).

With the aim of finding methodologies involving the use of harmless materials in place of toxic reagents such as phosgene, Feroci et al. used carbon dioxide in the presence of 2-pyrrolidone electrogenerated base and tosyl chloride (Scheme 6.35) [76]. In this methodology, the absolute configuration of all chiral atoms is retained; this result excludes the tosylation of the hydroxyl group.

Carboxylation followed by a Mitsunobu reaction has been investigated [77]. These reactions occurred with high yields. The stereochemical course of the Mitsunobu reaction is dependent on the N-substituent: primary amines gave oxazolidinones **51** with retention of configuration at the oxygen-bearing center, while secondary amines led to oxazolidinones **52** with inversion of configuration at the oxygen-bearing center (Scheme 6.36). Experiments with ^{18}O-labeled carbon dioxide evidenced two distinct mechanisms during the cyclization step [77b].

Chiral oxazolidinones were synthesized by using the selenium-catalyzed carbonylation of the corresponding 2-amino alcohols by bubbling CO at atmospheric pressure without racemization (Scheme 6.37) [78].

An alternative strategy involves the intramolecular displacement of a leaving group by the nitrogen atom of a N-benzoyl or N-tosyl carbamate (Scheme 6.38). The leaving group can be a bromine [79], a phenylselenide substituent [80]. The

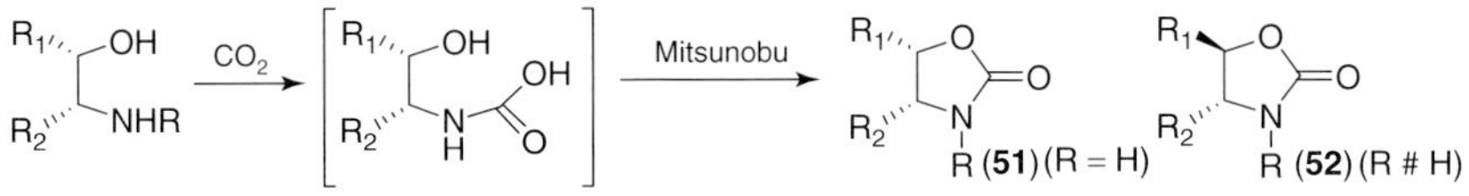

6.1 Five-Membered Heterocycles with N and O Atoms

Scheme 6.36

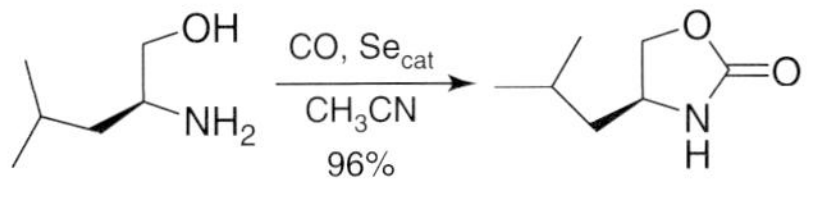

Scheme 6.37

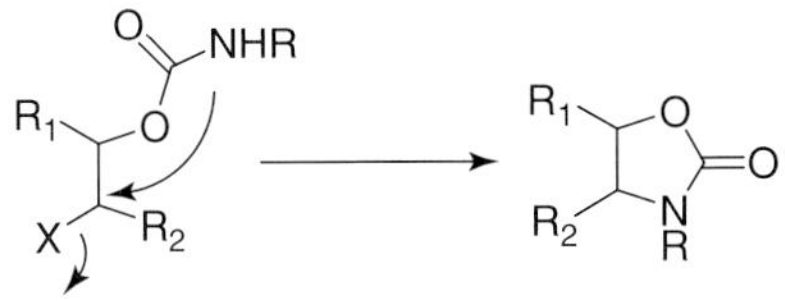

Scheme 6.38

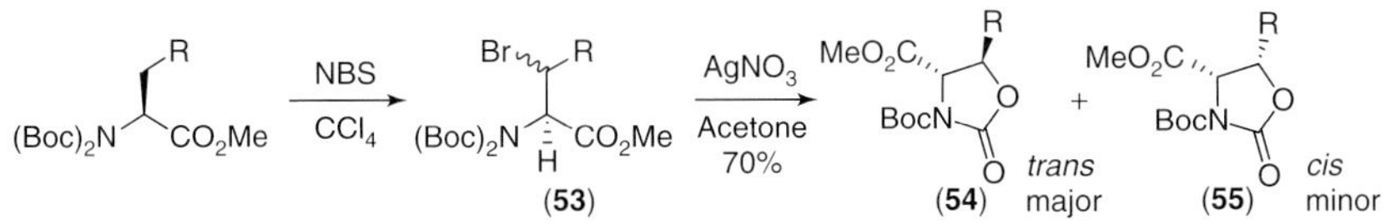

Scheme 6.39

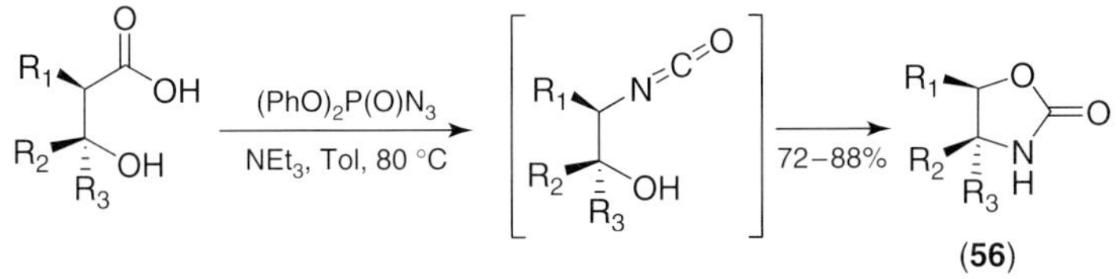

Scheme 6.40

oxazolidinone ring was also generated by a similar intramolecular nucleophilic addition, in basic media, to a vinyl sulfoxide moiety [81] or an epoxide [82].

Crich and Banerjee reported an expedient synthesis of oxazolidinones from α-amino acids, as precursors of β-hydroxy-α-amino acids (Scheme 6.39). [83] The N-bromosuccinimide (NBS)-mediated radical bromination of N,N-di-tert-butoxycarbonyl-protected α-amino acids afforded a 1:1 mixture of the diastereoisomeric bromide **53**, which was treated with silver nitrate to give predominantly the *trans*-oxazolidinones **54**.

Formation of acyl azide from chiral β-hydroxy-α-amino acids, Curtius rearrangement and internal cyclization afforded 4-substituted and 4,5-disubstituted oxazolidin-2-ones **56** in good yields and with complete stereochemical control, as described in Scheme 6.40 [84].

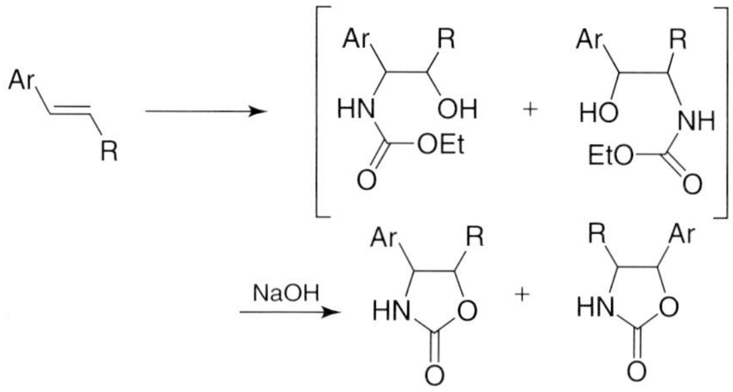

Scheme 6.41

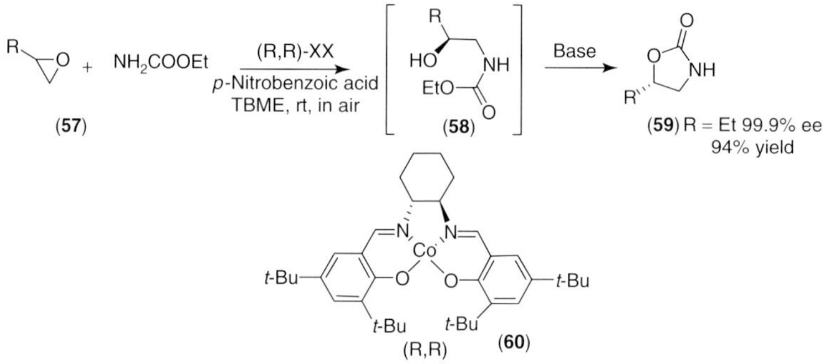

Scheme 6.42

In the following approach, the authors envisioned the use of the Sharpless asymmetric amino hydroxylation of a styrene derivative followed by a base-mediated selective ring closure of the resulting carbamate to afford oxazolidinones in a single step (Scheme 6.41). The optimized aminohydroxylation conditions were used on several β-substituted styrene derivatives [85].

Bartoli et al. reported a methodology that provided a simple and practical tool for synthesizing a collection of varied 5-substituted oxazolidinones in almost enantiomerically pure form from racemic terminal epoxides [86]. The asymmetric carbamate-based aminolytic kinetic resolution (AKR) of racemic epoxides **57** catalyzed by Jacobsen's (salen)-Co^{III} chiral complex **60**, which was generated in situ by oxidation of the catalytically inactive (salen)-Co^{II} complex, gave N-protected 1-amino-2-ol **58** in almost enantiomerically pure form (Scheme 6.42). The best base to promote the subsequent cyclization step was sodium hydride. This one-pot reaction sequence appeared as an easy and very convenient protocol to access a wide range of enantiopure 5-substituted oxazolidinones **59**.

5-Functionalized enantiomerically pure oxazolidin-2-ones **64** were also prepared in one pot from commercially available chiral aziridines **61** bearing an EWG at C-2 (R = ester, vinyl or acyl, Scheme 6.43) [87]. The reaction was performed with methyl chloroformate to give a high yield (up to 85%) of the 5-functionalized oxazolidinones. This reaction occurred with retention of the configuration through an S$_N$2-type double inversion process: acylation on the nitrogen to form the

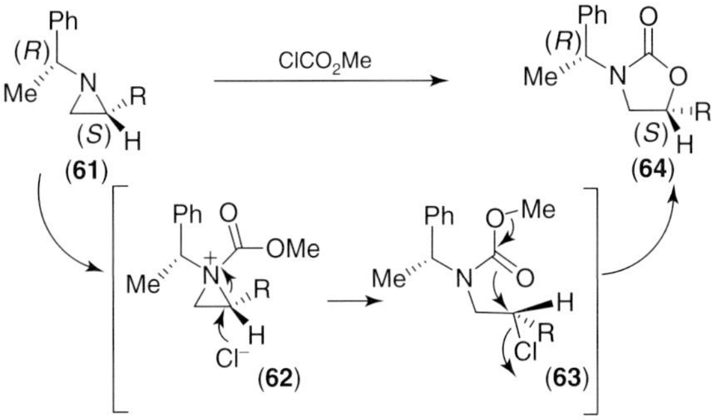

Scheme 6.43

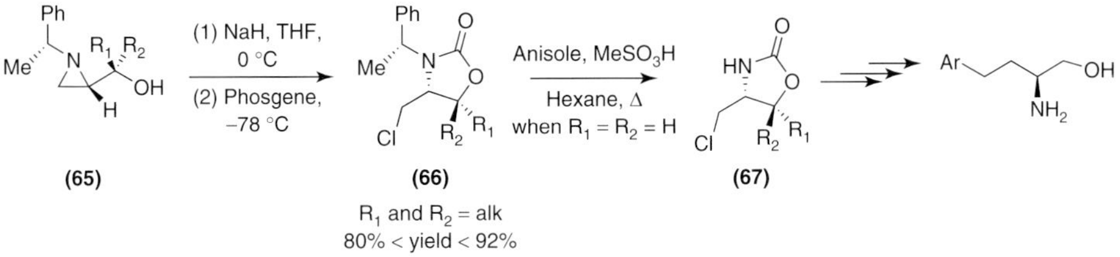

Scheme 6.44

activated aziridinium species **62**, which was regioselectively cleaved by chloride ion via an S_N2 process, followed by intramolecular cyclization by the carbamate moiety.

The same authors reported a novel and efficient pathway to oxazolidinones from aziridines **65** [88]. Treatment of these compounds with sodium hydride and phosgene led to enantiopure (chloromethyl)-5-substituted-oxazolidin-2-ones **66**. Phosgene was selected as the intramolecular cyclizing agent of the amino alcohol moiety in order to form the cyclic carbamate with regioselective ring opening of the aziridine by the chloride ion (Scheme 6.44). The oxazolidinone **67** ($R_1 = R_2 = H$) was then transformed into enantiopure (L)-homophenylalaninol derivatives.

Trost and coworkers have investigated the palladium-catalyzed desymmetrization of meso biscarbamates. [89] The oxazolidinone-forming reaction was easily carried out by preparing the biscarbamate substrate *in situ* from the *cis*-diol **68**, and the biscarbamate solution was added to a solution of 5 mol% catalyst, prepared by stirring a mixture of ligand (L) and tris(dibenzylideneacetone)dipalladium–chloroform complex. The best yields and enantiomeric excesses (84% yield and >99% ee for $n = 1$) were obtained by using L* with $R = R_1 = H$ and by adding triethylamine to the mixture (Scheme 6.45). The origin of the enantioselectivity was explained by a model suggesting that the difference in diastereotopic transition states (ionization of the pro-*S* leaving group vs. ionization of the pro-*R* leaving group) was the result of steric interactions between the chiral ligand and the cycloalkene.

Larksarp and Alper reported a highly enantioselective method for the synthesis of oxazolidinones by palladium(0) catalyzed cycloaddition of vinyloxiranes with heterocumulenes such as carbodiimides or isocyanates by using chiral phosphine ligands [90]. Oxidative addition of the chiral phosphine–palladium complex

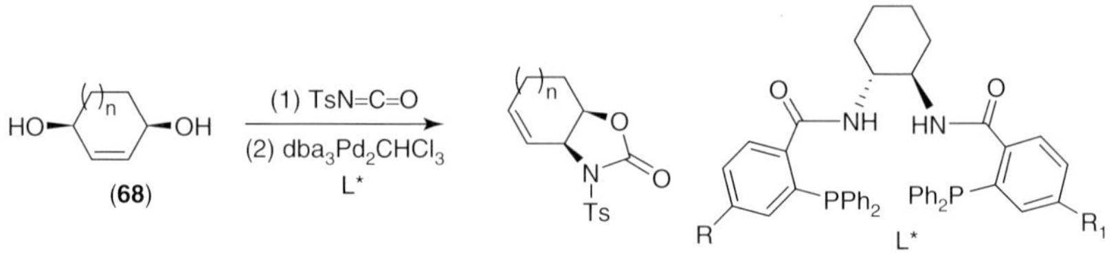

Scheme 6.45

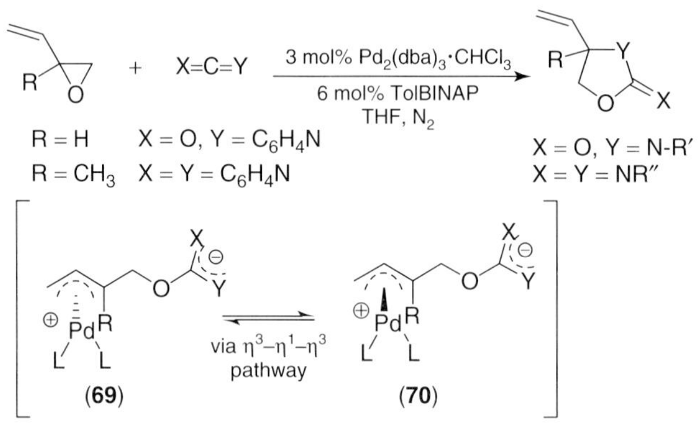

Scheme 6.46

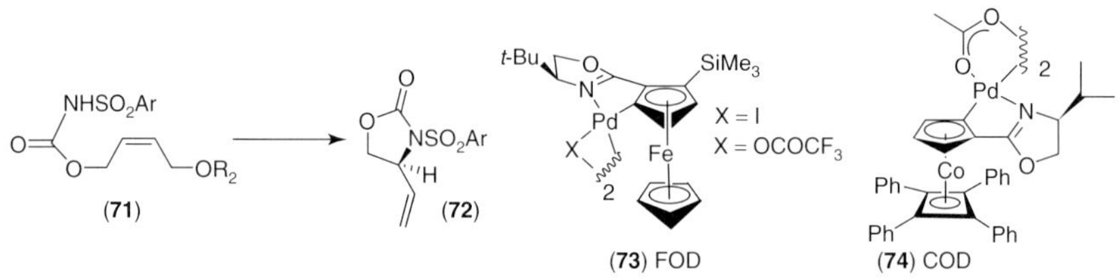

Scheme 6.47

to racemic vinyl oxirane followed by heterocumulene complexation led to a diastereoisomeric π-allyl palladium complexes **69** and **70**, which are equilibrated via an $\eta^3-\eta^1-\eta^3$ mechanism (Scheme 6.46). This equilibrium is much faster than the nucleophilic addition of the nitrogen.

Cyclization of the allylic N-arylsulfonylcarbamates **71** (prepared from allylic alcohol and arylsulfonylisocyanate) catalyzed by ferrocenyloxazoline palladacycles (FOP) afforded 4-vinyloxazolidin-2-ones in high yields and enantioselectivities. A better catalyst was recently found by the same authors: the cobalt oxazoline palladacyclic complex **74** that does not require preactivation like the FOD complexes (Scheme 6.47) [91].

Lespino and Du Bois described the catalytic oxidative cyclization of various carbamates to oxazolidinones [92]. The reactions were performed optimally using $[Rh_2(OAc)_4]$ as catalyst, $PhI(OAc)_2$ as oxidant and MgO as base additive

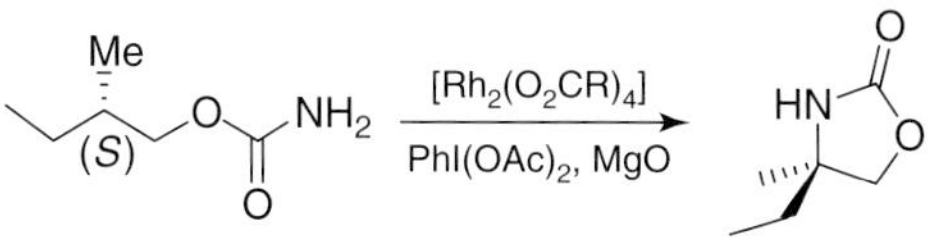

Scheme 6.48

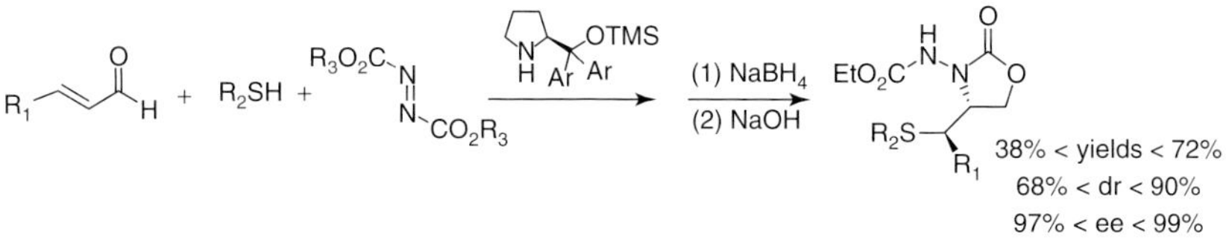

Scheme 6.49

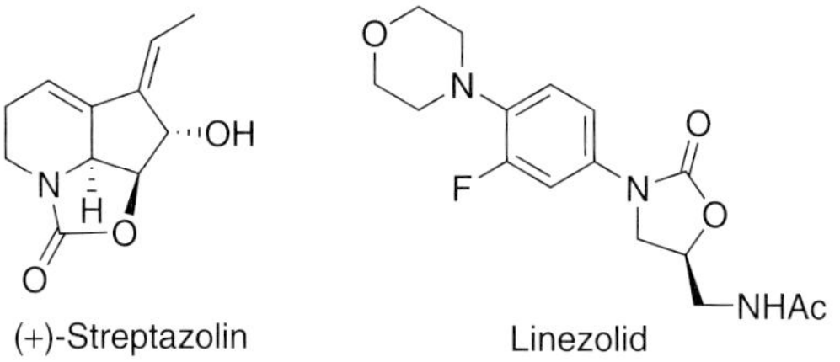

38% < yields < 72%
68% < dr < 90%
97% < ee < 99%

(+)-Streptazolin   Linezolid

Scheme 6.50

(Scheme 6.48). The authors showed that the C–N bond formation is totally stereoselective and postulated a metal-directed N-atom insertion process. This method is significant because it gives access to quaternary α-amino acids.

A very efficient asymmetric multicomponent reaction [93] involving both enamine and iminium activation gave access to highly functionalized oxazolidinones (Scheme 6.49). The domino-conjugated nucleophilic addition of thiols to α-β-unsaturated aldehydes in the presence of 2-[bis(3,5-bistrifluoromethylphenyl)trimethylsilyloxymethyl]pyrrolidine as catalyst, followed by electrophilic amination reaction to the enamine intermediate, resulted in the formation of nearly enantiopure oxazolidinones.

Many recent works have reported the asymmetric synthesis of oxazolidinones via asymmetric desymmetrization of prochiral diols. Serinol derivatives were used to give enantiomerically enriched monoacetates, which were transformed by appropriate reactions into oxazolidinones [94].

Besides the importance of oxazolidinones as chiral auxiliaries, these compounds also constitute a new class of antimicrobial compounds, which were discovered by researchers at Dupont in the 1980s [95]. The oxazolidinone moiety was found in the structure of few biologically active natural products such as (−)-cytoxazone [96] or (+)-streptazolin [97], and much effort has been devoted to the development of analogs (Scheme 6.50).

Actually, these compounds exert potent *in vitro* and *in vivo* antibacterial activity against multiple antibiotic-resistant strains of Gram-positive bacteria. Linezolid [98] was the first oxazolidinone that was introduced into clinical trials and approved

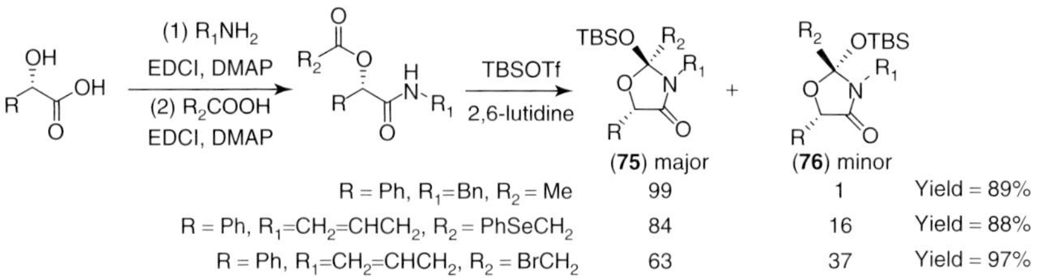

6 Asymmetric Synthesis of Five-Membered Ring Heterocycles with More Than One Heteroatom

Scheme 6.51

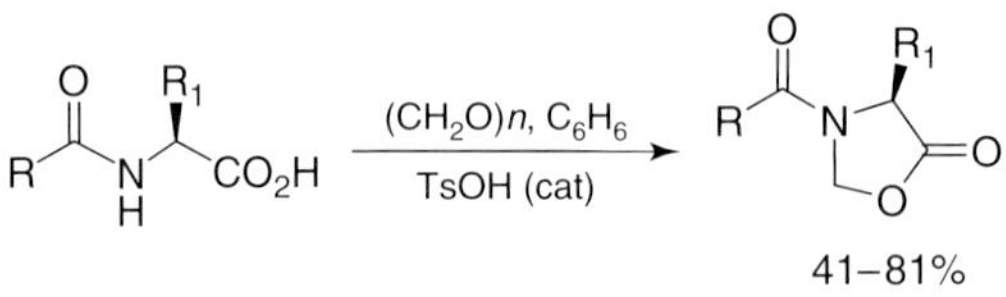

	(75) major	(76) minor	
R = Ph, R_1=Bn, R_2 = Me	99	1	Yield = 89%
R = Ph, R_1=CH$_2$=CHCH$_2$, R_2 = PhSeCH$_2$	84	16	Yield = 88%
R = Ph, R_1=CH$_2$=CHCH$_2$, R_2 = BrCH$_2$	63	37	Yield = 97%

Scheme 6.52

41–81%

in 2000 by the Food and Drug Administration. Many enantioselective syntheses of these products and their analogs have been described by using the classical reactions, cited above, to generate the oxazolidinone ring.

6.1.3.2 Oxazolidin-4-ones and 5-ones

Only a few reports are devoted to the asymmetric synthesis of 4-oxazolidinones. Kamimura et al. [99] described in 2002 the synthesis of a new type of heterocycles: the oxyoxazolidinones **75** and **76** (Scheme 6.51), which are readily prepared from O-acylmandelamides on treatment with TBSOTf.

These compounds were used by the same authors as chiral auxiliaries for heterocyclic synthesis [100].

Although much less used than the oxazolidin-2-ones, 5-oxazolidinones are attractive precursors for the synthesis of various substituted amino acids [101]. Ben-Ishai [102] was the first to describe the reaction between N-Cbz-amino acids with paraformaldehyde, under acid catalysis, that led to 5-oxazolidinones via an intramolecular cyclization. With a view to synthesizing N-methyl amino acids, several chemists extended the range of substrates that can be transformed into 5-oxazolidinones, and different protecting groups (N-Fmoc [103], N-Cbz, N-Boc [104] were used to develop this methodology (Scheme 6.52). Recently, Hughes et al. [105] described the syntheses of N-methyl amino acids of the common 20 amino acids in high yields and without racemization via 5-oxazolidinone intermediates.

The synthesis of α,α'-disubstituted amino acids from a bicyclic oxazolidinone was first developed by Seebach and coworkers [106]. Reaction between proline and pivalaldehyde under acidic catalysis gave the single product **77**, in which the *tert*-butyl group is on the exo face, in a cis relationship with the bridgehead hydrogen. No epimerization occurred during this condensation. Deprotonation and alkylation led to the products **78** in moderate to excellent yields and with essentially complete diastereoselection (Scheme 6.53). Recently, Vartak et al. [107] described a stereoselective synthesis of α,α'-biprolines by using this procedure.

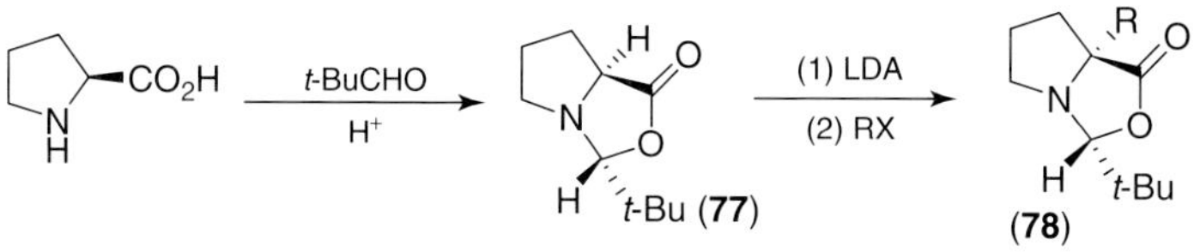

6.1 Five-Membered Heterocycles with N and O Atoms

Scheme 6.53

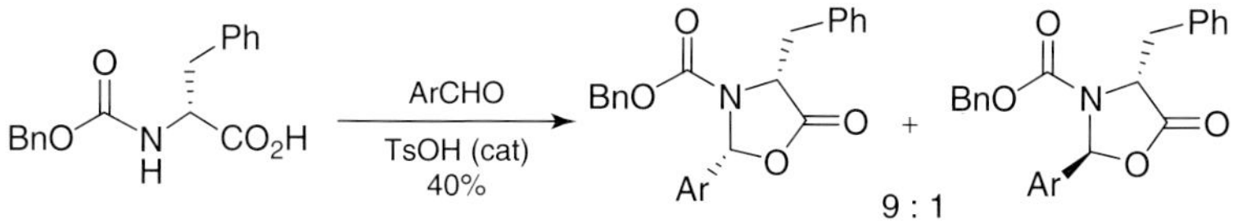

Scheme 6.54

This methodology has been applied to acyclic amino acids with moderate yields. The procedure involves an aromatic aldehyde [108, 109] or pivalaldehyde [110] and afforded a separable mixture of oxazolidinones, whose major isomer showed a cis configuration, as depicted in Scheme 6.54.

Most of the syntheses described here are based on stereoselective synthesis using chiral precursors (amino alcohols, amino acids, hydroxy acids, aziridines, etc.). But, during the last few years, new strategies using asymmetric catalysis to obtain enantioenriched oxazolidinones have emerged. They have contributed to widening the field of application of these compounds as chiral auxiliaries and as biologically active compounds.

6.1.4
Isoxazolines and Isoxazolidines

As clearly suggested by the presence of their N–O bond, isoxazolidines **79** and isoxazolines **80** are commonly prepared (Scheme 6.55) through 1,3-dipolar cycloaddition reactions [62, 111] from alkenes and nitrones **81** or nitrile oxides **82**, respectively.

As it will be detailed below, isoxazolidines **79** are especially valuable reagents for synthetic purposes, whereas isoxazolines **80** essentially constitute the basis

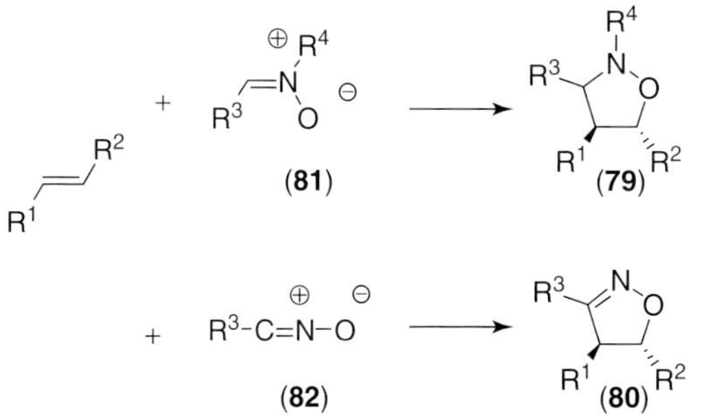

Scheme 6.55

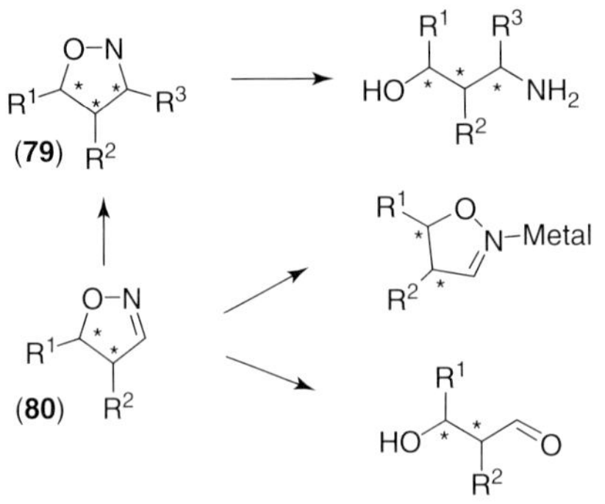

Scheme 6.56

for meaningful ligands that are used during various metal-catalyzed reactions (Scheme 6.56).

6.1.4.1 **Isoxazolidines**

Chiral isoxazolidines are known for a long time as valuable intermediates in the synthesis of many important classes of molecules: alkaloids, β-amino acids, β-lactams, amino sugars, and so on. This is due to the labile nature of their N–O bond; isoxazolidines can be viewed as masked forms of γ-amino alcohols and they are easily prepared in an enantiopure form from chiral nitrones or chiral alkenes or also, to a lesser extent, from both chiral nitrones and alkenes [112]. Intermolecular as well as intramolecular [3 + 2] cycloadditions have been performed and the recent use of metal catalysis was especially examined [62, 111, 113].

First, it should be underscored that such reaction creates up to three stereogenic centers and, irrespective of the enantioselectivity issue, this fact requires that regio- and diastereoselectivities are controlled [114]. The classical endo/exo problem adding further to these difficulties, much work has been devoted to circumvent these points [111, 115].

For instance, MacMillan [116] (Scheme 6.57) described an example of remarkable control of all these selectivities by using crotonaldehyde **83** (the presence of the EWG on the alkene clearly directs the regioselectivity on the dipolarophile) reacting with nitrone **84** in the presence of the imidazolidinone salt **85**, which acts as the chiral organocatalyst.

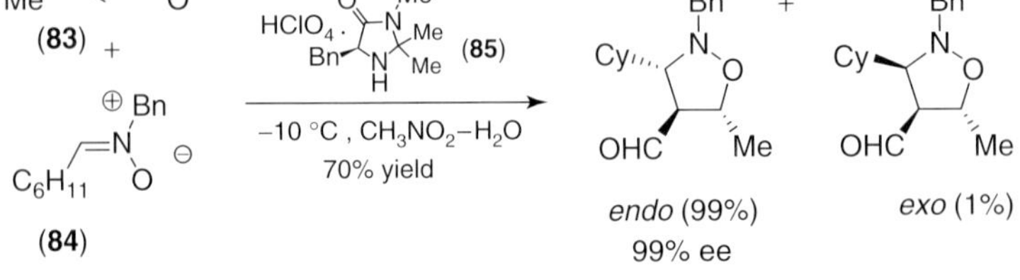

Scheme 6.57

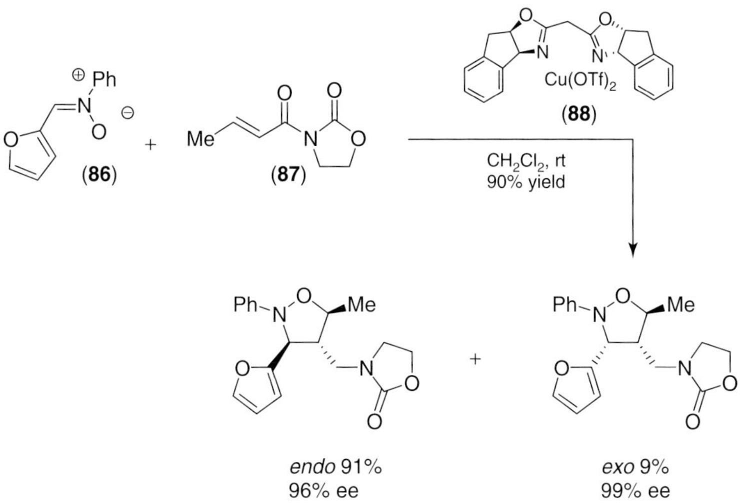

6.1 Five-Membered Heterocycles with N and O Atoms

Scheme 6.58

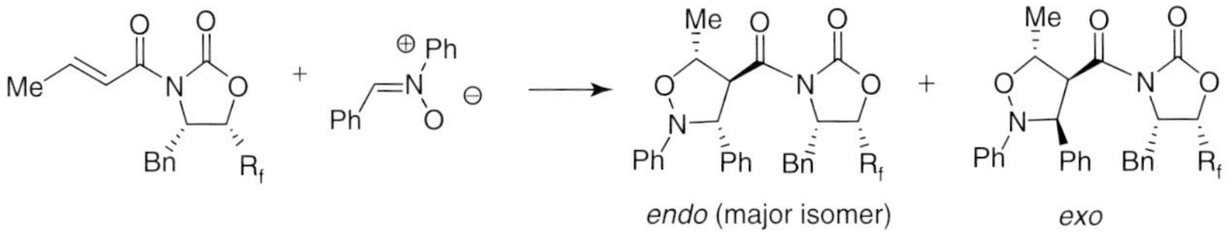

Scheme 6.59

The improvement of the diastereofacial selectivity provoked by the presence of a Lewis acid is common in the cycloaddition field. This can be also substantiated in the present case as shown in Scheme 6.58, which depicts the enantioselective reaction of nitrone **86** with dipolarophile **87** under the catalytic effect of a chiral bisoxazoline/Lewis acid complex **88** [117a].

Nitrone cycloaddition of a fluorous oxazolidinone chiral auxiliary catalyzed by Yb(Otf)$_3$ or Sc(Otf)$_3$ was reported [118] to give cycloadducts with excellent yields and stereoselectivities (Scheme 6.59). This initial work stimulated intensive research in the field of fluorous-supported auxiliaries [119] in order to make use of such reactions for specific synthesis purposes [73, 120] or for elaborating combinatorial libraries [121].

Interestingly, N-oxide derivative **90** was produced via a reverse-Cope elimination from hydroxylamine **89** [122]; on the other hand, after being oxidized by air, hydroxylamine **89** reacted intramolecularly to afford the spiro tricyclic isoxazolidine **91** with excellent diastereoselectivity. Eventually, the N–O bond was cleaved with zinc, yielding a hydroxylamine, as shown in Scheme 6.60. Similar strategies were described by others later on [123].

As might be anticipated, nitrones derived from natural products have given rise to numerous studies. Two of these (Scheme 6.61) may serve as examples of this fruitful area. First, highly regio- and stereoselective cycloaddition of methyl acrylate

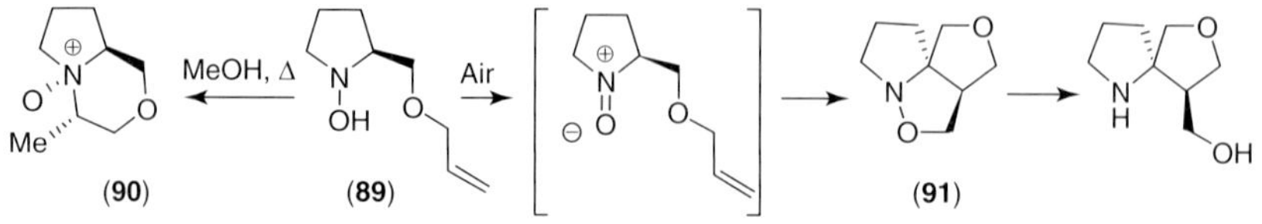

Scheme 6.60

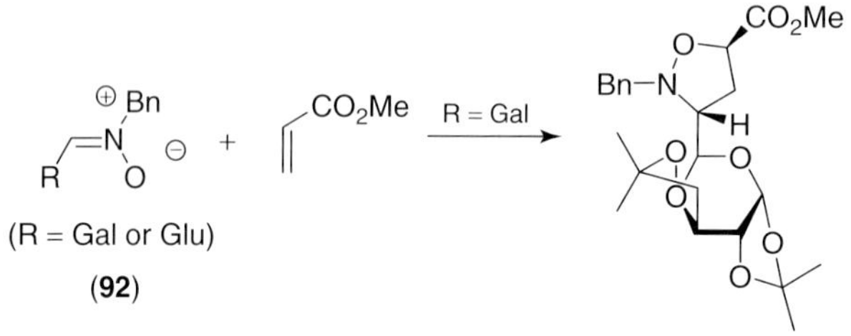

Scheme 6.61

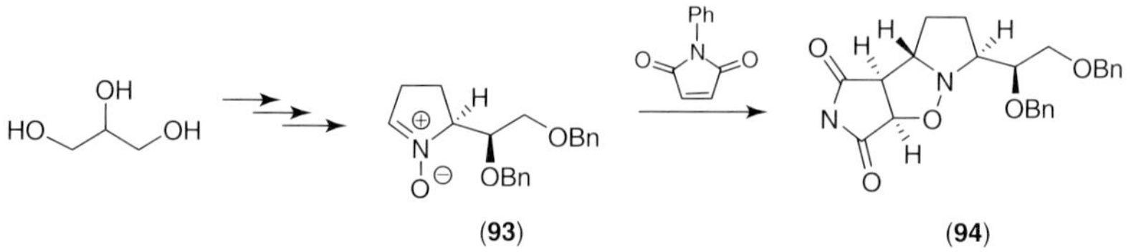

Scheme 6.62

to nitrones **92** derived from D-galactose and D-glucose afforded starting material for the synthesis of complex molecules [124].

The second example [125] is related to nitrone **93**, synthesized from a D-glyceraldehyde derivative, which shows complete regio- and stereoselectivities when it reacts with a large range of dipolarophiles. For instance (Scheme 6.62), its reaction with N-phenyl succinimide produced compound **94** in an exclusive exo/anti cyclization mode.

An attractive organocatalytic synthesis of 5-hydroxyisoxazolidines has been described by Cordova et al. [126]. These authors made use of a tandem reaction involving hydroxylamines and α,β-ethylenic aldehydes as reagents (Scheme 6.63); the target compounds offer a versatile entry to the synthesis of different β-amino acids and γ-amino alcohol derivatives.

6.1.4.2 Isoxazolines

2-Isoxazolines are also prepared from a [3 + 2] cycloaddition reaction to alkenes, the reagents here being nitrile oxides (Scheme 6.55) [127]. They can be prepared as single stereoisomers when the chiral allylic alcohol is a dipolarophile [128]. Synthesis of such heterocycles as well as their applications have been intensively reviewed [62, 111] and this can be easily understood given the usefulness of such compounds, in particular, as starting material for the synthesis of amino acids [129] and as chiral ligands in metal-catalyzed reactions [130].

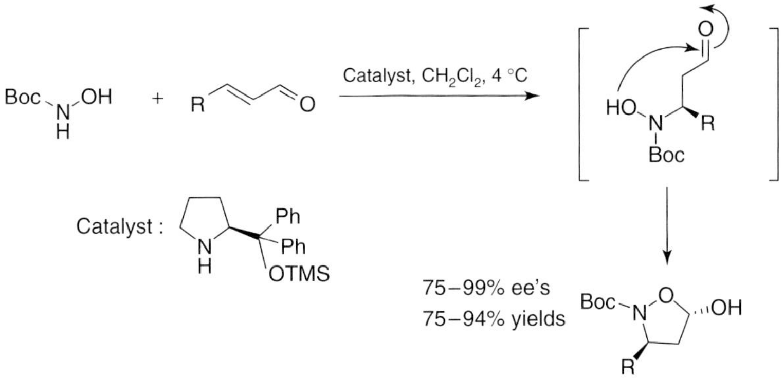

6.1 Five-Membered Heterocycles with N and O Atoms | 247

Scheme 6.63

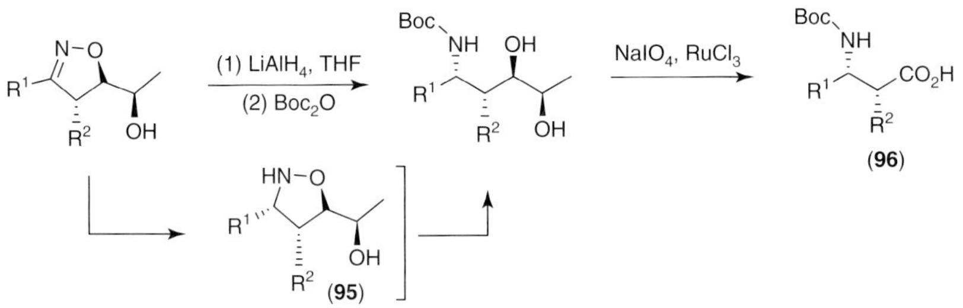

Scheme 6.64

It could be first mentioned that isoxazoline can be transformed into their saturated homologs, namely, the isoxazolidines (see above), with advantageous control of the created stereogenic center at the C-3 position. For instance, Mapp et al. [129] reported stereoselective hydride addition to the C-3 C=N bond of the unsaturated heterocycle in order to obtain intermediate isoxazolidines **95**, which were then transformed into β-amino acids **96** (Scheme 6.64). Similar reactions were also described by using Grignard and organolithium reagents.

Though enantiopure isoxazolidines are commonly produced when starting from either chiral reagents or chiral catalyst, racemic isoxazolidine can be resolved by using lipases. In this way, condensation of achiral nitrile oxide **97** and isobutyl vinylacetate **98** afforded racemic isoxazoline **99**. However, enzymatic hydrolysis with lipase PS30 operates smoothly a kinetic resolution, yielding the enantiopure isoxazoline acid **100** (Scheme 6.65) [131].

The effect of Lewis acids linked to a bisoxazoline ligand gave rise to a recent, interesting study about the control of regioselectivity which can be crucial during the nitrile oxide cycloaddition. Sibi et al. [132] recently made a careful examination of the different factors that affect this selectivity and delineated the role of various Z templates and Lewis acids in the relative production of the C-adduct and the O-adduct (Scheme 6.66). The higher regioselectivity as well as the best enantio-selectivity was obtained by using MgI_2 linked to the ligand **101** and the achiral pyrazolidinone template **102**.

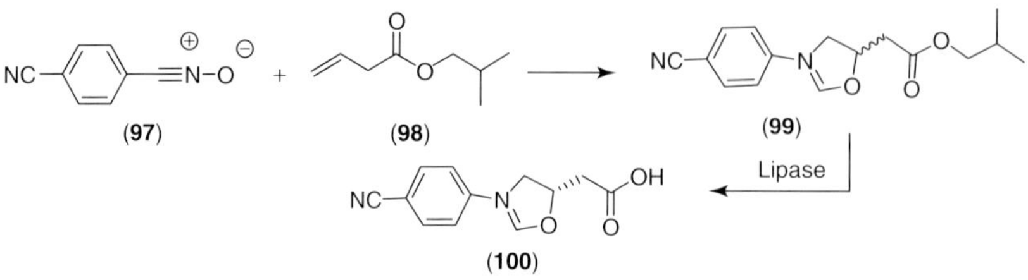

Scheme 6.65

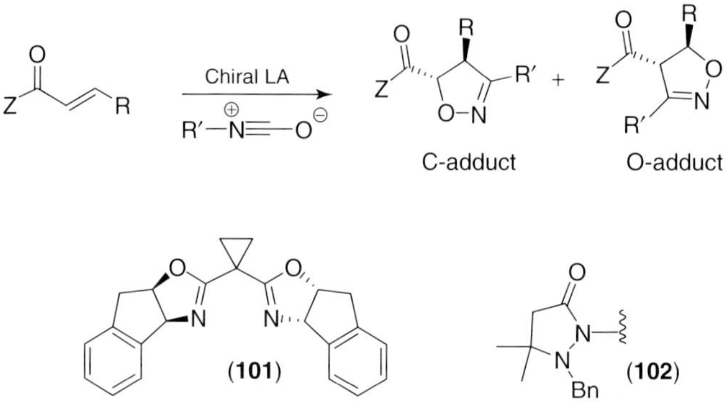

Scheme 6.66

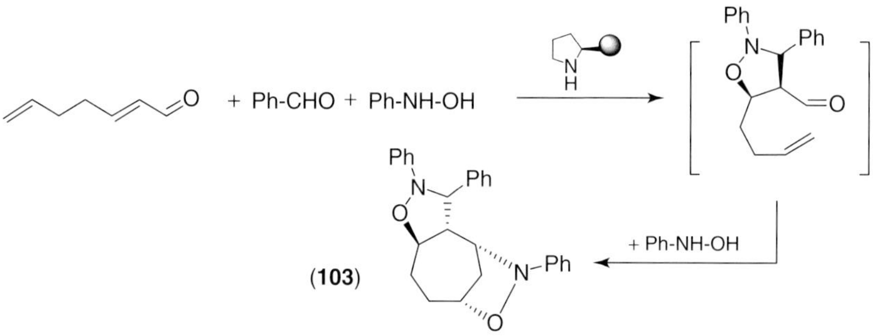

Scheme 6.67

Actually, very enantioselective processes have been described as the outcome of the 1,3-dipolar cycloaddition reactions that lead to isoxazolidines when mediated by a chiral Lewis acid, and it meaningful to note that Box ligands (see above) appear as the most appealing ones [133]. Apart from these processes involving chiral catalysts, the most common ones are definitely the use of chiral nitrile oxides or chiral alkenes; both processes have been extensively reviewed [111].

Recently, it has been reported [134] that a supported proline catalyst can be used for a three-component intermolecular [3 + 2] cycloaddition, which is displayed in Scheme 6.67. Tricyclic bis-isoxazoline **103** was thus obtained with an outstanding

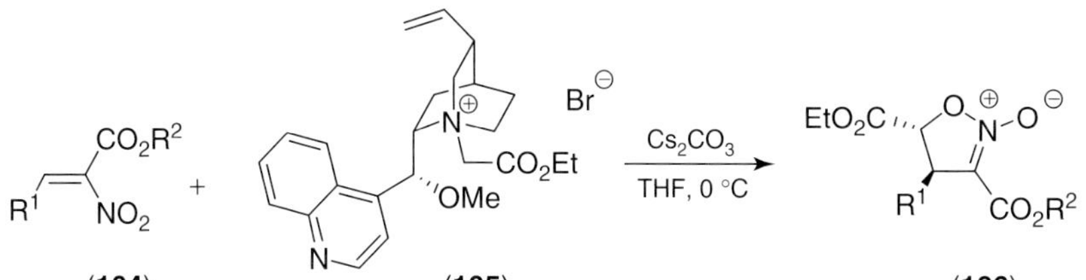

Scheme 6.68

stereocontrol (>25:1 dr and 99% ee) during the formation six new bonds and five new stereocenters.

In conclusion, it is interesting to note that a recent preliminary report describes a very stereoselective synthesis of isoxazoline N-oxides [135]. In the presence of cesium carbonate, nitroolefins **104** reacted with the ammonium salt **105** of cinchonidine to afford nitrile oxides **106** with utmost dia- and enantioselectivities (Scheme 6.68).

6.2
Five-Membered Heterocycles with Two N Atoms

6.2.1
Imidazolidines and Imidazolidinones

6.2.1.1 Imidazolidines

At the end of the 1970s, Mukaiyama [136] reported the first efficient asymmetric synthesis, which was based on the use of chiral diamino ligands (including heterocyclic ones), and since then this field is continuously growing [137]. As will be developed below, preparations of imidazolidine derivatives and oxazolidine (see above) are similar; however, it should be noted at the onset that the chiral pool does not furnish diamines as abundantly as the amino alcohols, which are the starting material for the synthesis of oxazolidines.

Actually, the condensation of chiral diamines with aldehydes produces imidazolidines smoothly, which are rather fragile compounds (as every aminal is), and care is needed during the hydrolysis process. Scheme 6.69 shows Mukaiyama's preparation of compounds **108–111** from diamine **107** derived from (S)-N-benzyloxycarbonylproline. These N-acylimidazolidines were reported to be very efficient chiral auxiliaries for remarkably enantioselective Grignard 1,2 or 1,4 additions. The 2R configuration of the aminals was produced exclusively, and this has been explained by its less crowded structure compared to that of its 2S epimer [138].

As mentioned above, enantiopure 1,2-diamines are not very commonly obtained from the chiral pool, and an original way for supplying for these heterocycles was reported (Scheme 6.70) [139]. It consists of asymmetric deprotonation of an achiral imidazolidine **112** carried out by butyl lithium in the presence of sparteine, and the resulting anion was then reacted with an alkyl halide, the eventual ring opening

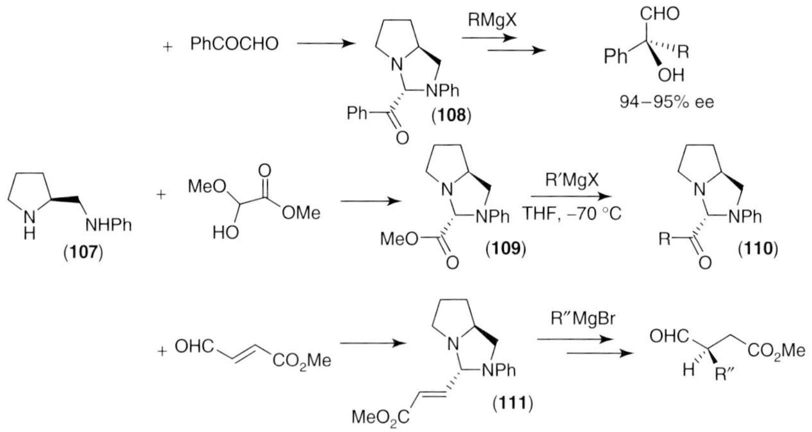

Scheme 6.69

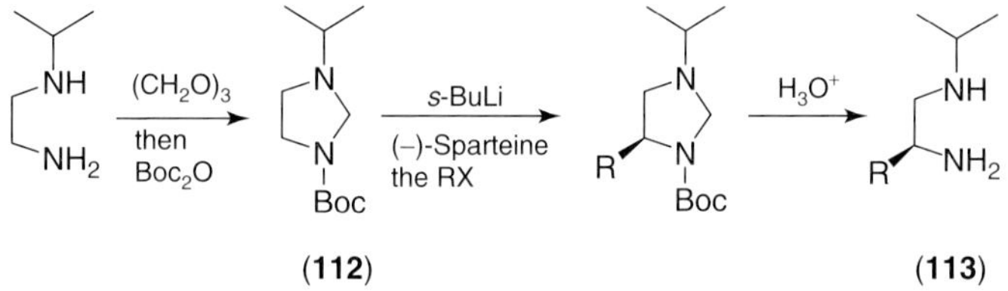

Scheme 6.70

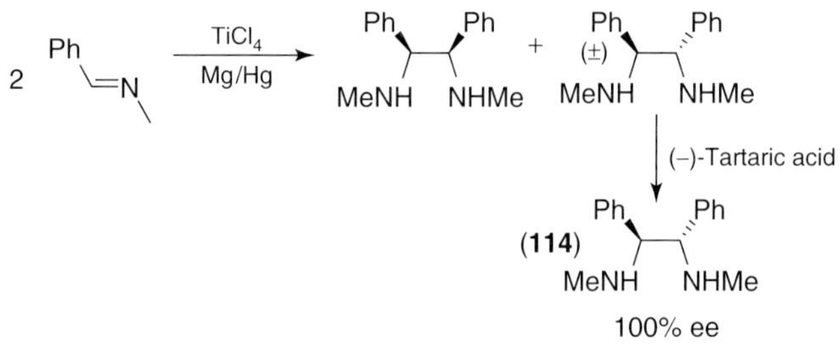

Scheme 6.71

and deprotection of the amino group allowing the synthesis of many optically active 2-alkyl diamines **113**.

The fact that such diamines are not easily attainable is also illustrated by the formation of N, N′-dimethyl-1,2-diphenylethylenediamine in the racemic form via a reductive coupling reaction with low-valent titanium species (Scheme 6.71). Flash chromatography allowed the separation of the chiral isomers from the meso isomers. This very attractive synthesis was reported by Normant's groups [140], who resolved the racemic mixture with (+)-tartaric acid and obtained large amounts R, R-N, N′-dimethyl-1,2-diphenylethylenediamine (DMPEDA) **114**.

The imidazolidines that result from the condensation of an aldehyde (ketones are much less reactive) with such a chiral diamine show the remarkable C_2 symmetry,

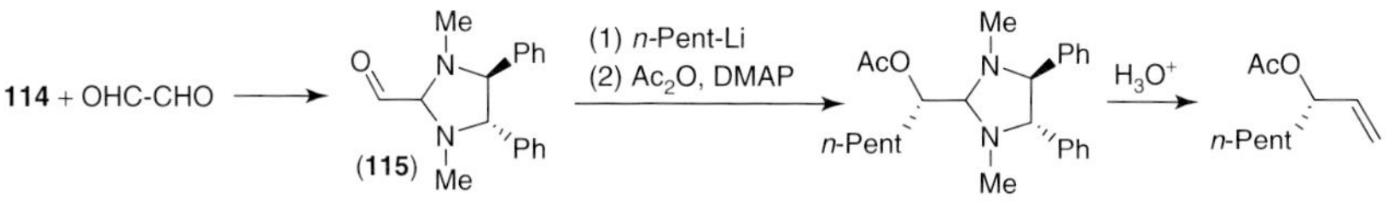

Scheme 6.72

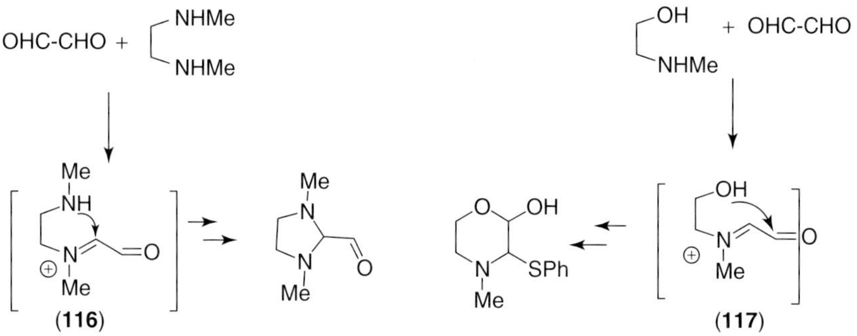

Scheme 6.73

whose interest since the pioneering work by Kagan [141] is widely recognized [142]. In the present case, it should be noted that, in imidazolidine **115**, the created C-2 center is nonstereogenic and this avoids the necessity of controlling the stereochemistry of this center (as was the case with oxazolidines, see above). Alexakis and Mangeney took advantage of this fact when they developed the use of such aminals as chiral auxiliaries for an impressive number of very original syntheses [143]. An example of such transformation is depicted in Scheme 6.72 [144].

Incidentally, Scheme 6.73 reveals another interesting feature: while condensation of glyoxal with 1,2-amino alcohols leads to a morpholine structure (see above) [17b], an analogous reaction leads to imidazolidines when 1,2-diamines come into play. Yet, both condensations involve iminium ions (**116** and **117**) as the key intermediate (Scheme 6.73), but its intramolecular cyclization affects either the C=N or the C=O bond according to the hydroxyl or amino nature of the nucleophile. To our knowledge, no explanation has been given to this opposite behavior.

As mentioned above, enantiopure, C_2-symmetric 1,2-diamines are not simply available. Corey et al. [145] have described another way for their production. It consists in condensing benzil with cyclohexanone and reducing the formed bis-imine **118** as shown in Scheme 6.74; the enantiopure diamine **119** was then obtained via optical resolution with tartaric acid.

This method was taken up many a time afterwards [146]. In particular, Kanemasa and Onimura [147] made use of such diamines to prepare new chirality-controlling auxiliaries such as imidazolidine and made an exhaustive conformational analysis in order to understand their behavior during dipolar cycloadditions and Diels–Alder reactions.

Aiming to synthesize isoindolones, which present many interesting biological activities, Katritzki et al. [148] realized the ingenious transformation depicted in

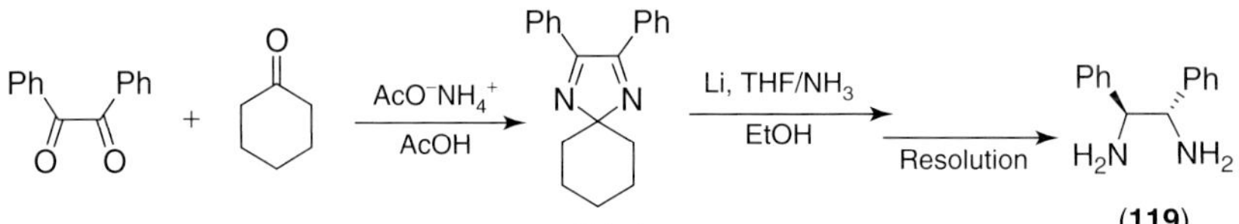

Scheme 6.74

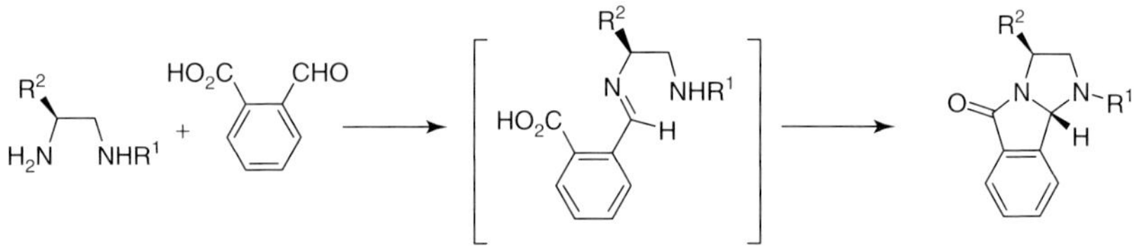

Scheme 6.75

Scheme 6.75. Condensation of enantiopure diamines **120** with 2-formylbenzoic acid **121** afforded, with both high yields and stereoselectivities, various tetrahydro-5H-imidazo[2,1-a]isoindol-5-one **123**. Imine **122** is a likely intermediate in which the addition of the amino moiety onto the imine bond occurs on the less crowded *re* diastereoface of the imine double bond.

Actually, imidazolidine and imidazolidinone heterocycles take part in many fruitful ligands for various metal-catalyzed reactions or for organocatalyzed reactions [149].

6.2.1.2 Imidazolidinones

4-Imidazolidinones Lithium enolates of 4-imidazolinones are the basis of the classical asymmetric transformation proposed by Seebach who founded the famous concept of "Self-generation of stereocenters" [150]. Imidazolidinones **125** were prepared in racemic form (Scheme 6.76) either by condensing the glycine-derived salt **124** with pivalaldehyde or via an indirect route from bromoacetyl bromide; mandelic acid was used in order to resolve the racemic mixtures and produce the

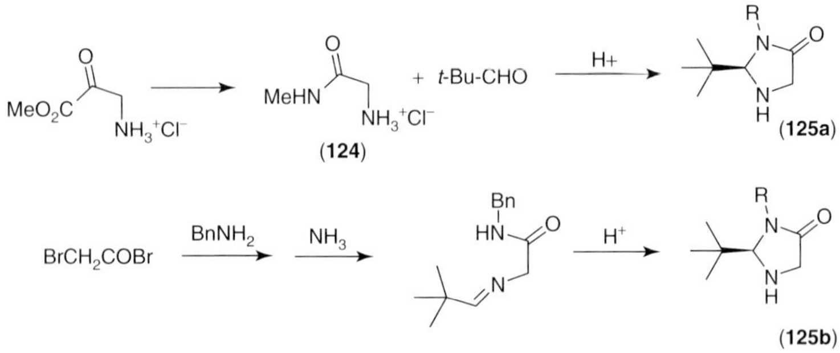

Scheme 6.76

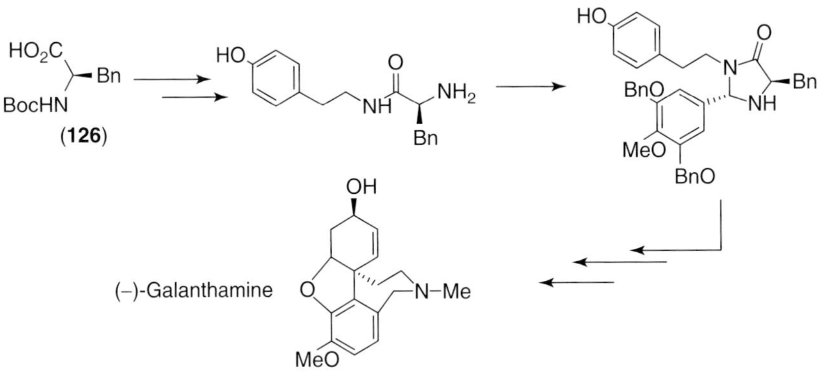

Scheme 6.77

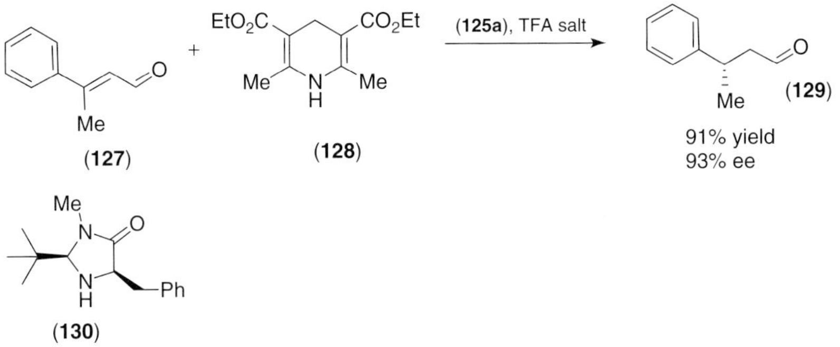

Scheme 6.78

enantiopure imidazolidinones [151]. Numerous other imidazolidinones were prepared in this way and these works mark out a new method for the enantioselective syntheses of many amino acids from such an asymmetric derivatization of glycine [152].

Clearly, every α-amino acid, other than glycine, can be used, and in such a case the produced imidazolidinone shows one more stereocenter. In this fashion, the synthesis of (−)-galanthamine, performed by Node et al. [153], made use, as a starting material, of the imidazolidinone **126** formed from Boc-D-alanine (Scheme 6.77).

MacMillan recently introduced a new concept in the realm of organocatalysis: the lowest unfilled molecular orbital (LUMO) lowering iminium activation which has found many rewarding applications for various transformations [154]. For instance (Scheme 6.78), imidazolidine **125a** catalyzes an enantioselective transfer hydrogenation from Hantzsch ester **128**, and enal **127** was reduced to the optically active aldehyde **129**. Imidazolidinone **130** was also tested but the enantioselectivities thus obtained were lower than with **125a**.

An original route to chiral oxazolidin-2-ones has been described by Kunieda et al. [155] through the smooth cycloaddition of 1,3-diacetyl-2-imidazolone **131** and

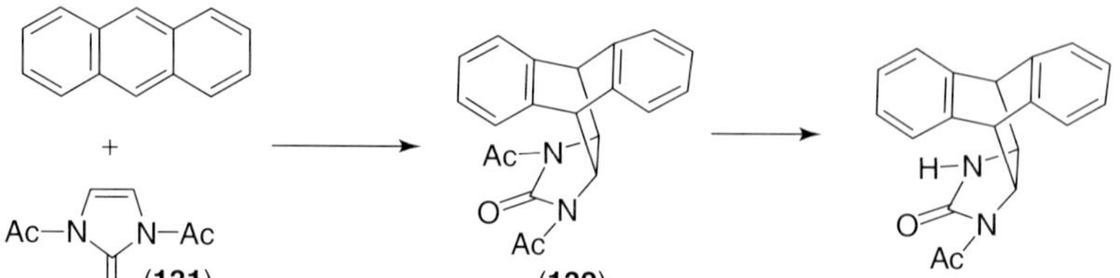

Scheme 6.79

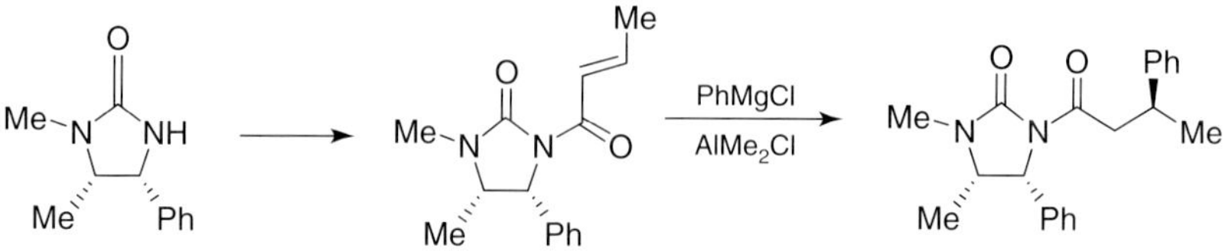

Scheme 6.80

anthracene (Scheme 6.79). This reaction leads to a meso adduct **132**, which was enantioselectively N-deacylated by reaction with an enantiopure amino alcohol.

Compound **133** was used as an exceptionally rigid and congested chiral auxiliary for various asymmetric syntheses such as Diels–Alder reactions and enolate methylations [156].

2-Imidazolidinones 2-Imidazolidinone **134** is easily prepared following a Helmchen procedure [157] from ephedrine and urea. The chiral α, β-unsaturated imide **135**, which was derived from the heterocycle **134**, was submitted (Scheme 6.80) to an asymmetric 1,4-addition in the presence of Lewis acids and this led to the formation of the adduct **136** [158].

Another original application of a 4-imidazolidinone derivative in catalysis was found in the metallocarbene field. Complex **138**, which is formed from imidazolidinone **137** (Scheme 6.81), has been used for an original "aldol reaction" that affords the N-acyl imidazolidinone **139** with excellent stereoselectivity, as shown in Scheme 6.82 [159].

2-Imidazolidinones can also be prepared from an original cycloaddition to phenyl isocyanate [160]. As a matter of fact, Alper reported (Scheme 6.82) that this isocyanate adds to (S)-(+)-n-butyl-2-phenylaziridine in the presence of a Pd^{II} catalyst. It is worth noting that the configuration of the aziridine stereogenic center is retained during this process.

It can be concluded that, owing to their straightforward availability, imidazolidines as well as imidazolidinones play an important role in the asymmetric synthesis as chiral auxiliaries, a field whose expansion has been at lightning speed. However, it seems that their use as chiral ligands in metal-catalyzed reactions or as organocatalysts constitutes at the present time a major subject of research in this area.

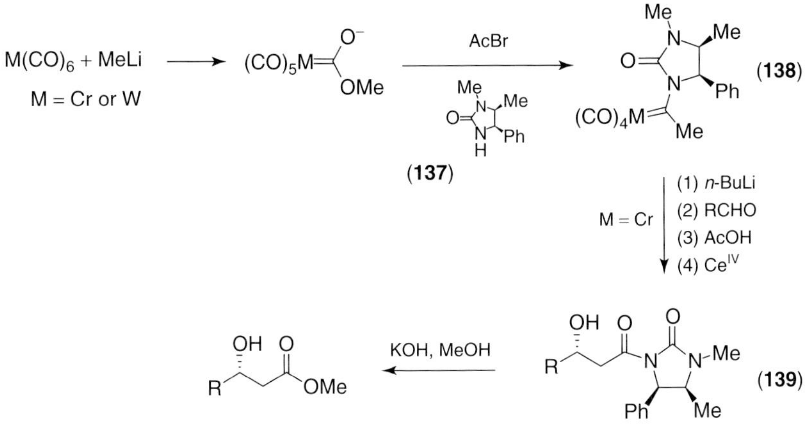

6.2 Five-Membered Heterocycles with Two N Atoms

Scheme 6.81

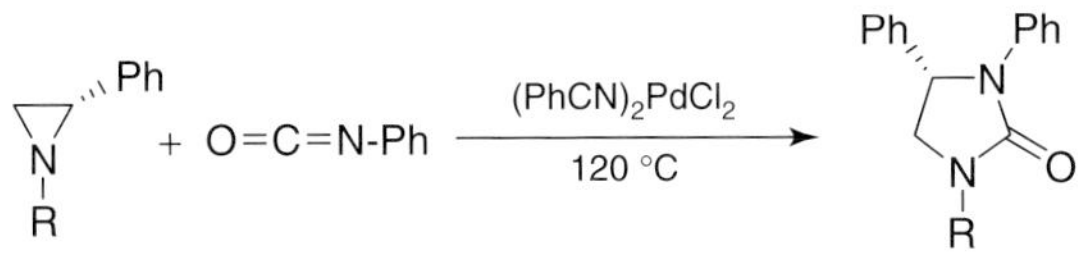

Scheme 6.82

6.2.2
Pyrazolidines and Pyrazolines

6.2.2.1 Pyrazolidines

As was commonly observed in this series, pyrazolidines exhibit very interesting biological properties [161]. On the other hand, from a purely chemical point of view, chiral pyrazolidines can be used for the synthesis of optically active 1,3-diamines by the N–N bond cleavage [162].

Here again, as in the aforementioned case of oxazines, most synthetic methods resort to dipolar cycloadditions to various dipolarophiles. For instance, the azomethine imine dipole **142** is an intermediate during the totally regio- and diastereoselective reaction (Scheme 6.83) of the substituted hydrazine **141** to the substrate **140**, which gave pyrazolidine **143** [163].

Kobayashi et al. described an enantioselective approach to such structures. They made use of a chiral zirconium catalyst to perform the [3 + 2] cycloaddition of

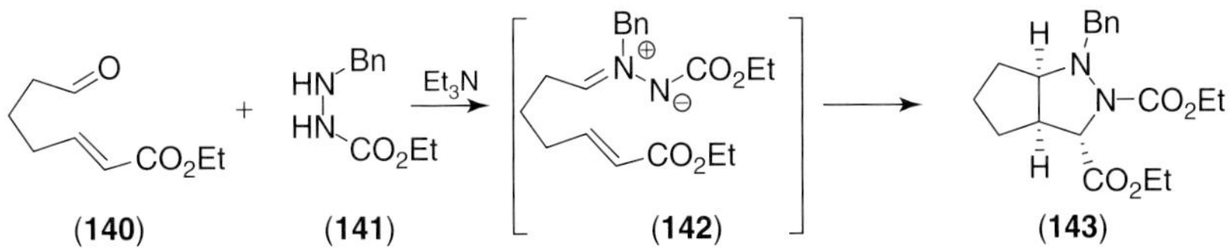

Scheme 6.83

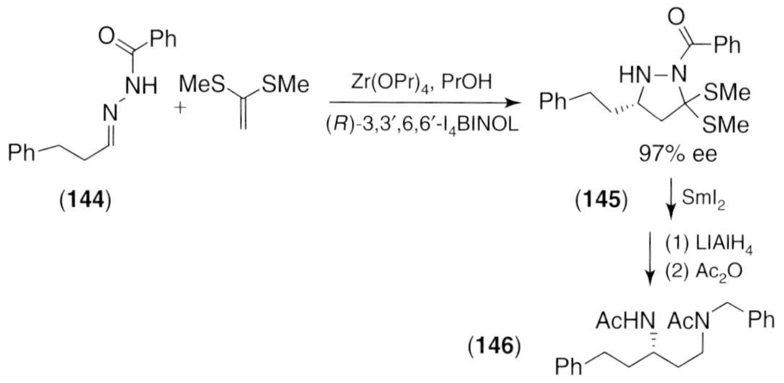

Scheme 6.84

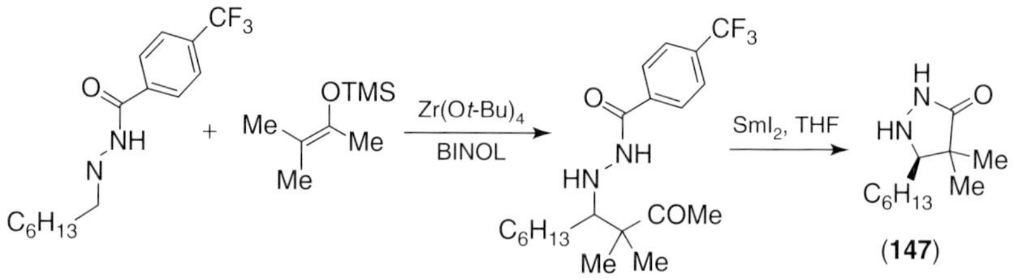

Scheme 6.85

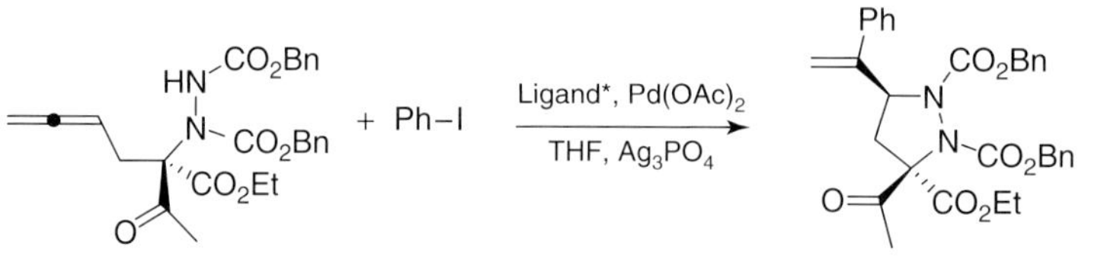

Scheme 6.86

acylhydrazones onto achiral substrates (Scheme 6.84). With tetraiodo-(R)-4-binaphtol (BINOL) as the enantiopure ligand, they were able to synthesize, for instance, pyrazolidine **145**, whose N–N bond cleavage mediated by SmI_2 allowed the formation of the diamino compound **146** [164]. The observed enantioselectivity enabled the authors to discard a possible stepwise pathway in favor of a concerted one.

Acylhydrazones were also used [165] as imine equivalents during a chiral-zirconium-catalyzed Mannich-type reaction (Scheme 6.85). Here again, a BINOL derivative (3,3′-BrBINOL) was chosen as the chiral ligand. The produced hydrazide treated by SmI_2 eventually gave pyrazolidine **147**.

In the same way, a palladium-based chiral catalyst was reported to allow the cyclization between optically active allenylic hydrazines and organic halides: that is, with a double asymmetric induction process (Scheme 6.86). Many asymmetric ligands were tested, and finally the (R, R)-box ligand **ent-43** (R = Bn) gave the best results [166].

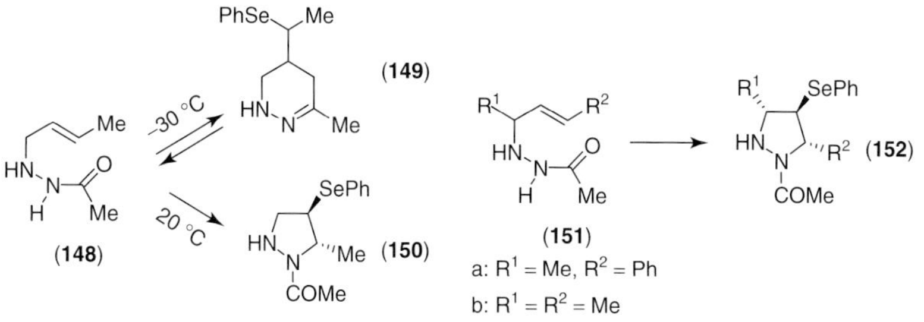

6.2 Five-Membered Heterocycles with Two N Atoms

Scheme 6.87

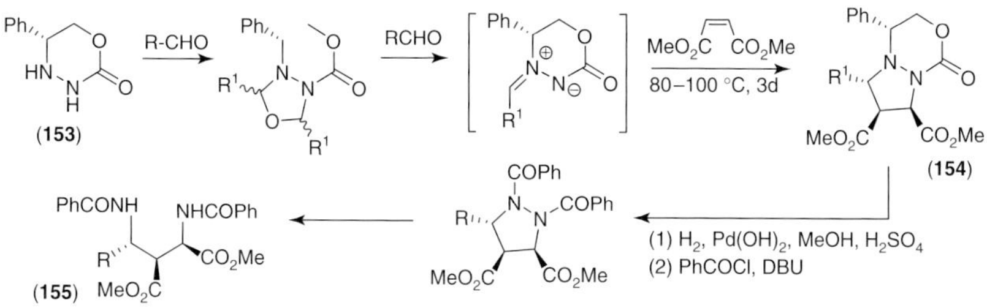

Scheme 6.88

Organoselenium-induced intramolecular cyclization of N-allyl acetylhydrazide **148** affords either oxadiazine **149** or pyrazolidine **150** (Scheme 6.87). It was shown that the five-membered heterocycle was the result of a thermodynamic control, whereas its six-membered isomer was formed under kinetically controlled conditions. Interestingly, substrates **151a–b** were cyclized to produce compounds **152a–b** as the sole diastereomers [167].

Husson et al. [162, 168] described the formation of pyrazolidines **154** via a tandem process from carbazate **153**. The last step in Scheme 6.88 involves a cycloreversion–cycloaddition process; various dipolarophiles (methyl maleate, fumarate and crotonate) were successfully employed. A chemoselective electroreduction of the functionalized pyrazolidines allowed the synthesis of enantiopure 1,3-diamines such as **155**, which was obtained when dimethyl maleate was chosen as the dipolarophile.

Carreira [169] described the asymmetric synthesis of Δ^2-pyrazolines **157** from dipolar cycloaddition reaction of trimethylsilyl diazomethane and camphorsultam-derived acrylates and showed afterwards that various Lewis acid-catalyzed additions to these pyrazolines afforded very smoothly the corresponding pyrazolidines. TiCl$_4$ proved to be the optimal Lewis acid and the adducts were isolated in many cases as single diastereomers.

6.2.2.2 Pyrazolines

The preceding report initiates here a section devoted to the synthesis of chiral pyrazolines. In the field of the scarcely explored dipolar cycloadditions with metal carbene complexes, Barluenga and coworkers [170] described the reaction of

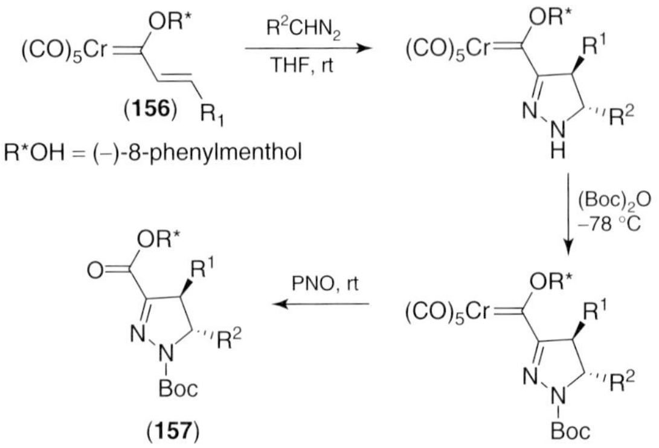

Scheme 6.89

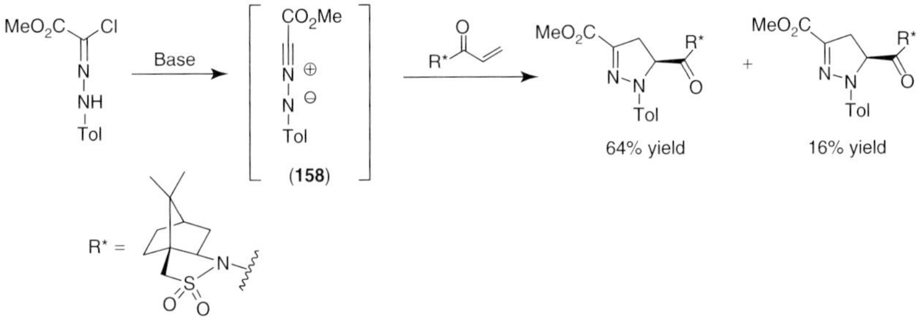

Scheme 6.90

diazo compounds with alkenyl Fischer carbenes **156** derived from chiral alcohols (Scheme 6.89). In this way, enantiopure Δ^2-pyrazolines **157** were obtained.

The process displayed in Scheme 6.89 can be still improved because the formation of Δ^2-pyrazolines **157** can be performed in a one-pot reaction. It is worth noting that stereoselectivity was excellent, whereas an analogous reaction from menthyl *trans*-cinnamate alone (that is, the ligand linked to Cr in **156**) showed a much lesser stereoselectivity when directly condensed with TMSCHN$_2$. Barluenga also showed that treatment of such pyrazolines with LiBEt$_3$H leads to pyrazolidines (see above). Barluenga also described a similar reaction in which metal carbene complexes reacted with nitrile oxides; however, in this case, the stereoselectivity was not so high as in the preceding case [171]. Apart from this single use, Δ^2-pyrazolines were examined for their physical applications [172a] and their pharmacological activity [172b].

Cycloadditions with nitrile imine intermediates **158** were studied by Molteni *et al.* (Scheme 6.90), who tested various chiral auxiliaries R* in such condensations [173]. Intramolecular cycloadditions with nitrile imine intermediates **159** were also described [174], but in this case too the stereoselectivity was rather poor as shown in Scheme 6.91; however, the resulting diastereomers can be easily separated through column chromatography and appeared to be enantiomerically pure.

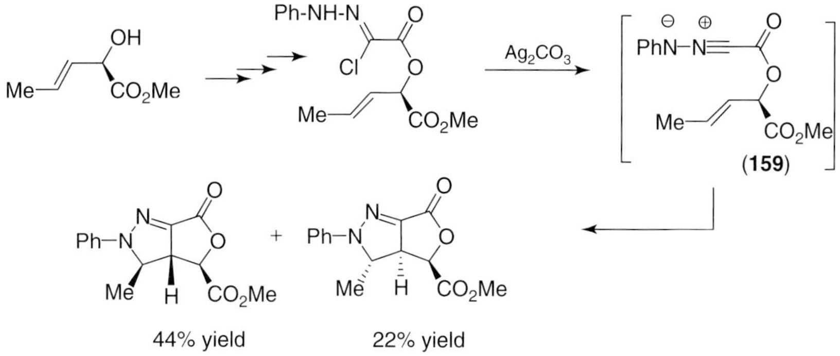

6.2 Five-Membered Heterocycles with Two N Atoms

Scheme 6.91

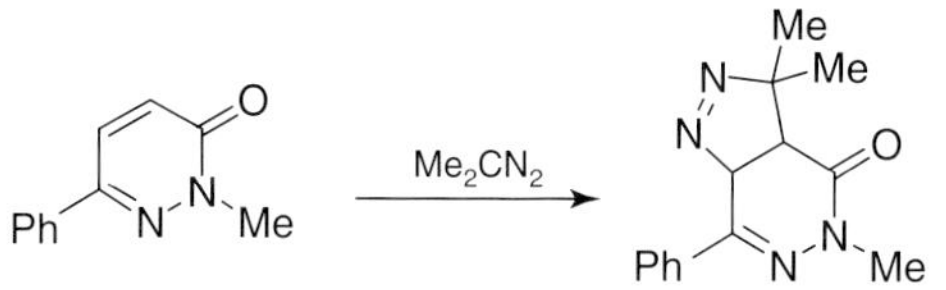

Scheme 6.92

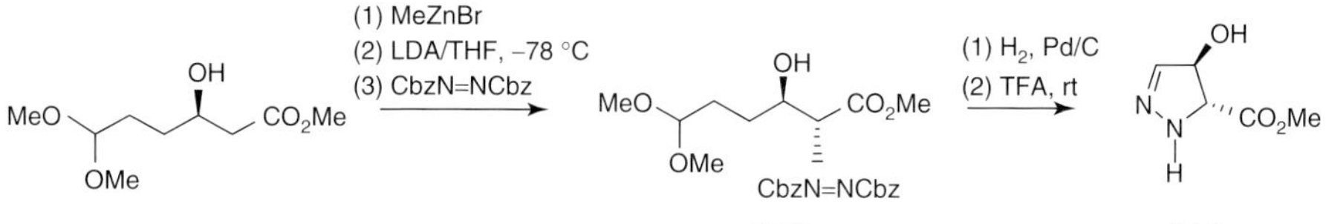

Scheme 6.93

Actually, it was already shown by Stanovnick and Svete [175] that dipolarophiles can be the substrates of such cycloaddition: Scheme 6.92 displays one example of such reaction with diazomethane as the dipolar constituent.

Another intramolecular cycloaddition leading to Δ^2-pyrazolines **161** was described by Genet and Greck [176]. The hyrazino derivative **160** was first obtained (Scheme 6.93) from the zinc enolate of the hydroxyester by treatment with dibenzylazodicarboxylate, and it was then N-deprotected and cyclized in the presence of trifluoroacetic acid.

A cooperative chirality control was observed by Kanemasa [177] during a Lewis acid-catalyzed cycloaddition of a diazoalkane (Scheme 6.94). Thus, several chiral catalysts were screened, and ligand **162** was found to be the most effective to perform the cycloaddition that led to Δ^2–pyrazolines **163** with excellent yields and enantioselectivity.

Unsaturated pyrazolidinone imides, though less reactive than the usual dipolarophiles, were found to be efficient substrates for cycloadditions involving diazoacetates [178]. The produced pyrazoline **165** was obtained in an enantiopure form, and Sibi took advantage of this reaction in order to synthesize (−)-manzazidin

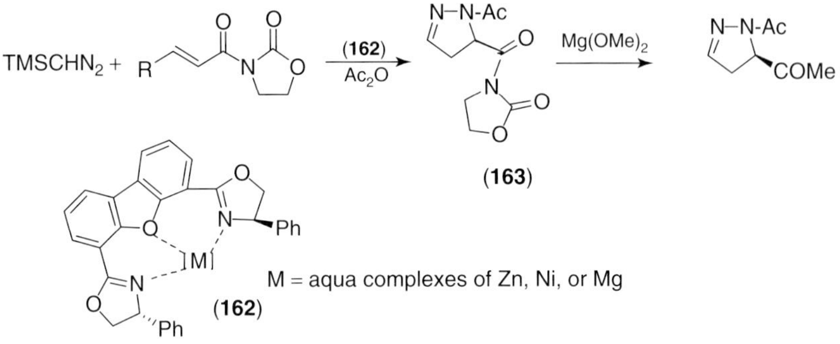

Scheme 6.94

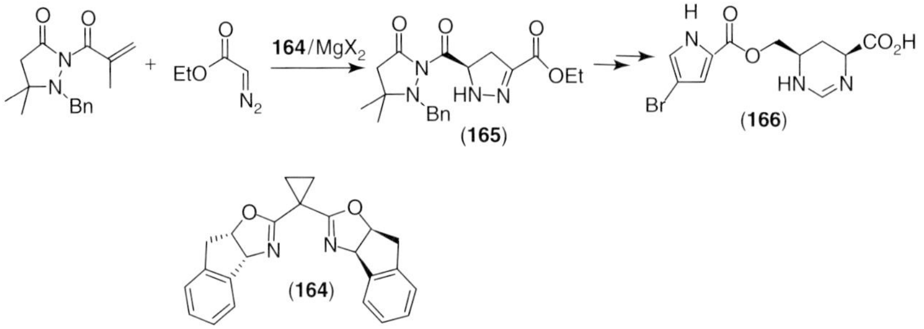

Scheme 6.95

A **166** (Scheme 6.95). In this case, as in many others reviewed above, a chiral Box ligand **164** was used in the catalyst system.

6.2.3
Pyrazolidinones

One of the first preparations of a chiral pyrazolidinone was performed by Elly's group [179]; it concerns compound **168**, a key intermediate for the synthesis of carbapenem analogs that have useful antibacterial properties [180]. The Mitsunobu reaction (Scheme 6.96) applied to the serine-derivative **167** allowed the preparation of large quantities of pyrazolidinone **168**.

An interesting synthesis was described by Matsuyama [181], who showed that the conjugate addition of pyrazolidine **169** to optically active vinylsulfoxide **170** led to

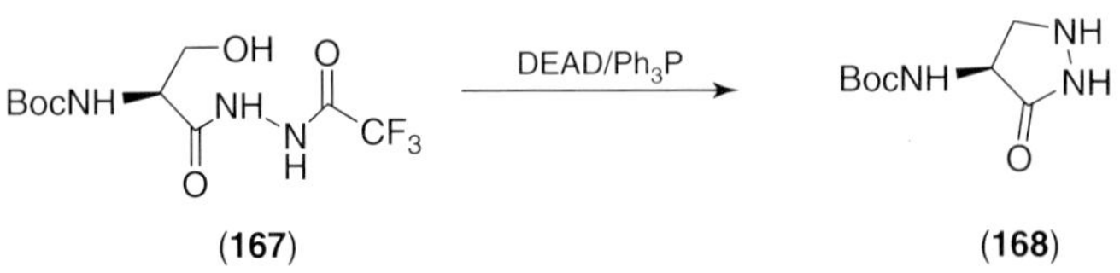

Scheme 6.96

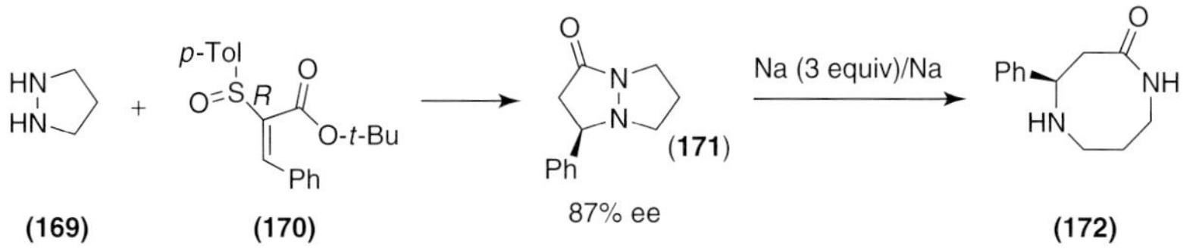

Scheme 6.97

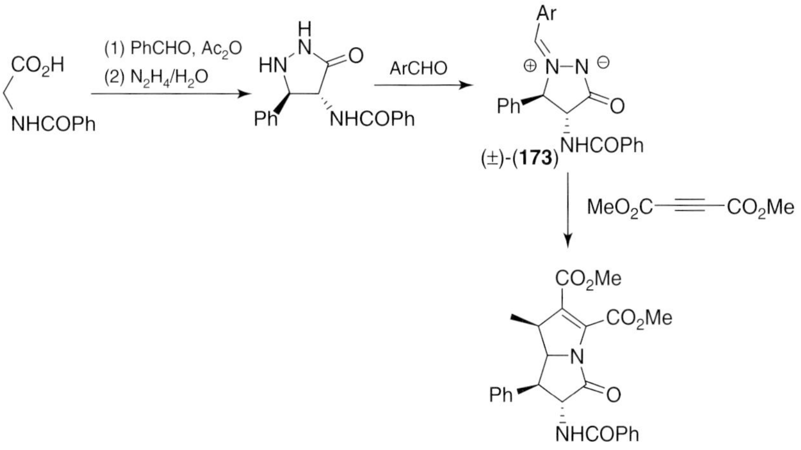

Scheme 6.98

pyrazolidinone **171** with a high level of enantioselectivity (Scheme 6.97). Breaking the N–N bond with Na/NH$_3$ furnishes diazacyclooctanone **172**.

Another strategy was used by Svete's group [182]; however, this method affords pyrazolidinones but only as racemic mixtures. Scheme 6.98 depicts this process, which furnishes an heterocycle that can be conveniently transformed into azomethine imine **173**. This dipole reagent is clearly very reactive toward various cycloadditions; one of these reactions with dimethylacetylenedicarboxylate is illustrated here.

Actually, the fact that pyrazolidinones can be easily transformed into such dipolar reagents is known from the pioneering work by Dorn [183], and Oppolzer already made use of the resourceful reactivity of azomethine imines [184].

Sharpless [185] described a versatile access to these heterocycles from an enantiopure oxirane as shown in Scheme 6.99, improving greatly the above-reported Svete's approach since nonracemic compounds are now obtained. Transformation of the oxirane into the aziridinium structure **174** is the key step of this process, and it should be noted that pyrazolidinone **175** could be converted into azomethines **176** which were used as starting materials for cycloadditions leading to more complex pyrazolidinones.

During his studies on the effect of pyrazolidinones as chiral templates on the reactivity and selectivity of many enantioselective reactions [186], Sibi described recently [187] the condensation of benzylhydrazine **177** to various benzimides **178**

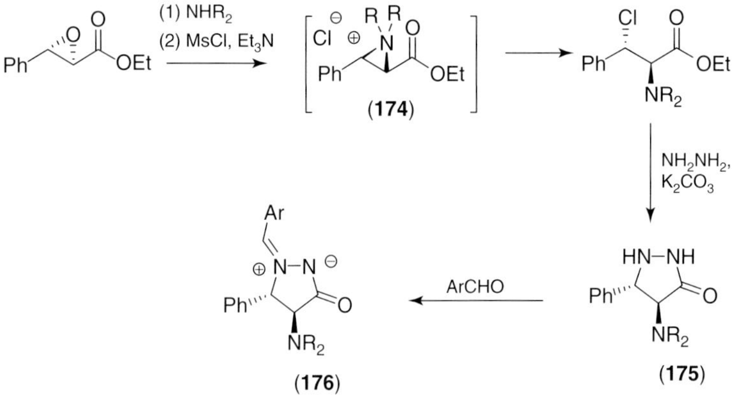

Scheme 6.99

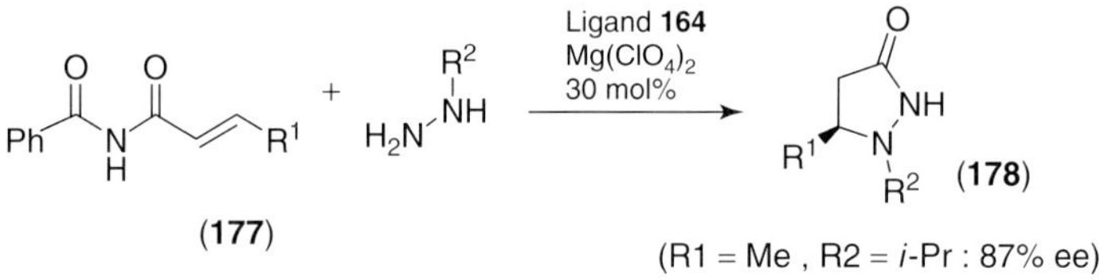

Scheme 6.100

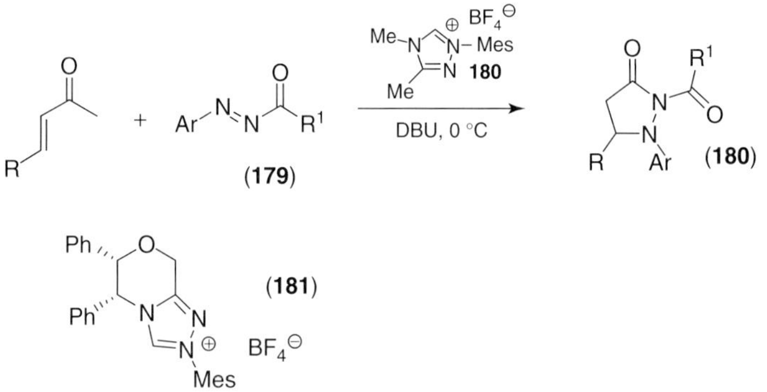

Scheme 6.101

in the presence of a catalytic system composed by a Lewis acid linked to the Box ligand **164** (Scheme 6.100). In this way, many pyrazolidinones (for instance **179**) were stereoselectively obtained.

A new method that looks very promising has been described by Chan and Scheidt [188]. The heterocyclic carbene derived from the triazolium salt is an excellent catalyst for a completely regioselective cycloaddition (Scheme 6.101) between enones and acylaryldiazenes **179** in order to produce various pyrazolidinones **180**.

Moreover, these authors reported also that the chiral carbene formed from the triazolium salt **181** produces enantioselectively pyrazolidinones. The mechanism of

6.3 Five-Membered Heterocycles with N and S Atoms

these transformations involves the addition of the carbene to the α, β-unsaturated aldehyde to give a homoenolate intermediate, which then undergoes a formal [3 + 2] cycloaddition with the diazene.

6.3.1 Thiazolidines

Thiazolidines have received particular attention owing to their pharmacological activity. Among the large number of compounds containing a thiazolidine ring, penicillins are the most popular ones (Scheme 6.102). The thiazolidine ring in these structures was commonly introduced from D-penicillamine hydrochloride, which was cyclized in the presence of an aldehyde [189]. In this manner, a countless number of penicillin derivatives have been prepared [190].

L-Cysteine (and 1,2-aminothiols in general) reacts with simple aromatic aldehydes in a nonstereoselective way to produce mixtures of the two epimers at C-2. Under specific conditions, N-acetylation of the diastereoisomeric mixture leads to either 2-4-*cis*- or 2-4-*trans*-2-aryl-3-acetyl-1,3-thiazolidine-4-carboxylic acids [191]. For instance, the acylation of a diastereoisomeric mixture of (2S, 4R) and (2R, 4R)-2-phenyl-carbomethoxy-1,3-thiazolidine with chloroacetylchloride in the presence of potassium carbonate led to the selective formation of the cis isomer **183** [192], which was used to synthesize the chiral thiazolopyrazine **184** (Scheme 6.103).

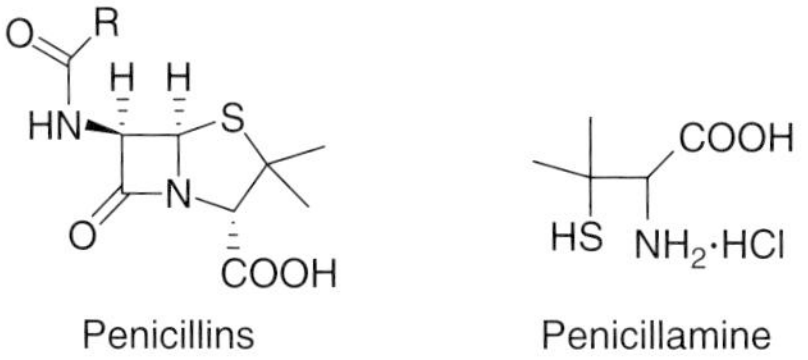

Scheme 6.102

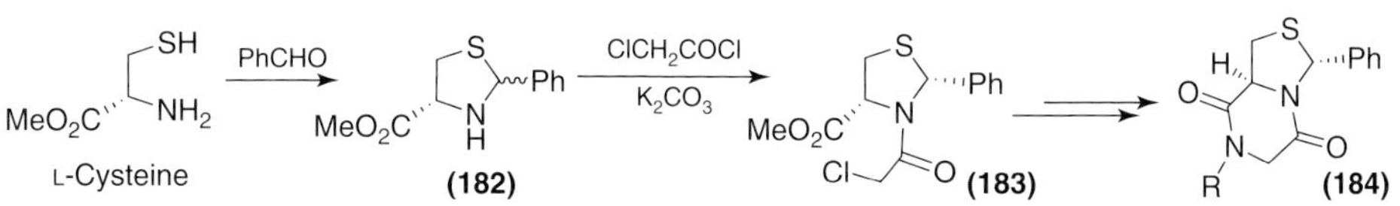

Scheme 6.103

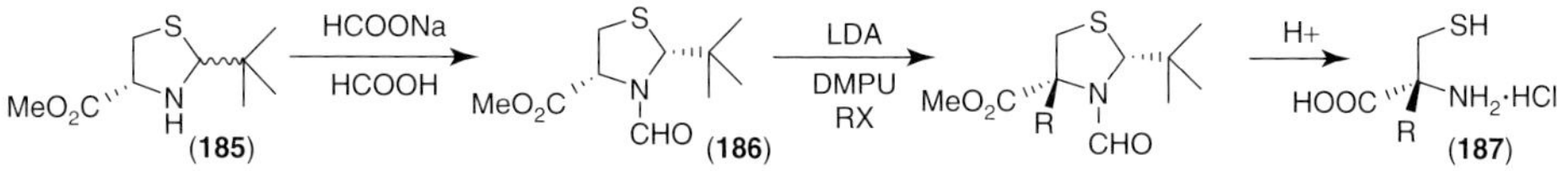

Scheme 6.104

N-Formylation of the thiazolidine **185**, derived from (R)-cysteine methyl ester hydrochloride and pivalaldehyde, led only to the syn diastereoisomer of the compound **186**. Alkylation of compound **186** in a basic medium with alkyl bromide occurred exclusively anti to the bulky t-butyl substituent (Scheme 6.104). This method was applied to the synthesis of enantiopure 2-alkyl cysteines **187** [193].

Baldwin and coworkers [194] synthesized several γ-lactam analogs of penems possessing antibacterial activity (Scheme 6.105). Condensation of aspartic acid semialdehyde with L-cysteine methyl ester **188** in pyridine afforded a 1:1 mixture of the diastereoisomeric thiazolidines **189**, which was refluxed in pyridine to give the bicyclic lactam **190** (45% yield).

An analogous approach was reported to synthesize spiro bicyclic thiazolidine lactams **191** [195] and **192** [196] (Scheme 6.106), which are β-turn mimetics and therefore valuable tools in order to study the bioactive conformation of biologically active peptides.

Geyer and Moser [197] prepared the first bicyclic peptidomimetic, the thiazolidine lactam **195** derived from the carbohydrate precursor **193** (Scheme 6.107). After deprotection, the amino group of **195** was elongated with a glycine derivative.

Several chiral catalysts containing a thiazolidine ring have been reported [198]. Natural L-cysteine and derivatives were often used in this way. For example, chiral thiazolidines were easily prepared from the condensation of L-cysteine ester with aldehydes or ketones; they were used as starting materials for the synthesis of a wide range of ligands **196**, which catalyzed an enantioselective addition of diethylzinc to aldehydes [199] and an arylation of aldehydes using aryl boronic acids [200], as shown in Scheme 6.108.

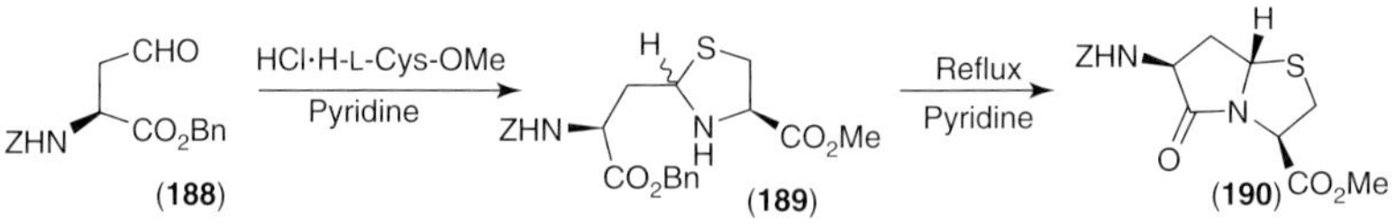

Scheme 6.105

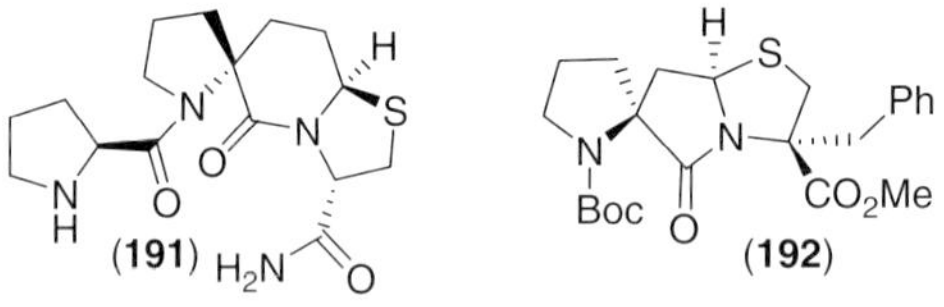

Scheme 6.106

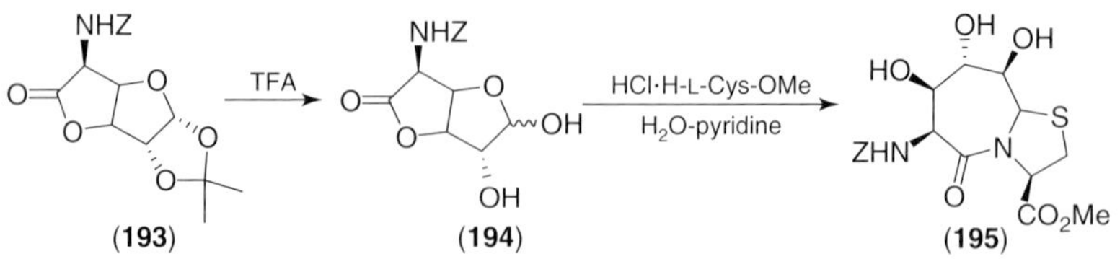

Scheme 6.107

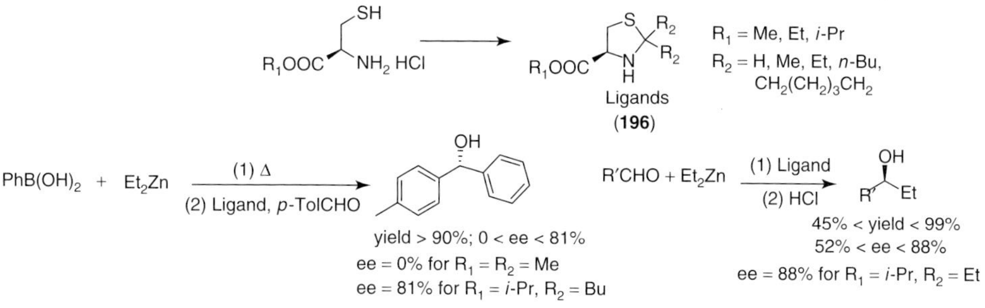

6.3 Five-Membered Heterocycles with N and S Atoms

$R_1OOC-CH(NH_2 \cdot HCl)-CH_2SH \longrightarrow$ thiazolidine with R_1OOC, R_2, R_2 substituents

Ligands (196)

R_1 = Me, Et, i-Pr
R_2 = H, Me, Et, n-Bu, $CH_2(CH_2)_3CH_2$

PhB(OH)$_2$ + Et$_2$Zn → (1) Δ; (2) Ligand, p-TolCHO → diarylmethanol

yield > 90%; 0 < ee < 81%
ee = 0% for $R_1 = R_2$ = Me
ee = 81% for R_1 = i-Pr, R_2 = Bu

R'CHO + Et$_2$Zn → (1) Ligand; (2) HCl → R'CH(OH)Et

45% < yield < 99%
52% < ee < 88%
ee = 88% for R_1 = i-Pr, R_2 = Et

Scheme 6.108

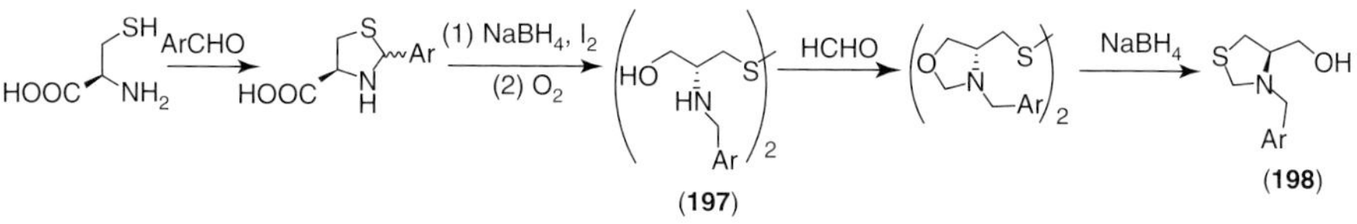

HOOC-CH(NH$_2$)-CH$_2$SH → ArCHO → HOOC-thiazolidine-Ar → (1) NaBH$_4$, I$_2$; (2) O$_2$ → disulfide (197) → HCHO → bicyclic → NaBH$_4$ → (198)

Scheme 6.109

Structures (199) and (200): thiazolidines with Ph, Ph, OH, NH, R_1, R_2 substituents

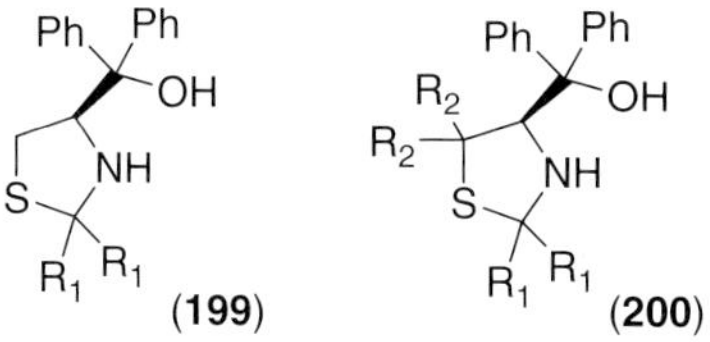

Scheme 6.110

From these works, it appears that the thiazolidine derivatives have higher enantioselectivity when they have a larger steric bulk in the thiazolidine rings.

Braga et al. [201] synthesized the new thiazolidine ligand **198** derived from disulfide **197**, obtained in an easy three-step synthesis from L-cysteine, as described in Scheme 6.109. This ligand was used in asymmetric palladium-catalyzed allylations and provided the allylation product in high yield and with high enantiomeric excess (80%).

The chiral 2,2-disubstituted thiaprolinol derivatives **199** [202] and **200** [203] were used as ligands for diethylzinc addition to aldehydes and for borane reduction of aromatic ketones. These catalysts led to alcohols with moderate to good enantioselectivities (Scheme 6.110).

Thiazolidines were also used as new chiral formyl anion equivalents. Several silylated thiazolidines were synthesized, generally as a mixture of the cis and trans diastereoisomers **201** and **202**, which could be separated by chromatography to give enantiopure compounds [204]. Diastereoselectivity was not observed during the reaction with aldehydes in the presence of fluoride ion; however, functionalization occurred with a total retention of configuration of the starting C–Si bond (Scheme 6.111).

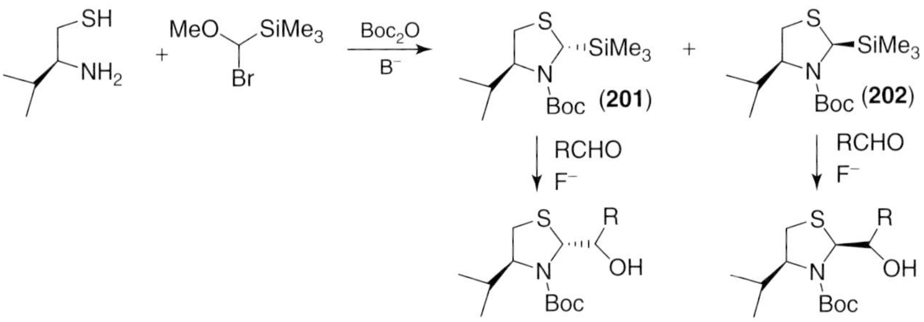

Scheme 6.111

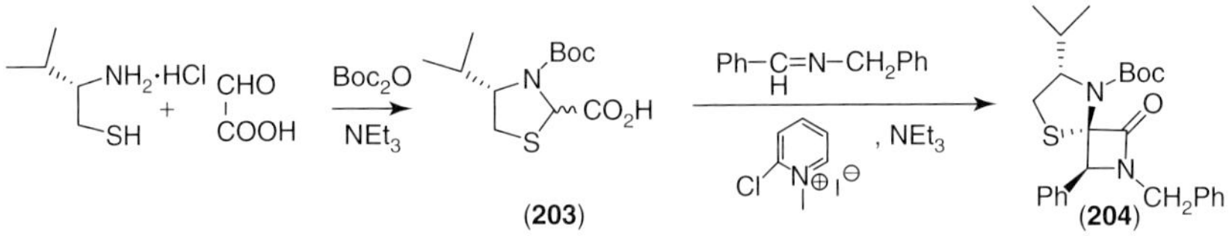

Scheme 6.112

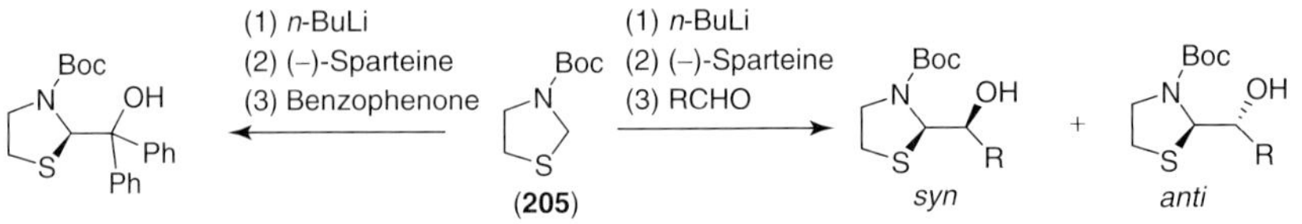

Scheme 6.113

A Staudinger reaction between chiral ketenes and imines was reported to afford enantiopure spiro-β-lactams derived from 1,3-thiazolidines [205]. The best result, as regards diastereoselectivity, was obtained with the 4-*iso*-propyl-1,3-thiazolidine-2-carboxylicacid **203**, which led to only one enantiomerically pure spiro-β-lactam **204** in 46% yield (Scheme 6.112). The authors stressed the importance of the C-4 substitution of the thiazolidine ring on the stereochemical outcome of this reaction.

Toru and coworkers used the simple *N*-Boc-thiazolidine **205** as a formyl anion equivalent [206]. As shown in Scheme 6.113, the reaction of the lithiated **205** with benzophenone in the presence of (−)-sparteine afforded the products with up to 93% ee. This reaction was expanded to various aromatic and aliphatic aldehydes to afford the products of addition with high enantioselectivity but moderate diastereoselectivity (syn/anti = 55/45).

Recently, Kwon and coworkers [207] prepared functionalized thiazolidines and oxazolidines with a high yield, one-step biphosphine-catalyzed mixed double-Michael reaction. Diphenylphosphinopentane (DPPP) was the best catalyst used for this reaction compared to diphenylphosphinomethane (DPPM), diphenylphosphinoethane (DPPE), and diphenylphosphonobutane (DPPB). The importance of the tether length between the two phosphine moieties was explained by invoking the stabilization of the intermediate phosphonium ion **207**, as represented in Scheme 6.114.

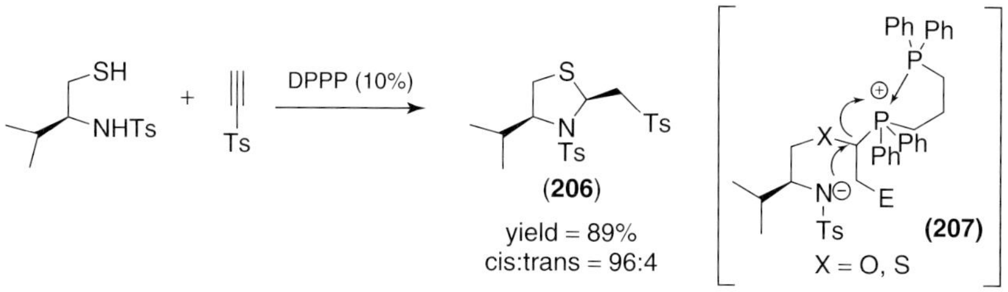

Scheme 6.114

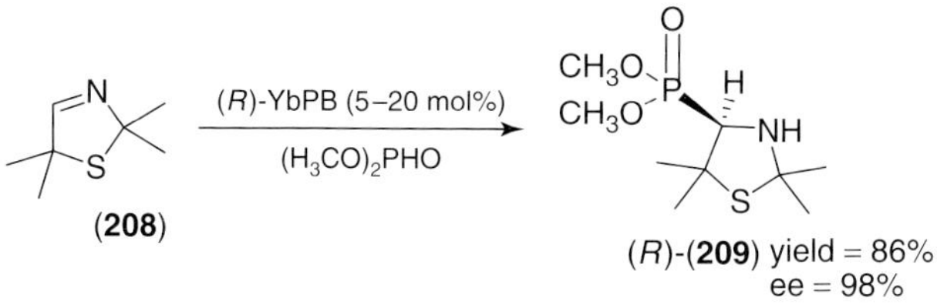

Scheme 6.115

An efficient enantioselective approach to the pharmaceutically interesting 4-thiazolidinephosphonate (R)-209 was described [208] using 2,2,5,5-tetramethyl-3-thiazoline 208 and various heterobimetallic lanthanoid catalysts (Scheme 6.115). The best results were obtained with (R)-YbKB as catalyst (where B = (R)-(+)-binaphtol).

6.3.1.1 Iminothiazolidines

Being present in a large variety of biologically active compounds, 2-iminothiazolidines have been intensively studied [209]. Ring expansion of aziridines, which leads to five-membered heterocycles, appears to be a powerful tool for the construction of iminothiazolidines. Palladium-catalyzed ring expansions of aziridines have thus been described by Alper et al., who reported an enantioselective version of this reaction (Scheme 6.116) [210]. Treatment of aziridines 210 with isocyanates in the presence of bis(acetonitrile)palladium dichloride afforded thiazolidines 211 in both regio- and stereoselective manner. The enantioselectivity of the palladium-catalyzed cycloaddition was unequivocally demonstrated; consequently, the relative and absolute stereochemistry was conserved throughout this cycloaddition reaction. An analogous ring expansion of thiiranes with carbodiimide catalyzed

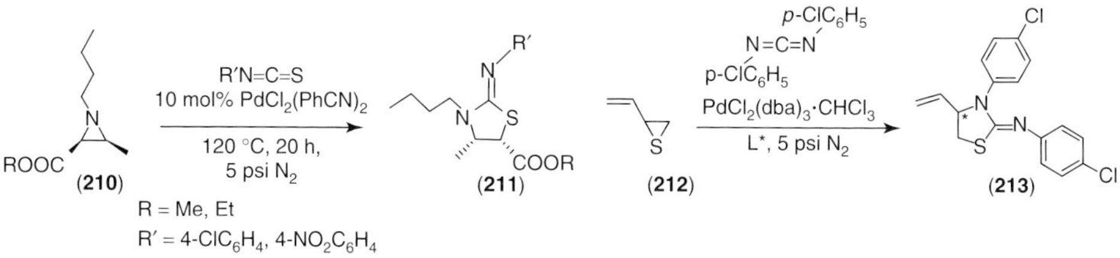

Scheme 6.116

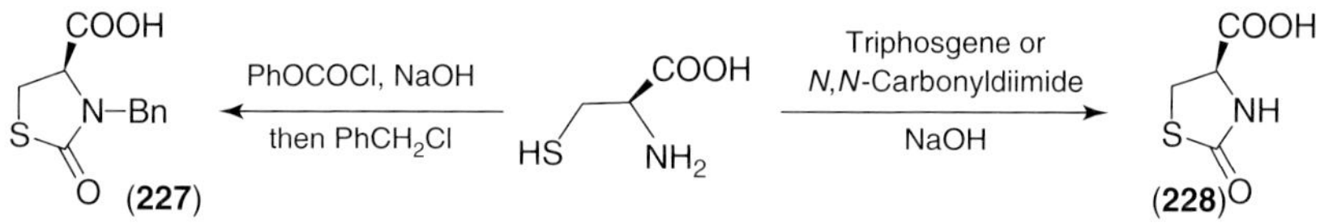

Scheme 6.121

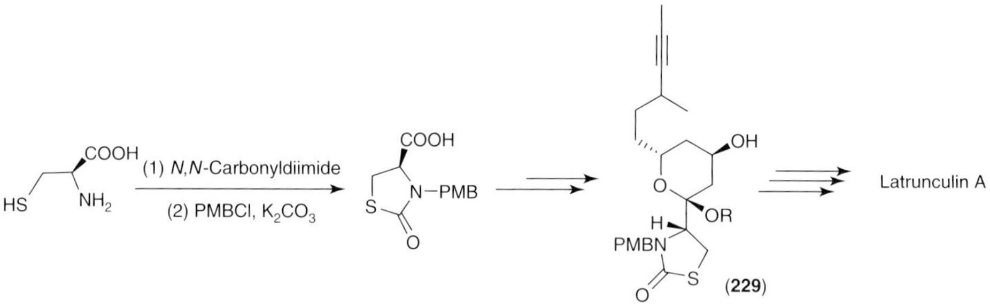

Scheme 6.122

Compound **229**, possessing a thiazolidine moiety, is a key building block for the synthesis of latrunculin A. It was synthesized from cysteine in a classical way (Scheme 6.122) [228].

6.3.2
Thiazolines

6.3.2.1 2-Thiazolines

Chiral thiazolines are mainly known as subunits of biologically active macromolecules, which exhibit antitumor and anti-HIV activities, but they are less familiar than their oxygenated analogs, namely the oxazolines. As a matter of fact, the 2-aminothiols, which can be viewed as the most obvious precursors, are not easily available compared to the corresponding amino alcohols. Nevertheless, the asymmetric synthesis of thiazolines has received renewed attention because thiazolines and bis-thiazolines have been proved to possess great potential as a new class of ligands in enantioselective synthesis.

Until 1990, desferrithiocyn **230**, a natural ferric ion chelator isolated from *Streptomyces antibioticus*, was the only example of a naturally occurring 4-methyl substituted Δ2-thiazoline; since then, a very wide range of new and biologically active linear and cyclic peptides containing at least one thiazoline ring have been isolated and characterized. For example, thiangazole B **231**, isolated from blue green algae, is a linear, fused 2,4-disubstituted poly-thiazoline/oxazole ring, and lissoclinamide **232** is a marine cyclopeptide alkaloid with two thiazolidine rings, as shown in Scheme 6.123.

In structures **230** and **231**, the thiazoline rings are derived from 2-methylcysteine, and different authors have developed different strategies for the elaboration of this thiazoline ring. With the aim of synthesizing desferrithiocyn, Pattenden

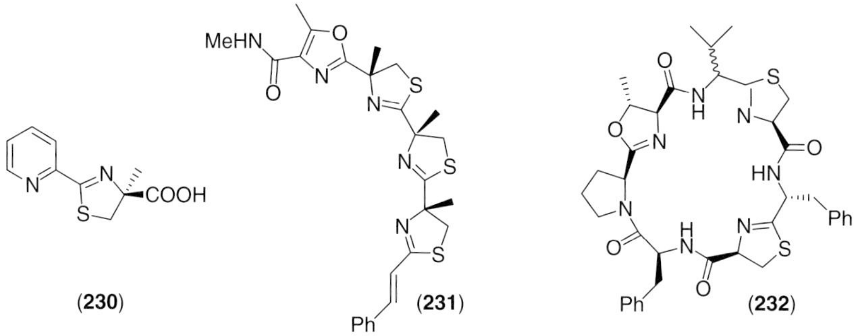

6.3 Five-Membered Heterocycles with N and S Atoms | 271

(230) Desferrithiocyn

(231) Thiangazole

(232) Lissoclinamide

Scheme 6.123

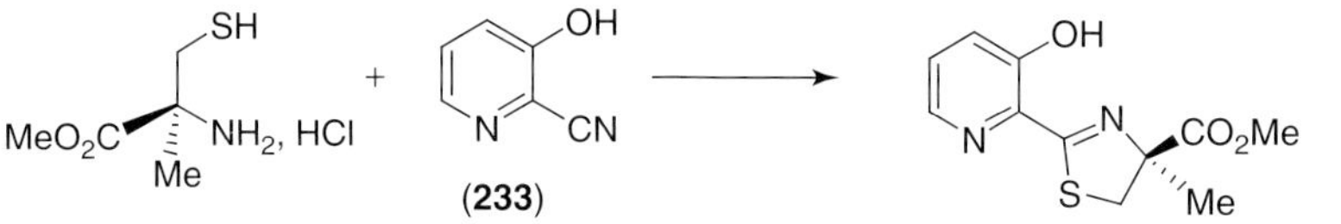

Scheme 6.124

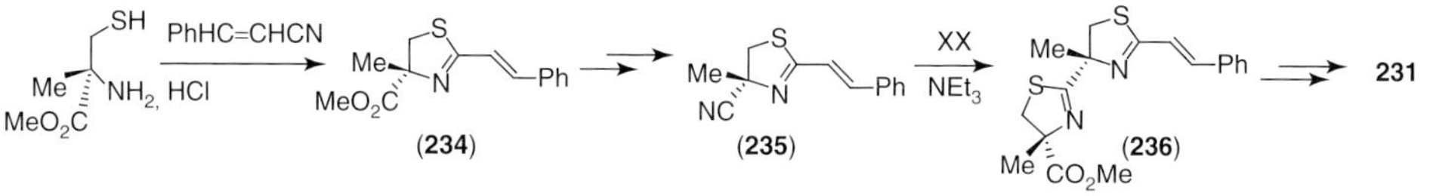

Scheme 6.125

and coworkers elaborated first the thiazoline ring from the hydrochloride of (S)-(2)-methylcysteine methyl ester [193], which reacted with the pyridine nitrile derivative 233 to give the (S)-desferrithiocyn 230 in high yield (Scheme 6.124).

To synthesize thiangazole, the same authors synthesized the thiazoline ring as described above. The resulting ester 234 was converted into the nitrile 235, and a second cyclocondensation reaction produced the bis-thiazoline 236 as shown in the Scheme 6.125 [229].

Thiazolines containing amino acids 238 could be obtained efficiently in an optically pure form from simple condensation reactions between cysteine derivatives and N-protected imino ethers 237, which are derived from natural amino acids (Scheme 6.126) [230].

Chiral centers attached to the C-2 position of thiazolines are prone to epimerization; therefore racemization-free methods have been developed. The Mitsunobu reaction [231a] or treatment with Burgess' reagent (methyl N-(triethylammoniosulfonyl)carbamate) [231b] applied to thiopeptides were efficiently used to introduce thiazolines in the peptide backbone and proceeded without loss of stereochemical integrity.

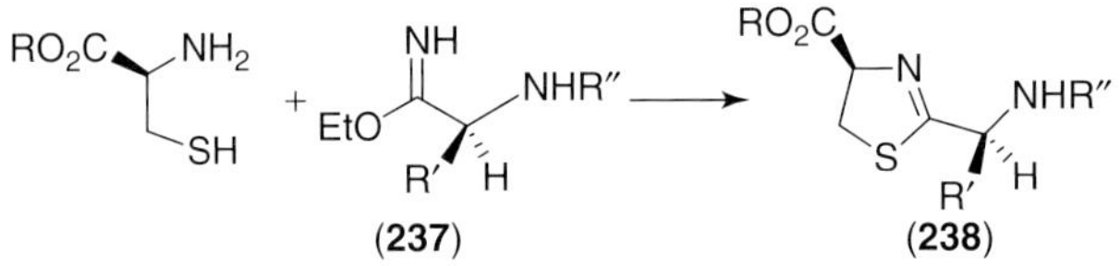

Scheme 6.126

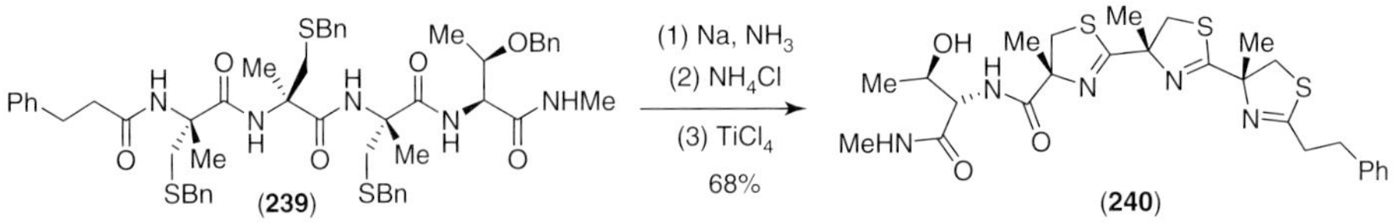

Scheme 6.127

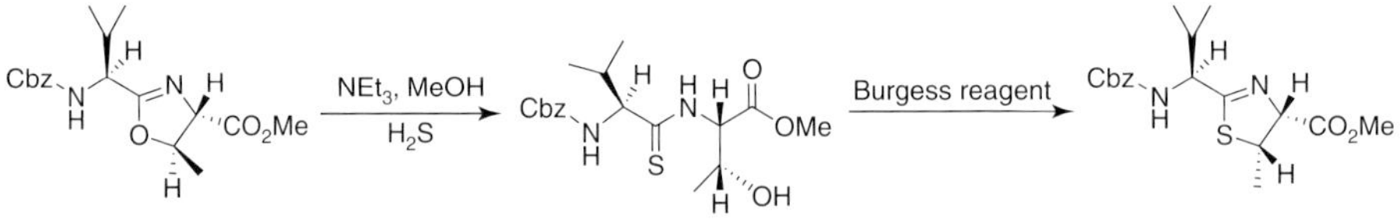

Scheme 6.128

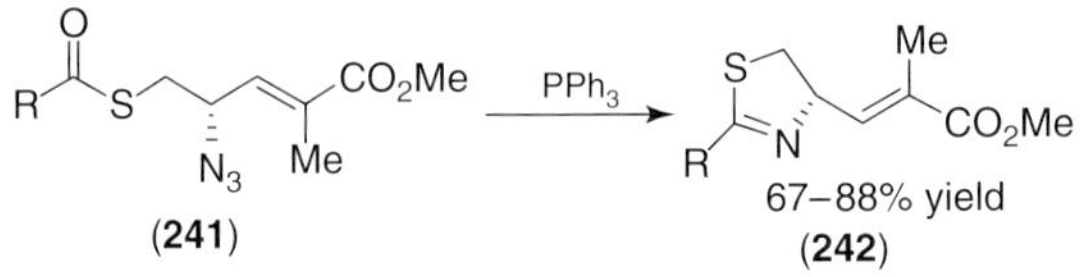

Scheme 6.129

As shown in Scheme 6.127, Heathcock and Parsons [232] made use of the tetrapeptide **239** that was reductively debenzylated and was cyclized by treatment with TiCl$_4$ in order to provide the tris-thiazoline **240**, a precursor of thangazole **231**.

Wipf and coworkers [233] used also Burgess' reagent in a general method for the oxazoline–thiazoline conversion. This reaction was stereoselective and essentially free of racemization (Scheme 6.128).

A one-pot Staudinger reduction/aza-Wittig reaction was used as a mild and versatile process for the conversion of azido-thiolesters **241** into enantiopure 2,4-disubstituted thiazolines **242** (Scheme 6.129) [234].

Recently, the dehydrative cyclization of N-acylcysteine derivatives was carried out in the presence of a molybdenum oxide catalyst to afford thiazolines in high yield, as shown in Scheme 6.130 [235].

Xu and Ye [236] used amino alcohols to synthesize 2,4,5-trisubstituted thiazolines **246** with high diastereoselectivity via the intramolecular conjugate reaction described in Scheme 6.131. The unstable intermediate thioamide **245** readily undergoes an intramolecular conjugate addition reaction in a weak acidic medium (silica gel).

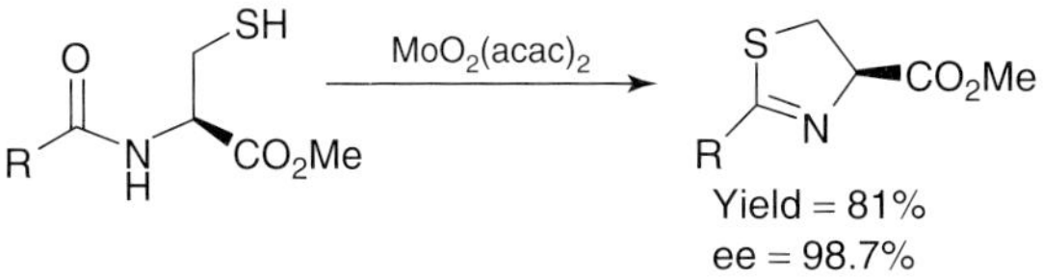

6.3 Five-Membered Heterocycles with N and S Atoms

Scheme 6.130

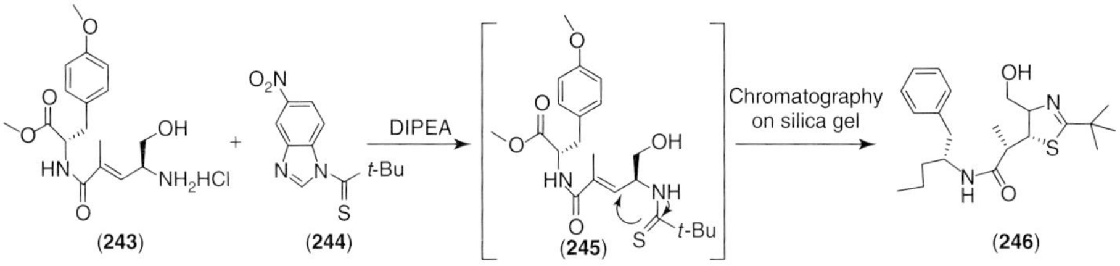

Scheme 6.131

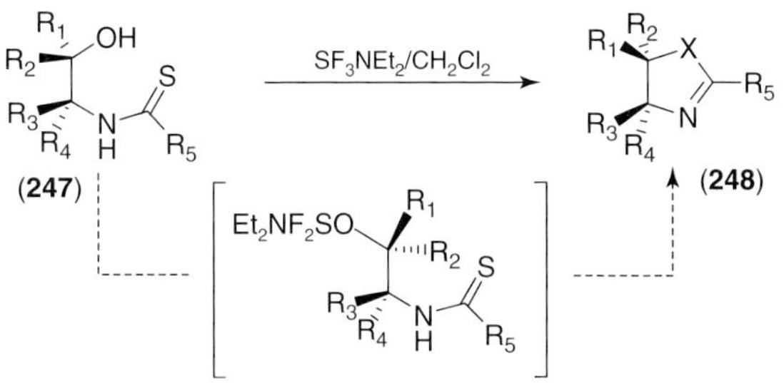

Scheme 6.132

Diethylamidosulfur trifluoride (DAST) was used as a powerful hydroxyl activating agent for the intramolecular cyclization of (1,2)-thioamidoalcohols to 2-thiazolines, as shown in Scheme 6.132; this reaction was totally diastereoselective with the inversion of configuration at C-5 [237].

The amino sugar **249** was allowed to react with para-substituted phenyl isothiocyanates, and thiazoline derivatives **250** were thus obtained (Scheme 6.133) [238].

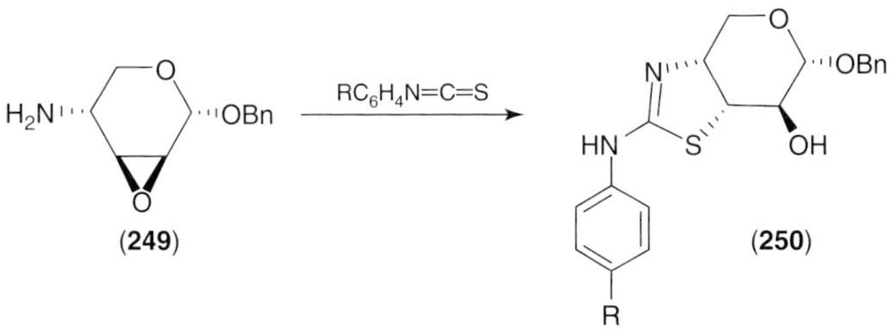

Scheme 6.133

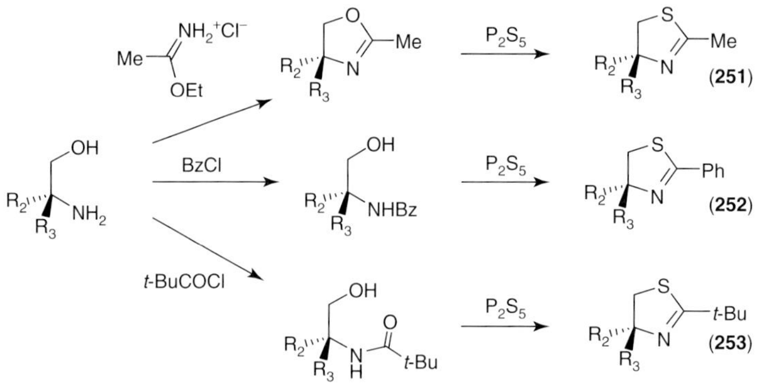

Scheme 6.134

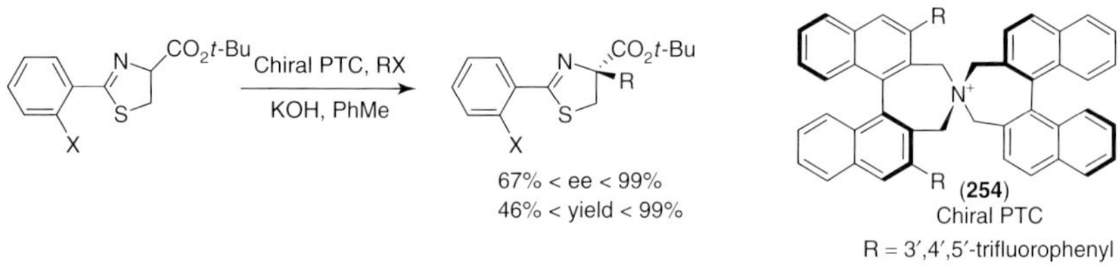

Scheme 6.135

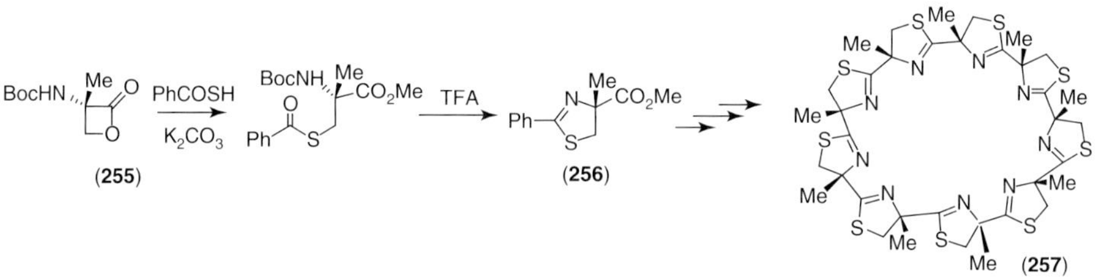

Scheme 6.136

With the aim to study the oxidation of thiazolines into thiazolines dioxide, thiazolines **251–253** have been prepared from β-amino alcohols in two steps (Scheme 6.134); the cyclization occurred in the presence of the simple reagent phosphorus pentasulfide [239].

Park et al. described an enantioselective synthesis of alkylcysteines via phase-transfer catalytic alkylation of 2-aryl-2-thiazoline-4-carboxylate ester, by using the chiral catalyst **254**. This methodology seems very efficient, given the mild reaction conditions and the high enantioselectivity obtained in the most cases (Scheme 6.135) [240].

In a very elegant synthesis of cyclic oligothiazolines, Fukuyama and coworkers [241] have built the thiazolidine ring from β-lactone **255** with thiobenzoic ester to provide in two steps the monothiazoline unit **256**, which was submitted to repetitive elongation to give ultimately the cyclic oligothiazoline **257** (Scheme 6.136).

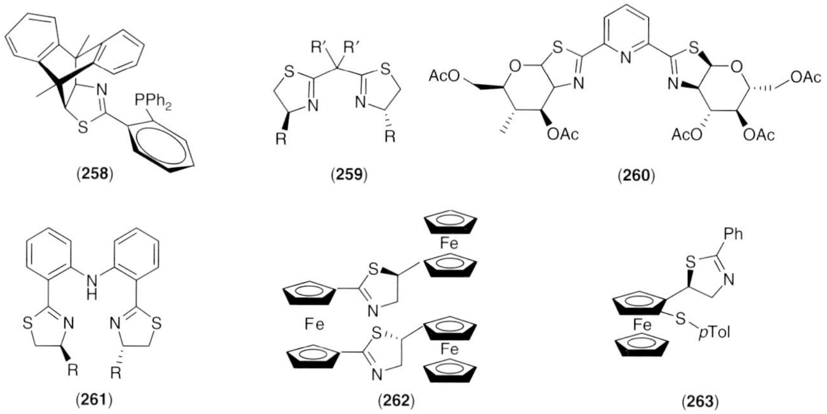

6.3 Five-Membered Heterocycles with N and S Atoms

Scheme 6.137

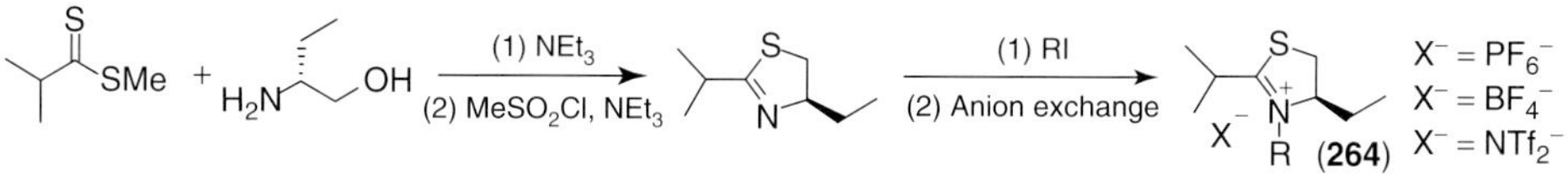

Scheme 6.138

The first application of a chiral bis(thiazoline) ligand for an asymmetric hydrosilylation reaction was reported by Helmchen et al. [242]. Since then, a few promising examples have been reported for asymmetric catalytic reactions with thiazolines or bis-thiazolines as ligands, such as Diels–Alder reactions (with ligands **258** and **259**) [243], cyclopropanation (in the presence of the carbohydrate-derived pyridyl bis(thiazoline) **260**) [244] and allylic substitution (with ligands **259**) [245] (Scheme 6.137).

C_2-symmetric tridentate bis(thiazoline) ligands **261** were used in catalytic asymmetric Henri reaction (nitroaldol reaction); an interesting metal-controlled reversal of enantioselectivity was obtained by replacing Cu(II) with Zn(II) [246]. In order to use them as ligands for asymmetric catalysis, new chiral ferrocene-thiazolines with (ligand **262**) [247] and without (ligand **263**) [248] C_2 symmetry have been synthesized.

A novel class of chiral ionic liquids based on the thiazolium cation **264** was prepared from amino alcohols in a multigram scale, as shown in Scheme 6.138. Their properties (low melting point, water tolerance and stable under acidic or basic conditions) make them interesting potential candidates for new chiral solvents [249].

6.3.2.2 3-Thiazolines

Very few enantioselective syntheses of 3-thiazolines have been reported in the literature. The first synthesis of enantiopure 3-thiazolines via an Asinger reaction was described by Martens and coworkers [250]. The easily accessible galactose

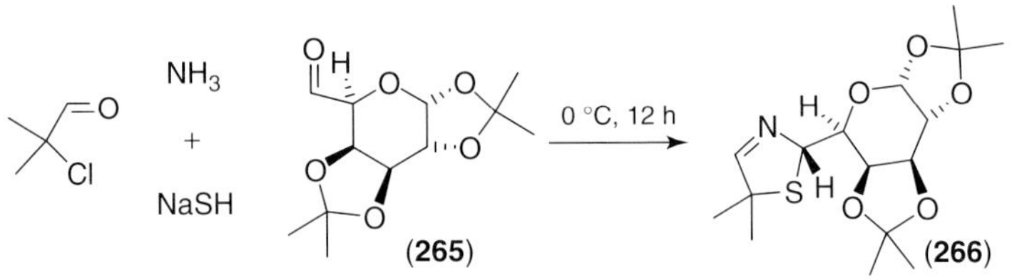

Scheme 6.139

derivative **265** reacted under Asinger conditions to give 3-thiazolines **266** in good yields and in a highly diastereoselective way (Scheme 6.139).

6.3.3
Sultams

Chiral γ-sultams are useful heterocycles for asymmetric synthesis and medicinal chemistry. The most important application concerned the use of these sultam derivatives as chiral auxiliaries. In 1984, Oppolzer described the synthesis of the camphor-derived sultam **269** (named Oppolzer's camphor sultam auxiliary), which was easily prepared from (+)-camphor-10-sulfonyl chloride **267**, as shown in Scheme 6.140 [251].

Many asymmetric syntheses of chiral sultams that are not derived from camphor have been developed [252]. Oppolzer and coworkers [253] have performed the synthesis of compounds **272a** and **272b**, which are structurally simpler than the compound **269** (Scheme 6.141). The product **272a** was obtained in two steps from saccharine **270**. The key step was the asymmetric hydrogenation of imine **271** catalyzed by $Ru_2Cl_4[(R)-(+)-BINAP]_2(NEt_3)$. While the reduction of **271a** furnished, after crystallization, the enantiomerically pure sultam **272a**, reaction from imine **271b** gave only traces of the racemic **272b**. An alternative approach consisted in the resolution of the racemic **272b**.

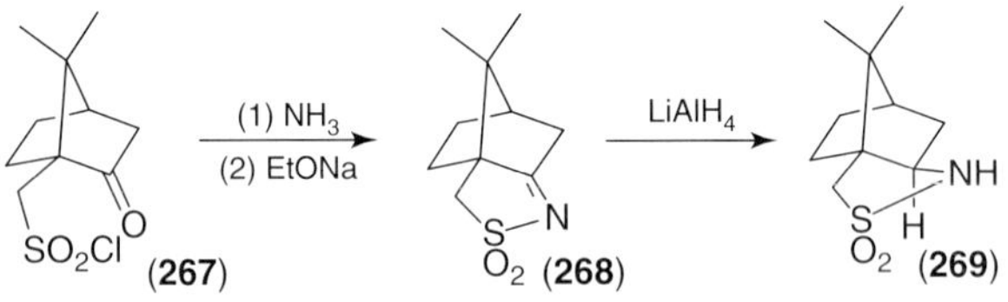

Scheme 6.140

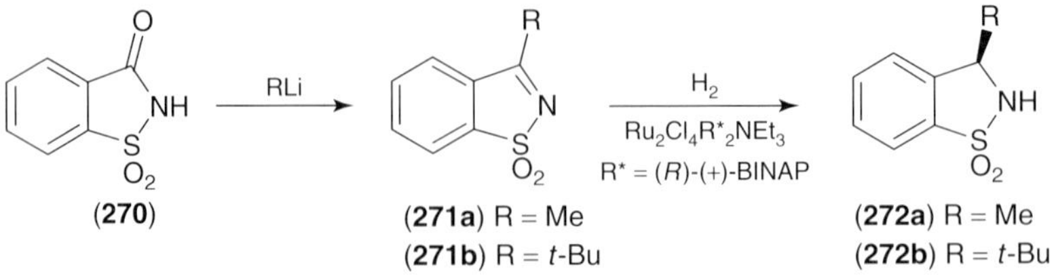

Scheme 6.141

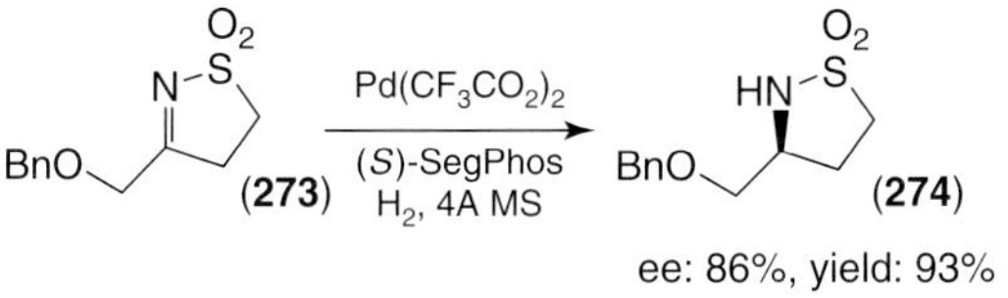

6.3 Five-Membered Heterocycles with N and S Atoms | 277

Scheme 6.142

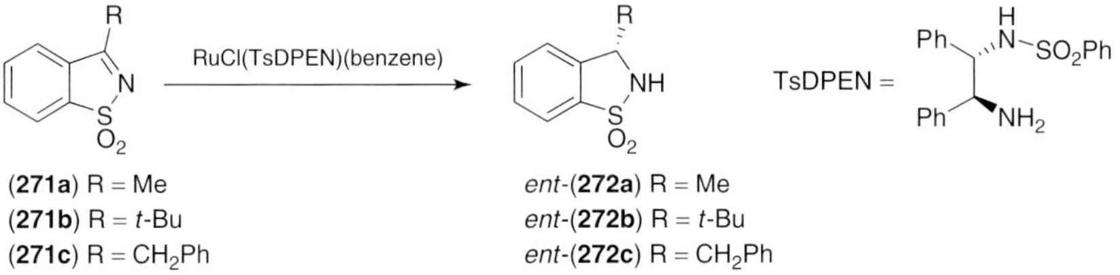

Scheme 6.143

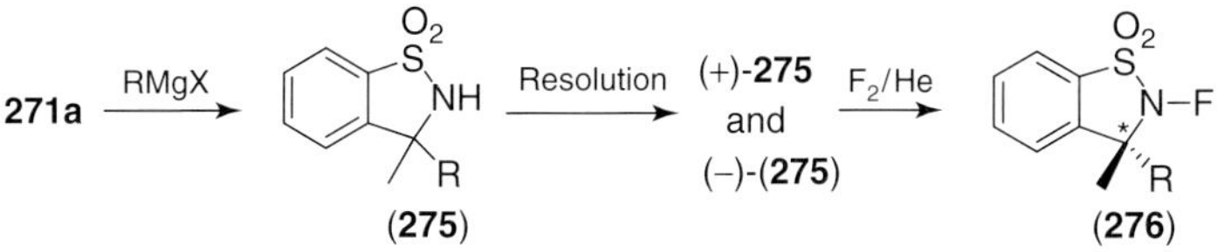

Scheme 6.144

The Pd/phosphine complexes were found to be valuable catalysts for the asymmetric hydrogenation of a series of cyclic sulfonamides **271** [254]. For example, Scheme 6.142 shows the first enantioselective synthesis of the chiral sultam **274** from the sulfonamide **273**.

The asymmetric transfer hydrogenation of imines **271** catalyzed by ruthenium and rhodium complexes led to the corresponding sultams in good to excellent enantiomeric excesses (Scheme 6.143) [255]. Ahn et al. reported a practical synthesis of the sultams *ent*-**272a–c** by using Noyori's RuCl(*N*-(4-toluenesulfonyl)-1,2-diphenylethylenediamine (TsDPEN))(benzene) catalyst as hydrogen donor. The sultam auxiliaries *ent*-**272** were obtained in high enantiomeric excess (93%> ee >91%) [256].

All these sultams have been shown to be highly selective chiral auxiliaries in asymmetric alkylations [257], aldol reactions [258], radical additions [259], Diels–Alder reactions [260], addition reactions [261], cyclopropanation [262], epoxidation [263], azidation [264], ene-reaction [265], asymmetric Bayliss – Hillman reaction [266], and so on.

With a view to synthesizing new chiral fluorinating agents, some authors reported the synthesis of chiral sultams with a stable but reactive N–F bond [267], such as compound **276** obtained by the fluorination of enantiopure compound **275** (Scheme 6.144).

Chiral β-amino alcohols were used as precursors for the synthesis of chiral halosulfonamides and sultams [268]. In the case of the *cis*-aminoindanol **277**, the hydroxyl group was converted to the *trans*-β-halosulfinamide **278** with inversion

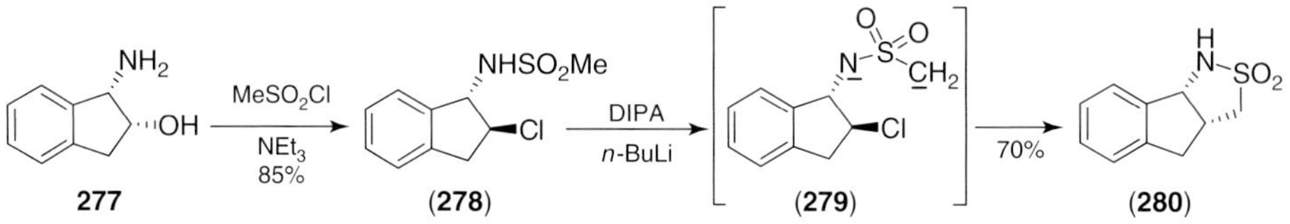

Scheme 6.145

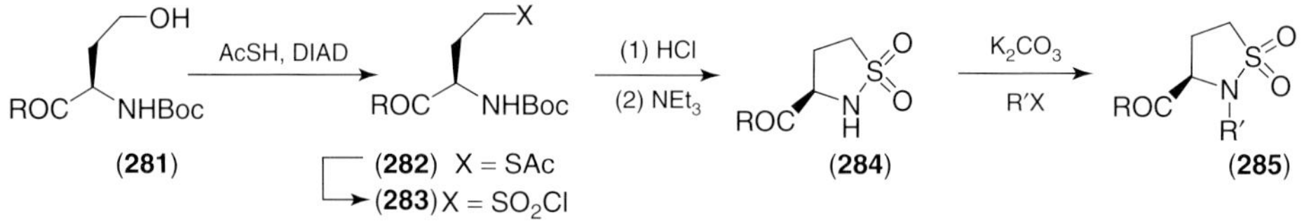

Scheme 6.146

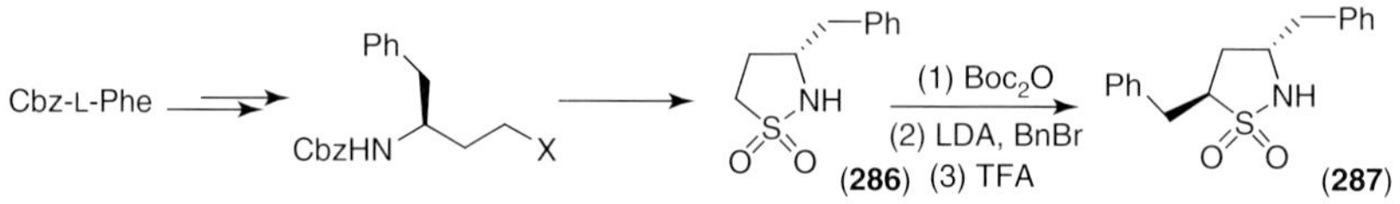

Scheme 6.147

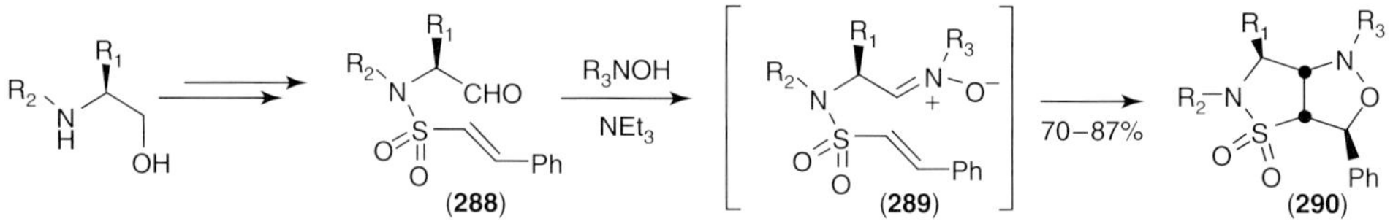

Scheme 6.148

of configuration. Sultam **180** was obtained via the intramolecular sulfonamide dianion **279** alkylation, as shown in Scheme 6.145.

The synthesis of enantiopure sultam **285**, a potential anti-inflammatory agent [269, 270], was accomplished from chiral alcohols **281** that were converted to the thioacetate **282**. Chlorine oxidation followed by selective N-Boc removal and cyclization gave **284**, which is a precursor of the target compound **285** (Scheme 6.146).

A novel series of HIV protease inhibitors containing the sultam scaffold **287** has been synthesized from the enantiopure Cbz-L-phenylalanine [271]. Alkylation of sultam **286** led to the trans isomer **287** as the sole isomer (Scheme 6.147).

Bicyclic sultams **290**, including an isoxazoline or isoxazolidine moiety, have been synthesized by intramolecular cycloaddition of sulfonamide oxime **289** [272]. This occurs with a complete diastereoselectivity: the stereochemical information present in the dipolarophile was completely preserved in the cycloadducts and the stereochemistry was predetermined by the alkene geometry, as outlined in Scheme 6.148.

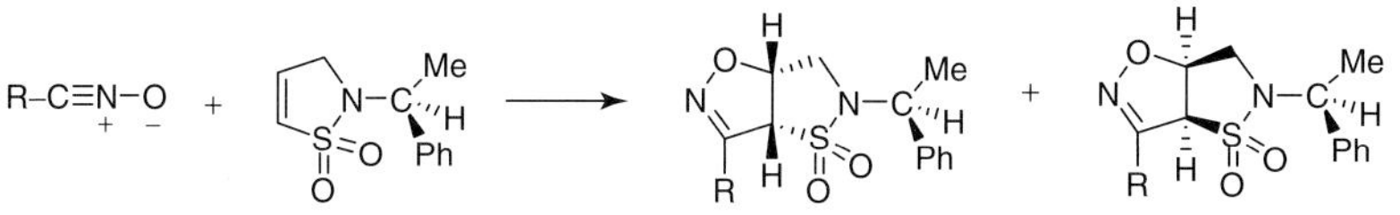

6.3 Five-Membered Heterocycles with N and S Atoms | 279

Scheme 6.149

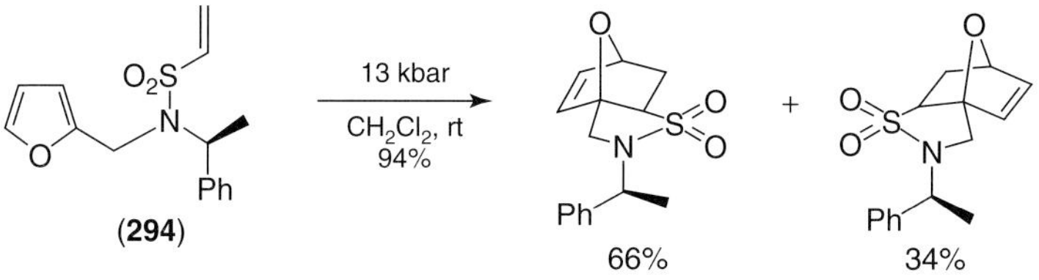

Scheme 6.150

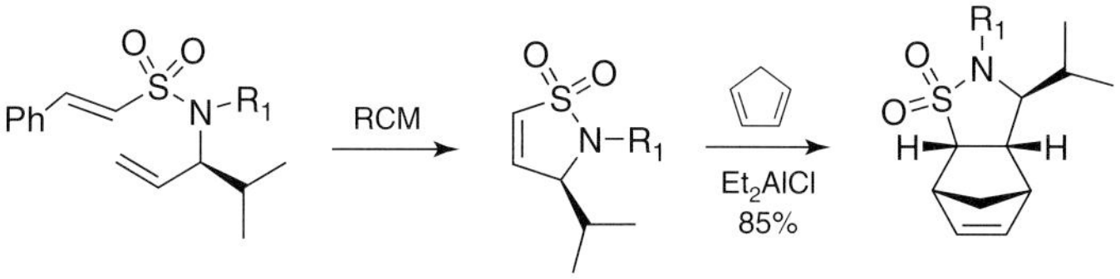

Scheme 6.151

Scheme 6.149 describes an intermolecular version of this 1,3-dipolar cycloaddition by using the chiral sultam **291** and nitrile oxide as starting materials. The reaction led to a separable 1:1 diastereoisomeric mixture of the bicyclic sultams **292** and **293** [273].

Enantiopure sultams have been readily prepared by intramolecular Diels–Alder reaction of the furan-containing vinylsulfonamides **294**, but with a relatively low diastereoselectivity, as shown in Scheme 6.150 [274].

α-β-Unsaturated-γ-sultams **296** were generated via ring-closing metathesis (RCM) [275]; these sultams reacted with cyclopentadiene under Lewis acid catalysis to yield the tricyclic sulfonamides **297** with complete facial and endo selectivity (Scheme 6.151).

With the aim of synthesizing a new class of nucleoside analogs with a specific inhibition against the HIV-1, Postel and coworkers reported the synthesis and the treatment of new cyanomethanesulfonamides of monosaccharidic and nucleosidic substrates under carbanion-mediated sulphonamide intermolecular coupling (CSIC) conditions [276]. This reaction, depicted in Scheme 6.152, led to the dihydroisothiazole 1,1-dioxide **299** in good yields.

Recently, Combs and coworkers [277] described the first asymmetric synthesis of compound **302** from the enantiopure sulfoxide **300**. The key step of this synthesis was the reduction of the heterocycle with excellent regiochemical and stereochemical controls (Scheme 6.153). The hydride attacked the ethylenic double

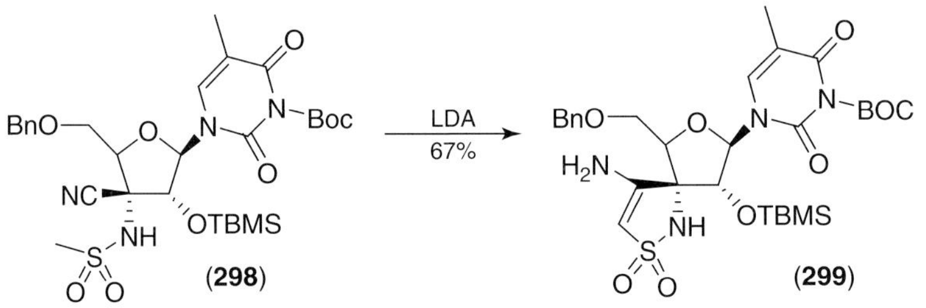

Scheme 6.152

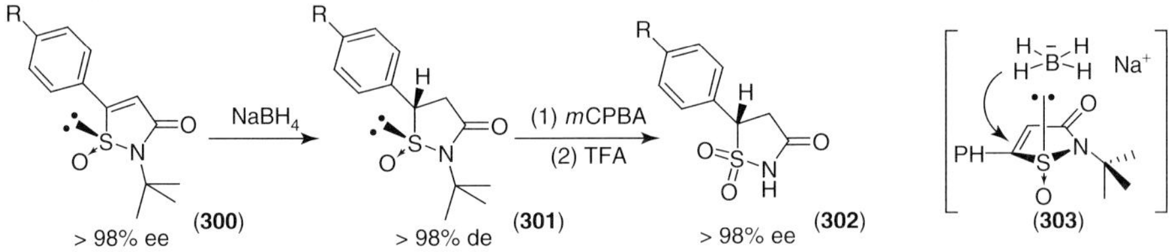

Scheme 6.153

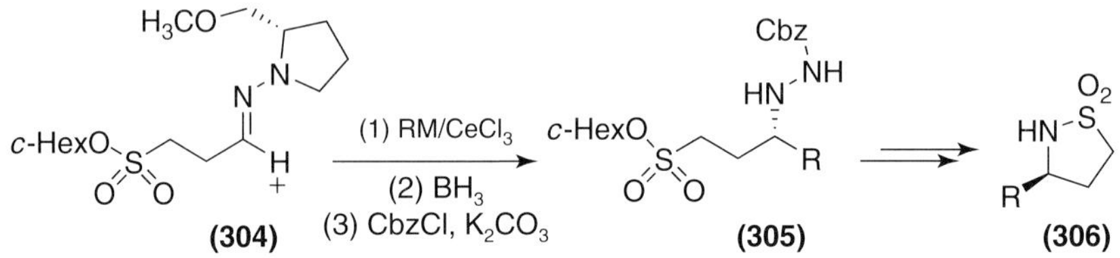

Scheme 6.154

bond adjacent to the sulfonamide on the opposite face of the heterocycle that bears the sulfonamide oxygen, as depicted on model **303**.

Enders and coworkers [278] have developed an efficient asymmetric synthesis of 3-substituted γ-lactams through the diastereoselective nucleophilic 1,2-addition of various organocerium compounds to the CN double bond of ω-S-1-amino-2-methoxymethylpyrrolidine (SAMP)-hydrazonosulfonates **304**. Compounds **306** were obtained in good overall yields and with high enantiomeric excesses (Scheme 6.154).

As can be seen from the above discussion, five-membered heterocycles with N, O and S atoms constitute a large category of highly valuable reagents. Their value is clearly noticeable in the field of asymmetric synthesis as well as biologically active compounds. As a matter of fact, these compounds not only act as chiral auxiliaries (some of them even belonging to very classical stereoselective reactions) but also as chiral ligands linked to metal catalysts, which allow various remarkable enantioselective transformations in many diverse topics. Though revealed more recently as important pharmaceutical tools, their usefulness in this area cannot be underestimated and the ongoing extensive research should lead to still more promising results.

References

1. Agami, C. and Couty, F. (**2004**) *Eur. J. Org. Chem.*, 677–85.
2. Royer, J., Bonin, M. and Micouin, L. (**2004**) *Chem. Rev.*, **104**, 2311–52.
3. Baldwin, J.E. (**1976**) *J. Chem. Soc., Chem. Commun.*, 736–38.
4. Fulop, F., Bernath, G., Mattinen, J. and Pihlaja, K. (**1989**) *Tetrahedron*, **45**, 4317–24.
5. Neelakantan, L. (**1971**) *J. Org. Chem.*, **36**, 2261–62.
6. (a) Baudet, M. and Gelbcke, M. (**1979**) *Anal. Lett.*, **12B**, 325–38. (b) Santiesteba, F., Grimaldo, C., Contreras, R. and Wrackmeyer, B. (**1983**) *J. Chem. Soc., Chem. Commun.*, 1486–87.
7. Just, G., Potvin, P., Uggowitzer, P. and Bird, P. (**1983**) *J. Org. Chem.*, **48**, 2923–24.
8. Agami, C. and Rizk, T. (**1985**) *Tetrahedron*, **41**, 537–40.
9. Beckett, H.A. and Jones, G.R. (**1977**) *Tetrahedron*, **33**, 3305–11.
10. Takahashi, H., Suzuki, Y. and Kametani, T. (**1983**) *Heterocycles*, **20**, 607–10.
11. (a) Takahashi, H., Hsieh, B.C.A. and Higashiyama, K. (**1990**) *Chem. Pharm. Bull.*, **38**, 2429–34. (b) Higashiyama, K., Kyo, H. and Takahashi, H. (**1998**) *Synlett*, 489–90. (c) Agami, C., Comesse, S. and Kadouri-Puchot, C. (**2002**) *J. Org. Chem.*, **67**, 1496–1500.
12. Schneider, P.H., Schrekker, H.S., Silveira, C.C., Wessjohann, L.A. and Braga, A.L. (**2004**) *Eur. J. Org. Chem.*, 2715–22.
13. Lu, Z. and Ma, S. (**2007**) *Angew. Chem. Int. Ed.*, **47**, 258–97.
14. (a) Gosselin, F., Roy, A., O'Shea, P.D., Chen, C. and Volante, R.P. (**2004**) *Org. Lett.*, **6**, 641–44. (b) Ishii, A., Higashiyama, K. and Mikami, K. (**1997**) *Synlett*, 1381–82.
15. (a) Tessier, A., Pytkowicz, J. and Brigaud, T. (**2006**) *Angew. Chem. Int. Ed.*, **45**, 3677–81. (b) Huguenot, F. and Brigaud, T. (**2006**) *J. Org. Chem.*, **71**, 2159–62.
16. Parrott, R.W. and Hithcock, S.R. (**2007**) *Tetrahedron: Asymmetry*, **18**, 377–82.
17. (a) Guerrier, L., Royer, J., Grierson, D.S. and Husson, H.P. (**1983**) *J. Am. Chem. Soc.*, **105**, 7754–55. (b) The simplest dialdehyde, that is glyoxal, leads to a hydroxymorpholine, instead of an oxazolidine, when it is reacted with aminoalcohols: Agami, C., Couty, F., Daran, J.C., Prince, B. and Puchot, C. (**1990**) *Tetrahedron Lett.*, **31**, 2889–92. (c) Agami, C., Couty, F., Hamon, L. and Puchot, C. (**1992**) *Tetrahedron Lett.*, **33**, 3645–46.
18. Husson, H.P. and Royer, J. (**1999**) *Chem. Soc. Rev.*, **28**, 383–94.
19. (a) Katrizky, A.R., Quiu, G., Yang, B. and Steel, P.J. (**1998**) *J. Org. Chem.*, **63**, 6699–703. (b) Katrizky, A.R., Cui, X.L., Yang, B. and Steel, P.J. (**1999**) *J. Org. Chem.*, **64**, 1979–85. (c) Katrizky, A.R., Rachwal, S. and Hitchkings, G.J. (**1991**) *Tetrahedron*, **47**, 2683–732.
20. (a) Mehmandoust, M., Marazano, C. and Das, B.C. (**1989**) *J. Chem. Soc., Chem. Commun.*, 1185–87. (b) Barbier, D., Marazano, C., Riche, C., Das, B.C. and Potier, P. (**1998**) *J. Org. Chem.*, **63**, 1767–72.
21. Roussi, G. and Zhang, J. (**1991**) *Tetrahedron Lett.*, **32**, 1443–46.
22. (a) Lavilla, R., Coll, O., Nicolas, M. and Bosch, J. (**1998**) *Tetrahedron Lett.*, **39**, 5089–92. (b) Diaba, F., Puigbo, G. and Bonjoch, J. (**2007**) *Eur. J. Org. Chem.*, 3038–44.
23. Meyers, A.I., Lefker, B.A., Wanner, K.T. and Aitken, R.A. (**1986**) *J. Org. Chem.*, **51**, 1936–38.
24. (a) Freville, S., Bonin, M., Celerier, J.P., Husson, H.P., Lhommet, G., Quirion, J.C. and Thuy, V.M. (**1997**) *Tetrahedron*, **53**, 8447–55. (b) Micouin, L., Quirion, J.C. and Husson, H.P. (**1996**) *Tetrahedron Lett.*, **37**, 849–52.
25. Pearson, A.J. and Kwak, Y. (**2005**) *Tetrahedron Lett.*, **46**, 3407–10.

26 (a) Colombo, L., Gennari, C., Poli, G. and Scolastico, C. (1985) *Tetrahedron Lett.*, **26**, 5459–62. (b) Bernardi, A., Caradani, S., Poli, G. and Scolastico, C. (1986) *J. Org. Chem.*, **51**, 5043–45. (c) Cardani, S., Poli, G., Scolastico, C. and Villa, R. (1988) *Tetrahedron*, **44**, 5929–38. (d) Bernardi, A., Cardani, S., Pilati, T., Poli, G., Scolastico, C. and Villa, R. (1988) *J. Org. Chem.*, **53**, 5499–502.

27 Belvisi, L., Carugo, O. and Poli, G. (1984) *J. Mol. Struct.*, **318**, 189–202.

28 (a) Bernardi, A., Cardani, S., Poli, G., Potenza, D. and Scolastico, C. (1992) *Tetrahedron*, **48**, 1343–52. (b) Bernardi, A., Poli, G., Scolastico, C. and Zanda, M. (1991) *J. Org. Chem.*, **56**, 6961–63.

29 (a) Conde-Friboes, K. and Hoppe, D. (1990) *Synlett*, 99–102. (b) Hoppe, I., Hoppe, D., Wolff, C., Egert, E. and Herbst, R. (1989) *Angew. Chem. Int. Ed.*, **28**, 67–69. (c) Hoppe, I., Hoffmann, H., Gärtner, I., Krettek, T. and Hoppe, D. (1991) *Synthesis*, 1157–62.

30 Conde Friboes, K., Harder, T., Aulbert, D., Strahringer, C., Bolte, M. and Hoppe, D. (1993) *Synlett*, 921–23.

31 Colombo, L., Di Giacomo, M., Brusotti, G. and Delogu, G. (1994) *Tetrahedron Lett.*, **35**, 2063–66.

32 (a) Agami, C., Couty, F. and Lequesne, C. (1994) *Tetrahedron Lett.*, **35**, 3309–12. (b) Agami, C., Couty, F. and Lequesne, C. (1995) *Tetrahedron*, **51**, 4043–56. (c) Agami, C., Couty, F., Lam, H. and Mathieu, H. (1998) *Tetrahedron*, **54**, 8793–96.

33 Agami, C., Amiot, F., Couty, F., Dechoux, L., Kaminsky, C. and Venier, O. (1998) *Tetrahedron: Asymmetry*, **9**, 3955–58.

34 (a) He, P. and Zhu, S. (2005) *Synthesis*, 2137–42. (b) Yamazaki, N., Suzuki, H. and Kibayashi, C. (1997) *J. Org. Chem.*, **62**, 8280–81. (c) Suzuki, H., Yamazaki, N. and Kibayashi, C. (2001) *Tetrahedron Lett.*, **42**, 3013–15.

35 (a) Lutomski, K.A. and Meyers, A.I. (1984) in *Asymmetric Synthesis*, (ed J.D., Morrison), Academic Press, Orlando, Vol. 3B, pp. 213–274. (b) Gant, T.G. and Meyers, A.I. (1994) *Tetrahedron*, **50**, 2297–360. (c) Meyers, A.I. (1998) *J. Heterocycl. Chem.*, **35**, 991–1002.

36 (a) Reuman, M. and Meyers, A.I. (1985) *Tetrahedron*, **41**, 837–60. (b) Meyers, A.I., Nelson, T.D., Moorlag, H., Rawson, D.J. and Meier, A. (2004) *Tetrahedron*, **60**, 4459–73.

37 Meyers, A.I. (2005) *J. Org. Chem.*, **70**, 6137–51.

38 (a) Zanoni, G., Castronovo, F., Franzini, M., Vidari, G. and Giannini, E. (2003) *Chem. Soc. Rev.*, **32**, 115–29. (b) Cardillo, G., Gentilucci, L. and Tolomelli, A. (2003) *Aldrichim. Acta*, **36**, 39–50. (c) Mc Manus, H.A. and Guiry, P.J. (2004) *Chem. Rev.*, **104**, 4151–202. (d) Desimoni, G., Faita, G. and Jorgensen, K.A. (2006) *Chem. Rev.*, **106**, 3561. (e) Hargaden, G.C. and Guiry, P.J. (2007) *Adv. Synth. Catal.*, **349**, 2407–24.

39 Meyers, A.I., Knaus, G. and Kamata, K. (1974) *J. Am. Chem. Soc.*, **75**, 268–70.

40 Meyers, A.I. and Slade, J. (1980) *J. Org. Chem.*, **45**, 2785–91.

41 Chandrasekhar, S. and Kausar, A. (2000) *Tetrahedron: Asymmetry*, **11**, 2249–53.

42 (a) Michael, J.P. and Pattenden, G. (1993) *Angew. Chem., Int. Ed. Chem.*, **32**, 1–23. (b) Davidson, B.S. (1993) *Chem. Rev.*, **93**, 1771–91. (c) Ma, D., Zou, B., Cai, G., Hu, X. and Liu, J.O. (2006) *Chem. Eur. J.*, **12**, 7615–26. (d) Hanessian, S., Vinci, V., Auzzas, L., Marzi, M. and Giannini, G. (2006) *Bioorg. Med. Chem. Lett.*, **16**, 4784–87.

43 Sakakura, A., Kondo, R. and Ishihara, K. (2005) *Org. Lett.*, **7**, 1971–74.

44 Ella-Menye, J.R. and Wang, G. (2007) *Tetrahedron*, **63**, 10034–41.

45 Nishimura, M., Minakata, S., Takahashi, T., Oderaotoshi,

Y. and Komatsu, M. (2002) *J. Org. Chem.*, **67**, 2101–10.
46 Brunner, H., Obermann, U. and Wimmer, P. (1986) *J. Organomet. Chem.*, **316**, C1–C3.
47 (a) Evans, D.A., Woerpel, K.A., Hinman, M.H. and Faul, M.M. (1991) *J. Am. Chem. Soc.*, **113**, 722–28. (b) Corey, E.J., Imai, N. and Zhang, H.Y. (1991) *J. Am. Chem. Soc.*, **113**, 728–29.
48 Denmark, S.E., Nakajima, N., Nicaise, O.J.C., Faucher, A.M. and Edwards, J.P. (1995) *J. Org. Chem.*, **60**, 4884–92.
49 Hanessian, S., Jnoff, E., Bernstein, N. and Simard, M. (2004) *Can. J. Chem.*, **82**, 306–13.
50 Annunziata, R., Benaglia, M., Cinquini, M., Cozzi, F. and Pozzi, G. (2003) *Eur. J. Org. Chem.*, 1191–97.
51 Fu, B., Du, D.M. and Wang, J. (2004) *Tetrahedron: Asymmetry*, **15**, 119–26.
52 (a) Rajaram, S. and Sigman, M.S. (2005) *Org. Lett.*, **7**, 5473–75. (b) Lee, J.Y., Miller, J.J., Hamilton, S.S. and Sigman, M.S. (2005) *Org. Lett.*, **7**, 1837–39. (c) Miller, J.J. and Sigman, M.S. (2007) *J. Am. Chem. Soc.*, **129**, 2752–53.
53 Hargaden, G.C., O'Sullivan, T.P. and Guiry, P.J. (2008) *Org. Biomol. Chem.*, **6**, 562–66.
54 Davies, I.W., Gerena, L., Cai, D., Larsen, R.D., Verhoeven, T.R. and Reider, P.J. (1997) *Tetrahedron Lett.*, **38**, 1145–48.
55 Sepac, D., Marinic, Z., Portada, T., Sinic, M. and Sinjic, V. (2003) *Tetrahedron Lett.*, **59**, 1159–67.
56 Benaglia, M., Cinquini, M., Cozzi, F. and Celentano, G. (2004) *Org. Biomol. Chem.*, **2**, 3401–7.
57 (a) Evans, D.A., Bartroli, J. and Shih, T.L. (1981) *J. Am. Chem. Soc.*, **103**, 2127–29. (b) Evans, D.A. (1982) *Aldrichim. Acta*, **15**, 22–32.
58 Ager, D.J., Prakash, I. and Schaad, D.R. (1997) *Aldrichim. Acta*, **30**, 3–12.
59 (a) Evans, D.A., Helmchen, G., Ruping, M. and Wolfgang, J. (2007) *Asymm. Synth.*, 3–9. (b) Zappia, G., Gacs-Baitz, E., Delle Monache, G., Misiti, D., Nevola, L. and Botta, B. (2007) *Curr. Org. Synth.*, **4**, 81–135.
60 (a) Phoon, C.W. and Abell, C. (1998) *Tetrahedron Lett.*, **39**, 2655–58. (b) Wu, Y., Shen, X.Y., Yang, Y.-Q., Hu, Q. and Huang, J.-H. (2004) *J. Org. Chem.*, **69**, 3857–3865. (c) Purandare, A.V. and Natarajan, S. (1997) *Tetrahedron Lett.*, **38**, 8777–8780.
61 (a) Evans, D.A., Chapman, K.T. and Bisaha, J. (1988) *J. Am. Chem. Soc.*, **110**, 1238–56. (b) Brailsford, J.A., Zhu, L., Loo, M. and Shea, K.J. (2007) *J. Org. Chem.*, **72**, 9402–5.
62 Gothelf, K.V. and Jørgensen, K.A. (1998) *Chem. Rev.*, **98**, 863–909.
63 (a) Sibi, M.P. and Ji, J. (1996) *Angew. Chem. Int. Ed.*, **35**, 190–92. (b) Desimoni, G., Faita, G., Galbiati, A., Pasini, D., Quadrelli, P. and Rancati, F. (2002) *Tetrahedron: Asymmetry*, **13**, 333–37.
64 (a) Mukhtar, T.A. and Wright, G.D. (2005) *Chem. Rev.*, **105**, 529–42. (b) Zappia, G., Menendez, P., Delle Monache, G., Misiti, D., Nevola, L. and Botta, B. (2007) *Mini Rev. Med. Chem.*, **7**, 389–409.
65 Ager, D.J., Prakash, I. and Schaad, D.R. (1996) *Chem. Rev.*, **96**, 835–75.
66 (a) Bonner, M.P. and Thornton, E.R. (1991) *J. Am. Chem. Soc.*, **113**, 1299–308. (b) Hamdach, A., El Hadrami, E.M., Gil, S., Zaragozá, R.J., Zaballos-García, E. and Sepúlveda-Arques, J. (2006) *Tetrahedron*, **62**, 6392–97.
67 (a) Pridgen, L.N., Prol Jr, J., Alexander, B. and Gillyard, L. (1989) *J. Org. Chem.*, **54**, 3231–33. (b) Shu, L., Wang, P., Gan, Y. and Shi, Y. (2003) *Org. Lett.*, **5**, 293–96.
68 Hamdach, A., El Hadrami, E.M., Gil, S., Zaragozá, R.J., Zaballos-García, E. and Sepúlveda-Arques, J. (2006) *Tetrahedron*, **62**, 6392–97. A difference of reactivity between phosgene and diphosgene was described.
69 Ritter, T., Kværnø, L., Werder, M., Hauser, H. and Carreira, E.M. (2005) *Org. Biomol. Chem.*, **3**, 3514–23.
70 (a) Badone, D., Bernassau, J.-M., Cardamone, R. and Guzzi, U. (1996)

Angew. Chem. Int. Ed., **35**, 535–38.
(b) Cutugno, S., Martelli, G., Negro, L. and Savoia, D. (**2001**) *Eur. J. Org. Chem.*, 517–22. (c) Kotake, T., Hayashi, Y., Rajesh, S., Mukai, Y., Takiguchi, Y., Kimura, T. and Kiso, Y. (**2005**) *Tetrahedron*, **61**, 3819–33.

71 Gage, J.R. and Evans, D.A. (**1990**) *Org. Synth.*, **68**, 77–82.

72 (a) Spino, C., Tremblay, M.-C. and Gobdout, C. (**2004**) *Org. Lett.*, **6**, 2801–4. (b) Trost, B.M., Chung, C.K. and Pinkerton, A.B. (**2004**) *Angew. Chem. Int. Ed.*, **43**, 4327–29. (c) Wu, Y. and Shen, X. (**2000**) *Tetrahedron: Asymmetry*, **11**, 4359–63. (d) Génisson, Y., Lamandé, L., Salma, Y., Andrieu-Abadie, N., André, C. and Baltas, M. (**2007**) *Tetrahedron: Asymmetry*, **18**, 857–64. (e) Green, R., Taylor, P.J.M., Bull, S.D., James, T.D., Mahon, M.F. and Merritt, A.T. (**2003**) *Tetrahedron: Asymmetry*, **14**, 2619–23. (f) Kim, J.D., Kim, I.S., Jin, C.H., Zee, O.P. and Jung, Y.H. (**2005**) *Org. Lett.*, **7**, 4025–28. (g) Mishra, R.K., Coates, C.M., Revell, K.D. and Turos, E. (**2007**) *Org. Lett.*, **9**, 575–78. (h) Cossy, J., Pévet, I. and Meyer, C. (**2001**) *Eur. J. Org. Chem.*, 2841–50.

73 Hein, J.E., Geary, L.M., Jaworski, A.A. and Hultin, P.G. (**2005**) *J. Org. Chem.*, **70**, 9940–46.

74 Agami, C., Couty, F., Hamon, L. and Venier, O. (**1993**) *Tetrahedron Lett.*, **34**, 4509–12.

75 (a) Benedetti, F. and Norbedo, S. (**2000**) *Tetrahedron Lett.*, **41**, 10071–74. (b) Davies, S.G., Hughes, D.G., Nicholson, R.L., Smith, A.D. and Wright, A.J. (**2004**) *Org. Biomol. Chem.*, **2**, 1549–53.

76 Casadei, M.A., Feroci, M., Inesi, A., Rossi, L. and Sotgiu, G. (**2000**) *J. Org. Chem.*, **65**, 4759–61.

77 (a) Kodaka, M., Tomohiro, T. and Okuno, H. (**1993**) *J. Chem. Soc.*, 81–82. (b) Dinsmore, C.J. and Mercer, S.P. (**2004**) *Org. Lett.*, **6**, 2885–88.

78 Li, P., Yuan, X., Wang, S. and Lu, S. (**2007**) *Tetrahedron*, **63**, 12419–23.

79 Knapp, S., Kukkola, P.J., Sharma, S., Murali Dhar, T.G. and Naughton, A.B.J. (**1990**) *J. Org. Chem.*, **55**, 5700–10.

80 Tiecco, M., Testaferri, L., Temperini, A., Bagnoli, L., Marini, F. and Santi, C. (**2004**) *Chem. Eur. J.*, **10**, 1752–64.

81 Bueno, A.B., Carreño, M.C., Ruano, J.L.G., Arrayás, R.G. and Zarzuelo, M.M. (**1997**) *J. Org. Chem.*, **62**, 2139–43.

82 (a) Roush, W.R. and Adam, M.A. (**1985**) *J. Org. Chem.*, **50**, 3752–57. (b) Clayden, J. and Warren, S. (**1998**) *J. Chem. Soc., Perkin Trans. 1*, 2923–31. (c) Cui, Y., Dang, Y., Yang, Y. and Ji, R. (**2006**) *J. Heterocycl. Chem.*, **43**, 1071–75.

83 Crich, D. and Banerjee, A. (**2006**) *J. Org. Chem.*, **71**, 7106–9.

84 (a) Takacs, J.M., Jaber, M.R. and Vellekoop, A.S. (**1998**) *J. Org. Chem.*, **63**, 2742–48. (b) Andruszkiewicz, R. and Wyszogrodzka, M. (**2002**) *Synlett*, 2101–3. (c) Bertau, M., Bürli, M., Hungerbühler, E. and Wagner, P. (**2001**) *Tetrahedron: Asymmetry*, **12**, 2103–7.

85 Barta, N.S., Sidler, D.R., Somerville, K.B., Weissman, S.A., Larsen, R.D. and Reider, P. (**2000**) *Org. Lett.*, **2**, 2821–24.

86 Bartoli, G., Bosco, M., Carlone, A., Locatelli, M., Melchiorre, P. and Sambri, L. (**2005**) *Org. Lett.*, **7**, 1983–85.

87 Sim, T.B., Kang, S.H., Lee, K.S., Lee, W.K., Yun, H., Dong, Y. and Ha, H.-J. (**2003**) *J. Org. Chem.*, **68**, 104–8.

88 Park, C.S., Kim, M.S., Sim, T.B., Pyun, D.K., Lee, C.H., Choi, D., Lee, W.K., Chang, J.W. and Ha, H.-J. (**2003**) *J. Org. Chem.*, **68**, 43–49.

89 (a) Trost, B.M., Van Vranken, D.L. and Bingel, C. (**1992**) *J. Am. Chem. Soc.*, **114**, 9327–43. (b) Trost, B.M. and Patterson, D.E. (**1998**) *J. Org. Chem.*, **63**, 1339–41.

90 Larksarp, C. and Alper, H. (**1997**) *J. Am. Chem. Soc.*, **119**, 3709–15.

91 (a) Overman, L.E. and Remarchuk, T.P. (**2002**) *J. Am. Chem. Soc.*, **124**, 12–13. (b) Kirch, S.F.

and Overman, L.E. (**2005**) *J. Org. Chem.*, **70**, 2859–61.

92. Lespino, C.G. and Du Bois, J. (**2001**) *Angew. Chem. Int. Ed.*, **40**, 598–600.
93. Marigo, M., Schulte, T., Franzén, J. and Jørgensen, K.A. (**2005**) *J. Am. Chem. Soc.*, **127**, 15710–11.
94. (a) Tsuji, T., Iio, Y., Takemoto, T. and Nishi, T. (**2005**) *Tetrahedron: Asymmetry*, **16**, 3139–42. (b) Sugiyama, S., Fukuchi, H. and Ishii, K. (**2007**) *Tetrahedon*, **63**, 12047–57. (c) Neri, C. and Williams, J.M.J. (**2003**) *Adv. Synth. Catal.*, **345**, 835–48. (d) Neri, C. and Williams, J.M.J. (**2002**) *Tetrahedron: Asymmetry*, **13**, 2197–99. (e) Allali, H., Tabti, B., Alexandre, C. and Huet, F. (**2004**) *Tetrahedron: Asymmetry*, **15**, 1331–33. (f) Sugiyama, S., Watanabe, S., Inoue, T., Kurihara, R., Itou, T. and Ishii, K. (**2003**) *Tetrahedron*, **59**, 3417–25. (g) Sugiyama, S., Watanabe, S. and Ishii, K. (**1999**) *Tetrahedron Lett.*, **40**, 7489–92.
95. Barbachyn, M.R. and Ford, C.W. (**2003**) *Angew. Chem. Int. Ed.*, **42**, 2010–23.
96. Grajewska, A. and Rozwadowska, M.D. (**2007**) *Tetrahedron: Asymmetry*, **18**, 803–13.
97. Kibayashi, C. (**2005**) *Chem. Pharm. Bull.*, **11**, 1375–86.
98. Mohsen, D. (**2006**) *Top. Heterocycl. Chem.*, **2**, 153–206.
99. Omata, Y., Kakehi, A., Shirai, M. and Kamimura, A. (**2002**) *Tetrahedron Lett.*, **43**, 6911–14.
100. (a) Kamimura, A., Omata, Y., Tnaka, K. and Shirai, M. (**2003**) *Tetrahedron*, **59**, 6291–99. (b) Kamimura, A., Tanaka, K., Hayashi, T. and Omata, Y. (**2006**) *Tetrahedron Lett.*, **47**, 3625–27.
101. Aurelio, L., Brownlee, R.T.C. and Hughes, A.B. (**2004**) *Chem. Rev.*, **104**, 5823–46.
102. Ben-Ishai, D. (**1957**) *J. Am. Chem. Soc.*, **79**, 5736–38.
103. Freidinger, R.M., Hinkle, J.S., Perlow, D.S. and Arison, B.H. (**1983**) *J. Org. Chem.*, **48**, 77–81.
104. Reddy, G.V., Rao, G.V. and Iyengar, D.S. (**1998**) *Tetrahedron Lett.*, **39**, 1985.
105. (a) Aurelio, L., Box, J.S., Brownlee, R.T.C., Hughes, A.B. and Sleebs, M.M. (**2002**) *J. Org. Chem.*, **68**, 2652–67. (b) Aurelio, L., Brownlee, R.T.C. and Hughes, A.B. (**2002**) *Org. Lett.*, **4**, 3767–69.
106. (a) Seebach, D. and Naef, R. (**1981**) *Helv. Chim. Acta*, **64**, 2704–8. (b) Seebach, D., Boes, M., Naef, R. and Schweiser, W.B. (**1983**) *J. Am. Chem. Soc.*, **105**, 5390–98.
107. Vartak, A.P., Young, V.G. and Jonhson, R.L. Jr (**2005**) *Org. Lett.*, **7**, 35–38.
108. Karady, S., Amato, J.S. and Weinstock, L.M. (**1984**) *Tetrahedron Lett.*, **25**, 4337–40.
109. (a) Abell, A.D., Taylor, J.M. and Oldham, M.D. (**1996**) *J. Chem. Soc., Perkin Trans. 1*, 1299–304. (b) Abell, A.D., Edwards, R.A. and Oldham, M.D. (**1997**) *J. Chem. Soc., Perkin Trans. 1*, 1655–62.
110. Procopiou, P.A., Ahmed, M., Jeulin, S. and Perciaccante, R. (**2003**) *Org. Biomol. Chem.*, **1**, 2853–58.
111. Pellissier, H. (**2007**) *Tetrahedron*, **63**, 3235–85.
112. Frederickson, M. (**1997**) *Tetrahedron*, **53**, 403–25.
113. (a) Shirahase, M., Kanemasa, S. and Oderaotoshi, Y. (**2004**) *Org. Lett.*, **6**, 675–78. references therein. (b) Lemay, M., Trant, J. and Ogilvie, W.W. (**2007**) *Tetrahedron*, **63**, 11644–55. (c) Rios, R., Ibrahem, I., Vesely, J., Zhao, G.L. and Cordova, A. (**2007**) *Tetrahedron Lett.*, **48**, 5701–5.
114. (a) Cordero, F.M., Bonollo, S., Machetti, F. and Brandi, A. (**2006**) *Eur. J. Org. Chem.*, 3235–41. (b) Argyropoulos, N.G., Panagiotidis, T., Coutouli-Argyropoulou, E. and Raptopoulou, C. (**2007**) *Tetrahedron*, **63**, 321–30.
115. Gothelf, K.V., Hazell, R.G. and Jorgensen, K.A. (**1996**) *J. Org. Chem.*, **61**, 346–55.

116 Jen, W.S., Wiener, J.J.M. and Mac Millan, D.W.C. (2000) *J. Am. Chem. Soc.*, **122**, 9874–75.

117 (a) Sato, T., Yamada, T., Miyazaki, S. and Otani, T. (2004) *Tetrahedron Lett.*, **45**, 9581–84. (b) Suga, H., Nakajima, T., Itoh, K. and Kakehi, A. (2005) *Org. Lett.*, **7**, 1431–34. (c) Iwasa, S., Maeda, H., Nishiyama, K., Tsushima, S., Tsukamoto, Y. and Nishiyama, H. (2002) *Tetrahedron*, **58**, 8281–87. (d) Iwasa, S., Tsushima, S., Shimada, T. and Nishiyama, H. (2001) *Tetrahedron Lett.*, **42**, 6715–17.

118 Hein, J.E. and Hultin, P.G. (2005) *Tetrahedron: Asymmetry*, **16**, 2341–47.

119 Zhang, W. and Curran, D.P. (2066) *Tetrahedron*, **62**, 11837–65.

120 (a) Pedrosa, R., Andres, C., Nieto, J., Perez-Cuadrado, C. and San Francisco, I. (2006) *Eur. J. Org. Chem.*, **14**, 3259–65. (b) Tessier, A., Pytkowicz, J. and Brigaud, T. (2006) *Angew. Chem. Int. Ed.*, **45**, 3677–81.

121 Dolle, R.E., Le Bourdonnec, B., Morales, G.A., Moriarty, K.J. and Salvino, J.M. (2006) *J. Comb. Chem.*, **8**, 597–635.

122 O'Neil, I.A., Ramos, V.E., Ellis, G.L., Cleator, E., Chorlton, A.P., Tapolczay, D.J. and Kalindjian, S.B. (2004) *Tetrahedron Lett.*, **45**, 3659–61.

123 (a) Banerji, A., Gupta, M., Biswas, K.P., Prange, T. and Neuman, A. (2007) *J. Heterocycl. Chem.*, **44**, 1045–49. (b) Bainbridge, N.P., Currie, A.C., Cooper, N.J., Muir, J.C., Knight, D.W. and Walton, J.M. (2007) *Tetrahedron Lett.*, **48**, 7782–87. (c) Fabio, M., Roonzini, L. and Troisi, L. (2007) *Tetrahedron Lett.*, **63**, 12896–902.

124 Borrachero, P., Cabrera-Escribano, F., Gomez-Guillen, M. and Torres, M.I. (2004) *Tetrahedron Lett.*, **45**, 4835–39.

125 Alibes, R., Blanco, P., de March, P., Figueredo, M., Font, J., Alvarez-Larena, A. and Piniella, J.F. (2003) *Tetrahedron Lett.*, **44**, 523–25.

126 Ibrahem, I., Rios, R., Vesley, J., Zhao, G.L. and Cordova, A. (2007) *Chem. Commun.*, 849–51.

127 Romanski, J., Jozwik, J., Chapuis, C., Asztemborska, M. and Jurczark, J. (2007) *Tetrahedron: Asymmetry*, **18**, 865–72 and references therein.

128 (a) Kanemasa, S., Nishiuchi, M., Kamimura, A. and Hori, K. (1994) *J. Am. Chem. Soc.*, **116**, 2324–39. (b) Bode, J.W., Fraefel, N., Muri, D. and Carreira, E.M. (2001) *Angew. Chem. Int. Ed.*, **40**, 2082–85.

129 (a) Minter, A.R., Fuller, A.A. and Mapp, A.K. (2003) *J. Am. Chem. Soc.*, **125**, 6846–47. (b) Fuller, A.A., Chen, B., Minter, A.R. and Mapp, A.K. (2005) *J. Am. Chem. Soc.*, **127**, 5376–83.

130 Arai, M., Kuraishi, M., Arai, T. and Sasai, H. (2001) *J. Am. Chem. Soc.*, **123**, 2907–8.

131 Zhang, L.H., Chung, J.C., Costello, T.D., Valvis, I., Ma, P., Kauffman, S. and Ward, R. (1997) *J. Org. Chem.*, **62**, 2466–70.

132 Sibi, M.P., Itoh, K. and Jasperse, C.P. (2004) *J. Am. Chem. Soc.*, **126**, 5366–67.

133 Yamamoto, H., Hayashi, S., Kubo, M., Harada, M., Hasegawa, M., Noguchi, M., Sumimoto, M. and Hori, K. (2007) *Eur. J. Org. Chem.*, 2859–64.

134 Vesely, J., Rios, R., Ibrahem, I., Zhao, G.L., Eriksson, L. and Cordova, A. (2008) *Chem. Eur. J.*, **14**, 2693–98.

135 Zhu, C.Y., Deng, X.M., Sun, X.L., Zheng, J.C. and Tang, Y. (2008) *Chem. Commun.*, 738–40.

136 Mukaiyama, T. (1981) *Tetrahedron*, **37**, 4111–19.

137 Lemaire, M. and Mangeney, P. (2005) *Chiral Diaza Ligands for Asymmetric Synthesis*, Springer Verlag, Berlin.

138 Mukaiyama, T., Sakito, Y. and Asami, M. (1978) *Chem. Lett.*, 1253–56.

139 Coldham, I., Copley, R.C.B., Haxell, T.F.N. and Howard, S. (2001) *Org. Lett.*, **3**, 3799–801.

140 (a) Mangeney, P., Grojean, F., Alexakis, A. and Normant, J.F. (1988) *Tetrahedron Lett.*, **29**, 2675–76. (b) Betschart, C. and Seebach,

D. (1987) *Helv. Chim. Acta*, **70**, 2215–31.
141. Kagan, H.B. and Dang, T.P. (**1972**) *J. Am. Chem. Soc.*, **94**, 6429–33.
142. (a) Whitesell, J.K. (**1989**) *Chem. Rev.*, **89**, 1581–90. (b) Bowmick, K.C. and Joshi, N.N. (**2006**) *Tetrahedron: Asymmetry*, **17**, 1901–29.
143. (a) Alexakis, A., Mangeney, P., Lensen, N. and Tranchier, J.P. (**1996**) *Pure Appl. Chem.*, **68**, 531–34. (b) Alexakis, A. and Mangeney, P. (**1996**) in *Advanced Asymmetric Synthesis*, (ed G.R. Stephenson), Chapman & Hall, London, pp. 93–110.
144. (a) Alexakis, A., Tranchier, J.P., Lensen, N. and Mangeney, P. (**1995**) *J. Am. Chem. Soc.*, **117**, 10767–768. (b) Frey, L.F., Tillyer, R.D., Caille, A.S., Tschaen, D.M., Dolling, U.F., Grabovski, E.J. and Reider, P.J. (**1998**) *J. Org. Chem.*, **63**, 3120–24.
145. Corey, E.J., Imwinkelried, R., Pikul, S. and Xiang, Y.B. (**1989**) *J. Am. Chem. Soc.*, **111**, 5493–95.
146. (a) Yoshida, S., Sugihara, Y. and Nakayama, J. (**2007**) *Tetrahedron Lett.*, **48**, 8116–19.
(b) Kull, T. and Peters, R. (**2007**) *Adv. Synth. Catal.*, **349**, 1647–52.
(c) Somfai, P. and Panknin, O. (**2007**) *Synlett*, **8**, 1190–202.
147. Kanemasa, S. and Onimura, K. (**1992**) *Tetrahedron*, **48**, 8631–44.
148. Katritzky, A.R., He, H.Y. and Verma, A.K. (**2002**) *Tetrahedron: Asymmetry*, **13**, 933–38.
149. (a) Halland, N., Hazell, R.G. and Jorgensen, K.A. (**2002**) *J. Org. Chem.*, **67**, 8331–38. (b) Braga, A.L., Vargas, F., Silveira, C.C. and de Andrade, L.H. (**2002**) *Tetrahedron Lett.*, **43**, 2335–37. (c) Lee, E.K., Kim, S.H., Jung, B.H., Ahn, W.S. and Kim, G.J. (**2003**) *Tetrahedron Lett.*, **44**, 1971–74. (d) Prieto, A., Halland, N. and Jorgensen, K.A. (**2005**) *Org. Lett.*, **7**, 3897–900. (e) Jin, M.J., Takale, V.B., Sarkar, M.S. and Kim, Y.M. (**2006**) *Chem. Commun.*, 663–64. (f) Lee, S. and MacMillan, D.W.C. (**2006**) *Tetrahedron*, **62**, 11413–24. (g) Fournier, P.A. and Collins, S.K. (**2007**) *Organometallics*, **26**, 2945–49. (h) Uria, U., Vicario, J.L., Badia, D. and Carillo, L. (**2007**) *Chem. Commun.*, 2509–11. (i) Gordillo, R. and Houk, K.N. (**2006**) *J. Am. Chem. Soc.*, **128**, 3543–53.
150. Seebach, D., Sting, A.R. and Hoffmann, M. (**1996**) *Angew. Chem. Int. Ed. Engl.*, **35**, 2708–48 and references therein.
151. (a) Fitzi, R., Seebach, D. and Fitzi, R. (**1986**) *Angew. Chem. Int. Ed. Engl.*, **25**, 345–46. (b) Seebach, D. (**1988**) *Tetrahedron*, **44**, 5277–92.
152. Williams, R.M. (**1989**) *Synthesis of Optically Active α–Amino Acids*, Pergamon, Oxford, pp. 63–78 and 81–84.
153. Node, M., Kodama, S., Hamashima, Y., Katoh, T., Nishide, K. and Kajimoto, T. (**2006**) *Chem. Pharm. Bull.*, **54**, 1662–79.
154. Ouelet, S.G., Walji, A.M. and MacMillan, D.W.C. (**2007**) *Acc. Chem. Res.*, **40**, 1327–39.
155. Yokohama, K., Ishizuka, T., Ohmachi, N. and Kunieda, T. (**1998**) *Tetrahedron Lett.*, **39**, 4847–50.
156. Abdel-Aziz, A.A.M., Okuno, J., Tanaka, S., Ishizuka, T., Matsunaga, H. and Kunieda, T. (**2000**) *Tetrahedron Lett.*, **41**, 8533–37.
157. Roder, H., Helmchen, G., Peters, E.M. and Schneering, H.G.V. (**1984**) *Chem. Int. Ed. Engl.*, **23**, 898–99.
158. Bongini, A., Cardillo, G., Mingardi, A. and Tomasini, C. (**1996**) *Tetrahedron: Asymmetry*, **7**, 1457–66.
159. Wulff, W.D. (**1998**) *Organometallics*, **17**, 3116–14 and references therein.
160. Baeg, J.O., Bensimon, C. and Alper, H. (**1995**) *J. Am. Chem. Soc.*, **117**, 4700–1.
161. Ahn, J.H., Shin, M.S., Jun, M.A., Kang, S.K., Kim, K.R., Rhee, S.D., Kang, N.S., Kim, S.Y., Sohn, S.K., Kim, S.G., Jin, M.S., Lee, J.O., Cheon, H.G. and Kim, S.S. (**2007**) *Bioorg. Med. Chem. Lett.*, **17**, 2622–28.
162. Chauveau, A., Martens, T., Bonin, M., Micouin, L. and Husson, H.P. (**2002**) *Synthesis*, 1885–90.

163 Gallos, J.K., Koumbis, A.E. and Apostolakis, N.E. (1997) *J. Chem. Soc., Perkin Trans. 1*, 2457–59.

164 (a) Kobayashi, S., Shimizu, H., Yamashita, Y., Ishitani, H. and Kobayashi, J. (2002) *J. Am. Chem. Soc.*, **124**, 13678–79. (b) Yamashita, Y. and Kobayashi, S. (2004) *J. Am. Chem. Soc.*, **126**, 11279–82.

165 Kobayashi, S., Hasegawa, Y. and Ishitani, H. (1998) *Chem. Lett.*, 1131–32.

166 Yang, Q., Jiang, X. and Ma, S. (2007) *Chem. Eur. J.*, **13**, 9310–16.

167 Tiecco, M., Testaferri, L. and Marini, F. (1996) *Tetrahedron*, **36**, 11841–48.

168 (a) Roussi, F., Bonin, M., Chiaroni, A., Micouin, L., Riche, C. and Husson, H.P. (1999) *Tetrahedron Lett.*, **40**, 3727–30. (b) Roussi, F., Chauveau, A., Bonin, M., Micouin, L. and Husson, H.P. (2000) *Synthesis*, 1170–79.

169 Guerra, F.M., Mish, M.R. and Carreira, E.M. (2000) *Org. Lett.*, **2**, 4265–67.

170 Barluenga, J., Fernandez-Mari, F., Viado, A.L., Aguilar, E., Olano, B., Garcia-Grande, S. and Moya-Rubiera, C. (1999) *Chem. Eur. J.*, **5**, 883–96.

171 Barluenga, J., Fernandez-Mari, F., Aguilar, E., Viado, A.L. and Olano, B. (1998) *Tetrahedron Lett.*, **39**, 4887–90.

172 (a) De Silva, A.P., Gunaratne, H.Q.N., Gunnlaugsson, T. and Nieuwenhuizen, M. (1996) *Chem. Commun.*, 1967–68. (b) Johson, M., Younglove, B., Lee, L., LeBlanc, R., Holt Jr, H., Hills, P., Mackay, H., Brown, T., Mooberry, S.L. and Lee, M. (2007) *Bioorg. Med. Chem. Lett.*, **17**, 5897–901.

173 Garanti, L., Molteni, G. and Pilati, T. (2002) *Tetrahedron: Asymmetry*, **13**, 1285–89.

174 (a) Broggini, G., Garanti, L., Molteni, G. and Zecchi, G. (1999) *Tetrahedron: Asymmetry*, **10**, 487–92. (b) Broggini, G., Garanti, L., Molteni, G., Pilati, T., Ponti, A. and Zecchi, G. (1999) *Tetrahedron: Asymmetry*, **10**, 2203–12. (c) Broggini, G., Garanti, L., Molteni, G. and Pilati, T. (2001) *Synth. Commun.*, **31**, 2649–56.

175 Stanovnik, B., Jelen, B., Turk, C., Zlicar, M. and Svete, J. (1998) *J. Heterocycl. Chem.*, **35**, 1187–204 and references therein.

176 Poupardin, O., Greck, C. and Genet, J.P. (2000) *Tetrahedron Lett.*, **41**, 8795–97.

177 Kanemasa, S. and Kanai, T. (2000) *J. Am. Chem. Soc.*, **122**, 10710–11.

178 Sibi, M.P., Stanley, L.M. and Soeta, T. (2007) *Org. Lett.*, **9**, 1553–56.

179 (a) Holmes, R.B. and Neel, D.A. (1990) *Tetrahedron Lett.*, **31**, 5567–70. (b) Panfil, I., Urbanczyk-Lipkowska, Z., Suwinska, K., Solecka, J. and Chmielewski, M. (2002) *Tetrahedron*, **58**, 1199–212. (c) Dietrich, E. and Lubell, W.D. (2003) *J. Org. Chem.*, **68**, 6988–96 and references therein.

180 (a) Allen, N.E., Hobbs, J.N., Preston, D.A., Turner, J.R. and Wu, C.Y.E. (1990) *J. Antibiot.*, **43**, 92–99. (b) Panfil, I., Urbanczyk-Lipkowska, Z., Zuwinska, K., Solecka, J. and Chmielewski, M. (2002) *Tetrahedron*, **58**, 1199–212.

181 Itoh, N., Matsuyama, H., Yoshida, M., Kamigata, N. and Iyoda, M. (1995) *Bull. Chem. Soc. Jpn.*, **68**, 3121–30.

182 Svete, J., Preseren, A., Stanovnik, B., Golic, L. and Golic-Grdadolnik, S. (1994) *J. Heterocycl. Chem.*, **34**, 1323–28.

183 (a) Dorn, H. and Otto, A. (1968) *Chem. Ber.*, **101**, 3287–301. (b) Dorn, H. (1985) *Tetrahedron Lett.*, **26**, 5123–26.

184 Oppolzer, W. (1972) *Tetrahedron Lett.*, **17**, 1707–10.

185 Chuang, T.H. and Sharpless, K.B. (2000) *Helv. Chem. Acta*, **83**, 1734–43.

186 Sibi, M.P., Stanley, L.M., Nie, X., Venkatraman, L., Liu, M. and Jasperse, C.P. (2007) *J. Am. Chem. Soc.*, **129**, 395–405 and references therein.

187 Sibi, M.P. and Soeta, T. (2007) *J. Am. Chem. Soc.*, **129**, 4522–23.

188 Chan, A. and Scheidt, K.A. (2008) *J. Am. Chem. Soc.*, **130**, 2740–41.

189 (a) Sheehan, J.C. and Henery-Logan, K.R. (1959) *J. Am. Chem. Soc.*,

81, 3089–94. (b) Sheehan, J.C. and Yang, D.-D. (1958) *J. Am. Chem. Soc.*, **80**, 1158–64. (c) King, F.E., Clark-Lewis, J.W., Smith, G.R. and Wade, R. (1959) *J. Chem. Soc.*, 2264–66.

190 Sammes, P.G. (1976) *Chem. Rev.*, **76**, 113–55.

191 Szilágyi, L. and Györgydeák, Z. (1979) *J. Am. Chem. Soc.*, **101**, 427–32.

192 Pinho e Melo, T.M.V.D., Lopes, S.M.M., D'A Rocha Gonsalves, A.M., Paixão, J.A., Beja, A.M. and Silva, M.R. (2006) *Heterocycles*, **68**, 679–86.

193 Pattenden, G., Thom, S.M. and Jones, M.F. (1993) *Tetrahedron*, **49**, 2131–38.

194 Baldwin, J.E., Freeman, R.T., Lowe, C., Schofield, C.J. and Lee, E. (1989) *Tetrahedron*, **45**, 4537–50.

195 Khalil, E.M., Ojala, W.H., Pradham, A., Nair, V.D., Gleason, W.B., Mishra, R.K. and Jonhson, R.L. (1999) *J. Med. Chem.*, **42**, 628–37.

196 Subasinghe, N.L., Khalil, E.M. and Jonhson, R.L. (1997) *Tetrahedron Lett.*, **38**, 1317–20.

197 Geyer, A. and Moser, F. (2000) *Eur. J. Org. Chem.*, 1113–20.

198 Mellah, M., Voituriez, A. and Schulz, E. (2007) *Chem. Rev.*, **107**, 5133–209.

199 Meng, Q., Li, Y., He, Y. and Guan, Y. (2000) *Tetrahedron: Asymmetry*, **11**, 4255–61.

200 Braga, L., Milani, P., Vargas, F., Paixão, M.W. and Sehnem, J.A. (2006) *Tetrahedron: Asymmetry*, **17**, 2793–97.

201 (a) Schneider, H., Schrekker, H.S., Silveira, C.C., Wesjohann, L.A. and Braga, A.L. (2004) *Eur. J. Org. Chem.*, 2715–22. (b) Braga, A.L., Silveira, C.C., De Bolster, M.W.G., Schrekker, H.S., Wessjohann, L.A. and Schneider, P.H. (2005) *J. Mol. Catal. A: Chem.*, **239**, 235–38.

202 Huang, H.-L., Lin, Y.-C., Chen, S.-F., Wang, C.-L. J. and Liu, L.T. (1996) *Tetrahedron: Asymmetry*, **7**, 3067–70.

203 Trentmann, W., Mehler, T. and Martens, J. (1997) *Tetrahedron: Asymmetry*, **8**, 2033–43.

204 (a) DegI'Innocenti, A., Pollicino, S. and Capperucci, A. (2006) *Chem. Commun.*, 4881–93. (b) Capperucci, A., DegI'Innocenti, A., Pollicino, S., Acciai, M., Castagnoli, G., Malesci, I. and Tiberi, C. (2007) *Heteroat. Chem.*, **18**, 516–26.

205 Cremonesi, G., Croce, P.D., Fontana, F., Forni, A. and La Rosa, c. (2005) *Tetrahedron: Asymmetry*, **16**, 3371–79.

206 (a) Wang, L., Nakamura, S. and Toru, T. (2004) *Org. Biomol. Chem.*, **2**, 2168–69. (b) Wang, L., Nakamura, S., Ito, Y. and Toru, T. (2004) *Tetrahedron: Asymmetry*, **15**, 3059–72.

207 Sriramurthy, V., Barcan, G.A. and Kwon, O. (2007) *J. Am. Chem. Soc.*, **129**, 12928–29.

208 Gröger, H., Saida, Y., Arai, S., Martens, J., Sasai, H. and Shibasaki, M. (1996) *Tetrahedron Lett.*, **37**, 9291–92.

209 D'hooghe, M. and De Kimpe, N. (2006) *Tetrahedron*, **62**, 513–35.

210 Baeg, J.-O., Bensimon, C. and Alper, H. (1995) *J. Am. Chem. Soc.*, **117**, 4700–1.

211 Larksarp, C., Sellier, O. and Alper, H. (2001) *J. Org. Chem.*, **66**, 3502–6.

212 Cruz, A., Macías-Mendoza, D., Barragan-Rodrígues, E., Tlahuext, H., Nöth, H. and Contreras, R. (1997) *Tetrahedron: Asymmetry*, **8**, 3903–11.

213 Ueda, S., Terauchi, H., Yano, A., Matsumoto, M., Kubo, T., Kyoya, Y., Suzuki, K., Ido, M. and Kawasaki, M. (2004) *Bioorg. Med. Chem.*, **12**, 4101–16.

214 Velazquez, F. and Olivo, H.F. (2002) *Curr. Org. Chem.*, **6**, 303–40.

215 Delauney, D., Toupet, L. and Le Corre, M. (1995) *J. Org. Chem.*, **60**, 6604–7.

216 (a) Nagao, Y., Yamada, S., Kumagai, T., Ochiai, M. and Fujita, E. (1985) *J. Chem. Soc., Chem. Commun.*, 1418–19. (b) Nagao, Y., Hagiwara, Y., Kumagai, T., Ochiai, M., Inoue, T., Hashimoto, K. and Fujita, E. (1986) *J. Org. Chem.*, **51**, 2391–93. (c) Velàzquez, F. and Olivo, H.F. (2002) *Curr. Org. Chem.*, **6**, 303–40. (d) Ortiz, A. and Sansinenea, E.

(2007) *J. Sulfur Chem.*, **28**, 109–47.
(e) Dang, T.V., Miyamoto, M., Sano, S., Shiro, M. and Nagao, Y. (2005) *Heterocycles*, **65**, 1139–56.

217 Yan, T.H., Hung, A.-W., Lee, H.-C., Chang, C.-S. and Liu, W.-H. (1995) *J. Org. Chem.*, **60**, 3301–6.

218 Guz, N.R. and Phillips, A.J. (2002) *Org. Lett.*, **4**, 2253–56.

219 Zhang, Y., Phillips, A.J. and Sammakia, T. (2004) *Org. Lett.*, **6**, 23–25.

220 Zhang, Y. and Sammakia, T. (2004) *Org. Lett.*, **6**, 3139–41.

221 Crimmins, M.T. and Shamzad, M. (2007) *Org. Lett.*, **9**, 149–52.

222 Osorio-Lozada, A. and Olivo, H.F. (2008) *Org. Lett.*, **10**, 617–20.

223 Singh, S.P., Parmar, S.S., Raman, K. and Stenberg, V.I. (1981) *Chem. Rev.*, **81**, 175–203.

224 Prabhakar, Y.S., Solomon, V.R., Gupta, M.K. and Katti, S.B. (2006) *Top. Heterocycl. Chem.*, **4**, 161–249.

225 (a) Maclaren, J.A. (1968) *Aust. J. Chem.*, **21**, 1891–96. (b) White, J.D. and Kawasaki, M.J. (1990) *J. Am. Chem. Soc.*, **112**, 4991–93.

226 Falb, E., Nudelman, A. and Hassner, A. (1993) *Synth. Commun.*, **23**, 2839–44.

227 Seki, M., Kimura, M., Hatsuda, M., Yoshida, S.I. and Shimizu, T. (2003) *Tetrahedron Lett.*, **44**, 8905–7.

228 (a) Fürstner, A. and Turet, L. (2005) *Angew. Chem. Int. Ed.*, **44**, 3462–66. (b) Fürstner, A., De Souza, D., Turet, L., Fenster, M.D.B., Parra-Rapado, L., Wirtz, C., Mynott, R. and Lehmann, C.W. (2007) *Chem. Eur. J.*, **13**, 115–34.

229 Boyce, R.J., Mulqueen, G.C. and Pattenden, G. (1995) *Tetrahedron*, **51**, 7321–30.

230 (a) North, M. and Pattenden, G. (1990) *Tetrahedron*, **46**, 8267–90. (b) Meyers, A.I. and Tavares, F.X. (1996) *J. Org. Chem.*, **61**, 8207–15.

231 (a) Galéotti, N., Montagne, C., Poncet, J. and Jouin, P. (1992) *Tetrahedron Lett.*, **33**, 2807–10. (b) Wipf, P. and Miller, C.P. (1992) *Tetrahedron Lett.*, **33**, 6267–70.

232 (a) Walker, M.A. and Heathcock, C.H. (1992) *J. Org. Chem.*, **57**, 5566–68. (b) Parsons Jr, R.L. and Heathcock, C.H. (1994) *J. Org. Chem.*, **59**, 4733–34.

233 (a) Wipf, P., Miller, C.P., Venkatraman, S. and Fritch, P.C. (1995) *Tetrahedron Lett.*, **36**, 6395–98. (b) Wipft, P. and Fritch, P.C. (1994) *Tetrahedron Lett.*, **35**, 5397–400. (c) Wipft, P. and Venkatraman, S. (1997) *Synlett*, 1–10.

234 Chen, J. and Forsyth, C.J. (2003) *Org. Lett.*, **5**, 1281–83.

235 Sakakura, A., Kondo, R. and Ishihara, K. (2005) *Org. Lett.*, **7**, 1971–74.

236 Xu, Z. and Ye, T. (2005) *Tetrahedron: Asymmetry*, **16**, 1905–12.

237 Lafargue, P., Guenot, P. and Lellouche, J.-P. (1995) *Synlett*, 171–72.

238 Abdel-Jalil, R.J., Saeed, M. and Voelter, W. (2001) *Tetrahedron Lett.*, **42**, 2435–37.

239 Aitken, R.A., Armstrong, D.P., Galt, R.H.B. and Mesher, S.T.E. (1997) *J. Chem. Soc., Perkin Trans. 1*, 935–43.

240 Kim, T.-S., Lee, Y.-J., Jeong, B.-S., Park, H.-G. and Jew, S.-S. (2006) *J. Org. Chem.*, **71**, 8276–78.

241 (a) Han, F.S., Osajima, H., Cheung, M., Tokuyama, H. and Fukuyama, T. (2006) *Chem. Commun.*, 1757–59. (b) Han, F.S., Osajima, H., Cheung, M., Tokuyama, H. and Fukuyama, T. (2007) *Chem. Eur. J.*, **13**, 3026–38.

242 Helmchen, G., Krotz, A., Ganz, K.-T. and Hansen, D. (1991) *Synlett*, 257–59.

243 (a) Nishio, T., Kodama, Y. and Tsurumi, Y. (2005) *Phosphorus, Sulfur, and Silicon*, **180**, 1449–50. (b) Yamakuchi, M., Matsunaga, H., Tokuda, R., Ishizuka, T., Nakajima, M. and Kunieda, T. (2005) *Tetrahedron Lett.*, **46**, 4019–22.

244 Irmak, M., Lehnert, T. and Boysen, M.K. (2007) *Tetrahedron Lett.*, **48**, 7890–93.

245 (a) Abrunhosa, I., Gulea, M., Levillain, J. and Masson, S. (2001) *Tetrahedron: Asymmetry*, **12**, 2851–59.

(b) Abrunhosa, I., Delain-Bioton, L., Gaumont, A.-C., Gulea, M. and Masson, S. (**2004**) *Tetrahedron*, **60**, 9263–72. (c) Fu, B., Du, D.-M. and Xia, Q. (**2004**) *Synthesis*, 221–26.

246 (a) Du, D.-M., Lu, S.-F., Fang, T. and Xu, J. (**2005**) *J. Org. Chem.*, **70**, 3712–15. (b) Lu, S.-F., Du, D.-M., Xu, J. and Zhang, S.-W. (**2006**) *J. Am. Chem. Soc.*, **128**, 7418–19. (c) Lu, S.-F., Du, D.-M., Zhang, S.-W. and Xu, J. (**2004**) *Tetrahedron: Asymmetry*, **15**, 3433–41.

247 (a) Molina, P., Tárraga, A. and Curiel, D. (**2002**) *Synlett*, 435–38. (b) Tárraga, A., Molina, P., Curiel, D. and Bautista, D. (**2002**) *Tetrahedron: Asymmetry*, **13**, 1621–28.

248 Bernardi, L., Bonini, B.F., Comes-Franchini, M., Femoni, C., Fochi, M. and Ricci, A. (**2004**) *Tetrahedron: Asymmetry*, **15**, 1133–40.

249 Levillain, J., Dubant, G., Abrunhosa, I., Gulea, M. and Gaumont, A.-C. (**2003**) *Chem. Commun.*, 2914–15.

250 Schlemminger, I., Janknecht, H.-H., Maison, W., Saak, W. and Martens, J. (**2000**) *Tetrahedron Lett.*, **41**, 7289–92.

251 (a) Oppolzer, W., Chapuis, C. and Bernardinelli, G. (**1984**) *Helv. Chem. Acta*, **67**, 1397–401. (b) Oppolzer, W. (**1987**) *Tetrahedron*, **43**, 1969–2004. (c) Oppolzer, W. (**1990**) *Pure Appl. Chem.*, **62**, 1241–50.

252 Lee, A.W.M., Chan, W.H., Zhang, S.-J. and Zhang, H.-K. (**2007**) *Curr. Org. Chem.*, **11**, 213–28.

253 (a) Oppolzer, W., Wills, M., Kelly, M.J., Signer, M. and Blagg, J. (**1990**) *Tetrahedron Lett.*, **31**, 4117–20. (b) Oppolzer, W., Wills, M., Kelly, M.J., Signer, M. and Blagg, J. (**1990**) *Tetrahedron Lett.*, **31**, 5015–18. (c) Oppolzer, W., Rodriguez, I., Starkemann, C. and Walther, E. (**1990**) *Tetrahedron Lett.*, **31**, 5019–22.

254 Wang, Y.-Q., Lu, S.-M. and Zhou, Y.-G. (**2007**) *J. Org. Chem.*, **72**, 3729–34.

255 (a) Liu, P.-N., Gu, P.-M., Deng, J.-G., Tu, Y.-Q. and Ma, Y.-P. (**2005**) *Eur. J. Org.Chem.*, 3221–27. (b) Mao, J. and Baker, D.C. (**1999**) *Org. Lett.*, **1**, 841–43. (c) Wu, J., Wang, F., Ma, Y., Cui, X., Cun, L., Zhu, J., Deng, J. and Yu, B. (**2006**) *Chem. Commun.*, 1766–68. (d) Chen, Y.-C., Wu, T.-F., Deng, J.-G., Liu, H., Ciu, X., Zhu, J., Jiang, Y.-Z., Choi, M.C.K. and Chan, A.S.C. (**2002**) *J. Org. Chem.*, **67**, 5301–6.

256 Ahn, K.H., Ham, C., Kim, S.-K. and Cho, C.-W. (**1997**) *J. Org. Chem.*, **62**, 7047–48.

257 Lin, J., Chan, W.H., Lee, A.W.M. and Wong, W.Y. (**1999**) *Tetrahedron*, **55**, 13983–98.

258 (a) Oppolzer, W., Blagg, J., Rodriguez, I. and Walther, E. (**1990**) *J. Am. Chem. Soc.*, **112**, 2767–72. (b) Kumaraswamy, G., Padmaja, M., Markondaiah, B., Jena, N., Sridhar, B. and Kiran, M.U. (**2006**) *J. Org. Chem.*, **71**, 337–40. (c) Kumaraswamy, G. and Markondaiah, B. (**2008**) *Tetrahedron Lett.*, **49**, 327–30.

259 Curran, D.P., Shen, W., Zhang, J. and Heffner, T.A. (**1990**) *J. Am. Chem. Soc.*, **112**, 6738–40.

260 Chan, W.H., Lee, A.W.M., Jiang, L.S. and Mak, T.C.W. (**1997**) *Tetrahedron: Asymmetry*, **8**, 2501–4.

261 (a) Oppolzer, W., Kingma, A.J. and Poli, G. (1989) *Tetrahedron*, **45**, 479–88. (b) Oppolzer, W., Kingma, A.J. and Pillai, S.K. (1991) *Tetrahedron Lett.*, **32**, 4893–96. (c) Kim, K.S., Kim, B.H., Park, W.M., Cho, S.J. and Mhin, B.J. (**1993**) *J. Am. Chem. Soc.*, **115**, 7472–77.

262 Vallgârda, J., Appelberg, U., Csöregh, I. and Hacksell, U. (**1994**) *J. Chem. Soc., Perkin Trans. 1*, 461–70.

263 Zhang, S.-J., Chen, Y.-K., Li, H.-M., Huang, W.-Y., Rogatchov, V. and Metz, P. (**2006**) *Chin. J. Chem.*, **24**, 681–88.

264 (a) Ahn, K.H., Kim, S.-K. and Ham, C. (**1998**) *Tetrahedron Lett.*, **39**, 6321–22. (b) Ku, H.-Y., Jung, J., Kim, S.-H., Kim, H.Y., Ahn, K.H. and Kim, S.-G. (**2006**) *Tetrahedron: Asymmetry*, **17**, 1111–15.

265 Adam, W., Degen, H.-G., Krebs, O. and Saha-Möller, C.R. (2002) *J. Am. Chem. Soc.*, **124**, 12938–39.

266 Brzezinski, L.J., Rafel, S. and Leahy, J.W. (1997) *J. Am. Chem. Soc.*, **119**, 4317–18.

267 (a) Takeuchi, Y., Suzuki, T., Satoh, A., Shiragami, T. and Shibata, N. (1999) *J. Org. Chem.*, **64**, 5708–11. (b) Liu, Z., Shibata, N. and Takeushi, Y. (2002) *J. Chem. Soc., Perkin Trans. 1*, 302–3. (c) Kakuda, H., Suzuki, T., Takeuchi, Y. and Shiro, M. (1997) *Chem. Commun.*, 85–86.

268 Lee, J., Zhong, Y.-L., Reamer, R.A. and Askin, D. (2003) *Org. Lett.*, **5**, 4175–77.

269 Clerici, F., Gelmi, M.L., Pellegrino, S. and Pocar, D. (2007) *Top. Heterocycl. Chem.*, **9**, 179–264.

270 Cherney, R.J., King, B.W., Gilmore, J.L., Liu, R.-Q., Covington, M.B., Duan, J. J.-W. and Decicco, C.P. (2006) *Bioorg. Med. Chem. Lett.*, **16**, 1028–31.

271 Spaltenstein, A., Almond, M.R., Bock, W.J., Cleary, D.G., Furfine, E.S., Hazen, R.J., Kazmierski, W.M., Salituro, F.G., Tung, R.D. and Wright, L.L. (2000) *Bioorg. Med. Chem. Lett.*, **10**, 1159–62.

272 (a) Chiacchio, U., Corsaro, A., Rescifina, A., Bkaithan, M., Grassi, G., Piperno, A., Privitera, T. and Romeo, G. (2001) *Tetrahedron*, **57**, 3425–33. (b) Chiacchio, U., Corsaro, A., Gambera, G., Rescifina, A., Piperno, A., Romeo, R. and Romeo, G. (2002) *Tetrahedron: Asymmetry*, **13**, 1915–21.

273 Zhang, H.-K., Chan, W.-H., Lee, A.W.M., Wong, W.-Y. and Xia, P.-F. (2005) *Tetrahedron: Asymmetry*, **16**, 761–71.

274 (a) Rogatchov, V.O., Bernsmann, H., Schwab, P., Fröhlich, R., Wibbeling, B. and Metz, P. (2002) *Tetrahedron Lett.*, **43**, 4753–56. (b) Rogatchov, V.O. and Metz, P. (2007) *Arkivoc*, 167–90.

275 (a) McReynolds, M.D., Dougherty, J.M. and Hanson, P.R. (2004) *Chem. Rev.*, **104**, 2239–58. (b) Wanner, J., Harned, A.M., Probst, D.A., Poon, K.W.C., Klein, T.A., Snelgrove, K.A. and Hanson, P.R. (2002) *Tetrahedron Lett.*, **43**, 917–21.

276 (a) Postel, D., Nhien, A.N.V. and Marco, J.L. (2003) *Eur. J. Org. Chem.*, 3713–26. (b) Nhien, A.N.V., Tomassi, C., Len, C., Marco-Contelles, J.L., Balzarini, J., Pannecouque, C., De Clercq, E. and Postel, D. (2005) *J. Med. Chem.*, **48**, 4276–84.

277 Combs, A.P., Glass, B., Galya, L.G. and Li, M. (2007) *Org. Lett.*, **9**, 1279–82.

278 Enders, D., Moll, A. and Bats, J.W. (2006) *Eur. J. Org. Chem.*, 1271–84.

7
Asymmetric Synthesis of Six-Membered Ring Nitrogen Heterocycles with More Than One Heteroatom

Péter Mátyus and Pál Tápolcsanyi

7.1
Six-Membered Rings with Another Heteroatom in the Same Ring

This chapter covers the synthesis of chiral partially or fully saturated substituted pyridazines, pyrimidines, piperazines, oxadiazines, and morpholines with defined configuration of the substituents. Methods considering the formation of the heterocyclic ring are discussed separately from the stereoselective transformations by the involvement of the already existing ring.

A literature search was performed using the SciFinder program by using both keyword search (combination of the name of the corresponding ring with the keyword "asymmetric synthesis") and structure search. In the latter case, for the corresponding ring given as "product", unlimited substitution was allowed, but ring tools were locked out. The hits were refined (i) by omitting patents, (ii) by selecting the stereoselective reactions, and (iii) by the keyword "asymmetric synthesis". The literature is surveyed covering the period from 1957 till 2007.

7.1.1
Pyridazines

Partially saturated pyridazine derivatives with defined chirality have so far been obtained via two different routes.

In the first, most frequently employed pathway, the key step is the ring closure of enantiomerically pure precursors to pyridazines. Such precursors have generally been (i) a lactone or lactam, (ii) a hydroxy ester or acid, (iii) an amino acid, (iv) a protected hydroxy aldehyde, (v) keto ester, or (vi) a chiral heterocycle.

The second route, which may be considered an important variant of the first, is based on the Diels–Alder reaction, which involves the cycloaddition of a chiral diene and azodicarboxylic ester or triazolo derivative as dienophile.

Asymmetric Synthesis of Nitrogen Heterocycles. Edited by Jacques Royer
Copyright © 2009 WILEY-VCH Verlag GmbH & Co. KGaA, Weinheim
ISBN: 978-3-527-32036-3

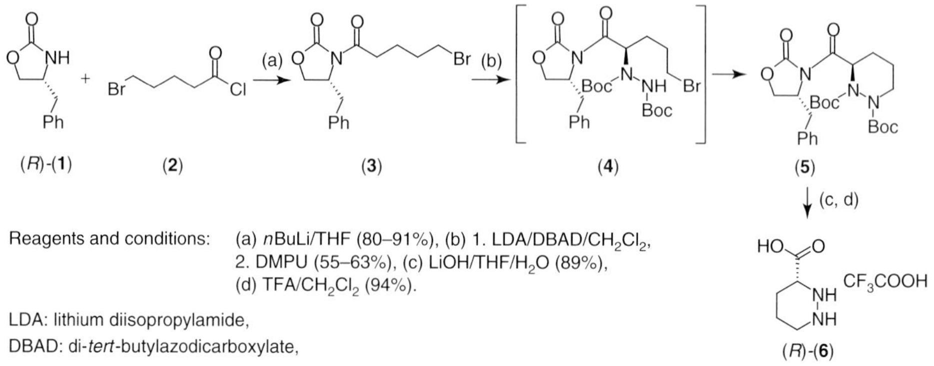

Reagents and conditions: (a) nBuLi/THF (80–91%), (b) 1. LDA/DBAD/CH$_2$Cl$_2$, 2. DMPU (55–63%), (c) LiOH/THF/H$_2$O (89%), (d) TFA/CH$_2$Cl$_2$ (94%).

LDA: lithium diisopropylamide,
DBAD: di-tert-butylazodicarboxylate,
DMPU: 1,3-dimethyl-3,4,5,6-tetrahydro-2(1H)-pyrimidinone

Scheme 7.1

7.1.1.1 Ring Closure of Optically Active Precursors

Piperazic acid (hexahydropyridazine-3-carboxylic acid) is a typical building block for a number of pharmacologically active compounds. Moreover, one of its enantiomers (3S) has been found to exert γ-aminobutyric acid (GABA) uptake inhibitory activity in rat cerebral cortex slices. Subsequently, enantioselective syntheses of such compounds have been thoroughly studied. Hale and coworkers described the synthesis of both enantiomers. The synthesis of (3R)-piperazic acid ((R)-6) starts with the N-acylation of (4R)-phenylmethyl-2-oxazolidinone (1) with 5-bromovaleryl chloride (2), followed by diastereoselective α-hydrazination of 3 with di-tert-butylazodicarboxylate and LiNiPr$_2$. When the reaction was carried out in the presence of DMPU (1,3-dimethyl-3,4,5,6-tetrahydro-2(1H)-pyrimidinone), the cyclized product 5 could be obtained directly without isolation of the intermediate bromovaleryl hydrazide 4. Interestingly, without the addition of DMPU, 4 was isolated as the major product. These findings were explained by the formation of a highly aggregated lithium-aza anion, which is transformed to a more reactive species in the presence of DMPU. After removal of the protecting groups from 5, (3R)-piperazic acid ((R)-6) was obtained as trifluoroacetate salt with 96% ee [1, 2]. The method has also been used for the preparation of (S)-piperazic acid [3, 4].

An analogous, convenient procedure was developed for the preparation of the S-enantiomer of 6 by Hale's group, starting from an oxazolidinone precursor derived from norephedrine. The product in this case was obtained with a lower enantiomeric excess (88–93%).

Three other pathways for the synthesis of enantiomerically pure pyridazinecarboxylic acid derivatives have been described by Schmidt's group, which are based on cyclocondensation with hydrazone formation or cyclization with alkylation of α- or δ-hydrazinocarboxylic acids.

In method I, as the key step, asymmetric hydrogenation of dehydro amino acid derivative 7 in the presence of a chiral rhodium catalyst ((R,R)-(Rh(1,5-COD)(DIPAMP)$^+$BF$_4^-$) afforded amino ester 8 with high enantioselectivity. By treatment with NaNO$_2$ in acetic acid, 8 was transformed into hydrazine derivative 9, in which

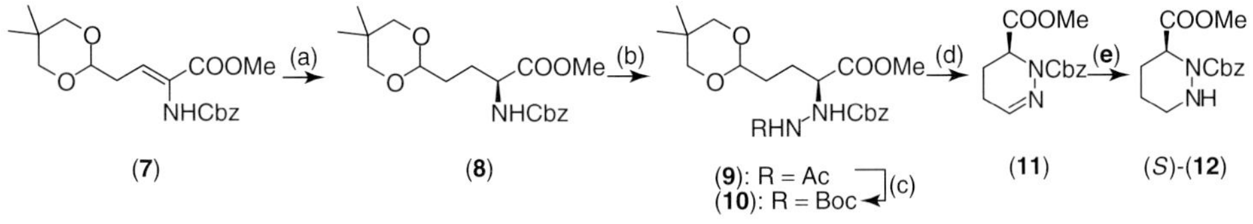

7.1 Six-Membered Rings with Another Heteroatom in the Same Ring | 295

Reagents and conditions: (a) (R,R)-(R)h(1,5-COD)(DIPAMP)⁺BF₄⁻/H₂/MeOH (100%, 97% ee), (b) 1. NaNO₂/Ac₂O/AcOH, 2. Zn/Ac₂O/AcOH (50%), (c) 1. Boc₂O/DMAP/MeCN, 2. K₂CO₃/MeOH (84%), (d) 6 N HCl/dioxane (87%), (e) NaCNBH₃/AcOH (81%).

COD: cyclooctadiene
DIPAMP: (R,R)-1,2-bis[2-(methoxyphenyl)phenylphosphanyl]ethane
Cbz: benzyloxycarbonyl

Scheme 7.2

the protecting group was replaced to obtain **10**. Its ring closure to **11** was achieved with HCl/dioxane; **11** could then be smoothly reduced to (S)-**12** [5].

Methods II and III used enantiomerically pure (2S)-pyroglutamic acid isopropyl ester or (2R)-5-oxotetrahydrofuran-2-carboxylic acid *tert*-butyl ester, respectively, which were transformed to properly protected α-hydrazino-δ-hydroxy or δ-hydrazino-α-hydroxy ester, and subsequently cyclized via intramolecular nucleophilic substitution of the corresponding mesylate or triflate to pyridazinecarboxylic ester [5].

Tetrahydropyridazine derivative **16**, which can be regarded as a useful intermediate for synthesis of the luzopeptins (natural products possessing inhibitory activity against reverse transcriptase), was prepared from β-hydroxy ester **13**. The Gennari–Evans–Vederas reaction of **13** with dibenzyl azodicarboxylate afforded an 18:1 mixture of the *anti* (**14**) and *syn* diastereomers of the corresponding α-hydrazino ester. Sequential O-acetylation and catalytic debenzylation in the presence of di-*tert*-butyl dicarbonate, followed by acylation of the mono-Boc-protected derivative thus formed with (4R)-3-acetyl-2-oxo-1,3-oxazolidine-4-carbonyl chloride, resulted in **15**, which could be smoothly cyclized with trifluoroacetic acid to **16** [6].

Enzyme-assisted synthetic methods have been described for the preparation of tetrahydropyridazines. Lipase Triacylglycerol lipase (TL)-mediated kinetic

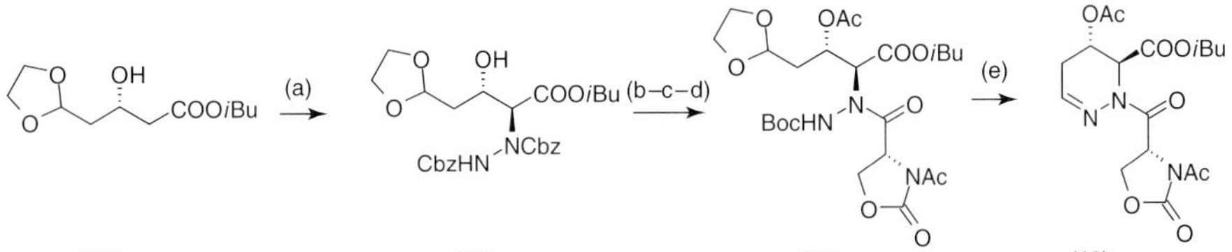

Reagents and conditions: (a) CbzN=NCbz/LDA/THF (61%, *anti*:*syn* = 18:1), (b) Ac₂O/pyridine (95%), (c) Boc₂O/H₂/Pd–C/MeOH (97%), (d) sym-collidine/(4R)-3-acetyl-2-oxo-1,3-oxazolidine-4-carbonyl chloride/CH₂Cl₂ (60%), (e) TFA/H₂O (97%).

Scheme 7.3

resolution of protected alcohol **17** gave the (*R*)-acetate (*R*)-**18** and the corresponding (*S*)-alcohol with high enantiomeric purity. The reaction conditions were thoroughly studied: application of *n*-hexane as solvent led to the best chemical yields and enantiomeric ratios, whereas basic additives such as pyridine, 4-*N*,*N*-dimethylaminopyridine, 2,4- and 2,6-lutidine were found to increase reaction rates considerably. Compound (*R*)-**18** was debenzylated and subsequently oxidized to aldehyde (*R*)-**19** with Dess–Martin reagent. Treatment of (*R*)-**19** with *tert*-butyl carbazate and NaBH$_3$CN, and then protection with benzyloxycarbonyl group, led to compound (*R*)-**20**. For its cyclization to (*S*)-**21**, an intramolecular Mitsunobu reaction was carried out. (*S*)-**33** was transformed to piperazic acid derivative (*S*)-**22** with 98% ee [7].

Four possible stereoisomers of 5-hydroxypiperazic acid were synthesized starting from chiral compounds **23** and **27**, which were transformed to the corresponding O-protected 3-butyraldehyde derivatives **24** and **28**, respectively, possessing a good leaving group at position 4. Ring closure with hydrazine hydrate and subsequent N-protection with benzoyl chloride gave tetrahydropyridazine derivatives (*S*)-**25** and (*R*)-**25**. Their diastereoselective Strecker reaction was investigated with trimethylsilyl cyanide in the presence of a Lewis acid. Interestingly, when Zn(OTf)$_2$ was used as Lewis acid, together with NaOAc/AcOH as additives, the *syn* cyano products (3*S*,5*S*)-**26** or (3*R*,5*R*)-**26** were formed in higher amounts. However, when Mg(OAc)$_2$ was used in combination with acetic acid, the *anti* isomers (3*R*,5*S*)-**26** or (3*S*,5*R*)-**26**, respectively, were obtained as the major product [8].

Another series of dihydropyridazines, the 6-aryl-5-methyl-4,5-dihydropyridazin-3(2*H*)-ones (*R*)-**33** could be obtained in four steps from the respective arene. Friedel–Crafts acylation of arenes **29** with (2*R*)-2-chloropropanoyl chloride to **30**, followed by their S$_N$2 reactions with benzyl methyl malonate in the presence of NaH, afforded **31**. The reason for using benzyl methyl malonate instead of dimethyl malonate was to suppress possible epimerization in the subsequent step, in which removal of one of the ester groups was needed. Debenzylation of **31** and

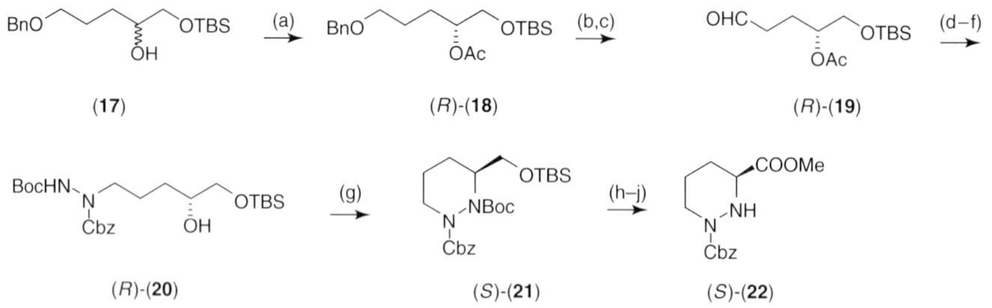

Reagents and conditions: (a) Lipase TL/CH$_2$CHOAc/solvent/additive (11–62%, 82–98% ee), b) H$_2$/Pd–C/MeOH (quant), (c) Dess–Martin periodinane/CHCl$_3$ (99%), (d) 1. BocNHNH$_2$/EtOH/AcOH, 2. NaCNBH$_4$ (67%), (e) CbzCl/NaHCO$_3$/CHCl$_3$ (97%), (f) K$_2$CO$_3$/MeOH (73%), (g) DEAD/PPh$_3$/THF (86%), (h) 1. TBAF/THF, 2. Jones reagent/acetone, (i) TMSCHN$_2$/MeOH (81%), (j) TFA/CH$_2$Cl$_2$ (84%).

Scheme 7.4

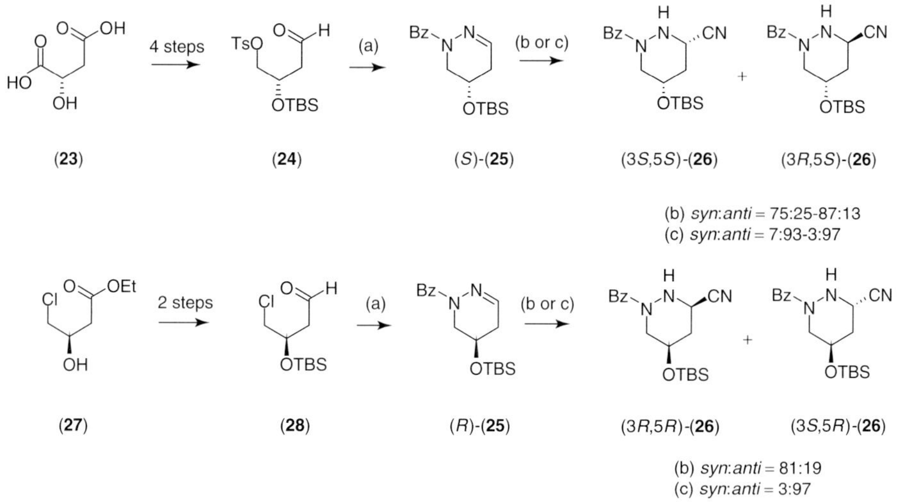

7.1 Six-Membered Rings with Another Heteroatom in the Same Ring | 297

Scheme 7.5

Reagents and conditions: (a) 1. NH$_2$NH$_2$·H$_2$O/EtOH, 2. BzCl/pyridine (89–93%), (b) Me$_3$SiCN/Zn(OTf)$_2$/AcOH/NaOAc/CH$_2$Cl$_2$ (42–70%), (c) Me$_3$SiCN/Mg(OAc)$_2$/CH$_2$Cl$_2$ (75–99%).

Scheme 7.6

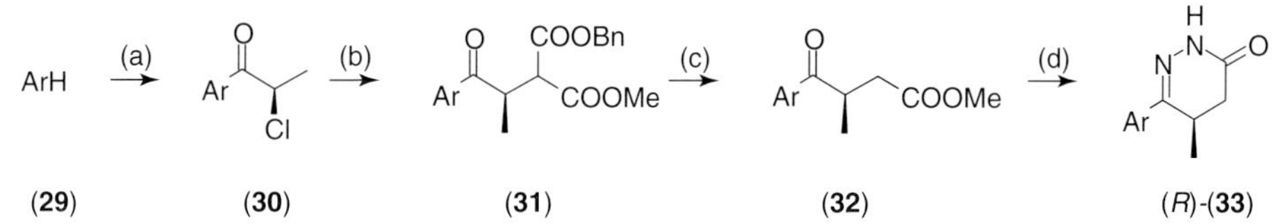

ArH: acetanilide, 2-isopropylpyrazolo[1,5-a]pyridine

Reagents and conditions: (a) (2R)-2-chloropropanoyl chloride/AlCl$_3$/1,2,4-trichlorobenzene (59–80%),
(b) NaH/CH$_2$(COOMe)(COOBn)/DMF (81–84%),
(c) 1. Pd–C/H$_2$/EtOAc, 2. diglyme/rfx (55–84%),
(e) N$_2$H$_4$·H$_2$O/AcOH/MeOH/H$_2$O (69–91%, 84–90.2% ee).

subsequent decarboxylation gave monoesters **32**, which cyclized to pyridazinones (R)-**33** in 84–90.2% ee on treatment with hydrazine hydrate [9, 10].

Diastereoselective synthesis of tetrahydropyridazinone **40** as a precursor of β-strand mimetics has been described starting from (S)-phenylalanine (**34**), which was first transformed to phenyloxazolidinone **35**. Alkylation of **35** with allyl bromide in the presence of lithium hexamethyl-disilazide (LiHMDS) selectively furnished allyl derivative **36** as the single isomer. Subsequent hydrolytic ring opening, followed by treatment with diazomethane, gave the α, α-dialkylated ester **37**. Ozonolysis of the double bond and the subsequent ring closure of aldehyde **38** led to dihydropyridazinone **39**, which could be hydrogenated to tetrahydroderivative **40** upon treatment with NaBH$_3$CN [11].

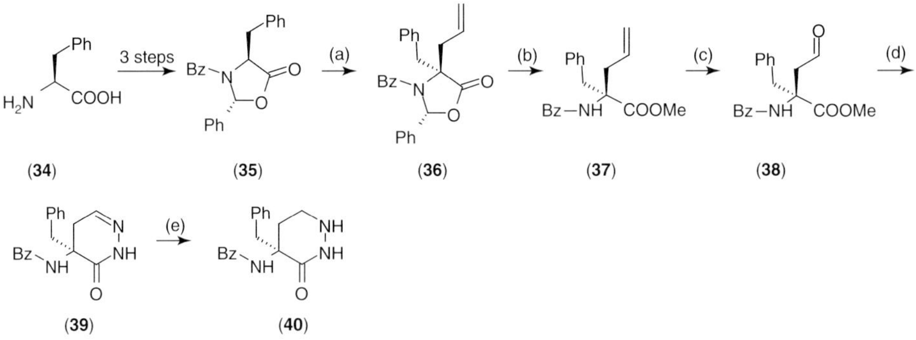

Scheme 7.7

Reagents and conditions: (a) LiHMDS/allyl bromide/THF (93%, 95% ee), (b) 1. NaOH/MeOH, 2. CH$_2$N$_2$ (99%), (c) O$_3$/CH$_2$Cl$_2$/MeOH (96%), (d) N$_2$H$_4$ (85%), (e) NaCNBH$_3$/MeOH (73%).

Asymmetric γ-alkylation of α,β-unsaturated glutamic acid derivatives **42** possessing a chiral auxiliary was the key step in the synthesis of 4-substituted 4,5-dihydropyridazinones **44**. Compounds **42** were obtained by selective Me$_3$Al-mediated acylation of methyl ester **41** with (2R)-bornane-10,2-sultam or 8-phenylmenthol. Alkylation of **42** was carried out under phase transfer-catalyzed conditions with various electrophiles, affording **43** with high diastereomeric excess (70–100% de) in favor of the 2R isomer. Reaction of the alkylated derivatives **43** with hydrazines afforded dihydropyridazinones **44** via N-deprotection and cyclization in one pot, with recovery of the chiral auxiliary [12].

Chiral pyridazine derivatives were synthesized by Young's group. They applied a "ring switching" method, starting from another optically active heterocyclic compound. Reaction of the β-lactam **45**, bearing a formylmethyl group, with hydrazine hydrate in methanol at room temperature gave a mixture of hydroxypyrrolidinones **46** (62%) and tetrahydropyridazine **47** (17%). Interestingly, reaction of the aldehyde

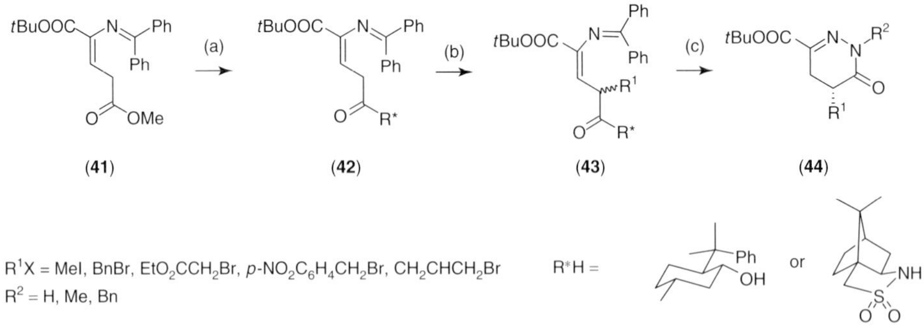

R^{1}X = MeI, BnBr, EtO$_2$CCH$_2$Br, p-NO$_2$C$_6$H$_4$CH$_2$Br, CH$_2$CHCH$_2$Br
R^2 = H, Me, Bn

Reagents and conditions: (a) Me$_3$Al/R*H (80%), (b) R^{1}X/NaOH/nBu$_4$NCl/MeCN (30–90%, 70–100% de), (c) R^{2}NHNH$_2$/EtOH (90–95%).

Scheme 7.8

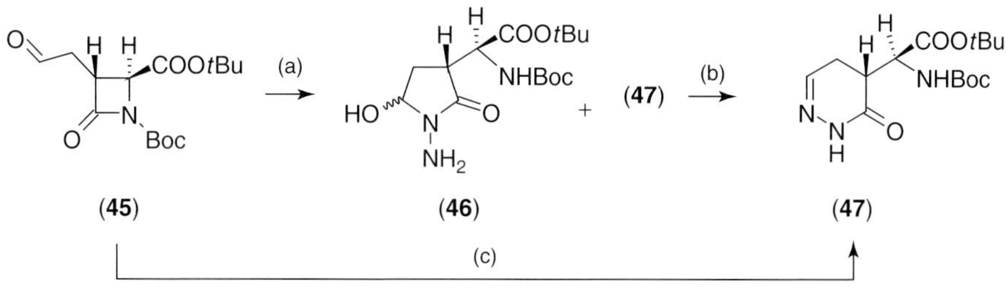

7.1 Six-Membered Rings with Another Heteroatom in the Same Ring

(45) → (a) → **(46)** + **(47)** → (b) → **(47)**
                                    ↑
                                   (c)

Reagents and conditions: (a) NH$_2$NH$_2$/MeOH/rt (46: 62%, 47: 17%), (b) heating, (c) NH$_2$NH$_2$/benzene/rfx (65%).

Scheme 7.9

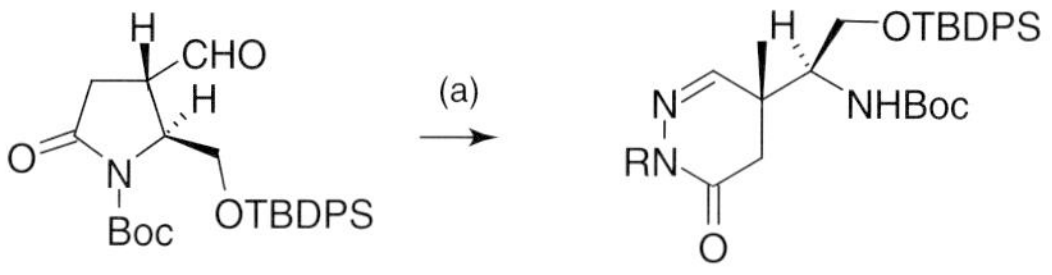

(48) → (a) → **(49)** R = H, Me

Reagents and conditions: (a) RNHNH$_2$/MeOH/rt (41–65%).

Scheme 7.10

with hydrazine hydrate in refluxing benzene afforded **47** in 65% yield as the sole product [13].

The ring switching method could also be successfully applied to functionalized pyrrolidinone derivative **48**, when dihydropyridazinones **49** were obtained in moderate to good yields [14].

An elegant one-pot method was recently described for the enantioselective synthesis of 3-substituted tetrahydropyridazines from achiral starting materials, using a chiral organocatalyst. In the first step, asymmetric catalytic α-amination of aldehydes **50** with azodicarboxylates **51** in the presence of (S)-proline or 5-[(2S)-pyrrolidin-2-yl]-5H-tetrazole as catalyst gave intermediates **52**. Base-promoted addition of the secondary nitrogen in **52** to vinylphosphonium bromide led to the ylide intermediate **53**, which could be cyclized to the product **54** in 69–99% ee [15].

7.1.1.2 Diels–Alder Reactions

In stereoselective hetero-Diels–Alder reactions, the stereochemical outcome of the reaction could be determined by chiral auxiliaries present either on the diene or on the dienophile, or as a further possibility, asymmetric catalysis could be employed.

The 2,3,4,6-tetra-O-acetyl-β-D-glucopyranosyl unit displayed an Re-face reactivity in the cycloaddition reactions of dienes **55** with azodicarboxylates **51**, furnishing cycloadducts **56** with high stereoselectivity. On removal of the protective groups, **56** can be transformed to tetrahydropyridazine ester (S)-**57**. The benzyl ester of **56**

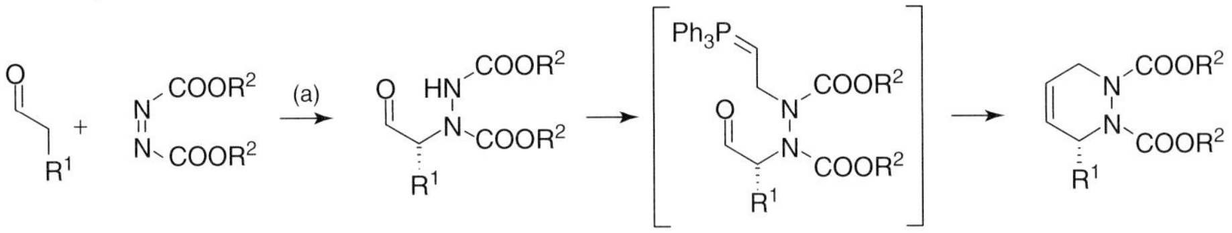

(50) **(51)** **(52)** **(53)** **(54)**

R^1 = iPr, tBu, allyl, Bn
R^2 = Et, tBu, Bn

Reagents and conditions: (a) 1. Proline or 5-[(2S)-pyrrolidin-2-yl]-5H-tetrazole/CH$_2$Cl$_2$,
2. CH$_2$CHPPh$_3$Br/NaH/DMSO or THF (63–89%, 69–99% ee).

Scheme 7.11

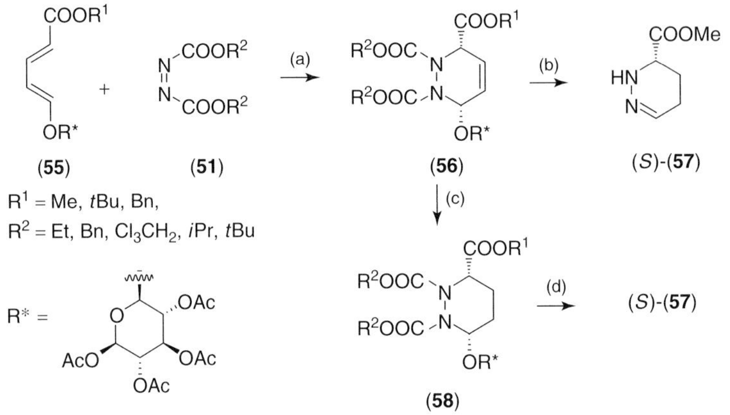

(55) **(51)** **(56)** **(S)-(57)**

R^1 = Me, tBu, Bn,
R^2 = Et, Bn, Cl$_3$CH$_2$, iPr, tBu

R* =

(58)

Reagents and conditions: (a) EtOAc/70 °C or toluene/100 °C (57–87%),
(b) H$_2$/Pd–C/EtOAc (R^1 = Me, R^2 = Bn) (37%, 98% ee),
(c) H$_2$/Pd–C/EtOAc (R^2 not Bn) (82–93%), (d) TFA (R^1 = Me, R^2 = tBu) (57%).

Scheme 7.12

furnished (S)-**57** directly, while hydrogenolysis of tert-butyl esters afforded saturated derivatives **58**, which were then treated with trifluoro-acetic acid (TFA) to yield (S)-**57** [16].

Oxazolidinethione moiety was applied as chiral auxiliary in an analogous Diels–Alder reactions of (4R)-3-[(1E)-buta-1,3-dien-1-yl]-4-phenyl-1,3-oxazolidine-2-thione with diethyl azodicarboxylate to furnish the product in moderate stereoselectivity (50% de) [17].

(3S,4R,5R)-4,5-Dihydroxy-3-methyl-2,3,4,5-tetrahydropyridazine **66** has been prepared from triazolocarbinol **63** derived from [(S)R]-(1E,3E)-1-p-tolylsulfinyl-1,3-pentadiene (**59**) and 4-methyl-1,2,4-triazoline-3,5-dione (**60**) in an asymmetric tandem hetero-Diels–Alder cycloaddition [2, 3] – sigmatropic rearrangement – sulfenate trapping reaction. After protection of the hydroxy group of **63**, the

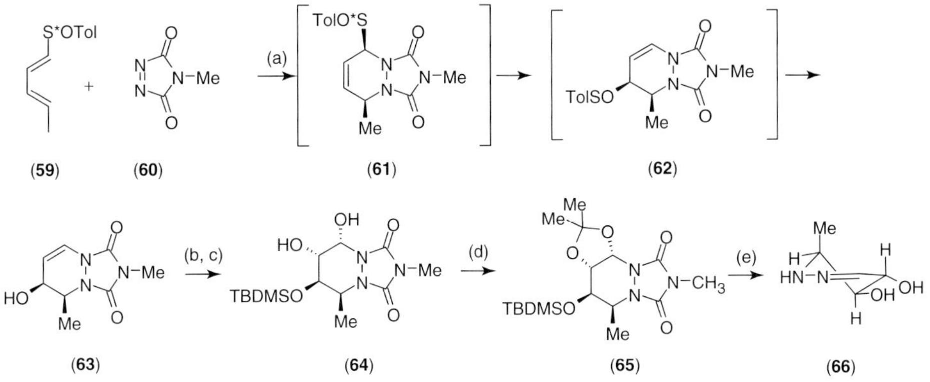

7.1 Six-Membered Rings with Another Heteroatom in the Same Ring

Reagents and conditions: (a) P(OMe)$_3$ (b) TBDMSOTf/Et$_3$N (57%), (c) OsO$_4$–NMO/acetone–H$_2$O (54%, 80% de), or (trifluoromethyl)methyldioxirane (64%, 98% de), (d) DMP/TsOH/acetone (75%), (e) NH$_2$NH$_2$ (quant).

TBDMS: tert-butyldimethylsilyl

Scheme 7.13

tert-butyldimethylsilyl derivative was dihydroxylated in the presence of OsO$_4$, yielding **64** with 80% de and 54% yield. When (trifluoromethyl)methyldioxirane was the oxidizing agent, 98% de and 64% yield were attained. Transformation of cis diol **64** to acetonide **65** and subsequent hydrazinolysis resulted in pyridazine derivative **66** [18].

The cycloadduct from the analogous diastereoselective aza-hetero-Diels–Alder reaction of 4-phenyl-1,2,4-triazoline-3,5-dione with (4S)-4-benzyl-3-[(2E)-penta-2,4-dienoyl]-1,3-oxazolidin-2-one in the presence of titanium tetrachloride (dr: 97:3) could be transformed to the trifluoroacetate of S-piperazic acid (S)-**6** via a series of steps [19].

According to the other approach, a chiral auxiliary was built in the dienophile. Thus, the asymmetric synthesis of piperidazine-3-phosphonic acid enantiomers was performed via a one-pot process of hetero-Diels–Alder reaction and subsequent phosphonylation. The reaction of diene **67** with dimenthyl azodicarboxylate in the presence of trimethyl phosphite and trimethylsilyl triflate as Lewis acid gave an isomeric mixture of 3-(dimethoxyphosphoryl)-3,6-dihydropyridazine-1,2-dicarboxylate **68**. After catalytic hydrogenation, the saturated derivatives could be separated by chromatography, affording (S)-**69** and (R)-**69** in the ratio of 66:34. Subsequent hydrolysis gave (S)-**70** and (R)-**70**, respectively [20].

The reactions of variously functionalized dienes **71** with azopyridine **72** have been investigated in the presence of chiral catalysts. A combination of (R)-2,2′-Bis(diphenylphosphino)-1,1′-binaphthalene (BINAP) and AgOTf as catalyst and propionitrile as solvent provided the most effective reaction conditions, affording products **73** in good yields and high enantiomeric purity for most substrates [21].

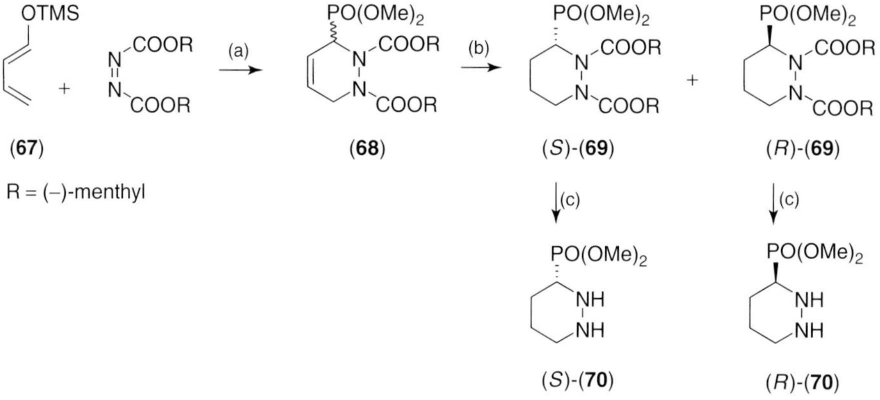

Scheme 7.14

Reagents and conditions: (a) P(OMe)$_3$/TMSOTf/CH$_2$Cl$_2$, (b) H$_2$/Pd–C/MeOH (99%, 66:34 dr), (c) 1. 6 N HCl/AcOH, 2. propylene oxide (64–67%).

(67) R = (−)-menthyl

Scheme 7.15

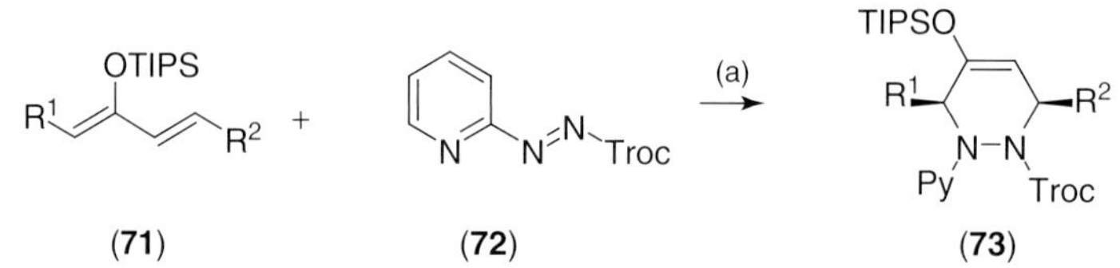

R^1 = Me, Bn, 4-MOMO-Bn, (CH$_2$)$_3$OTBS
R^2 = Me, Ph, iBu, iPr, BOM, 2-furyl

Reagents and conditions: (a) 5 mol% (R)-BINAP/10 mol% AgOTf/EtCN (65–87%, 55–99% ee).

7.1.2
Pyrimidines

For the synthesis of substituted pyrimidines with defined stereochemistry, condensation of an enantiomerically pure β-amino acid amide derivative with an aldehyde followed by the intramolecular addition of the amino nitrogen to the imine bond formed, in one pot or via separate steps, is a frequently applied method. The other most frequently used approach involves an intramolecular nucleophilic substitution as the ring closure step. Besides, stereoselective transformation such as alkylation, halogenation, or hydroxylation of the pyrimidine ring could also be employed.

7.1.2.1 Formation of the Pyrimidine Ring
Ring Formation by Addition to Imine Bond [5 + 1] Types Synthesis of (2R, 4S)-2-tert-butyl-6-oxohexahydropyrimidine-4-carboxylic acid and its protected derivatives (2R,4S)-3 has been reported in several papers. Thus, treatment of

7.1 Six-Membered Rings with Another Heteroatom in the Same Ring

(S)-asparagine (**1**) with pivaldehyde in the presence of KOH resulted in pyrimidinone **2** with *cis* orientation of the substituents as the single isomer. Protection of the amine nitrogen in **2** with benzyl chloroformate gave compound **3** [22–28].

In an analogous way, (6S)-1-benzoyl-3,6-dimethyltetrahydropyrimidin-4(1H)-one ((S)-**6**) was prepared from methyl (S)-3-aminobutanoate (**4**) by conversion to amide with methylamine and subsequent Schiff's base formation to **5** followed by ring closure and subsequent benzylation [29].

The same method was applied for the synthesis of the phenyl derivative (S)-**10** by amide formation of methyl (S)-3-amino-3-phenylpropionate (**9**) with methylamine, followed by cyclization with formaldehyde. Compound (S)-**9** was prepared via diastereoselective addition of lithium (R)-benzyl(α-methylbenzyl)amide to *tert*-butyl cinnamate (**7**), followed by hydrogenolysis of **8** and subsequent esterification [30].

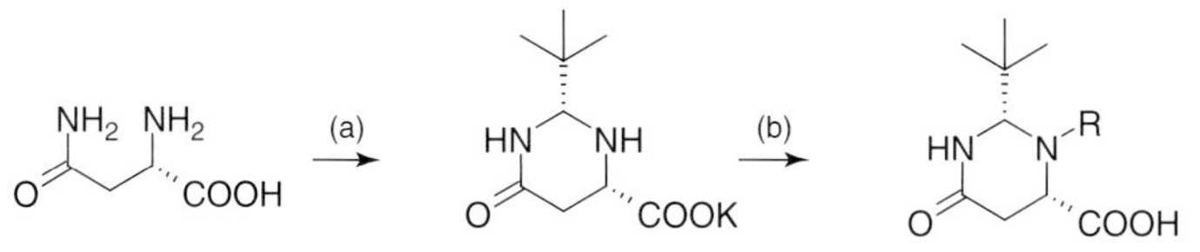

Reagents and conditions: (a) KOH/tBuCHO, (b) RCl/NaHCO$_3$/H$_2$O (67–80% from **1**).

Scheme 7.16

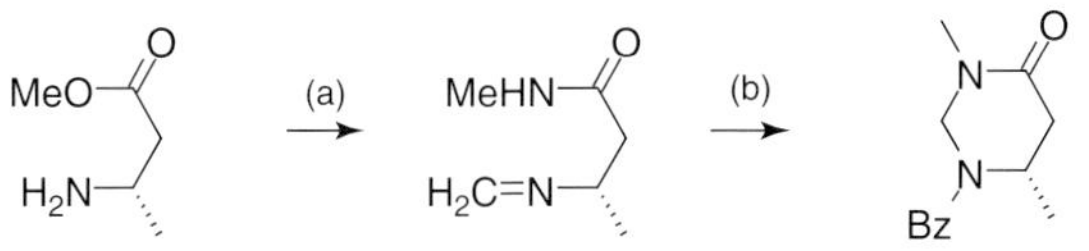

Reagents and conditions: (a) 1. MeNH$_2$/MeOH, 2. (CH$_2$O)$_n$ (79%),
(b) BzCl/DMAP/benzene (92%).

Scheme 7.17

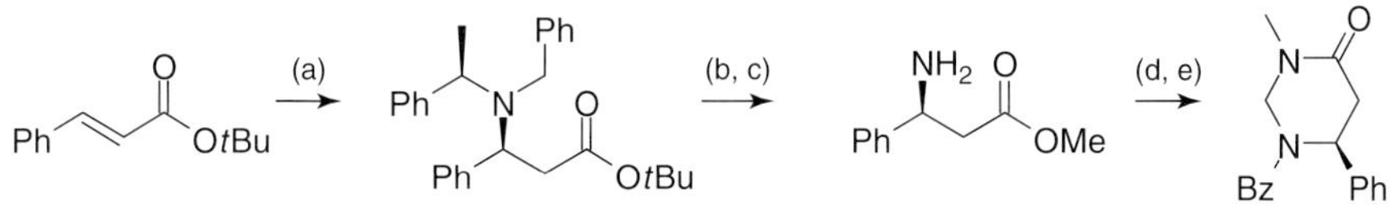

Reagents and conditions: (a) Li (R)-benzyl(α-methylbenzyl)amide, (b) H$_2$/Pd-C/AcOH,
(c) Me$_3$SiCl/MeOH (69% from **7**), (d) 1. MeNH$_2$/MeOH,
2. (CH$_2$O)$_n$, (e) BzCl/DMAP/benzene (82% from **9**).

Scheme 7.18

Optically active oxazoline **11** derived from D-valinol could be stereoselectively transformed to β-aminoalkanamide **15** and its substituted derivatives by addition of the lithium salt of **11** to N-cumyl nitrone **12**. The addition afforded a mixture of the equilibrating spirocycle **13** and hydroxylamine **14**. Hydrogenation of the mixture gave **15** in a highly enantioselective way (dr: 93:7), which was subsequently converted to tetrahydropyrimidinone **16** by treatment with paraformaldehyde [31].

Ring Closure by Intramolecular Nucleophilic Displacement 1,3,6-Trisubstituted 2,4-dioxohexahydropyrimidines (S)-**21** could be synthesized via intramolecular nucleophilic attack of the terminal amino group to methyl ester in anilinocarbonylamino diester intermediate (S)-**20**, which was prepared from partially protected aspartic acid (S)-**17**. Its activation with isobutyl chloroformate in the presence of N-methylmorpholine and the subsequent reaction of the mixed anhydride with diazomethane gave diazo compound (S)-**18**, which underwent the Arndt–Eistert rearrangement upon treatment with silver benzoate with methanol, leading to the homoaspartic acid derivative (S)-**19**. Hydrogenolysis, followed by addition to phenyl isocyanate, afforded the linear urea intermediate (S)-**20**, which, under basic conditions, cyclized regioselectively through the methyl ester and subsequently alkylated on the nitrogen, providing the 2,4-dioxohexahydropyrimidine derivative (S)-**21** [32].

(S)-1-*tert*-Butyl-4-methyl 2-aminobutanedioate ((S)-**22**) was reacted with diphenyl cyanocarbonimidate (**23**) to give (S)-**24**, which was then treated with benzylamine, affording a 1:1 mixture of pyrimidine (S)-**25** and imidazole **26**. The latter showed little optical activity, but (S)-**25** was obtained with 80% ee. The same reaction sequence starting from (R)-1-*tert*-butyl-4-methyl 2-aminobutanedioate gave the enantiomeric (R)-**25** with 90% ee [33].

Dihydropyrimidinone derivative (S)-**30**, a precursor of the antibiotics TAN-1057A, has been synthesized from N-protected diamino acid (S)-**27**, which was coupled with S-methylisothiobiuret (**28**). Removal of the Boc-protecting group from (S)-**29**, followed by ring closure and subsequent hydrogenolysis, gave (S)-**30** in 87% ee [34].

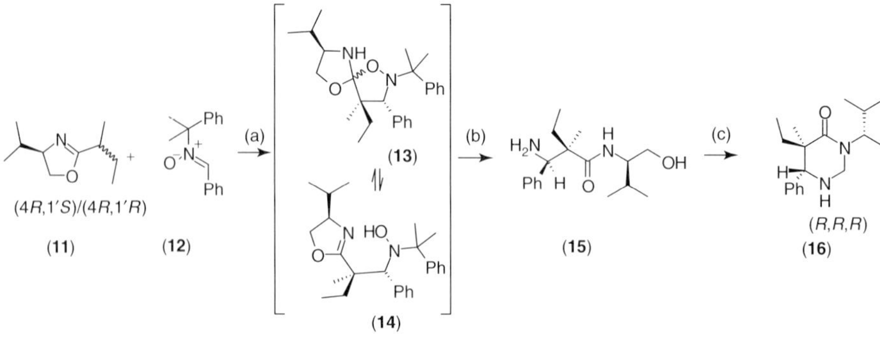

Reagents and conditions: (a) sBuLi/THF, (b) H_2/Pd–C/MeOH (98%, 93:7 dr), (c) $(CH_2O)_n$/MeOH (60%).

Scheme 7.19

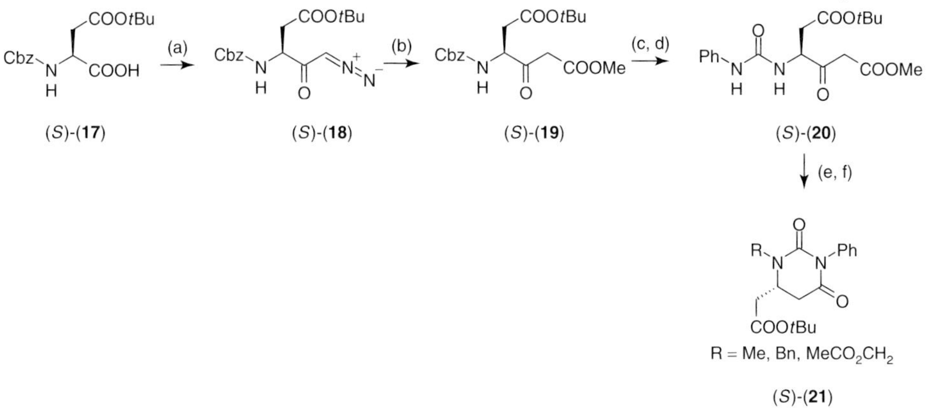

7.1 Six-Membered Rings with Another Heteroatom in the Same Ring | 305

Reagents and conditions: (a) iBuOCOCl/NMM/(THF)/CH$_2$N$_2$/Et$_2$O, (b) PhCOOAg/Et$_3$N/MeOH, (c) H$_2$/Pd–C/MeOH, (d) PhNCO/THF, (e) KOtBu/THF (70%), (f) RX/KOtBu/THF (80–93%).

NNM: N-methylmorpholine

Scheme 7.20

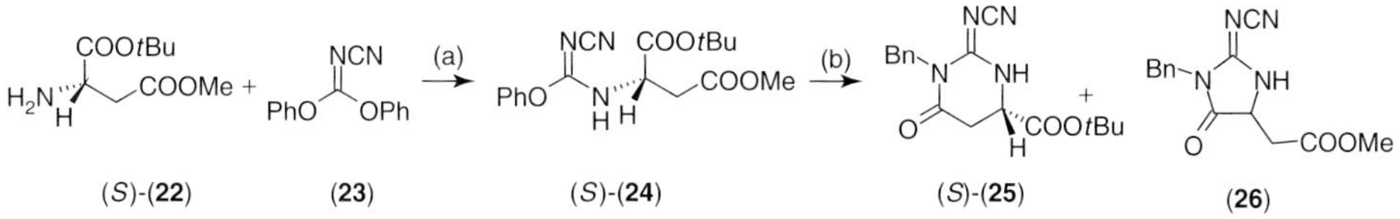

Reagents and conditions: (a) iPrOH (84%, 100% ee),
(b) BnNH$_2$/iPrOH (68% 1:1 mixture of (S)-**25** and **26**, 80% ee for (S)-**25**).

Scheme 7.21

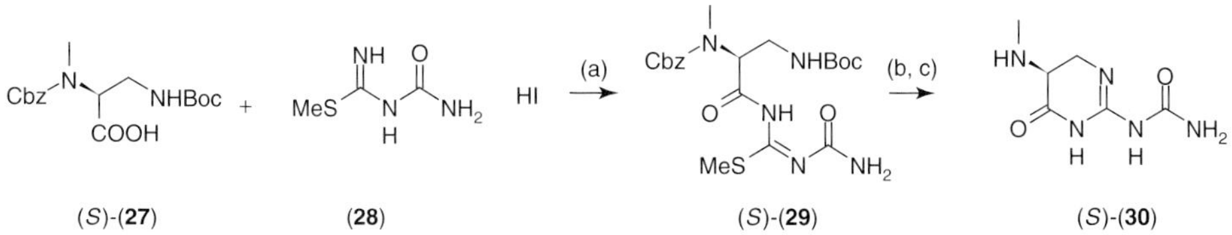

Reagents and conditions: (a) HOBt, EDC, DIEA (100%), (b) 1. TFA/anisole/CH$_2$Cl$_2$, 2. Et$_3$N/AcOH (55%, 92% ee),
(c) H$_2$/Pd–C/DMA (96%, 87% ee).

HOBt: 1-hydroxybenzotriazole
EDC: 1-ethyl-3-(3-dimethylaminopropyl) carbodiimide hydrochloride
DIEA: N,N-diisopropylethylamine

Scheme 7.22

For the asymmetric synthesis of (2S,3R)-capreomycidine (**35**) as a structural unit of pentapeptides possessing tuberculostatic properties, the pyrimidine ring was elaborated by intramolecular Mitsunobu reaction. The synthesis started with the

7 Asymmetric Synthesis of Ring Nitrogen Heterocycles

preparation of the aluminum enolate of chiral morpholinone **31** by treatment with LiHMDS and subsequent transmetalation with MeAl$_2$Cl. Addition of O-protected N-benzyl-3-hydroxypropylidene amine to the aluminum enolate gave a mixture of two diastereomers **32**, in which the imine is situated on the opposite face of the morpholine ring rather than the phenyls. Guanidinylation of **32** with N,N'-di-*tert*-butoxycarbonyl-S-methylisothiourea in the presence of HgCl$_2$ afforded a single diastereomer, which was deprotected to **33**. Formation of the pyrimidine ring was achieved by Mitsunobu reaction. Hydrogenolysis of the cyclized product **34** and subsequent acidic hydrolysis resulted in (2S,3R)-capreomycidine (**35**) in over 99% ee [35].

Further Ring Closure Methods 1,2-Dihydropyrimidines **38** were synthesized by the reaction of 3-aminoalkyl-2-enimines **36** with chiral aldehydes **37** in the presence of zinc chloride. Subsequent stereoselective reduction of **38** gave tetrahydroderivatives **39** in 99% de [36, 37].

Three-component Biginelli reactions between glycosylated aldehydes **40**, keto esters **41**, and urea (**42**) have been carried out in the presence of a mixture of CuCl, acetic acid and BF$_3$·OEt$_2$ as additives for the synthesis of mono- and bis-C-glycosylated chiral dihydropyrimidinones **43** (due to asymmetric induction of the sugar moiety) with moderate to good (35–80%) stereoselectivities [38].

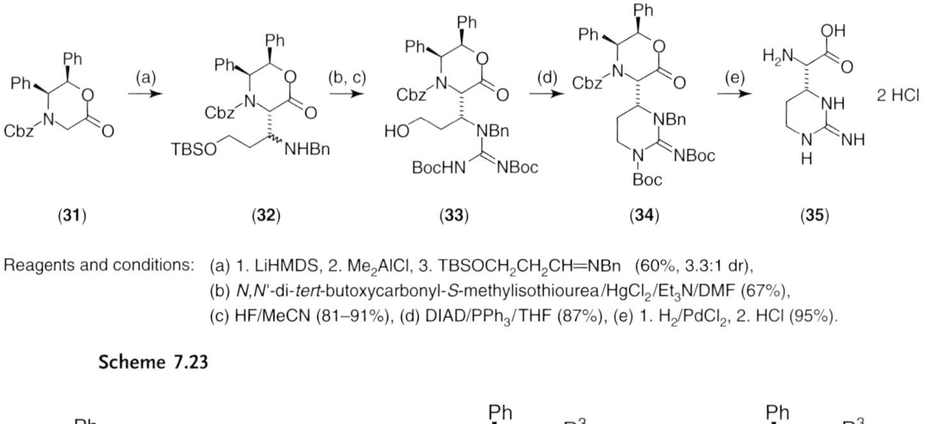

Reagents and conditions: (a) 1. LiHMDS, 2. Me$_2$AlCl, 3. TBSOCH$_2$CH$_2$CH=NBn (60%, 3.3:1 dr), (b) N,N'-di-*tert*-butoxycarbonyl-S-methylisothiourea/HgCl$_2$/Et$_3$N/DMF (67%), (c) HF/MeCN (81–91%), (d) DIAD/PPh$_3$/THF (87%), (e) 1. H$_2$/PdCl$_2$, 2. HCl (95%).

Scheme 7.23

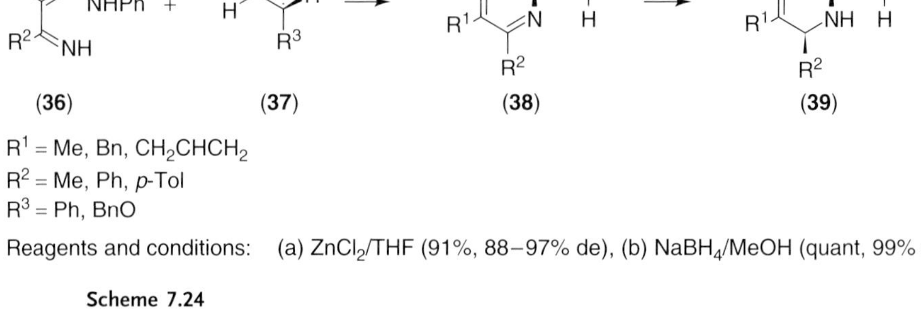

R^1 = Me, Bn, CH$_2$CHCH$_2$
R^2 = Me, Ph, *p*-Tol
R^3 = Ph, BnO

Reagents and conditions: (a) ZnCl$_2$/THF (91%, 88–97% de), (b) NaBH$_4$/MeOH (quant, 99% de).

Scheme 7.24

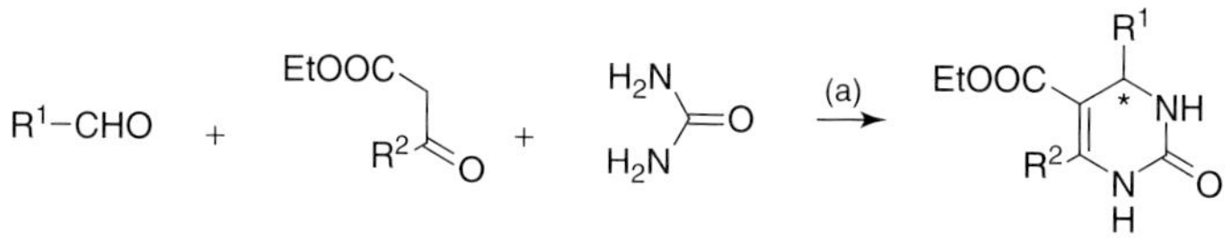

7.1 Six-Membered Rings with Another Heteroatom in the Same Ring

R^1–CHO + [EtOOC–CH(R^2)–C(=O)] (**40**) (**41**) + [H$_2$N–C(=O)–NH$_2$] (**42**) $\xrightarrow{(a)}$ **43** (EtOOC, R^1, R^2, NH, N–H, =O pyrimidinone)

(**40**)   (**41**)   (**42**)   (**43**)

R^1 = ribosyl, galactosyl, mannosyl, Ph, 2-(CF$_3$)-C$_6$H$_4$
R^2 = Me, ribosyl, galactosyl, mannosyl

Reagents and conditions: (a) CuCl/AcOH/BF$_3$·Et$_2$O (35–92%, 35–80% de).

Scheme 7.25

(*R*)-*N*-Boc-2,2-dimethyloxazolidine-4-carbaldehyde was subjected to the Biginelli reaction with ethyl acetoacetate and urea, with Yb(OTf)$_3$ as Lewis acid, furnishing the corresponding pyrimidinones as a 5:1 mixture of 4*R*,4′*S* and 4*S*,4′*S* diastereomers [39].

Asymmetric Biginelli reactions could also be performed starting from achiral components (benzaldehyde, ethyl acetoacete, and urea) in the presence of CeCl$_3$ or InCl$_3$ as Lewis acid, using chiral diamine or amide ligands as additives. In this way, 4-phenyldihydropyrimidinone derivative **43** (R^1 = Ph, R^2 = Me) could be obtained with moderate enantioselectivity (8–40%) [40].

7.1.2.2 Stereoselective Transformation by the Involvement of the Pyrimidine Ring

Stereoselective Alkylations In the presence of chiral amines, achiral pyrimidinone **44** was enantioselectively alkylated at 5-position in good yields with moderate (47:53–70:30) enantiomeric ratio. Lithium amides of the chiral amines were prepared with *n*BuLi, and the pyrimidinone **44** was treated with the chiral base obtained in this way, affording the lithium enolate of the heterocycle (**45**), which was then reacted with the corresponding electrophile. The enantioselectivity of the alkylation was the highest when (−)-sparteine was used as chiral amine, in toluene or tetrahydrofuran (THF) as solvent, and could be increased by the addition of LiBr. Moreover, a racemic mixture of alkylated pyrimidinones **46** could be enantiomerically enriched by deprotonation with lithium diisopropylamide (LDA), and subsequently reprotonated with a chiral proton source, achieving 53–68% ee [41].

A substituent with definite stereochemistry can determine the position of the entering electrophile in the C-alkylation of chiral substituted pyrimidinones. Compounds (*S*)-**6** could be stereoselectively alkylated at position 5 by deprotonation with

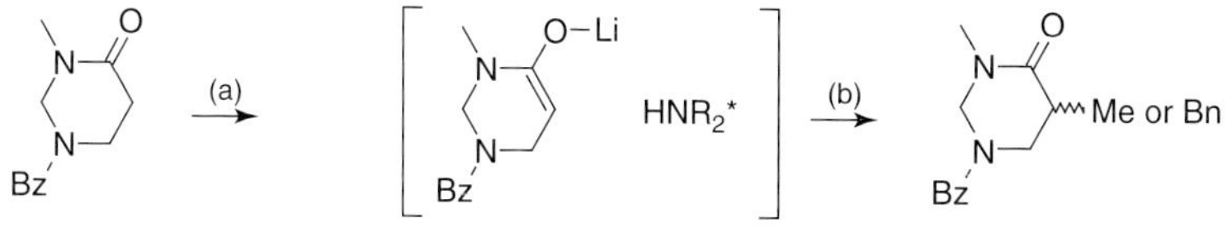

44 $\xrightarrow{(a)}$ [**45** · HNR$_2$*] $\xrightarrow{(b)}$ **46** (Me or Bn)

(**44**)   (**45**)   (**46**)

Reagents and conditions: (a) LiNR$_2$*/toluene, (b) MeI or BnBr (70–92%, er (*S*:*R*) = 47:53 to 70:30).

Scheme 7.26

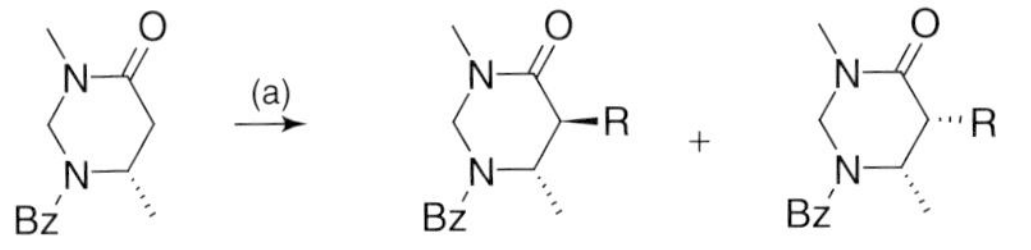

(S)-(6) → (5S,6S)-(47) + (5R,6S)-(47)

R = Me, Bn

Reagents and conditions: (a) LDA/THF/MeI or BnBr/−75 °C (71–91%, dr trans:cis = 4:1).

Scheme 7.27

LDA and subsequent treatment with alkyl halides, furnishing a 4:1 mixture of the trans and cis diastereomers 47 [29].

Alkylation of (2S)-1-benzoyl-2-isopropyl-3-methyltetrahydropyrimidin-4(1H)-one was carried out in an analogous way, affording the products with the same stereoselectivity [27].

Considerably higher stereoselectivity (de > 95%) could be achieved in the C-alkylation of 5-alkyl-1-benzoyl-2(S)-tert-butyl-3-methylpyrimidinone at position 5 with various alkyl halogenides. Again, addition of the electrophile to the corresponding enolate preferably took place from the face opposite the tert-butyl group, leading to the 5-dialkylated product with the entering group trans to the tert-butyl group [42].

Asymmetric alkylation of the pyrimidine ring of 48 could be achieved under SAMP ((2S)-1-amino-2-(methoxymethyl)pyrrolidine) control as a temporary chiral auxiliary providing 50 via SAMP-hydrazone intermediates 49. The hydrazone moiety was cleaved off, and the alkylated ketones 51 were then reduced to the corresponding 5-hydroxypyrimidin-2-ones 52 as potential HIV protease inhibitors [43].

Asymmetric induction via 1,5-radical translocation is used for the highly diastereoselective transformation of racemic and enantiomerically pure N-(o-bromo- and iodobenzoyl)-2-tert-butylperhydropyrimidinones 53 with electron-deficient alkenes

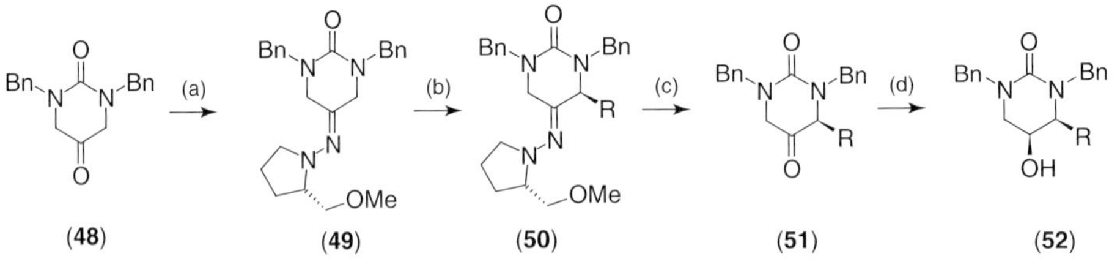

(48) → (49) → (50) → (51) → (52)

R = Me, Bu, CH₂CHCH₂, Bn, BnCH₂

Reagents and conditions: (a) SAMP/CH₂Cl₂/MS (3A) (93%), (b) 2,2,6,6-tetramethylpiperidine/nBuLi/THF/RX, (c) dimethyldioxirane/acetone/MeCN/H₂O or CuCl₂/THF/H₂O (44–59%, 2 steps), (d) LiAlH₄/Et₂O (76–84%, 96% de, 76–96% ee).

SAMP: (2S)-1-amino-2-(methoxymethyl)pyrrolidine.

Scheme 7.28

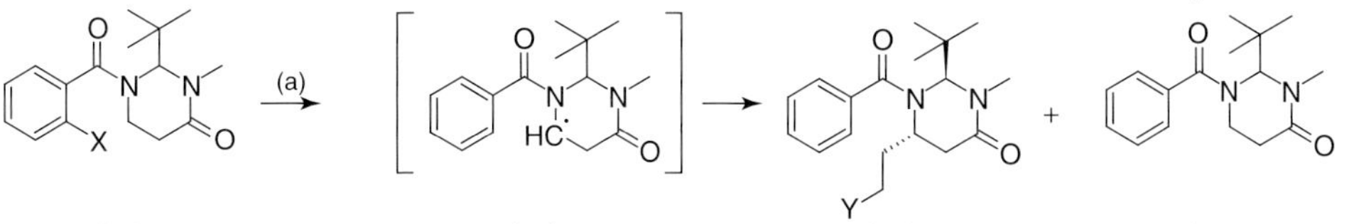

7.1 Six-Membered Rings with Another Heteroatom in the Same Ring | 309

(53)
X = Br, I
Y = COOMe, CN, SO$_2$Ph

Reagents and conditions: (a) nBu$_3$NCl/NaCNBH$_3$/AIBN/CH$_2$=CH–Y/tBuOH
(27–64% for 55, 16–70% for 56, over 95:5 dr).

AIBN: azobisisobutyronitrile.

Scheme 7.29

to substituted derivatives **55**. Thus, reaction of racemic bromo- or iodoperhydropyrimidinone **53** with methyl acrylate, acrylonitrile or phenyl vinyl sulfone in the presence of a catalytic amount of nBu_3SnCl, NaCNBH$_4$, and AIBN (azobisisobutyronitrile) resulted in a mixture of addition product **55** and reduced derivative **56**, in moderate yields and high stereoselectivity. With bromo derivatives as starting material, the expected product **55** is obtained as the major product, whereas reduced derivative **56** is formed in higher amount from the iodo compound. From enantiomerically pure **53**, products **55** are obtained in similar yields and high enantiomeric excess [44].

The palladium-catalyzed asymmetric allylic alkylation of barbituric acid derivatives with chiral ligands has been described. The reaction of 1,5-dimethylbarbituric acid (**57**) with allyl acetate (**58**) resulted in the 5-allyl substituted product **59** in moderate enantiomeric excess in the presence of Pd(acac)$_2$ as catalyst and **60** as phosphine ligand [45].

Low enantioselectivities (0–37% ee) were achieved in the reaction of barbiturate **61** with allyl acetate (**58**) in the presence of chiral phosphine ligands **65** or **66**. Using cyclopentenyl carbonate as the allylating agent, the product **64** was obtained in higher yields and enantioselectivities; however, the diastereomeric ratio was low [45].

Asymmetric allylic alkylation was also used for preparation of stereoisomeric mixtures enriched with one of the isomers of the narcotic methohexital, when

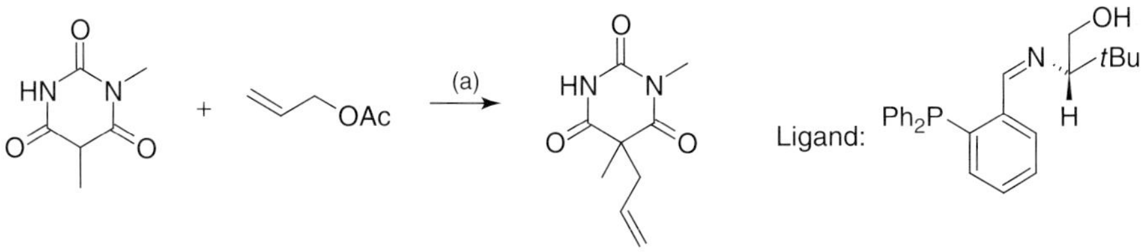

(57) (58) (59) (60)

Reagents and conditions: (a) Pd(acac)$_2$/DBU/toluene/CH$_2$Cl$_2$/ligand (77%, 34% ee).

Scheme 7.30

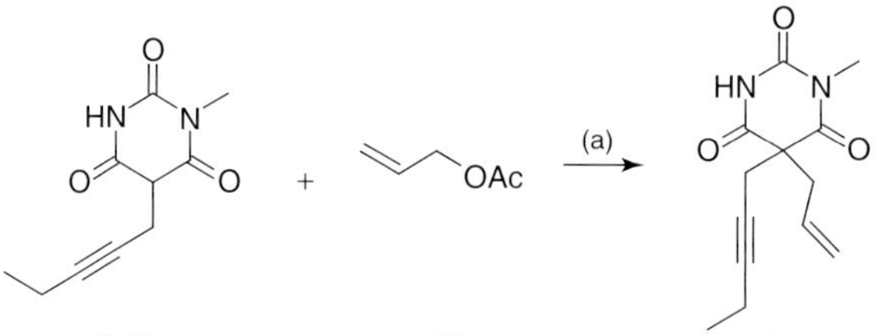

Scheme 7.31

Reagents and conditions: (a) $\eta^3C_3H_5PdCl_2$./CH_2Cl_2 or DMSO/ligand (42–72%, 0–37% ee).

Scheme 7.32

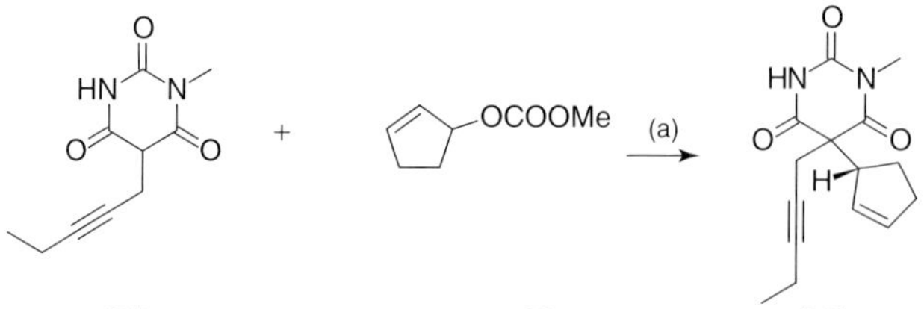

Reagents and conditions: (a) $Pd_2dba_3 \cdot CHCl_3$/CH_2Cl_2 or DMSO/ligand/54–93%, dr :1.1:1 to 2.51, 43–90% ee.

reactions of 5-allyl-5-(2′-hex-3′-ynyl)-1-methylbarbituric acid with allyl acetate were performed under various conditions. Ratio of the isomers mainly depended on the phosphine ligands used [46].

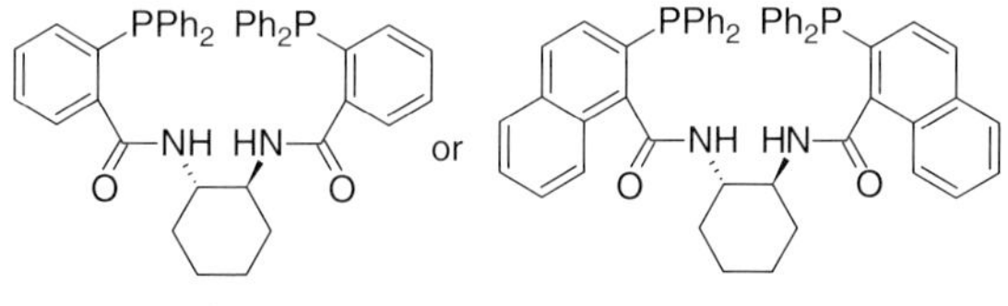

Halogenation, Hydroxylation Perhydropyrimidinone **67** could be halogenated via its enolate, formed by treatment with LiHMDS, using tosyl chloride, benzenesulfonyl bromide, or iodine as the halogenating agent. The chlorination reaction proved to be more selective than the bromination. Shorter reaction times (15 min) generally, favored the formation of the *cis* products **69**, while after longer reaction times (hours) a reversal of the stereoselectivity was observed for both chlorination and bromination. The *cis* adducts **69** are the kinetically preferred products, and the *trans* derivatives **68** are the thermodynamically preferred compounds. In contrast, iodination led to the *trans* isomer as the major product even after a short reaction time [47].

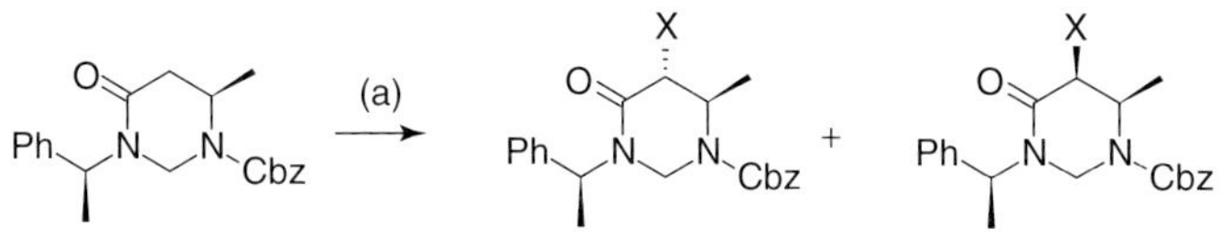

7.1 Six-Membered Rings with Another Heteroatom in the Same Ring | 311

(67) → **(68)** + **(69)**

X = Cl, Br, I

Reagents and conditions: (a) LiHMDS/THF/TsCl or PhSO$_2$Br or I$_2$ (74–98%, dr = 10:90 to 85:15).

Scheme 7.33

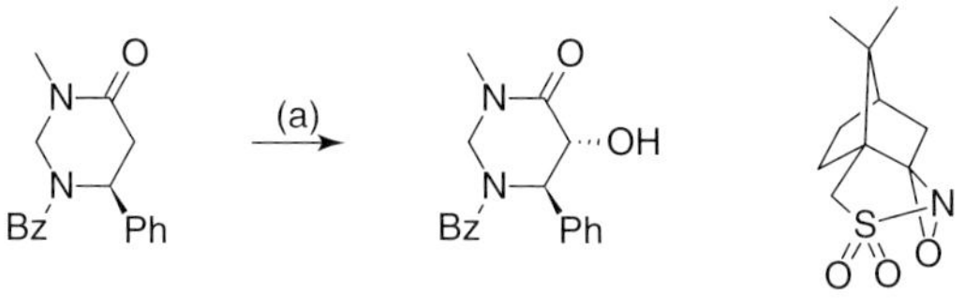

(S)-(**10**) → (5R,6R)-(**70**)   (**71**)

Reagents and conditions: (a) 1. LDA, 2. oxaziridine **71**.

Scheme 7.34

Enolization of pyrimidinone (S)-**10** with LDA followed by hydroxylation with oxaziridine **71** resulted in the exclusive formation of hydroxy compound (5R,6R)-**72** [30].

7.1.3
Piperazines

Substituted piperazines with defined configuration of the substituents are synthesized in most cases starting from enantiomerically pure starting materials. A method quite frequently applied for synthesis of piperazine ring is intramolecular amide coupling between amino and carboxylic acid ester moieties, resulting in ketopiperazines, which can be further functionalized. Further widely employed reactions for cyclization are intramolecular nucleophilic substitution, imine formation or cycloaddition. Asymmetric transformations of the piperazine ring could also be accomplished by alkylations with various electrophilic or nucleophilic reactants.

7.1.3.1 Formation of the Piperazine Ring
Ring Closure by Lactam Formation The following scheme presents three approaches for the synthesis of diketopiperazines (DKPs) via key intermediates N-monobenzylglutamate **2** (prepared from dimethyl glutamate **1**) and chloroacetamide **5**.

In the first approach, N-benzyl-protected derivative **2** was transformed to dipeptide **3**, and subsequently cyclized to piperazine derivative **4**. As an alternative pathway to **4**, compound **2** was acylated with chloroacetyl chloride to **5**, which was

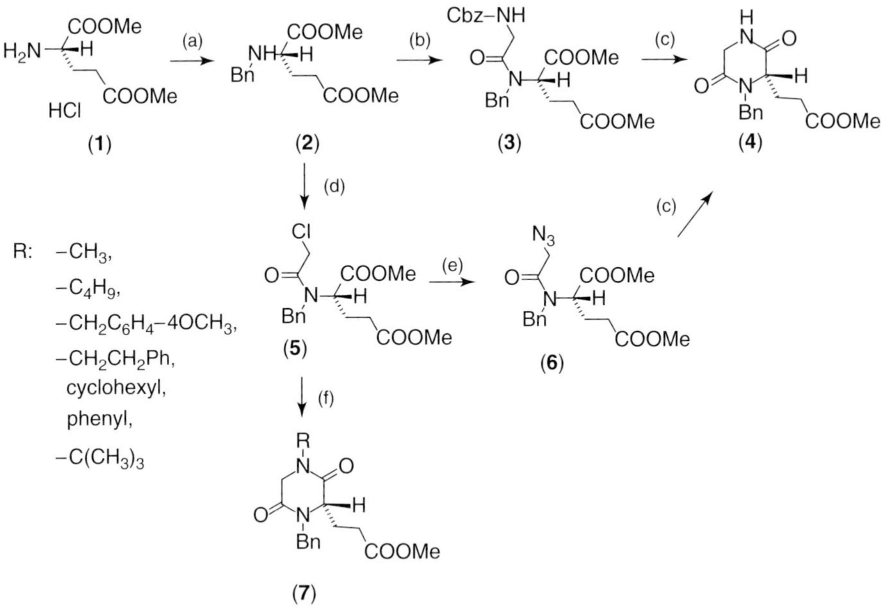

Scheme 7.35

R: –CH$_3$,
–C$_4$H$_9$,
–CH$_2$C$_6$H$_4$–4OCH$_3$,
–CH$_2$CH$_2$Ph,
cyclohexyl,
phenyl,
–C(CH$_3$)$_3$

Reagents and conditions: (a) 1. C$_6$H$_5$CHO/NEt$_3$/pentane/Na$_2$SO$_4$/rt, (91%),
2. NaBH$_4$/MeOH/ 0 °C (93%),
(b) Cbz-glycine/DCC/THF/CH$_2$Cl$_2$/rt,
(c) H$_2$/Pd–C/MeOH/rt (84% from 3, 87% from 6),
(d) ClCH$_2$COCl/NaHCO$_3$/CH$_2$Cl$_2$/rt (73%),
(e) NaN$_3$/acetone/rfx (99%),
(f) RNH$_2$/CH$_3$CN in case of cyclohexylamine: CH$_3$CN/48 h/reflux,
in case of aniline: DMF/16 h / 155 °C (52–97%).

reacted with NaN$_3$, and the azide **6** formed was hydrogenated to afford piperazine **4**. According to the third approach, the chloroacetyl derivative **5** was treated with various primary amines, resulting in the cyclized products **7** in one pot [48].

Principally in the same way as in the first approach above, a series of (2R,5S)- and (2S,5S)-2-hydroxymethyl-5-alkylpiperazines can be prepared starting from enantiomerically pure serine without any racemization. Commercial amino acids were converted into the corresponding N-benzyloxycarbonyl derivatives **8**, which were treated with (S)- or (R)-serine methyl ester hydrochloride (**9**) to give dipeptides **10** via the mixed anhydride coupling method. After deblocking of the benzyloxycarbonyl group, DKPs were formed in dry methanol by heating (65 °C) for five days. DKPs **11** were reduced to piperazines **12** with an excess of LiAlH$_4$ in refluxing THF [49].

DKP **18** was synthesized from N-sulfinyldiamino ester **13** via dipeptide **17**, which was prepared by coupling of the differently protected derivatives **15** and **16** of the same amino acid [50].

Dilactim ethers **22**, known as Schöllkopf chiral auxiliaries widely used for asymmetric synthesis of amino acids, were prepared from amio acids **19** in

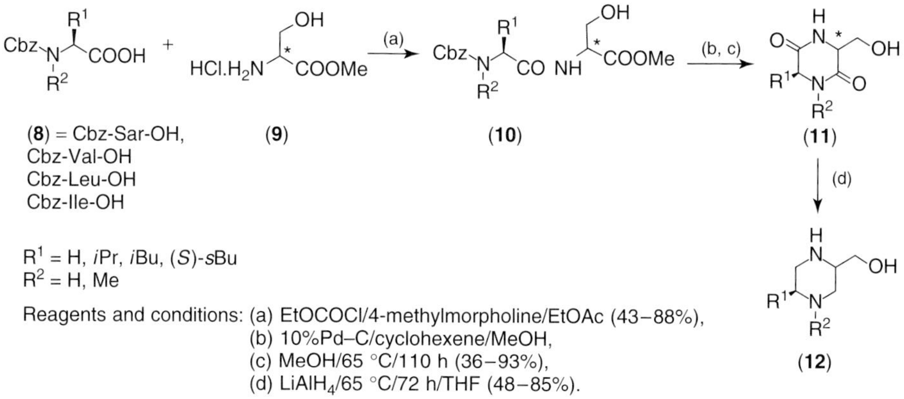

7.1 Six-Membered Rings with Another Heteroatom in the Same Ring | 313

(8) = Cbz-Sar-OH,
Cbz-Val-OH
Cbz-Leu-OH
Cbz-Ile-OH

R^1 = H, iPr, iBu, (S)-sBu
R^2 = H, Me

Reagents and conditions: (a) EtOCOCl/4-methylmorpholine/EtOAc (43–88%),
(b) 10%Pd–C/cyclohexene/MeOH,
(c) MeOH/65 °C/110 h (36–93%),
(d) LiAlH$_4$/65 °C/72 h/THF (48–85%).

Scheme 7.36

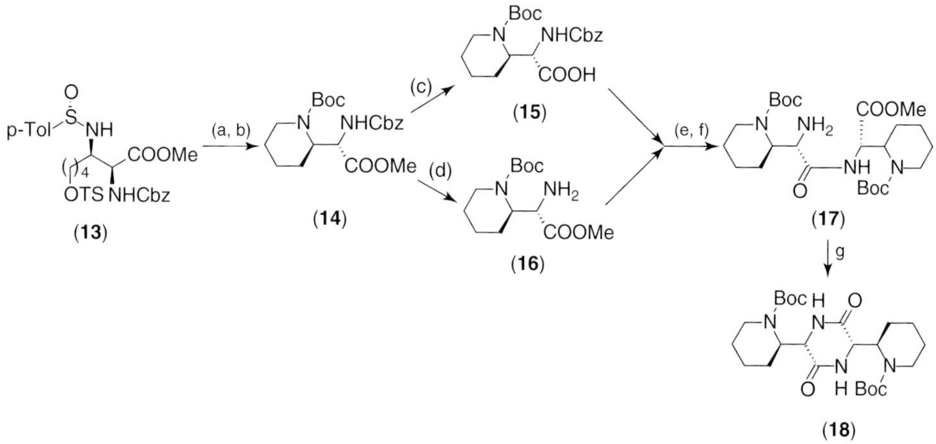

Reagents and conditions: (a) 1. H$_3$PO$_4$/MeOH/H$_2$O, 2. K$_2$CO$_3$, (b) Boc$_2$O/Et$_3$N/dioxane/H$_2$O (82%),
(c) LiOH/THF/H$_2$O (100%), (d) Pd–C/H$_2$ (100%), (e) BOP/DIPEA/CH$_2$Cl$_2$,
(f) Pd–C/H$_2$ (85% 2 steps), (g) DMF/rfx/40 h (22%)
or KCN/DMF/80 °C/4 d (29%).

Scheme 7.37

three steps. Reaction with triphosgene was followed by treatment of the resulting oxazolinedione **20** with ethyl glycinate to give piperazinedione **21**, which was subsequently transformed to dilactim ether **22** [51, 52]. The enantiomer of **22** could also be prepared via the same route starting from the enantiomer of the corresponding amino acid [53–55].

Synthesis of the naturally occurring DKPs (−)-phenylhistine (**28**, R = Bn) and (−)-aurantiamine (**28**, R = iPr) begins with the coupling of amino acid **23** and phosphinyl glycine **24**. Phosphinyl ester dipeptide **25** was then reacted with formylimidazole **26** under Horner–Emmons coupling conditions. Finally, *in situ*

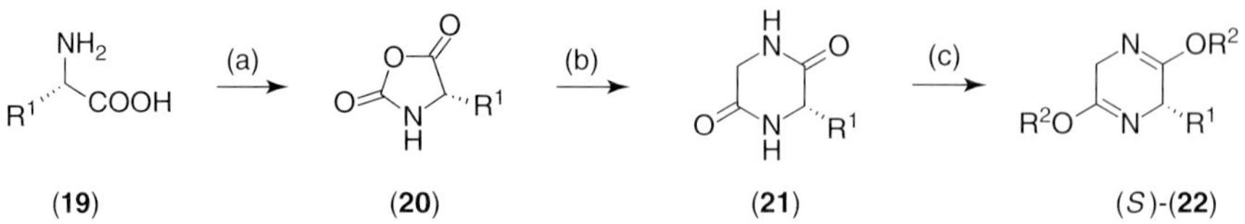

314 | *7 Asymmetric Synthesis of Ring Nitrogen Heterocycles*

Scheme 7.38

$R^1 = iPr, tBu$
$R^2 = Me, Et$

Reagents and conditions: (a) triphosgene/THF (99%), (b) 1. $NH_2CH_2CO_2Et/CHCl_3/EtN_3/THF/-70°C$, 2. toluene or xylene/rfx (70%), (c) $R_2^3OBF_4/CH_2Cl_2$ (90%).

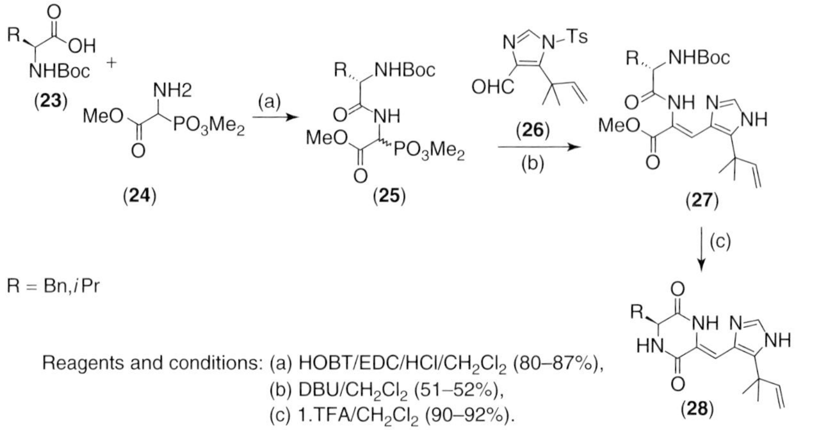

$R = Bn, iPr$

Reagents and conditions: (a) HOBT/EDC/HCl/CH_2Cl_2 (80–87%),
(b) DBU/CH_2Cl_2 (51–52%),
(c) 1.TFA/CH_2Cl_2 (90–92%).

Scheme 7.39

ring closure of **27**, during hydrolysis of the Boc-protecting group, resulted in the products **28** [56].

Analogous to the second approach shown in Scheme 7.35, ethyl or menthyl (2S)-5,6-dioxo-4-[(1R)-1-phenylethyl]piperazine-2-carboxylates were synthesized starting with ring opening of ethyl or menthyl (2R)-1-[(1R)-1-phenylethyl]aziridine-2-carboxylate with azide as nucleophile in aqueous acidic media in the presence of a catalytic amount of $AlCl_3 \cdot 6H_2O$, furnishing stereoselectively the corresponding (2S)-2-azido-3-[(1R)-1-phenylethyl]aminopropanoates, which were made to react with methyl oxalyl chloride, followed by catalytic hydrogenation. The absolute configuration, determined by X-ray crystallography, showed that the ring opening proceeded with complete inversion at the reaction center [57].

Monolactim ether **33**, widely used by Sandri et al. as a chiral auxiliary in syntheses of amino acids, was prepared according to the third approach shown in Scheme 7.35. L-Valine methyl ester (**29**) was benzylated to **30**, followed by N-acylation with chloroacetyl chloride. Ester hydrolysis of **31** and cyclization with ammonia gave **33** after treatment of **32** with Et_3OBF_4 [58].

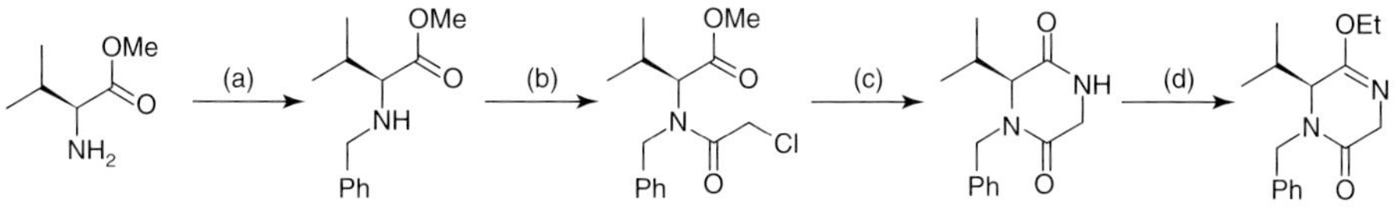

7.1 Six-Membered Rings with Another Heteroatom in the Same Ring

Scheme 7.40

Reagents and conditions: (a) PhCH$_2$Br/pyridine/CH$_2$Cl$_2$ (85%), (b) ClCH$_2$COCl/TEA/CH$_2$Cl$_2$ (96%), (c) 10 M NH$_3$ in EtOH (>98%), (d) EtO$_3$BF$_4$/CH$_2$Cl$_2$ (75%).

Applying the third approach shown in Scheme 7.35, piperazino-piperidine-based CCR5 antagonists were synthesized via reductive amination of 4-trifluoromethyl-acetophenone with various L-amino acids, followed by N-acylation with chloroacetyl chloride, and subsequent cyclization in the presence of 1-Boc-protected-4-substituted-4-aminopiperidines [59].

By the same strategy, the four possible stereoisomers of the N-acetyl derivative of the natural marine compound Etzionin were synthesized to carry out a stereochemical analysis by comparing the natural compound and the stereoisomers synthesized from enantiomerically pure starting materials. Thus, for synthesis of the (S,R) diastereomer, methyl (2S)-2-[(bromoacetyl)amino]-3-phenylpropanoate was submitted to nucleophilic displacement with methyl (3R)-3-aminododecanoate to give the corresponding peptide and subsequently cyclized [60].

DKPs **37** were prepared from protected amino acid amides **34** by the addition to methyl 3,3,3-trifluoropiruvate (**35**). The adduct dipeptides **36** were then converted to the corresponding 3-hydroxy-3-trifluoromethyl-2,5-diketopiperazines **37**, which are used as homochiral electrophilic synthons for α-trifluoromethyl amino acids [61].

The last step in the total synthesis of the cyclotryptophan alkaloid asperazine (**39**) is the formation of two DKP rings. A series of five reactions could be accomplished in a single step by heating **38** at 200 °C under argon. The reaction gave **39** and two stereoisomers in 70:3:2 ratio; after separation, **39** was obtained in 34% yield. However, the conversion of **38** to **39** was more efficient when two separate steps were applied. First, removal of the Boc-protecting groups, followed by heating of the crude deprotected reaction mixture in butanol in the presence of acetic acid, gave **39** in 59% yield [62].

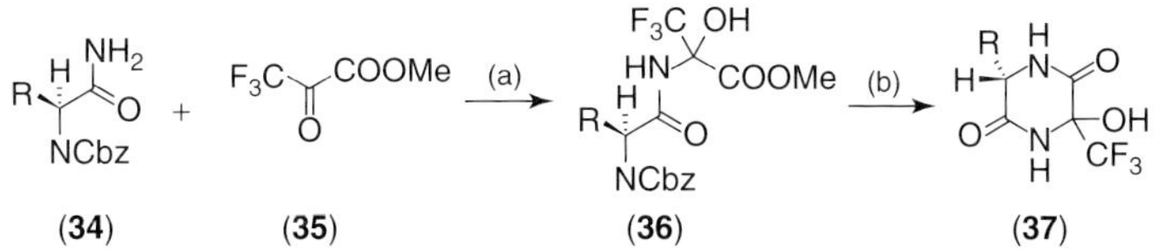

R = Me, iPr, iBu, Bzl

Reagents and conditions: (a) CH$_2$Cl$_2$, (b) H$_2$/Pd–C/MeOH (69–86%).

Scheme 7.41

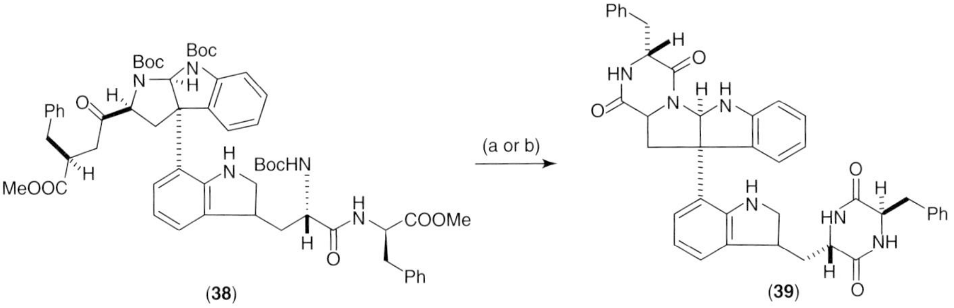

7 Asymmetric Synthesis of Ring Nitrogen Heterocycles

Reagents and conditions: (a) 200 °C, 4h (34%), (b) 1. HCOOH, 2. AcOH, nBuOH, 120 °C (59%).

Scheme 7.42

Analogous to DKP formation, monoketopiperazines are synthesized via intramolecular amide coupling of N-aminoethyl substituted amino acids. Thus, enantiomerically pure 2,6-dimethylpiperazin-5-ones **45** were synthesized by employing a diastereoselective triflate alkylation to set the required stereochemistry. The reaction sequence started with the conversion of N-Boc-L-alanine (**40**) to dibenzylamide (**41**) through the mixed anhydride. Removal of the Boc group, followed by borane-methyl sulfide reduction, gave diamine **42**, which was alkylated with the triflate (**43**) of methyl (R)-lactate. The alkylation proceeded with inversion to afford ester **44**. Hydrogenolysis of **44** in the presence of HCl resulted in monodebenzylation and partial cyclization, which was completed by heating in the presence of p-toluenesulfonic acid to ketopiperazine **45**. Reduction of the oxo group with LiAlH$_4$, followed by debenzylation with Pearlman's catalyst, gave

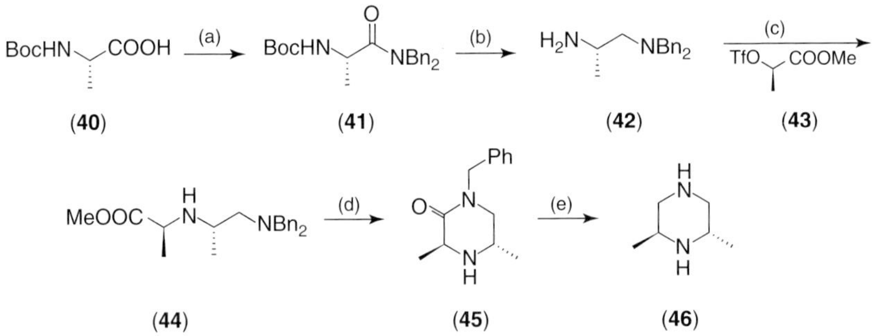

Reagents and conditions: (a) 1. ClCOOiBu/Et$_3$N/THF, 2. Bn$_2$NH (82%),
(b) 1. TFA/CH$_2$Cl$_2$,
2. BH$_3$·DMS/THF (84% for 2 steps, 98% ee),
(c) CH$_2$Cl$_2$/Et$_3$N (94%),
(d) 1. Pd–C/H$_2$/HCl/EtOH, 2. TsOH/EtOH/rfx (81%),
(e) LiAlH$_4$/THF (93%, 98% ee),
(f) Pd(OH)$_2$/H$_2$/MeOH (91%, 98% ee).

Scheme 7.43

(2S,6S)-2,6-dimethylpiperazine (**46**). The (R,R) enantiomer was prepared in an analogous way from N-Boc-D-alanine and methyl (S)-lactate [63, 64].

N-(2-Diallylamino-3-hydroxyalkyl)-substituted amino acid esters were deallylated to primary amines under mild conditions with (PPh$_3$)$_3$RhCl. While *tert*-butyl esters could be easily deprotected to the corresponding primary amine, methyl esters such as **47** spontaneously cyclized to piperazine **48** after deprotection [65].

As a building block for the synthesis of natural products TAN1251C and D, disubstituted piperazine **32** was synthesized from carbamate **49** obtained from L-tyrosine. Reduction of **49** with LiAlH$_4$, followed by Boc protection and subsequent Swern oxidation of the corresponding primary alcohol, gave aldehyde **50**, which was condensed with O-benzyl-L-tyrosine under reductive conditions. After deprotection of the Boc group, amine **51** was cyclized with NaOMe to piperazinone **52** [66].

Piperazine-2-carboxylic acid ester enantiomers **57** were prepared from L- or D-serine ester **53** via the azide intermediate **54**, obtained by Mitsunobu reaction. After deprotection of **54**, the amino compound was alkylated with the triflate (**55**) of dimethyl-D-malate to afford L-aspartic acid derivative **56** via an S$_N$2 reaction. Reduction of the azido group in **70** resulted in spontaneous lactam formation to piperazine derivative **57** [67].

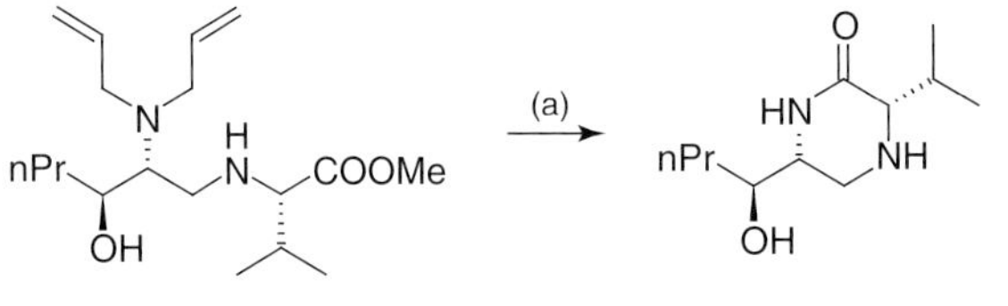

Reagents and conditions: (a) (PPh$_3$)$_3$RhCl/MeCN/H$_2$O (47%).

Scheme 7.44

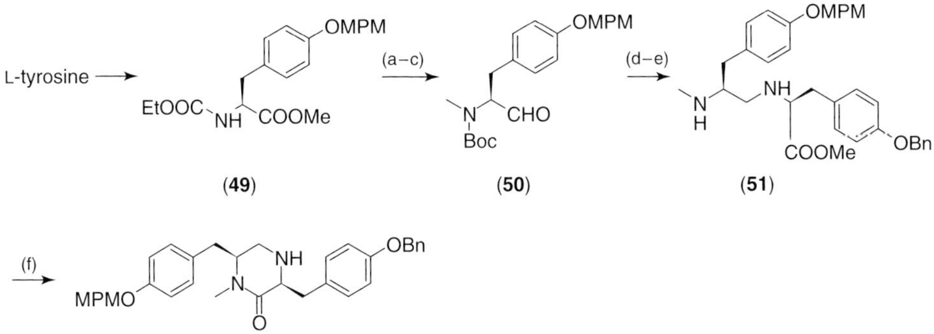

Reagents and conditions: (a) LiAlH$_4$/THF (82%), (b) Boc$_2$O/THF–H$_2$O (99%), (c) (COCl)$_2$/DMSO/Et$_3$N, (d) O-Bn-L-tyrosine methyl ester/NaBH$_3$CN/DMF (90%), (e) ZnBr$_2$/CH$_2$Cl$_2$ (99%), (f) NaOMe/THF (85%).

Scheme 7.45

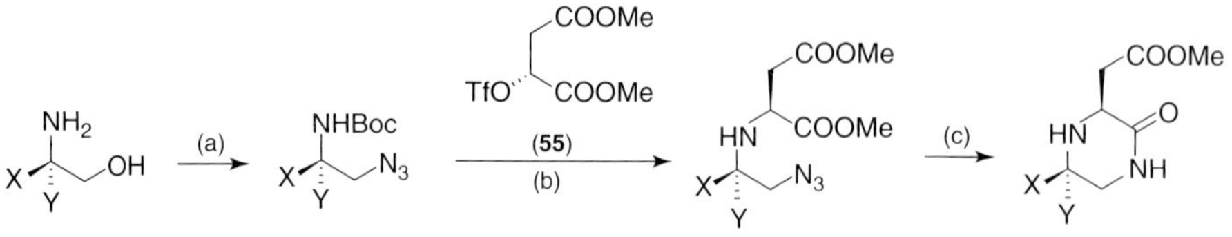

7 Asymmetric Synthesis of Ring Nitrogen Heterocycles

(53) X-CH(Y)-CH(NH$_2$)-OH →(a)→ **(54)** X-CH(Y)-CH(NHBoc)-N$_3$ →(b) [via TfO-CH$_2$-CH(COOMe)$_2$ **(55)**]→ **(56)** X-CH(Y)-CH(N$_3$)-NH-CH(COOMe)-CH$_2$-COOMe →(c)→ **(57)** piperazine-2,5-dione derivative

X = Cbz, Y = H or X = H, Y = Cbz

Reagents and conditions: (a) 1. Boc$_2$O/H$_2$O/dioxane, 2. HN$_3$/(iPrNOOC)$_2$/PPh$_3$ (39–67%),
(b) 1. TFA/CH$_2$Cl$_2$, 2. 55/CH$_2$Cl$_2$/2,6-lutidine (61–73%),
(c) PPh$_3$/THF (49–85%).

Scheme 7.46

Viso et al. employed chiral sulfinamides for the synthesis of enantiomerically pure substituted piperazines. 1-Benzyl-2,3-disubstituted piperazines **61** were obtained by treatment of N-sulfinyl-N-benzylamino alcohols **58** with diethyl oxalate and NaOMe, followed by borane reduction. Ring closure of **58** gave morpholine derivative **60** as a side-product, which could be transformed to piperazine **59** by NaOMe treatment [68, 69].

(2S,3S)-2,3-bis[(benzyloxy)methyl]piperazine was prepared from (S,S)-threitol-1,4-dibenzyl ether also via a 1,2-diamino intermediate by a five-step reaction

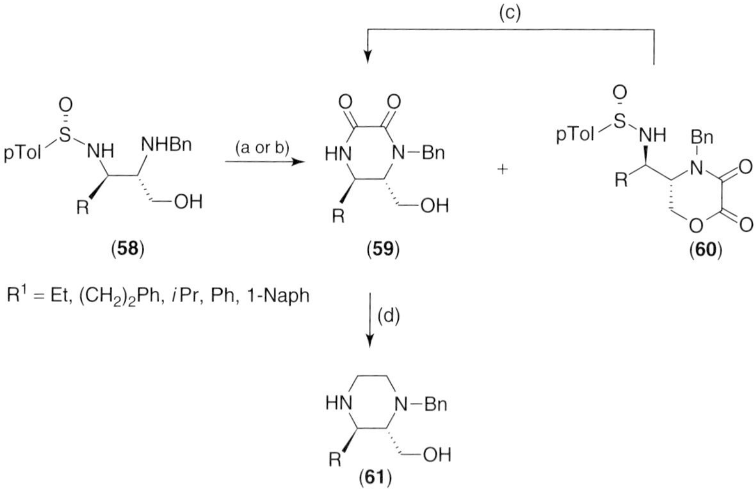

(58) pTol-S(O)-NH-CH(R)-CH(NHBn)-CH$_2$-OH →(a or b)→ **(59)** piperazine-2,3-dione (HN, N-Bn, R, -OH) + **(60)** pTol-S(O)-NH-CH(R)-CH(Bn)-morpholinedione →(c)→ **(59)**

R^1 = Et, (CH$_2$)$_2$Ph, iPr, Ph, 1-Naph

↓(d)

(61) HN—N-Bn piperazine with R and -OH substituents

Reagents and conditions: (a) (COOEt)$_2$/NaOMe/CH$_2$Cl$_2$/MeOH,
(b) 1. (COOEt)$_2$/CH$_2$Cl$_2$
2. NaOMe/MeOH (77–90% for 59, 10–15% for 60),
(c) NaOMe/MeOH,
(d) BH$_3$·DMS/THF (60–84%).

Scheme 7.47

7.1 Six-Membered Rings with Another Heteroatom in the Same Ring

sequence: (i) mesylation, (ii) nuclephilic displacement with azide, (iii) LiAlH$_4$ reduction, (iv) cyclization with diethyl oxalate, and finally (v) reduction of the oxo groups [70].

2,5-Diketopiperazine derivatives **65** were synthesized through Ugi four-center three-component reactions from commercially available dipeptides **62**, aldehydes **63**, and isocyanides **64** in 2,2,2-trifluoroethanol as solvent in moderate to good yields. A new chiral center was created at the carbon originating from the aldehyde; the relative stereochemistry of the resulting diastereomers was not determined. The best observed diastereomeric ratio was 6:1. The reaction time and yield could be improved by the application of microwave irradiation [71].

Trisubstituted DKPs were synthesized by intramolecular amidation of dipeptide esters, obtained in four-component Ugi reactions. Thus, the imine intermediate **67** formed *in situ* from (R)-leucine methyl ester (**66**) and benzaldehyde was reacted with *tert*-butylisonitrile and (R)-indanyl glycine **68**. The dipeptide ester **69** formed was then treated with 4 N HCl in dioxane, which resulted in removal of the Boc-protecting group and partial cyclization, which was completed by addition of triethylamine to give a 1:3 mixture of diastereomers **70** and **71**. The partial stereocontrol originates from the attack of the isonitrile on the less hindered face of the imine [72].

As an alternative route to isomeric DKPs, secondary amine **95** was prepared first via a four-center three-component Ugi reaction and subsequent hydrolysis. Then it was acylated with (R)-Boc-indanyl glycine **91** through mixed anhydride **96**. Dipeptide intermediate **97** was then cyclized to **98** [73].

Ring Closure by Intramolecular Nucleophilic Substitution O-protected N-sulfinyl diaminoalcohols **76** were acylated with chloroacetyl chloride to **77** and subsequently cyclized to N-sulfinylpiperazines **78**. The sulfinyl group was eliminated by treatment with NaH, affording unsaturated derivative **79** [69].

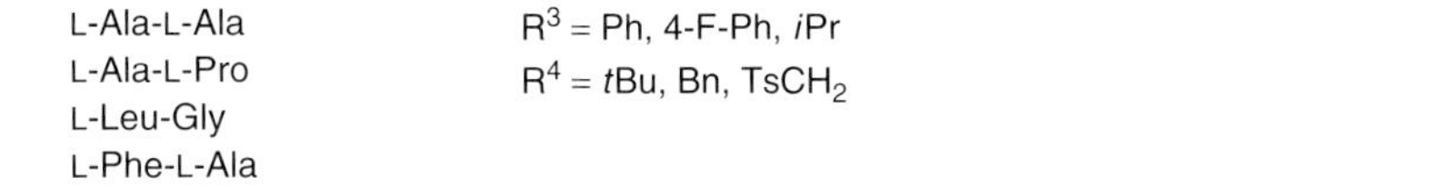

(62):
Gly-L-Leu
L-Ala-L-Ala
L-Ala-L-Pro
L-Leu-Gly
L-Phe-L-Ala

(63)
R^3 = Ph, 4-F-Ph, *i*Pr
R^4 = *t*Bu, Bn, TsCH$_2$

Reagents and conditions: (a) −40 °C to rt/ CF$_3$CH$_2$OH/33−52 h (24−87%, dr 2:1 to 6:1) or
(b) MW (30 W) 100 °C/CF$_3$CH$_2$OH-BmimPF$_6$/15 min (42−58%, dr 2:1 to 3:1).

Scheme 7.48

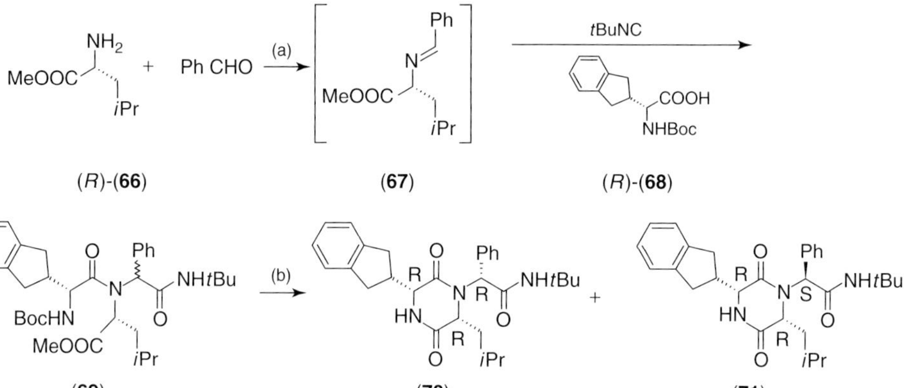

(R)-(66) (67) (R)-(68) (69) (70) (71)

Reagents and conditions: (a) Et_3N/MeOH, (b) 1. 4 N HCl/dioxane, 2. Et_3N/dioxane.

Scheme 7.49

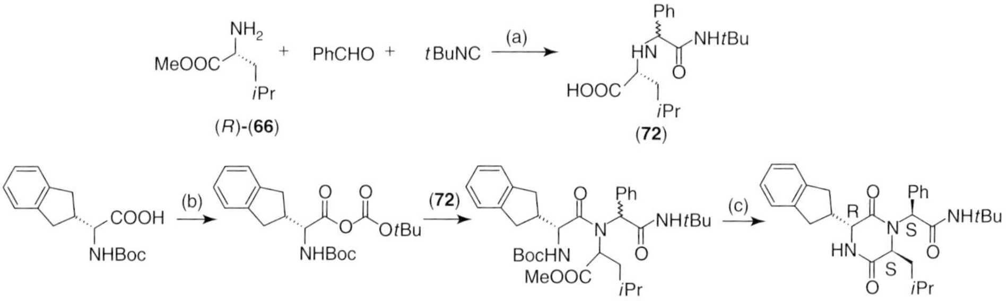

(R)-(66) (72) (R)-(68) (73) (74) (75)

Reagents and conditions: (a) MeOH/–30 °C to rt,
(b) iPrOCOCl/N-Me-morpholine/THF,
(c) 1.4 N HCl/dioxane, 2. HCl/MeOH (57%).

Scheme 7.50

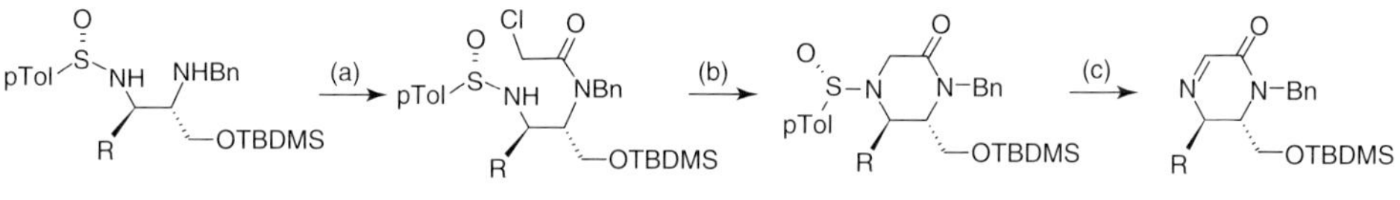

(76) (77) (78) (79)

R^1 = Et, $(CH_2)_2Ph$, iPr, Ph, 1-Naph

Reagents and conditions: (a) $ClCH_2COCl$/EtOAc:$NaHCO_3$ (1:1) satd, (52–90%),
(b) Cs_2CO_3/DMF (74–92%), (c) NaH/THF (73–87%).

Scheme 7.51

1-Benzyl 2-methyl (2*S*,3*R*)-3-alkyl-5-oxopiperazine-1,2-dicarboxylates were prepared in an analogous way from 1*N*-*p*-tolylsulfinyl derivative of methyl (2*R*, 5*R*)-5-methyl-2-phenylimidazolidine-4-carboxylate [50].

A diamino derivative was cyclized to piperazine in an unexpected way in the following reaction. The treatment of bistosylamide **80** with ethylene glycol ditosylate did not give the expected Richman–Atkins cyclization. Instead, chiral piperazine **81** was formed as a result of nucleophilic attack of the central benzylic amine on the electrophilic tosylate in ethylene glycol ditosylate, followed with expulsion of the C-3 tosamide unit either before or during the cyclization [73].

Although, generally, Mitsunobu reactions involving a nitrogen nucleophile require an acidic hydrogen ($pK_a < 13$), there are examples of intramolecular Mitsunobu reactions of amino alcohols to obtain heterocycles. Accordingly, cyclization of **85** by treatment with PPh_3/diethyl azodicarboxylate (DEAD) resulted in benzylated piperazine **86**. Amino alcohol **85** was prepared from (*R*)-1-aminopropan-2-ol (**82**) by reductive alkylation and the subsequent coupling of **83** with *N*-Boc-L-alanine providing **84**, which was reduced to **85** [63].

The method could be extended to the preparation of monomethylpiperazines. The first step was the coupling of *N*-Boc-D-alanine (**87**) with *N*-benzylethanolamine to amide **88**. Subsequent hydrolysis of the Boc group followed by the Mitsunobu reaction gave methylpiperazinone **89**, which was reduced to methylpiperazine **90** [63].

Cyclization via Imine/Enamine Formation The synthesis of tetrahydropyrazinone **93** was based on the coupling of α-aminoketone **91** with *N*-Boc-L-alanine pivalic acid mixed anhydride to **92**. Compound **91** was obtained from (*R*)-valine in a three-step reaction sequence (i) N-protection with Boc, (ii) transformation of the carboxylic group to dimethyl amide, followed by (iii) reaction with phenylmagnesium bromide. After acidic deprotection of **92**, the pyrazine ring was formed by intramolecular

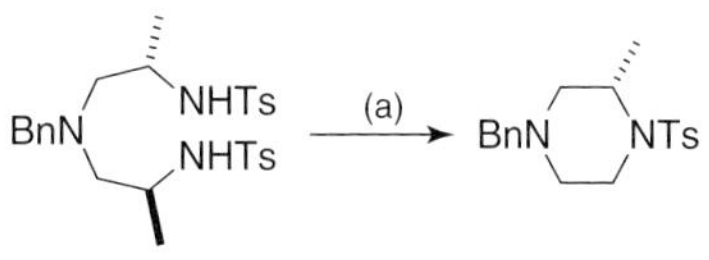

(**80**) → (**81**)

Reagents and conditions:
(a) 1. NaH/THF/Δ,
2. $(CH_2OTs)_2$/DMF/120 °C (32%).

Scheme 7.52

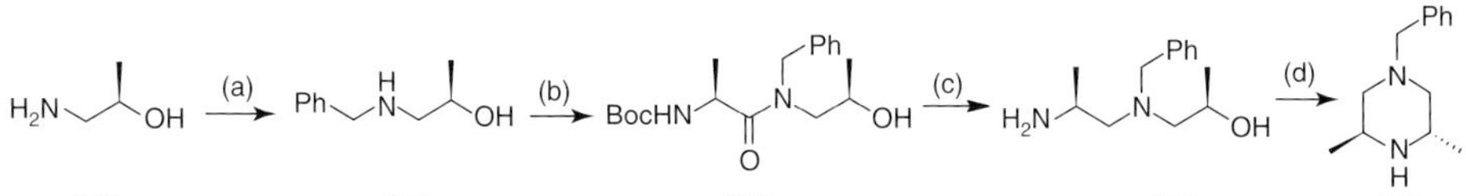

(**82**) → (**83**) → (**84**) → (**85**) → (**86**)

Reagents and conditions: (a) 1. $PhCHO/MgSO_4$/THF, 2. $NaBH_4$/EtOH (57%),
(b) CDI/CH_2Cl_2/*N*-Boc-L-Alanine (80%),
(c) 1. TFA/CH_2Cl_2, 2. $BH_3 \cdot DMS$/THF (76%, 2 steps), (d) $DEAD/PPh_3$/THF (66%).

Scheme 7.53

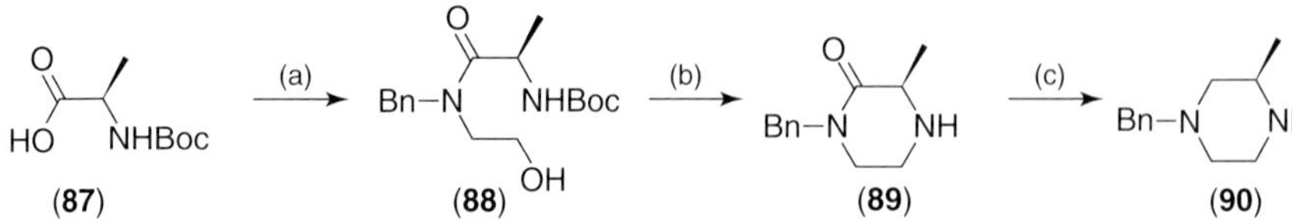

7 Asymmetric Synthesis of Ring Nitrogen Heterocycles

Reagents and conditions: (a) CDI/CH$_2$Cl$_2$/N-benzyl-ethanolamine (81%), (b) 1. TFA/CH$_2$Cl$_2$, 2. DEAD/PPh$_3$/THF (62%), (c) LiAlH$_4$/THF (88%).

Scheme 7.54

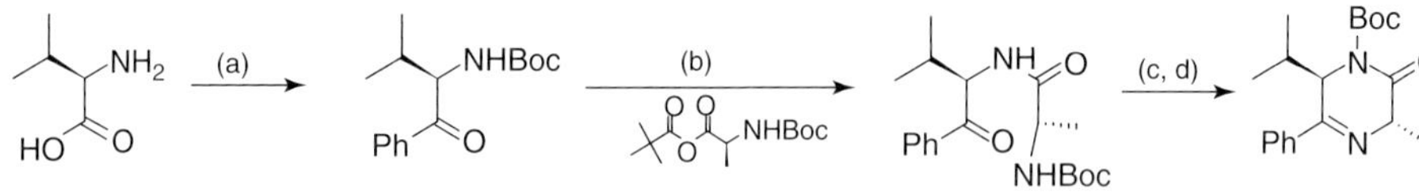

Reagents and conditions: (a) 1. (Boc)$_2$O, 2. Me$_2$NH·HCl/TBTU, 3. PhMgBr (78%), (b) HCl$_g$/EtOAc (95%), (c) 1. HCl$_g$/EtOAc, 2. K$_2$CO$_3$ (86%), (d) (Boc)$_2$O/0 °C/THF (84%, trans:cis = 20:1).

Scheme 7.55

condensation between the oxo and amino groups, which was transformed to Boc-protected derivative **93** as a 20:1 mixture of the *trans* and *cis* diastereomers [74, 75]. In an analogous way, a derivative of **121** without methyl substituent was also prepared using N-Boc-glycine pivalic acid mixed anhydride instead of alanine derivative as the coupling partner of **119** [76].

Synthesis of piperazinecarboxylic ester **98** was achieved by starting with the coupling of protected L-tyrosine **94** and 2-amino-3,3-diethoxypropionate (**95**). The dipeptide **96** obtained was then cyclized to unsaturated piperazine **97** by treatment with TFA. Subsequent catalytic hydrogenation under acidic conditions resulted selectively in the single diastereomer **98**. When the acid was omitted, no reduction of the double bond occurred, and **99** was isolated, which could be reduced to **98** with NaCNBH$_3$ in the presence of HCl [77].

Reaction of the primary amino moiety in **100** with aldehydes and subsequent reduction of the imine intermediate with NaBH$_4$ gave alkylamino derivatives **101**. Reductive cyclization of **101** with aqueous glyoxal solution in the presence of NaCNBH$_3$ as reducing agent provided 2-substituted-1,4-piperazines **102** [78].

Diastereomeric 3,5-disubstituted piperazines **104** and **105** were formed in high yields but in moderate ratios when the corresponding β-keto esters **103** were hydrogenated at 45 °C and 45 psi with Pd/C as catalyst. Removal of the Cbz protecting group and reductive amination took place in a one-pot reaction. Although in slightly lower yield, the piperazine ring was also formed in two steps involving

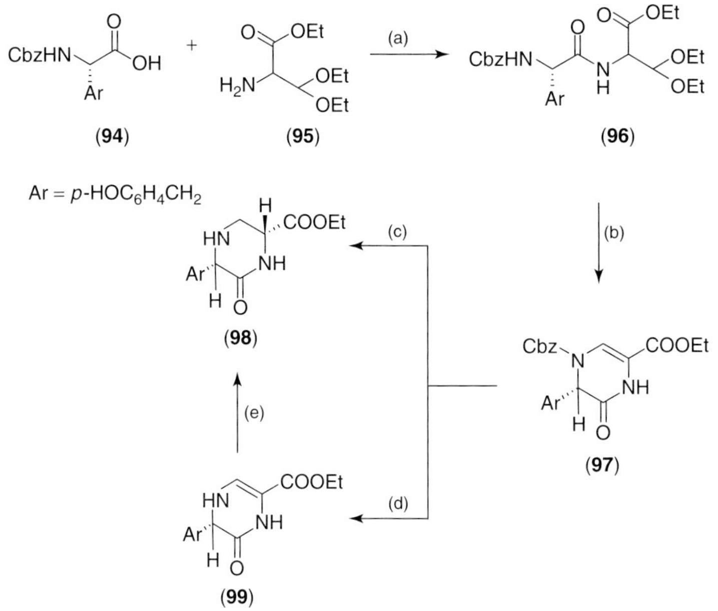

7.1 Six-Membered Rings with Another Heteroatom in the Same Ring

Ar = p-HOC$_6$H$_4$CH$_2$

Reagents and conditions: (a) DCC/HOBT/CH$_2$Cl$_2$ (100%),
(b) TFA/H$_2$O (70%),
(c) Pd(OH)$_2$/H$_2$/HCl then neutr. (70%),
(d) Pd(OH)$_2$/H$_2$/EtOH,
(e) NaCNBH$_3$/HCl/MeOH (70%).

Scheme 7.56

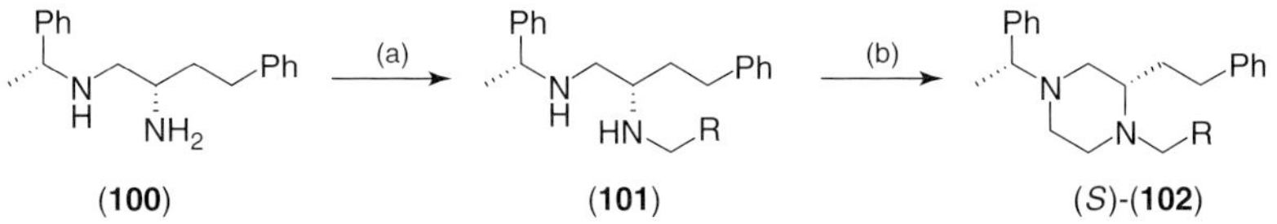

R = Ph, pentyl

Reagents and conditions: (a) 1. RCHO/MgSO$_4$/MeOH, 2. NaBH$_4$ (83–94%),
(b) (CHO)$_2$/NaCNBH$_3$/MeOH (85–87%).

Scheme 7.57

removal of the Cbz group, by catalytic hydrogenation and reduction of the resulting intermediates with NaBH$_3$CN in the presence of ZnCl$_2$ [79].

The enantioselective synthesis of the antipode (**111**) of the pyrazinone alkaloid hamacanthin A was based on Sharpless asymmetric dihydroxylation of vinylindole **106** with asymmetric dihydroxylation (AD)-mix-α to give (S)-indolyl-1,2-ethanediol **107**, which was then transformed to 3-indolylazidoethylamine **108** in seven steps. Compound **108** was coupled with 6-bromo-3-indolyl-α-oxoacetyl chloride to azide

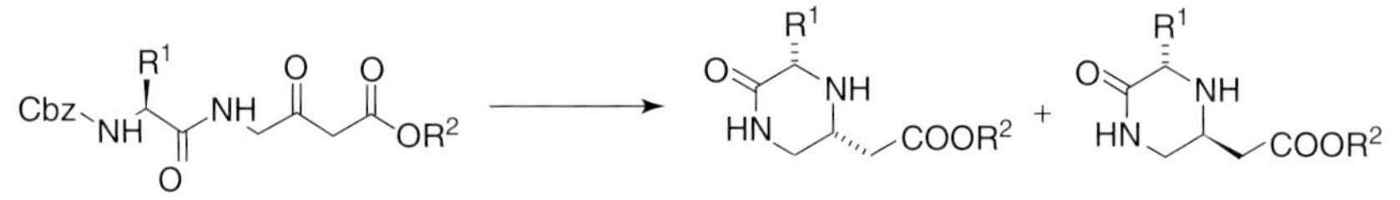

324 | *7 Asymmetric Synthesis of Ring Nitrogen Heterocycles*

(103) (104) (105)

R^1 = Me, Bn
R^2 = Me, Et

Reagents and conditions: Method A: H_2/Pd-C, 45 psi, 45 °C, 48–72 h (89–97%);
Method B: 1. H_2/Pd–C/25 °C/2 h, 2. $NaBH_3CN/ZnCl_2$/25 °C/3 h (64–80%)
(dr 1.7:1 to 5.7:1).

Scheme 7.58

109, followed by an intramolecular Staudinger-aza-Wittig cyclization through intermediate **110** [80]. For the synthesis of the related alkaloids hamacanthin A [81], hamacanthin B [81, 82], antipode of *cis*-dihydrohamachantin B [81, 83], and the antipode of *cis*- and *trans*-dihydrohamachantin A [84], a similar approach was applied.

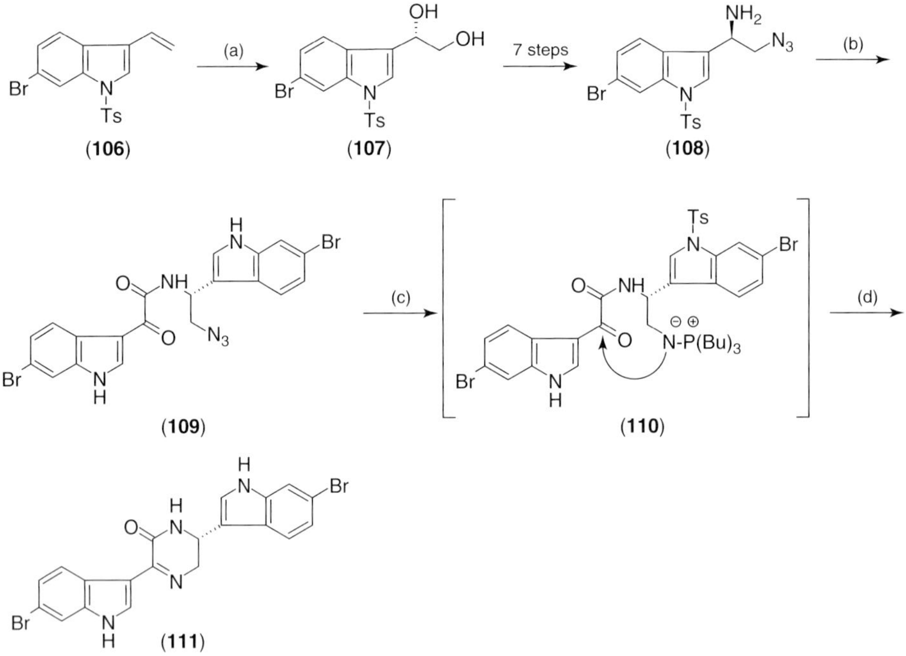

(106) (107) (108)

(109) (110)

(111)

Reagents and conditions: (a) AD-mix-α(92%),
(b) 6-bromo-3-indolyl-α-oxoacetyl chloride/Et_3N/DMF (74%),
(c) 1.nBu_3P/toluene/rfx, 2. reflux (97%), (d) NaOH/MeOH/rfx (88%).

Scheme 7.59

Piperazine-2-carboxylic acid is a conformationally restricted nonproteinogenic amino acid with medicinal applications. It is widely used as a building block in a number of bioactive compounds [85–87].

Synthesis of orthogonally protected (S)-piperazine-2-carboxylic acid **118** has been achieved in four steps from N-Boc-L-serine β-lactone **113**, obtained from protected L-serine **112** via a Mitsunobu reaction, by utilizing an extension of the Vederas' serine lactone ring-opening methodology. Thus, treatment of lactone **113** with allylamine gave a mixture of amino acid **114** and amide **115**. Cbz protection of **114** to **116** and subsequent ozonolysis gave piperazine **117** via the corresponding formyl intermediate. Compound **117** was then reduced chemoselectively to (S)-piperazinecarboxylic acid derivative **118** [88].

Cycloaddition The key step for the synthesis of the piperazinone ring system of pseudotheonamide A$_1$ (**123**) and A$_2$ (**124**) was intramolecular [3 + 2] cycloaddition of the appropriately oriented azido group and the α,β-unsaturated ester moiety in compound **121**, which was obtained from (R)-α-azido acid **119** and p-aminophenylalaninol **120** in three steps. The target molecules were then obtained as a 3:7 mixture of the corresponding diastereomers by reduction of **122** [89].

Similarly, enantiomerically pure piperazines were prepared via intermolecular [3 + 2] cycloaddition of an azide to a C–C double bond in N-allyl-1,2-azidoamines, which were prepared from 1,2-amino alcohols. The corresponding cycloadduct triazolopyrazine derivative was then reacted with acyl chlorides, carbamoyl chlorides, or alkyl halides to obtain halomethylpiperazines [90, 91].

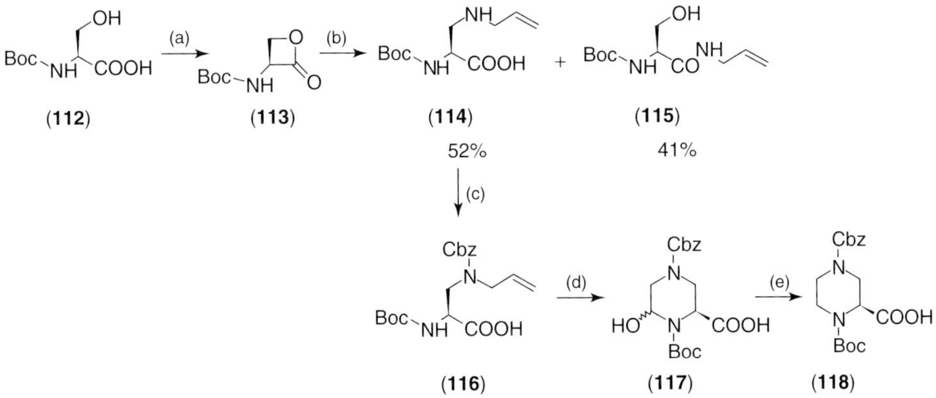

Reagents and conditions: (a) DEAD/Ph$_3$P/THF,
(b) CH$_2$=CHCH$_2$NH$_2$/MeCN,
(c) CbzCl/NaHCO$_3$/H$_2$O/acetone (97%),
(d) 1. O$_3$, 2. Me$_2$S, (e) Et$_3$SiH/BF$_3$(OEt)$_2$ (80%).

Scheme 7.60

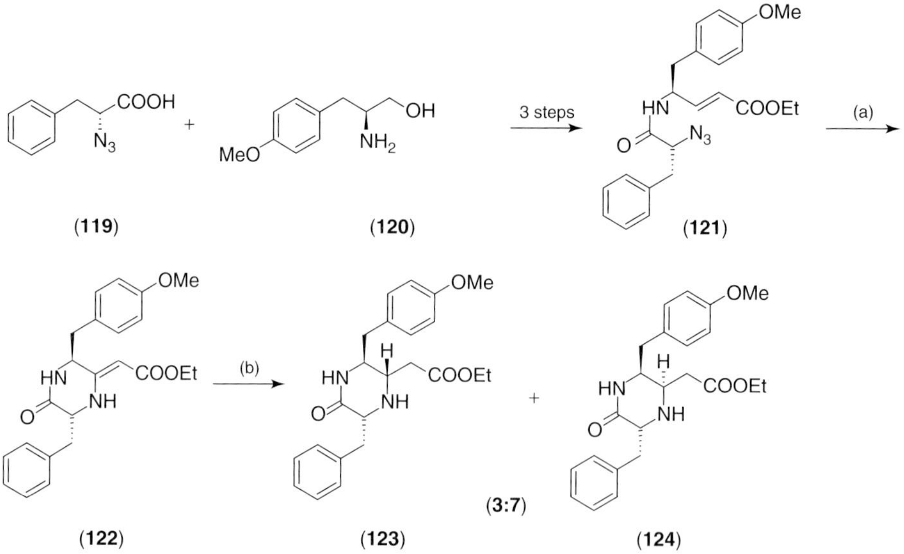

Reagents and conditions: (a) Et$_3$N (catalytic)/toluene/rfx (56%)
(b) NaCNBH$_3$/MeOH/5% HCl (90%).

Scheme 7.61

Miscellaneous Methods for Ring Closure The asymmetric synthesis of 2-vinyl-piperazines **127** has been described from olefins **125** and 1,2-bis[benzylamino]ethane **126** in the presence of a chiral Pd(0) complex. On using Pd$_2$(dba)$_3$ as the Pd source and [bicyclo[2.2.1]heptane-2,3-diylbis(methylene)]bis(diphenylphosphine) (BHMP) as chiral phosphine ligand, moderate yields (28–68%) and enantiose-lectivities (22–42% ee) were achieved [92]. The cyclized product was obtained in 87% yield and 60% ee by the application of BINAP as ligand [93], while the combination of [PdCl(η^3-C$_3$H$_5$)]$_2$ and phosphinooxazinane **162** gave vinylpiperazine **161** in 50% yield and 70% ee [94].

Kukula's method for synthesis of piperazine-(2S)-carboxylic acid **135** consists of three simple steps: (i) coupling of the precursor **129** with (S)-proline methyl ester (**130**) as a chiral auxiliary, (ii) heterogeneous diastereoselective hydrogenation of amide **131** via intermediate **132** to give diastereomers **133** and **134** in a ratio of 4:1, and (iii) acidic hydrolysis of the tricyclic amide **133**. The diastereoselectivity obtained in the hydrogenation of the substrate over a Pd catalyst was considerably higher than over Rh, Ru or Pt catalysts. Separation of the two diastereomers from the hydrogenation product by crystallization is possible [95].

With (R)-phenylglycinol (**137**) as the chiral inductor, 2,3-dialkyl-substituted piperazines were synthesized via hydroxyethylenediamine precursor **139**, which was obtained by reduction of oxazino-oxazine **138**. Compound **139** was condensed with α-dicarbonyl compounds (glyoxal or 1-phenyl-1,2-propanedienone), which afforded bisoxazolidine **140** or **143** as single diastereomers, while the reaction with

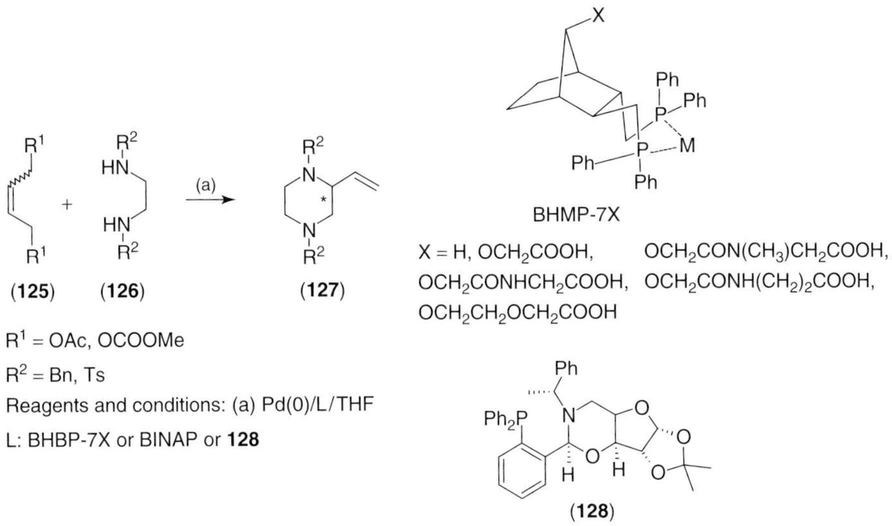

7.1 Six-Membered Rings with Another Heteroatom in the Same Ring | 327

Scheme 7.62

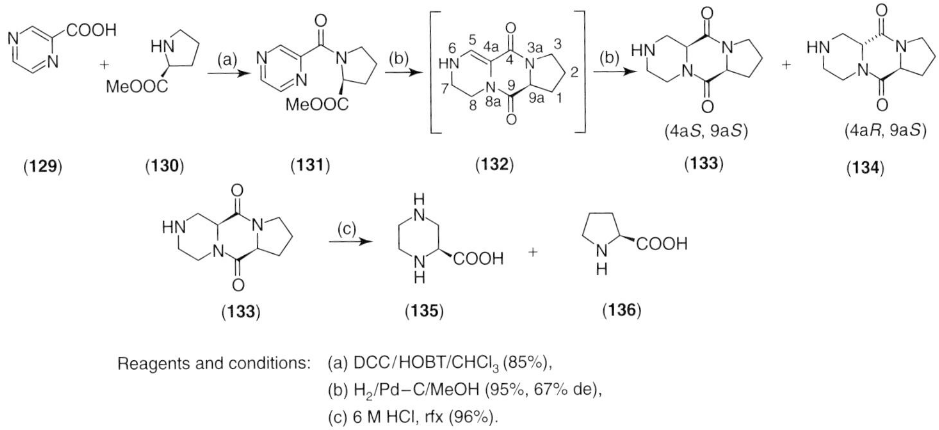

R¹ = OAc, OCOOMe
R² = Bn, Ts
Reagents and conditions: (a) Pd(0)/L/THF
L: BHBP-7X or BINAP or **128**

X = H, OCH$_2$COOH, OCH$_2$CON(CH$_3$)CH$_2$COOH,
OCH$_2$CONHCH$_2$COOH, OCH$_2$CONH(CH$_2$)$_2$COOH,
OCH$_2$CH$_2$OCH$_2$COOH

Scheme 7.63

Reagents and conditions: (a) DCC/HOBT/CHCl$_3$ (85%),
(b) H$_2$/Pd–C/MeOH (95%, 67% de),
(c) 6 M HCl, rfx (96%).

2,3-butanedienone gave a 1:1 diastereomeric mixture of **141** and **142**. The bisoxazolidines **140–143** obtained were reduced to give piperazine derivatives **144–147**, respectively [96].

7.1.3.2 Stereoselective Transformation of the Piperazine Ring

Alkylations Metalated bislactim ethers (known as Schöllkopf chiral auxiliaries) derived from DKPs are widely used as homochiral nucleophilic templates (homochiral glycine anion equivalents) for the synthesis of nonproteogenic α-amino acids. In this approach, enantiomerically pure bislactim ethers such as (S)-**22** can be deprotonated with nBuLi and the aza-enolate anions **148** formed are alkylated with an alkyl halogenide as electrophile. The alkyl group R³ enters trans to the R¹

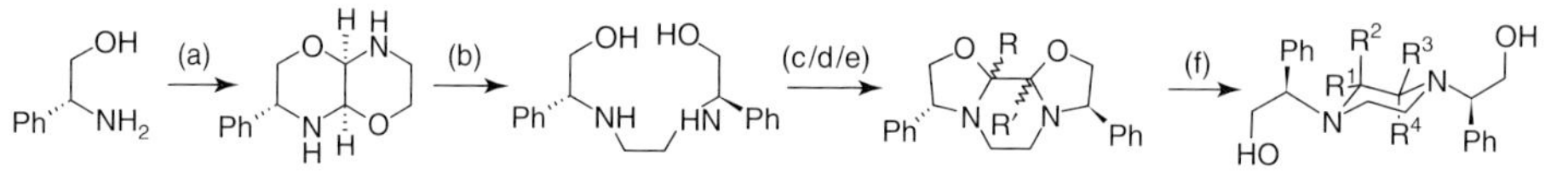

Scheme 7.64

Reagents and conditions: (a) glyoxal/H$_2$O/EtOH (70%),
(b) BF$_3$·THF (97%),
(c) glyoxal/H$_2$O/EtOH (68%),
(d) 2,3-butanedione/toluene (84%, 1:1 mixture),
(e) 1-phenyl-1,2-propanedienone/benzene (70%),
(f) BF$_3$·THF (70–99%).

(140): R = R' = H
(141+142): R = R' = Me
(143): R = Ph, R' = Me

(144): R^1 = R^2 = R^3 = R^4 = H
(145): R^1 = R^3 = H, R^2 = R^4 = Me
(146): R^1 = R^3 = Me, R^2 = R^4 = H
(147): R^1 = R^4 = H, R^2 = Ph, R^3 = Me

substituent of the piperazine ring, affording *trans*-alkylated bislactim ether **149** in high diastereomeric excess, which can be hydrolyzed to the mixture of L-valine or *tert*-L-leucine ester **150** and the target enantiomerically pure α-amino acid **151** [51, 97–113].

N,N-Bis(*p*-methoxybenzyl)piperazinedione derivative **152** was developed and used as chiral auxiliary for synthesis of enantiomerically pure amino acids. The methoxybenzyl derivative has the advantage over the dilactim ether Schöllkopf chiral auxiliaries in that it is crystalline and therefore easier to handle, and the *p*-methoxybenzyl (PMB) group can be removed under mild conditions. Thus, alkylation of the piperazine core with alkyl halides via the lithium or potassium enolate **152** afforded **153** with high diastereoselectivities, which could be deprotected to **154** [114–117].

Besides Schöllkopf chiral auxiliaries **22** and PMB derivative **152**, other piperazine precursors, such as monolactim ether **63** or piperazinones **155**, with a chiral N-substituent, can also be stereoselectively alkylated in essentially the same way [118–126].

It is not necessary to use a strong base for alkylation of the piperazine ring in all cases. Mild phase transfer catalysis (PTC) conditions in the presence of a weak base have been successfully employed for the diastereoselective alkylation

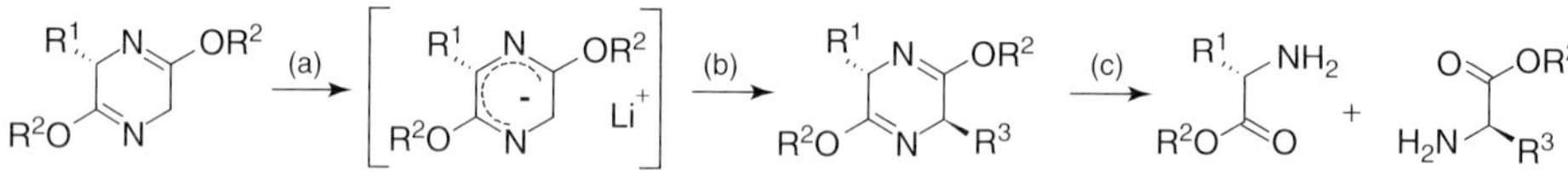

R^1 = *i*Pr, *t*Bu
R^2 = Me, Et
R^{3}X = alkyl halogenide

Reagents and conditions: (a) BuLi/THF/−78 °C,
(b) R^{3}X,
(c) 0.1 M HCl(aq)/MeCN.

Scheme 7.65

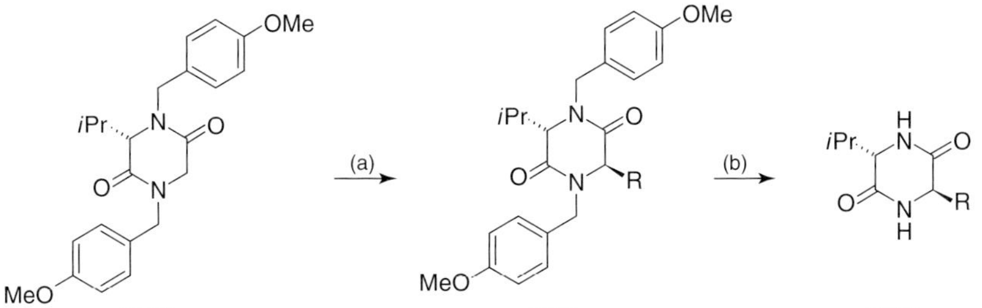

Scheme 7.66

Reagents and conditions: (a) LiHMDS or KHMDS/THF/R-X,
(b) Ce(NH$_4$)$_2$(NO$_3$)$_6$/MeCN/H$_2$O.

Scheme 7.67

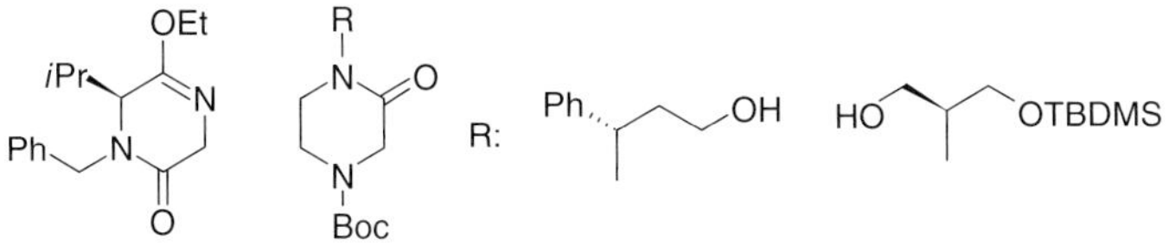

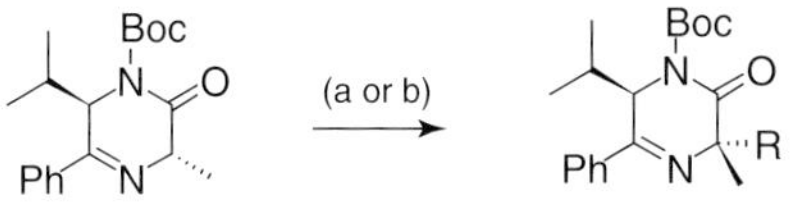

Scheme 7.68

R = allyl, propargyl, Bn, EtCOOCH$_2$

Reagents and conditions: (a) R-X/K$_2$CO$_3$/Bu$_4$NBr/MeCN or CH$_2$Cl$_2$/rt
(75–86%, >94% de)
(b) Allyl carbonates/Pd(OAc)$_2$/PPh$_3$/THF
(75–85%, >96% de).

of piperazinones **93** with activated halides, affording alkylated products **156**. Pd-catalyzed allylation of the same substrate was also performed by reacting **93** with allyl carbonates in the presence of Pd(OAc)$_2$ and PPh$_3$ [74, 75].

Reaction of the Piperazine Ring with Electrophiles Different from Alkyl Halides Not only alkyl halides but also other electrophiles, such as aldehydes, ketones, epoxides [127–129], and acyl halogenides [130], can react with enolate anions **148** to afford the corresponding alcohol.

Further possibilities include Michael conjugate addition of **148** to α,β-unsaturated esters [131–134], oxo compounds [135], or nitro compounds [136, 137] to result in the corresponding substituted piperazines. Even stereoselective arylation of the piperazine ring is possible by the reaction of lithium enolates **182** with arene-Mn(CO)$_3$ cations [138–140].

When aldehydes are reacted with **148** as such, the diastereoselectivity with regard to the heterocyclic carbon is high, whereas formation of the second chiral center at the secondary carbon in **157** (R^4 = H) proceeds with low stereoselectivity [141–145]. However, replacement of the lithium ion in complex **148** with another metal complex (trisdimethylaminotitanium group is used in most cases) leads to a strong increase in stereoselectivity as regards the formation of the alcoholic moiety. The predominant formation of one diastereomer can be rationalized on the basis of the two six-membered pericyclic transition states **A** and **B**. Transition state **A** is lower in energy because of the quasi-diaxial repulsion between the R moiety, the methoxy group and the ligands on the metal in transition state **B** [146–149].

Reaction of the Piperazine Ring with Nucleophiles As revealed by the previous examples, the Schöllkopf chiral auxiliary itself serves as a nucleophilic synthon, and therefore it can be attacked by electrophiles. Its chloro derivative **158**, however, has been used as an electrophilic synthon, and the chloride in **158** could be stereoselectively replaced by malonate anions as nucleophiles, furnishing the product **159** [150].

Another example of diastereoselective nucleophilic reactions on the piperazine ring is the reaction of organomagnesium or organocadmium reagents with cyclic acyl imines **161** *in situ* generated from the corresponding acetates **160**, resulting in **162** as the major and **163** as the minor diastereomer [61].

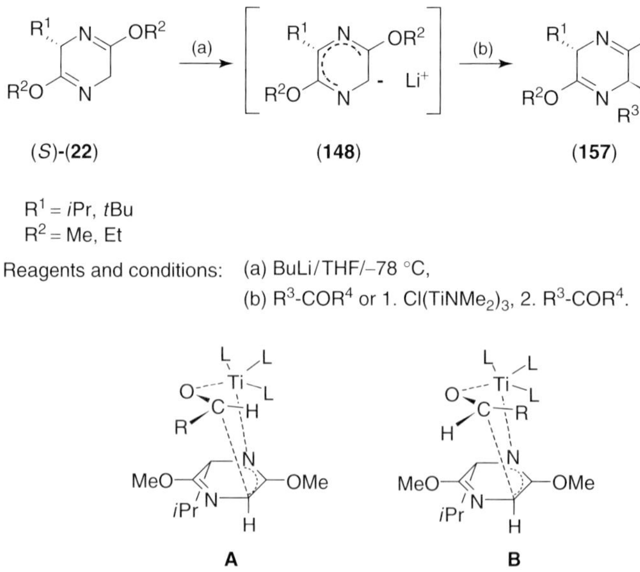

Scheme 7.69

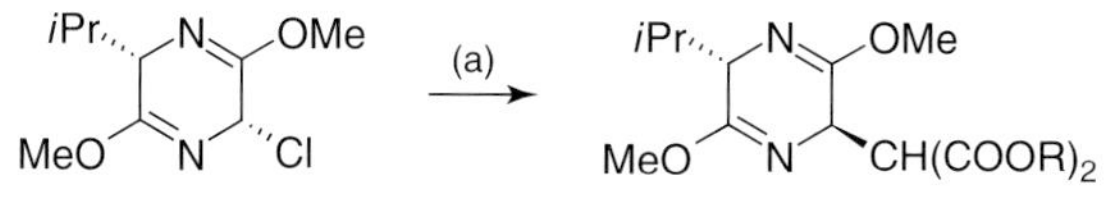

7.1 Six-Membered Rings with Another Heteroatom in the Same Ring | 331

(158) → **(159)**

R = Me, iPr, tBu

Reagents and conditions: (a) NaHC(COOR)$_2$/THF/18-crown-6 (65%, dr 25:1 to 60:1).

Scheme 7.70

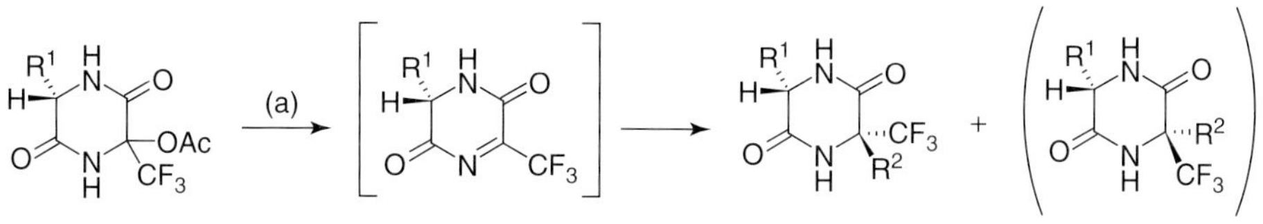

(160) → [**(161)**] → **(162)** + **(163)**

R^1 = Me, iPr, iBu, Bzl
R^2 = Me, Bz,

Reagents and conditions: (a) R^{2}MgX/THF or R$^2{}_2$Cd/THF (40–62%, de 54–99%).

Scheme 7.71

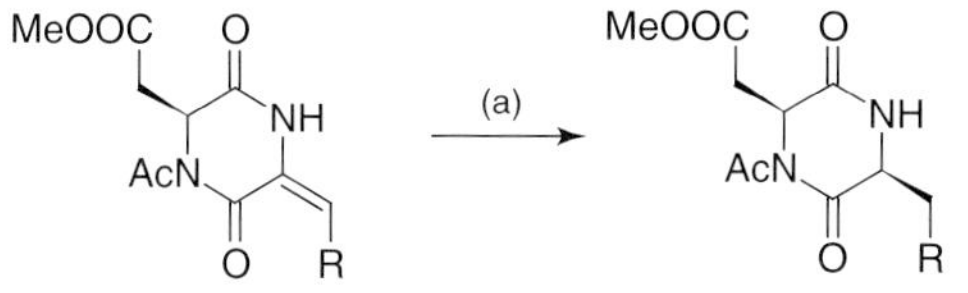

(164) → **(165)**

Reagents and conditions: (a) H$_2$/Pd–C (44–97%, >97% de).

Scheme 7.72

Asymmetric Hydrogenations Dehydropiperazines **164** could be stereoselectively hydrogenated by using Pd/C as the catalyst giving the *cis* saturated product **165** with high diastereoselectivity [151].

The highly diastereoselective conjugate reduction of benzylidene DKPs **166** by treatment with SmI$_2$ in THF and subsequent addition of deoxygenated water led to the *cis* 3,6-disubstituted derivative **167** on starting from either the *E* or the *Z* stereoisomer. Dideutero derivatives could also be synthesized using D$_2$O instead of H$_2$O [152, 153].

Tetrahydropyrazinecarboxamide **168** was reduced to the corresponding piperazine derivative **169** with moderate stereoselectivity (57% ee) by the application of rhodium catalysis using ferrocene-based phosphine ligand with planar chirality [154].

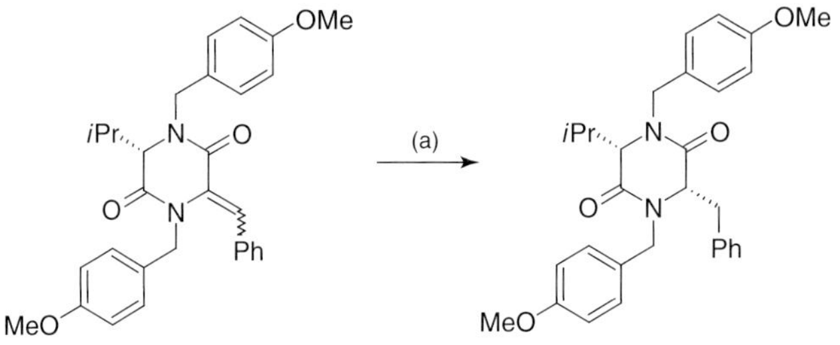

Scheme 7.73

Reagents and conditions: (a) SmI_2/THF/H_2O (89–93%, 95–96% de).

Scheme 7.74

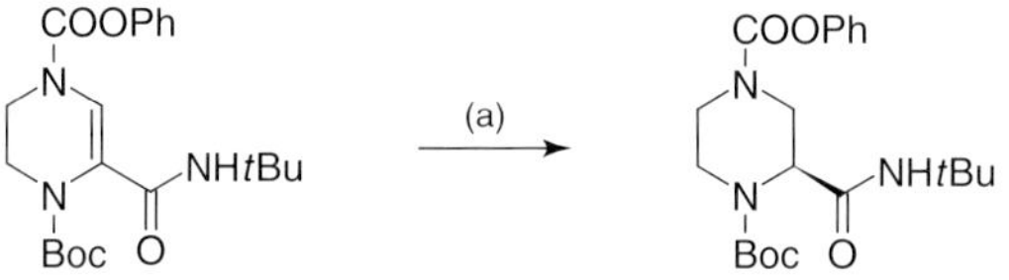

Reagents and conditions: (a) [Rh(nbd)$_2$]PF$_6$/(S,S)-EtTRAP-H/H_2/Cl(CH$_2$)$_2$Cl (92%, 57% ee).

nbd: norbornadiene
(S,S)-EtTRAP-H: (S,S)-2,2″-bis[(diethylphosphino)methyl]-1,1″-biferrocene.

7.1.4
Oxadiazines

Asymmetric syntheses of oxadiazines have been far less explored than those of other rings discussed in the previous sections. The methods will be classified according to the type of the oxadiazine ring: 1,2,5-oxadiazines and 1,3,4-oxadiazines.

7.1.4.1 1,2,5-Oxadiazines

1,2,5-Oxadiazines **4** were prepared from N-(benzotriazolylcarbonyl)-L-alanine or phenylalanine **1**, which was converted to the corresponding acyl chloride **2**. Its reaction with N-phenylhydroxylamine gave hydroxamic acid **3**, which readily underwent cyclization under basic conditions [155].

7.1.4.2 1,3,4-Oxadiazines

Synthesis of 1,3,4-oxadiazines **8** and **9** started from ethyl (S)-3-hydroxybutyrate (**5**). Diastereoselective amination with di-*tert*-butylazodicarboxylate gave an 84:16 mixture of hydrazino ester epimers **6** and **7**. Treatment of **6** with 2,2-dimethoxypropane afforded oxadiazine **8**, which underwent base-catalyzed equilibration to give the more stable *trans* product **9** (*trans/cis* ratio = 98:2). The latter compound

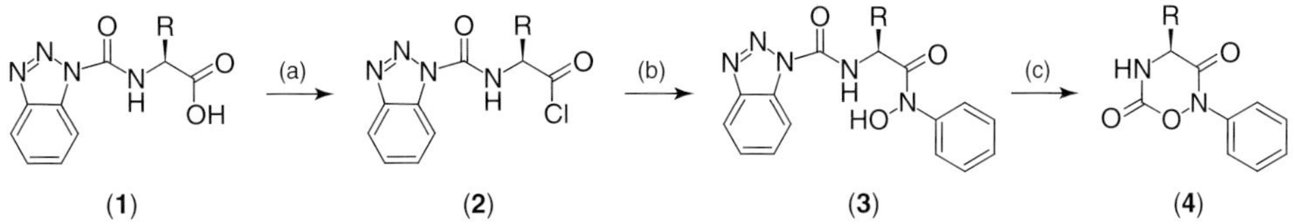

7.1 Six-Membered Rings with Another Heteroatom in the Same Ring | 333

(1) → (a) → **(2)** → (b) → **(3)** → (c) → **(4)**

R = Me, Bn

Reagents and conditions: (a) SOCl$_2$, (b) PhNHOH/N-methylmorpholine/toluene (15–31%), (c) Na$_2$CO$_3$/H$_2$O/acetone (28%).

Scheme 7.75

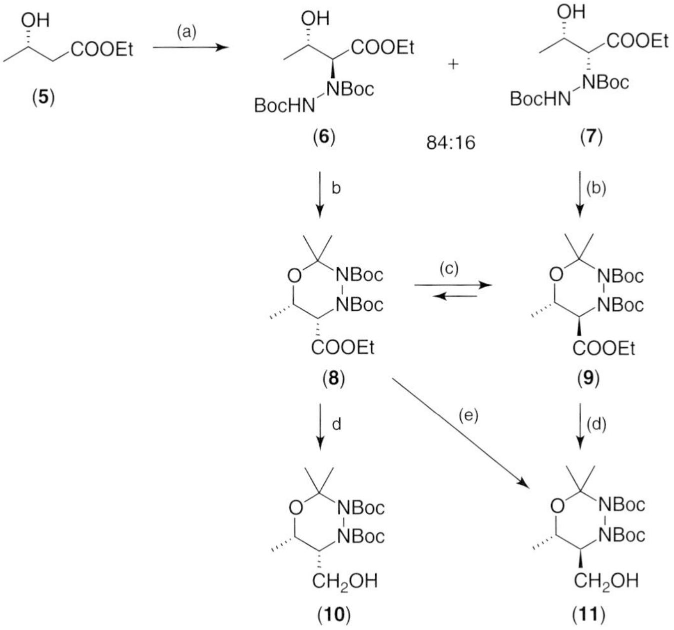

(5) → (a) → **(6)** + **(7)** (84:16)

(6) → b → **(8)**
(7) → (b) → **(9)**

(8) ⇌ (c) ⇌ **(9)**

(8) → d → **(10)**
(9) → (d) → **(11)**
(8) → (e) → **(11)**

Reagents and conditions: (a) LDA/CbzNNCBz, (b) (MeO)$_2$CMe$_2$/pTsOH/benzene/rfx (75% for **8**, 61% for (**9**), (c) NaH/EtOH/THF/rt/1d (78%), (d) Ca(BH$_4$)$_2$ (from 3.5 equiv CaCl$_2$ and 6 equiv NaBH$_4$), EtOH/THF (91–92%), (e) Ca(BH$_4$)$_2$ (from 3 equiv CaCl$_2$ and 6.5 equiv NaBH$_4$) NaOEt/EtOH/THF (86%).

Scheme 7.76

could also be obtained from the *syn* hydrazino derivative **7**. The esters **8** and **9** were reduced to the corresponding hydroxymethyl derivative **10** and **11** by Ca(BH$_4$)$_2$ *in situ* formed from CaCl$_2$ and NaBH$_4$. However, the *cis* isomer was sensitive to the conditions of the reduction: when CaCl$_2$ was used in a slight excess compared to NaBH$_4$, no epimerization took place; on the contrary, employing NaBH$_4$ in excess resulted in nearly complete conversion to the more stable *trans* alcohol **11** [156, 157].

7 Asymmetric Synthesis of Ring Nitrogen Heterocycles

Another method was applied by Hitchcock et al. for the synthesis of the 1,3,4-oxadiazine ring. They synthesized 3,4,5,6-tetrahydro-2H-1,3,4-oxadiazin-2-ones **16** starting from enantiomerically pure natural amino alcohols such as ephedrine, norephedrine or their stereoisomers. Introduction of alkyl moiety to the primary amino group in **12** could be carried out by condensation with a suitable oxocompound under reductive conditions. Subsequent treatment of **13** with NaNO$_2$/HCl gave N-nitrosamines **14**, which were reduced to β-hydrazinoalcohols **15**. Cyclization of **15** could be performed either by using 1,1′-carbonyldiimidazole or diethyl carbonate to afford oxadiazinones **16** (Scheme 7.77) [158–162].

Oxadiazines **17** and **18** could also be prepared principally by the same way starting from the corresponding amino alcohol [162–164].

3,4,5,6-Tetrahydro-2H-1,3,4-oxadiazin-2-ones **36** were used as chiral auxiliaries. Asymmetric aldol reactions of acylated oxadiazinone **19** with various aldehydes via titanium enolate afforded adducts **20**. When aldehydes without an α-hydrogen were employed, the products were obtained in high yields and diastereoselectivities; in contrast to this finding, by employment of aldehydes bearing an α-hydrogen, moderate diastereoselectivities were achieved.

The observed stereoselectivity could be explained by the conformation of the oxadiazinone ring, in which the N-4 substituent blocks the *si* face of the N-3 acyl substituent (see scheme) [165–169].

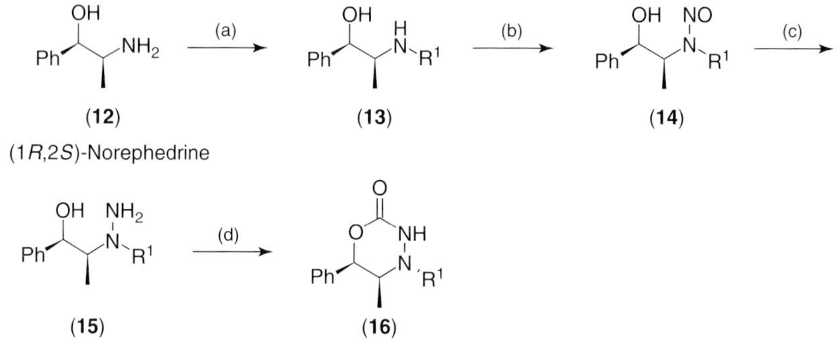

R^1 = iPr, nBn, Me
R^2 = Me, Bn, bornyl

Reagents and conditions: (a) Me$_2$CO or PhCHO/NaBH$_4$, (b) NaNO$_2$/HCl/THF (95–97%), (c) LiAlH$_4$/THF, (d) (EtO)$_2$CO/LiH/hexane or CDI/pTsOH (74%), (e) LiH/RCOCl/CH$_2$Cl$_2$ (43–95%).

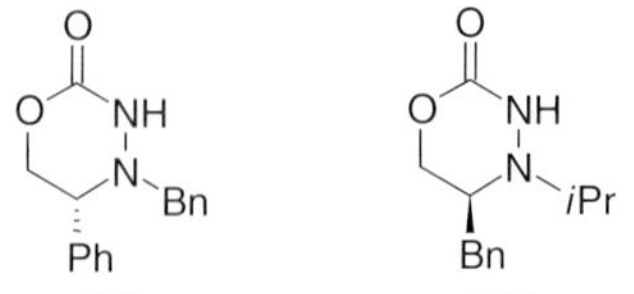

Scheme 7.77

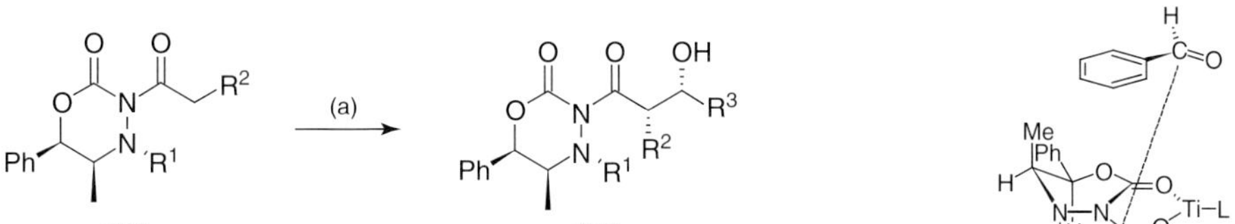

7.1 Six-Membered Rings with Another Heteroatom in the Same Ring

R^1 = Me, iPr, bornyl
R^2 = Me, PhSCH$_2$, MeO, BnO, PhO
R^3 = aryl, tBu, alkyl

Reagents and conditions: (a) 1. R^{3}CHO/THF, 2. Et$_3$N, 3. TiCl$_4$ (62–97%, dr 75:25 to 99:1 for aldehyde without α-hydrogen, 8:1 to 38:1 with α-hydrogen).

Scheme 7.78

7.1.5
Morpholines

Morpholine derivatives with defined configuration of the substituents can be obtained either by cyclization of enantiomerically pure starting material or by asymmetric transformation of the morpholine ring.

A review appeared in *Synthesis* in 2004 summarizing syntheses of C-substituted morpholine derivatives. It classifies the methods according to the type of starting material or reaction type; thus syntheses from amino alcohols, epoxides, olefins, and carboxylic acid derivatives and via organometallic reactions are described [170].

In the first part of this section, we follow a classification of the ring closures according to the method of cyclization: lactone or lactam formation, imine bond formation, nucleophilic substitution, conjugate addition, and by the involvement of organometallics. In the second part, asymmetric alkylations, arylations, radical additions, and hydrogenations on the morpholine ring are discussed.

7.1.5.1 Formation of the Morpholine Ring
Ring Closure by Lactone or Lactam Formation A frequent route to the synthesis of enantiomerically pure morpholinone derivatives is the cyclization of a suitably substituted enantiomerically pure N-(2-hydroxyethyl)-α-amino acid or ester by lactone formation. Thus, (1R,2S)-2-amino-1,2-diphenyl ethanol (**1**) was N-alkylated with ethyl bromoacetate in the presence of a weak base to afford **2**; subsequent N-protection and acidic cyclization gave 4-protected-5,6-diphenylmorpholin-2-ones **4** via intermediates **3** [171].

In the same pathway, differently substituted morpholinone derivatives were also synthesized [172–176].

Similar to the previous example, the hydroxy ester suitable for lactonization was prepared by N-alkylation of an amino alcohol; diastereoselective coupling between racemic α-halo acids **5** and N-benzylethanolamine (**6**) was performed in the presence of Bu$_4$NI, and **7** obtained with high diastereoselectivity was then cyclized to oxazinones **8** upon treatment with pTsOH [177].

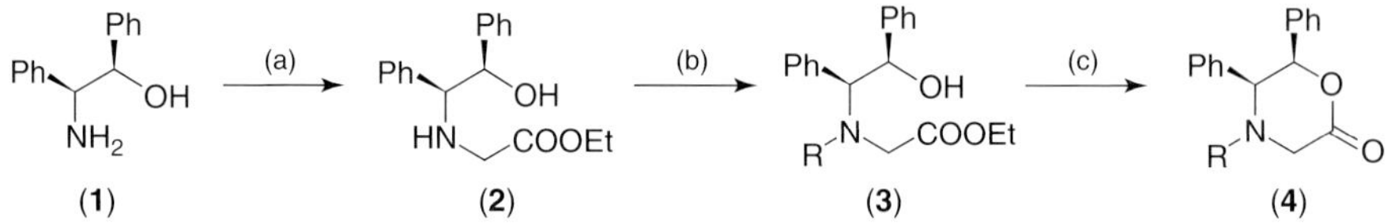

Scheme 7.79

R = Boc, Cbz
Reagents and conditions: (a) BrCH$_2$COOEt/Et$_3$N/THF, (b) Boc$_2$O/toluene or CbzCl/CH$_2$Cl$_2$/NaHCO$_3$ (aq), (c) pTsOH/toluene (75–86% overall).

Scheme 7.80

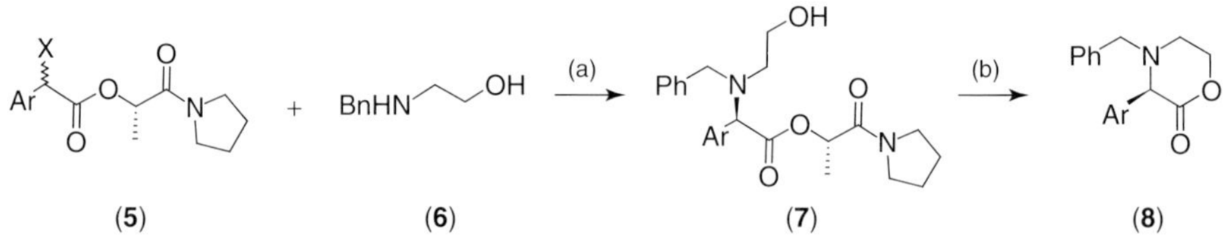

Reagents and conditions: (a) Et$_3$N/THF/Bu$_4$NI (84–90%, 76–96% de),
(b) toluene/pTsOH (82%, 86–91% ee).

Diphenylmorpholine (2R,3S)-**4a** was prepared in enantiomerically pure form in six steps starting from benzaldehyde. Enzymatic stereoselective HCN addition to the aldehyde was followed by O-protection to **9**, which was then converted to **11** via a one-pot Grignard addition–transamidation–reduction sequence. The reduction proceeded with complete diastereoselectivity to **11**, which upon O-deprotection gave **12**. Subsequent N-protection and acidic cyclization resulted in morpholine (2R,3S)-**4a** [178].

Asymmetric Strecker reaction of various aldehydes and (R)-2-phenylglycinol (**13**) gave nitriles **14**, which were esterified with methanolic HCl. The esters were cyclized

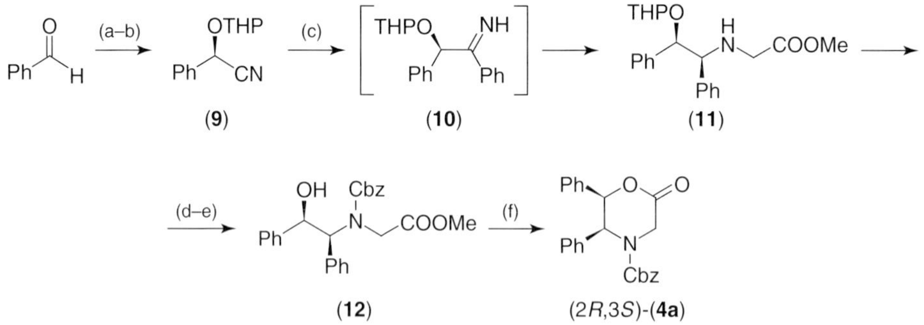

Reagents and conditions: (a) HCN/oxynitrilase (99% ee), (b) pTsOH/Et$_2$O/dihydropyran,
(c) 1. PhMgBr/Et$_2$O, 2. MeOH, 3. NH$_2$CH$_2$COOMe, 4. NaBH$_4$,
(d) pTsOH/MeOH, (e) CbzCl, (f) pTsOH/cyclohexane (48% overall).

Scheme 7.81

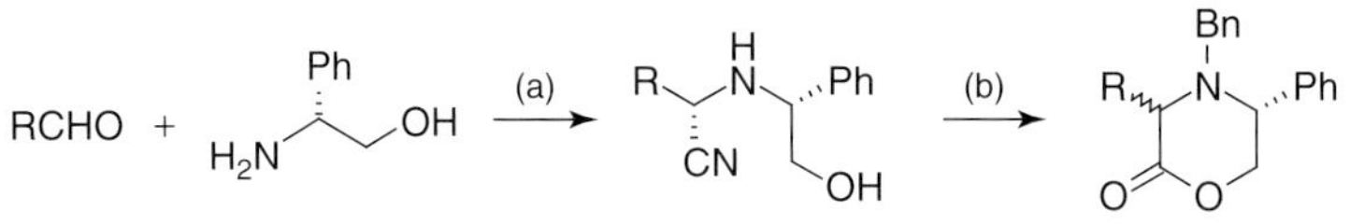

7.1 Six-Membered Rings with Another Heteroatom in the Same Ring

Scheme 7.82

R = Me, Bn, nPr, BnO(CH$_2$)$_2$

Reagents and conditions: (a) NaCN/NH$_4$Cl/MeOH/H$_2$O, (b) 1. HCl/MeOH, 2. toluene/pTsOH, 3. BnBr/K$_2$CO$_3$/DMF (overall: 33–48%, 66–80% de).

in the presence of pTsOH, and the intermediates obtained were N-benzylated to compounds **15**. Good diastereoselectivities were achieved [179].

The 3R,5S-disubstituted morpholinone **19** was elegantly obtained in the following pathway. A highly stereoselective (dr: 95:5) conjugate addition of amino alcohol (S)-**17** to unsaturated oxocarboxylic acid **16** in combination with crystallization-induced dynamic resolution (CIDR) provided the hydroxyethylamino carboxylic acid derivative (R,S)-**18** in high diastereoselectivity, which, by lactonization, led to (3R,5S)-**19**. In the CIDR process, the solvent plays an important role: the solubility of the arising amino acid should be as low as possible, but high enough to permit the equilibration processes [180].

Lactone formation for synthesis of compound (S)-**22** could be carried out either by cyclization between a hydroxy function and a carboxylic amide moiety in (S)-**21** derived from amino acid amide (S)-**20** or by reacting amino acid (S)-**23** with 1,2-dibromoethane [181].

Reacting ephedrine **24** with α-oxocarboxylic acid chlorides results in lactol formation, giving compounds **25** after N-acylation [182].

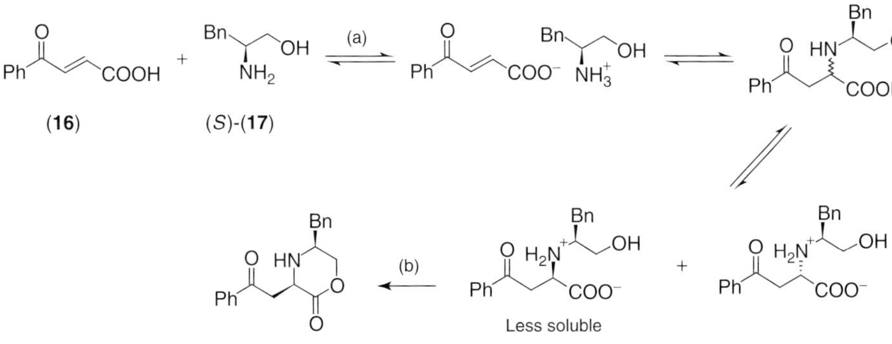

Reagents and conditions: (a) EtOH or CH$_2$Cl$_2$/25–30 °C/7d (for (R,S)-18: 81%, dr 95:5), (b) THF/H$_2$SO$_4$/rt (60%).

Scheme 7.83

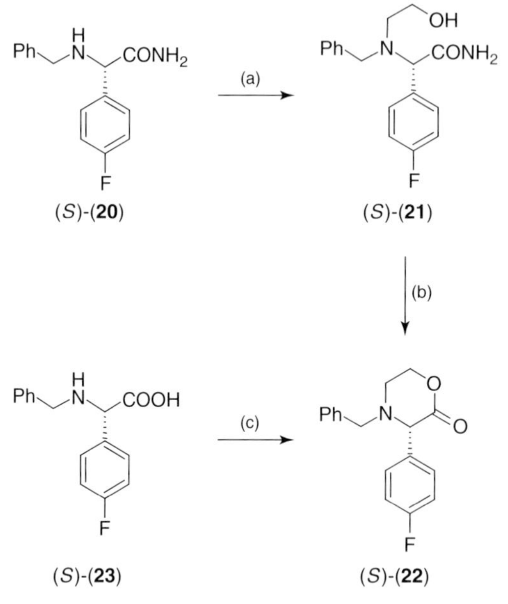

Scheme 7.84

Reagents and conditions: (a) BrCH$_2$CH$_2$OH/iPr$_2$NEt/DMF (75%),
(b) AcOH/60 °C (95%), (c) Br(CH)$_2$Br/iPr$_2$NEt.

Scheme 7.85

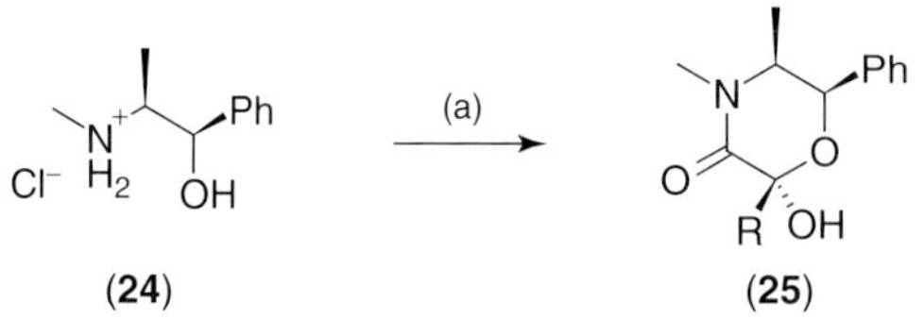

R = Me, Et, nPr, iPr
Reagents and conditions: (a) RCOCOCl/Et$_3$N/DMAP/CH$_2$Cl$_2$ (65–71%).

Lactone and lactam formation took place by reacting ephedrine **24** with oxalyl chloride, affording morpholine-2,3-dione (5S,6R)-**26** [183–185].

Reaction of amino alcohols with an α-halo ester in the presence of weak base results in N-alkylation (see above), while treatment of amino alcohols **27** with a strong base as NaH followed by addition of ethyl chloroacetate gave O-alkylated derivative of **27**, which spontaneously cyclized to form lactam **28** [186].

Morpholine-2,5-diones can be considered as lactones and lactams in the same ring. Two different approaches could be applied for their synthesis: (i) first ester formation of the hydroxy moiety of an α-hydroxy acid with an amino acid, followed by intramolecular amide formation as the ring closure step give the product or (ii) the amide formed in the first step is cyclized via lactonization.

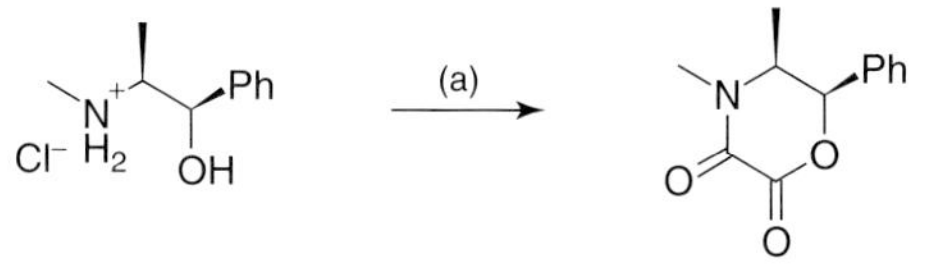

7.1 Six-Membered Rings with Another Heteroatom in the Same Ring

(24) → **(5S,6R)-(26)**

Reagents and conditions: (a) (COCl)$_2$/Et$_3$N/DMAP/CH$_2$Cl$_2$ (65–71%).

Scheme 7.86

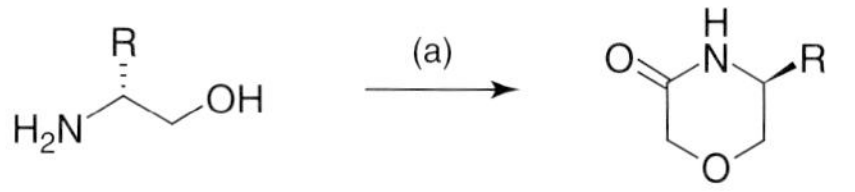

(27) → **(28)**

R = Me, Bn, iBu, tBu, sBu
Reagents and conditions: (a) NaH/ClCH$_2$COOEt/THF (69–88%).

Scheme 7.87

According to the first approach, synthesis of (S)-**32** started from hydroxy acid (S)-**29**, which was coupled with 1,1′-carbonyldiimidazole (CDI)-activated Cbz-glycine yielding amino acid (S)-**30** after deprotection. Cyclization of (S)-**30** took place upon treatment with Mukaiyama reagent. The lactam (S)-**31** could be transformed to lactim (S)-**32** using Me$_3$OBF$_4$ [187].

Employing the second approach, depending on the reaction conditions, two different stereoisomeric products were formed starting from the acid **33**. By treatment with diethyl azodicarboxylate and triphenyl phosphine, intramolecular Mitsunobu reaction with inversion of the configuration led to (6R)-**34**, while heating **33** under reduced pressure resulted in (6S)-**34** via spontaneous cyclization with retention [170–188].

The N-benzyl-protected amino alcohol **6** was cyclized with glyoxylic acid (**35**), then the product **36** activated with Tf$_2$O and reacted with the alcohol **37** followed by crystallization driven epimerization, affording (2R,1′R)-**38** in 99% diastereomeric excess [170, 189].

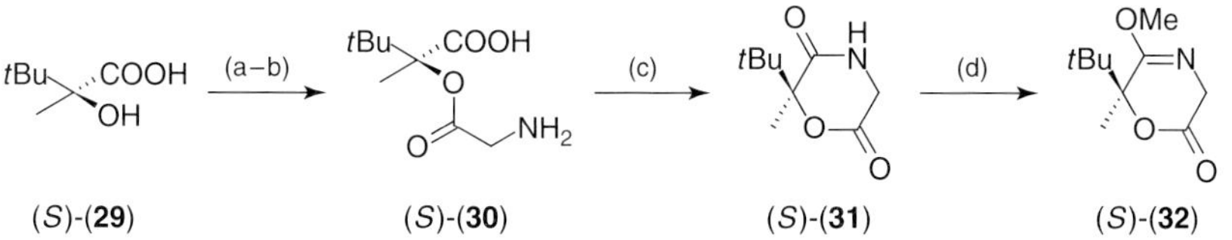

(S)-(29) → **(S)-(30)** → **(S)-(31)** → **(S)-(32)**

Reagents and conditions: (a) Cbz-glycine/CDI/THF (86%), (b) H$_2$/Pd–C/EtOH (96%), (c) 2-chloro-1-methylpyridinium iodide/iPr$_2$NEt/CH$_2$Cl$_2$ (84%), (d) Me$_3$OBF$_4$/CH$_2$Cl$_2$ (89%).

Scheme 7.88

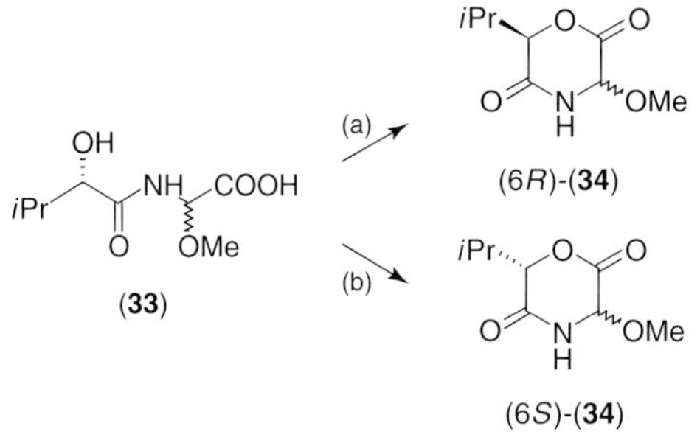

340 | 7 Asymmetric Synthesis of Ring Nitrogen Heterocycles

Scheme 7.89

Reagents and conditions: (a) DEAD/PPh$_3$ (30%), (b) 65 °C/0.2 bar (95%).

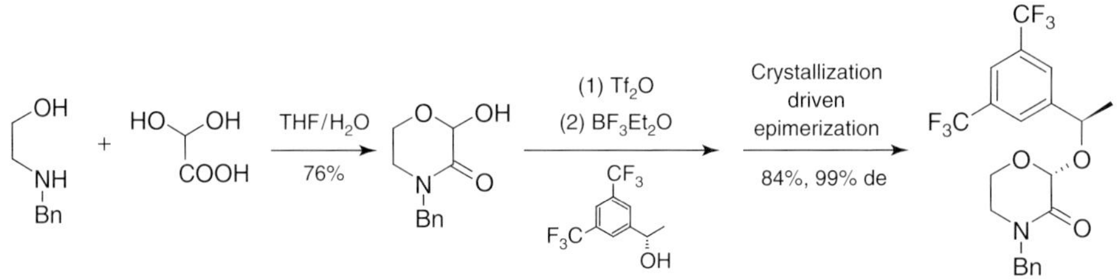

Scheme 7.90

Ring Closure via Imine/Enamine Bond Formation Intramolecular condensation of an oxo group with an amino moiety is also a suitable way of morpholine ring formation. Thus, for the synthesis of **43**, ketones **39** halogenated at α-position and potassium salts **40** of N-protected α-amino acids were condensed to give esters **41**, which were then deprotected to **42**. Cyclization to morpholinones **43** could be performed in acetate buffer to avoid pH decrease during the reaction, which would cause hydrolysis and slow down the cyclization [190].

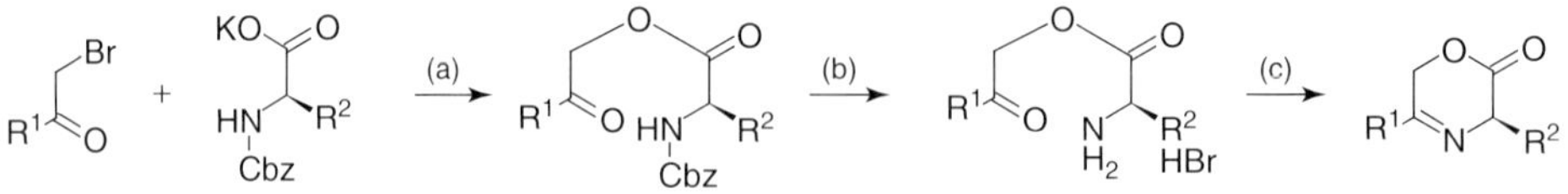

Scheme 7.91

R^1 = aryl, R^2 = Me, Bz, iPr

Reagents and conditions: (a) DMF/rt (74–100%), (b) HBr/AcOH/rt (80–99%), (c) 0.2 M acetate buffer/rt (65–90%).

7.1 Six-Membered Rings with Another Heteroatom in the Same Ring | 341

Via the corresponding ester intermediate analogous to **42** possessing oxo and amino group in proper position, synthesis of (6S)-6-isopropyl-5-phenyl-3,6-dihydro-2H-1,4-oxazin-2-one [191, 192], as well as (6R)-6-isopropyl-3-methyl-5-phenyl-3,6-dihydro-2H-1,4-oxazin-2-one diastereomers [193] were reported in similar way as above.

Principally the same method was applied for the synthesis of (3S,5R)-3,5-diphenylmorpholin-2-one ((3S,5R)-**46**) from (S)-**44** via (S)-**45**: esterification of an N-protected amino acid by reacting its potassium salt with an α-halogenated ketone, followed by N-deprotection of the formed ester and subsequent cyclization by imine formation. Finally, stereoselective hydrogenation of the imine bond gave the product [194].

Asymmetric Strecker reaction of the α-acyloxy ketone **48** afforded cyanomorpholine (3S,5S)-**50** as the major diastereomer through the iminium intermediate **49**. The stereochemistry is controlled by the configuration of the acyloxy amino chiral center: the relative configuration of the nitrile group is usually *anti* to that of the amino side chain. The Strecker precursor **48** was prepared from methyl 3-(chlorocarbonyl)propanoate (**47**) by addition of diazomethane followed by insertion of the resulting diazoketone to Boc-L-valine [195].

Enantiomerically pure morpholinone (5S,1′S)-**52** was prepared by condensation of amino alcohol (2S,1′S)-**51** with glyoxal [196].

When the 2-aminoethanol derivative (R)-**53** possessing N-homoallyl substituent was reacted with glyoxal in aqueous media, morpholine (3R,5R)-**56** was obtained from an aza-Cope rearrangement of the iminium intermediate **54**, directly formed by the condensation of amino alcohol (R)-**53** and glyoxal, followed by hydrolysis of intermediate **55**. In this reaction sequence, stereoselective formation of the new

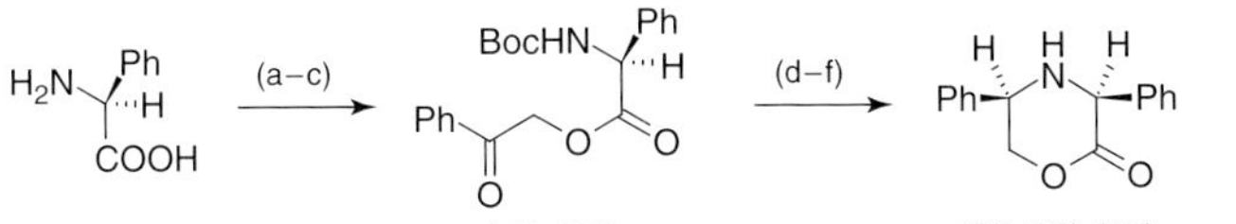

Reagents and conditions: (a) Boc₂O/NaOH/H₂O/dioxane (89%),
(b) KOH/MeOH (quant), (c) PhCOCH₂Br/DMF (74%),
(d) HBr/AcOH (81%), (e) NaOAc/AcOH buffer (89%),
(f) H₂/Pd–C/EtOAc (86%).

Scheme 7.92

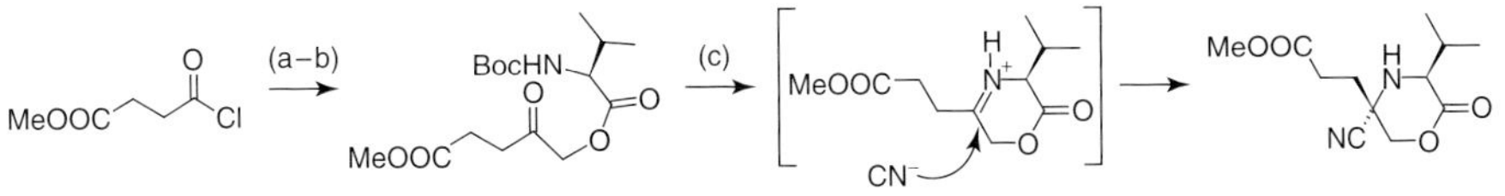

Reagents and conditions: (a) CH₂N₂ (90%), (b) Boc-L-valine/Cu(acac)₂/toluene (40%),
(c) 1. TFA/CH₂Cl₂, 2. NaCN/iPrOH (64%, dr 18:1).

Scheme 7.93

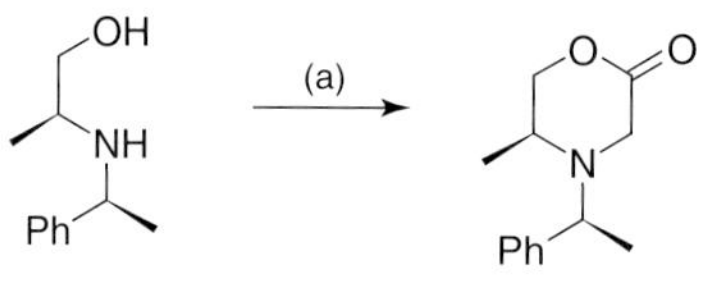

342 | *7 Asymmetric Synthesis of Ring Nitrogen Heterocycles*

(2S,1′S)-(**51**) → (5S,1′S)-(**52**)

Reagents and conditions: (a) OHC–CHO/THF/rfx (86%).

Scheme 7.94

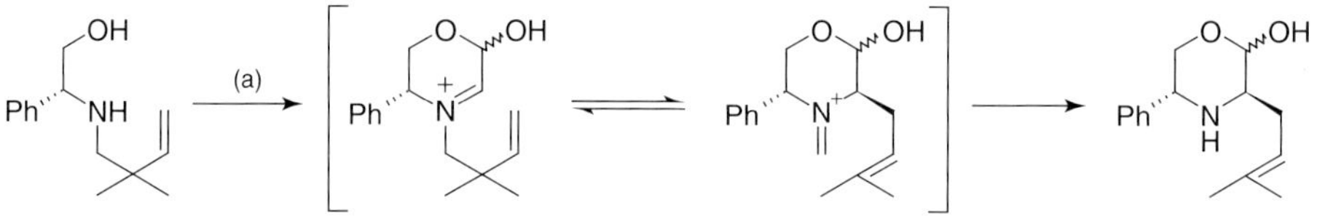

(R)-(**53**) → [(**54**) ⇌ (**55**)] → (3R,5R)-(**56**)

Reagents and conditions: (a) OHC–CHO/H$_2$O (54%).

Scheme 7.95

chiral center at 3 position was observed giving exclusively the 3R stereoisomer [197–199].

Reaction of amino alcohol (R)-**57** with 4-fluorophenylglyoxal **58** yielded a mixture (1:2) of the diastereomers **59**. However, the newly formed chiral center was labile under acidic conditions, and when the diastereomeric mixture was treated with HCl in isopropyl alcohol, the 3R isomer crystallized preferentially and nearly all of the other diastereomer was converted into the 3R isomer leading to formation of (3R,1′R)-**60** in excellent yield and stereoselectivity (98% de) [170, 200].

Ring Closure via Nucleophilic Substitution with C–O Bond Formation   Intramolecular displacement of a halogen with a hydroxy group in a suitable haloacetylaminoethanol derivative can be used for morpholine ring formation. Thus, compound (S,S)-**62** prepared by acylation of amino alcohol (S,S)-**61** with chloroacetyl chloride was treated with KOtBu to give morpholinone (S,S)-**63** [201, 202].

Optically active 2-hydroxymethylmorpholine derivatives (S)-**67** and (R)-**69** could be prepared from precursors possessing the stereogenic center from natural origin. Namely, the protected D-mannitol derivative **64** was cleaved with sodium

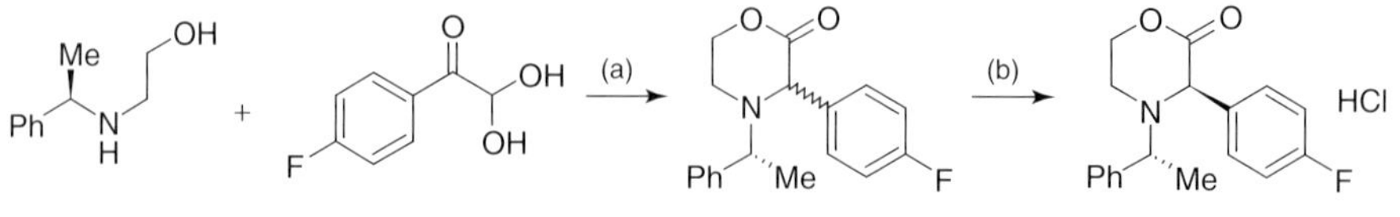

(R)-(**57**) + (**58**) → (**59**) → (3R,1′R)-(**60**)

Reagents and conditions: (a) AcOH, (b) HCl/iPrOAc (90%, 98% de).

Scheme 7.96

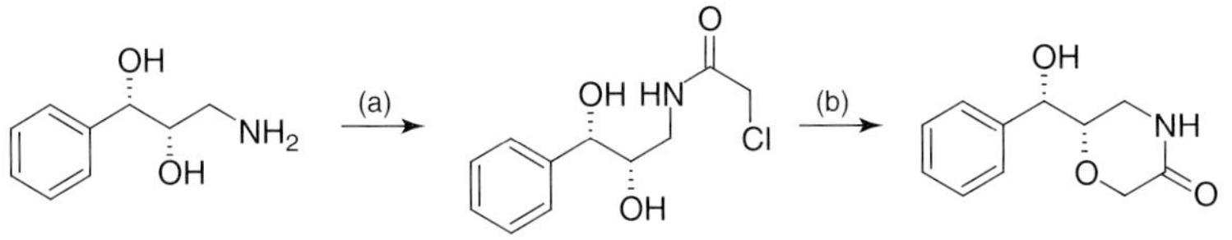

Scheme 7.97

(S,S)-(61) → (S,S)-(62) → (S,S)-(63)

Reagents and conditions: (a) ClCH$_2$COCl/Et$_3$N/CH$_2$Cl$_2$ (70%),
(b) KO*t*Bu/BuOH (80%).

metaperiodate to aldehyde (R)-65, and reductive amination with benzyl amine followed by acylation with chloroacetyl chloride and subsequent deprotection gave (S)-66, which was then cyclized to the product (S)-67 with S configuration by treatment with NaOEt. Since the enantiomer of (S)-67 could not be prepared in a stereochemically pure form by inversion of the configuration of (S)-66 as the key step, another route was applied. Treatment of the epoxide (R)-68 prepared from (R)-65 in five steps with aminoethyl hydrogensulfate gave the cyclized product, which was then debenzylated to (R)-69 [203].

Enantiomerically pure starting material, (S)-3-amino-1,2-propanediol ((S)-70), was also used for the preparation of (S)-2-hydroxymethylmorpholine ((S)-72) as a key intermediate for the synthesis of the selective norepinephrine inhibitor reboxetine. After N-acylation of (S)-70 with chloroacetyl chloride, (S)-71 could be cyclized to morpholine (S)-72 with KO*t*Bu [204].

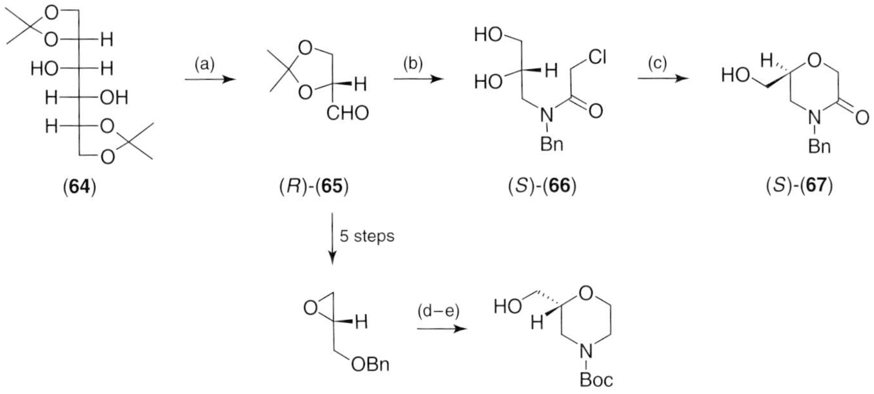

(64) → (R)-(65) → (S)-(66) → (S)-(67)

(R)-(65) → 5 steps → (R)-(68) → (R)-(69)

Reagents and conditions: (a) NaIO$_4$/H$_2$O, (b) 1. BnNH$_2$/EtOH/H$_2$O/Raney-Ni/H$_2$ (quant),
2. ClCH$_2$COCl/Et$_3$N/CH$_2$Cl$_2$, (86%), 3. 20% aq AcOH (quant),
(c) NaOEt/EtOH (88%), (d) 1. H$_2$N(CH$_2$)OSO$_3$H/NaOH/H$_2$O (62%),
2. Boc$_2$O/Et$_3$N/CH$_2$Cl$_2$, (88%), (e) Pd–C/H$_2$/MeOH (quant).

Scheme 7.98

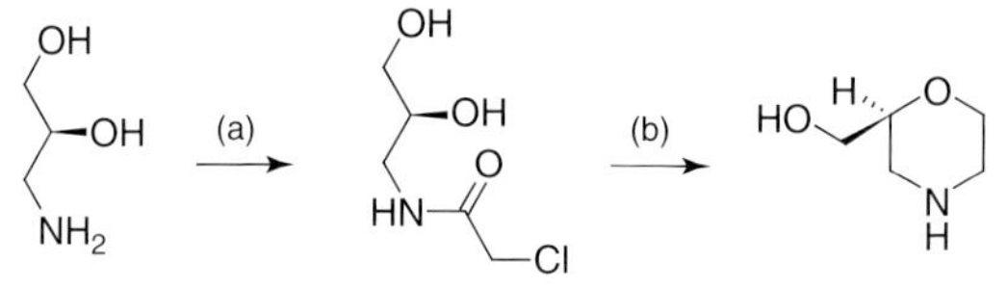

(*S*)-(**70**) (*S*)-(**71**) (*S*)-(**72**)

Reagents and conditions: (a) $ClCH_2COCl/Et_3N/MeCN/MeOH$ (94%), (b) KO*t*Bu/*t*AmOH (92%).

Scheme 7.99

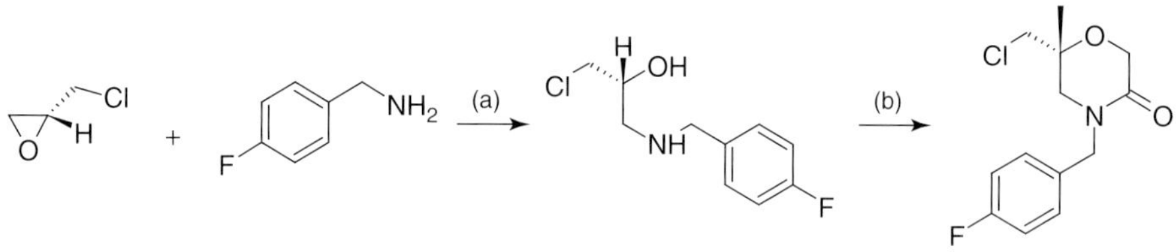

(*R*)-(**73**) (**74**) (*R*)-(**75**) (*R*)-(**76**)

Reagents and conditions: (a) cyclohexane/rt (70%, 98% ee), (b) $BrCH_2COBr/NaOH/H_2O/CHCl_3$ (86%).

Scheme 7.100

Synthesis of the morpholine derivative (*R*)-**76** as an intermediate for the gastro-protecting agent mosapride started from (*R*)-(−)-epichlorohydrin ((*R*)-**73**). Using hydrocarbons as cyclohexane as the solvent proved to be more efficient over other solvents in the reaction of (*R*)-**73** with 4-fluorobenzylamine (**74**) furnishing the product (*R*)-**75** in good yield and high enantiomeric purity. Reaction of (*R*)-**75** with bromoacetyl bromide in a mixture of chloroform and 30% aqueous NaOH solution resulted in the ring-closed product (*R*)-**76** without isolation of the bromoacetamide intermediate [205].

Enzymatic synthesis of enantiomerically pure morpholine (*R*)-**81** could be achieved through an epoxide intermediate, where the key step is the asymmetric reduction of the bromoaldehyde **77** to bromo alcohol (*S*)-**78** using baker's yeast.

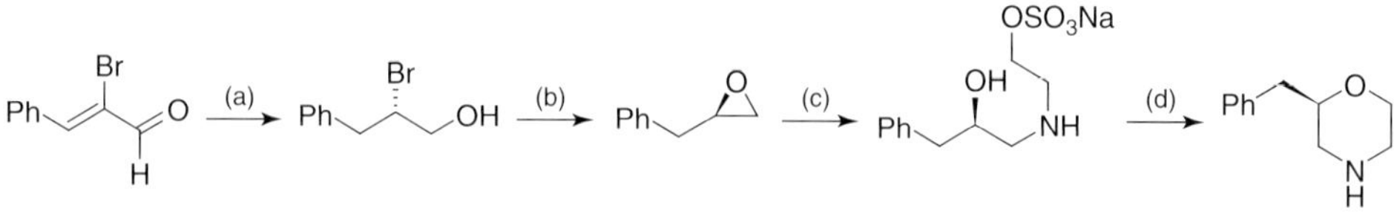

(**77**) (*S*)-(**78**) (*R*)-(**79**) (*R*)-(**80**) (*R*)-(**81**)

Reagents and conditions: (a) Baker's yeast/XAD1180 resin (100%, 98,6% ee), (b) $NaOH/H_2O$ (73%), (c) $H_2N(CH_2)_2OSO_3Na/NaOH/MeOH$, (d) NaOH/toluene (66%).

Scheme 7.101

7.1 Six-Membered Rings with Another Heteroatom in the Same Ring

Ring opening of the epoxide (R)-**79** with ethanolaminesulfonate to (R)-**80** and subsequent basic cyclization gave (R)-**81** [170].

For the synthesis of N,N'-dibenzyl derivative of (2S,2'S)-2, 2'-bimorpholine (2S,2'S)-**84**, tetraol **83** prepared from tartaric acid derivative **82** was cyclized in a one-step procedure with NaH and p-toluenesulfonyl imidazole [206].

trans-2,5-Disubstituted morpholine derivative (2S,5R)-**88** was prepared starting from enantiomerically pure epoxide (S)-**85** and D-alaninol (R)-**86** via diol **87** applying the same cyclization method. Thus, refluxing (S)-**85** and (R)-**86** in n-propanol and subsequent N-tosylation yielded selectively diol **87**, which was then cyclized via deprotonation with NaH in the presence of p-toluenesulfonyl imidazole. Removal of the tosyl group with sodium in ethanolic ammonia provided (2S,5R)-**88** in pure form [207].

However, adaptation of the method bearing larger substituents was not unambiguous. When attempting the same reaction sequence for synthesis of analogous morpholine derivatives with larger 5-substituents (e.g. phenyl and benzyl), ring opening proceeded smoothly, but instead of selective N-tosylation, O-tosylation occurred as a result of sterical hindrance. For the preparation of the N-tosylated diol (analog of **87**), first, transient protection of hydroxy groups was necessary with trimethylsilyl groups. Ring closure of the phenyl analog of diol **87** was also unsuccessful using the same procedure (NaH/TsIm) as used for **87**, but O-tosylation and subsequent treatment with K_2CO_3 in tBuOH resulted in the cyclized product [207].

An enantioselective route to reboxetine analog (S,S)-**94** with S,S configuration was described, where the key step was the Sharpless asymmetric epoxidation of the cinnamyl alcohol derivative **89** to epoxide (R,R)-**90** in 85% ee. The stereochemical outcome was highly dependent on the oxidating agent: the best enantiomeric excess was achieved by using cumene hydroperoxide. The (R,R) enantiomer of the product

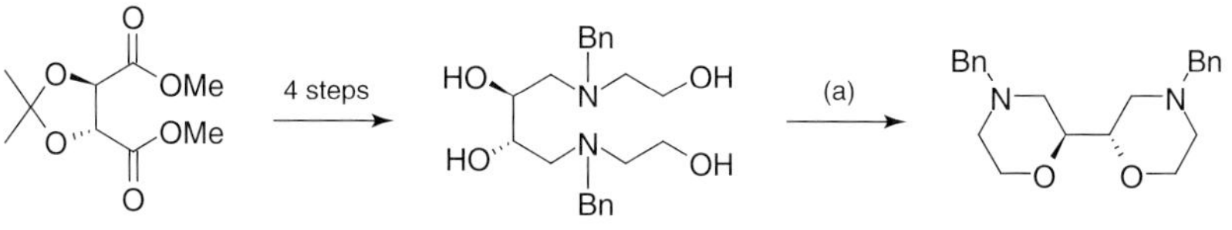

Reagents and conditions: (a) NaH/THF/TsIm (82%, 50% overall, 99% ee).

Scheme 7.102

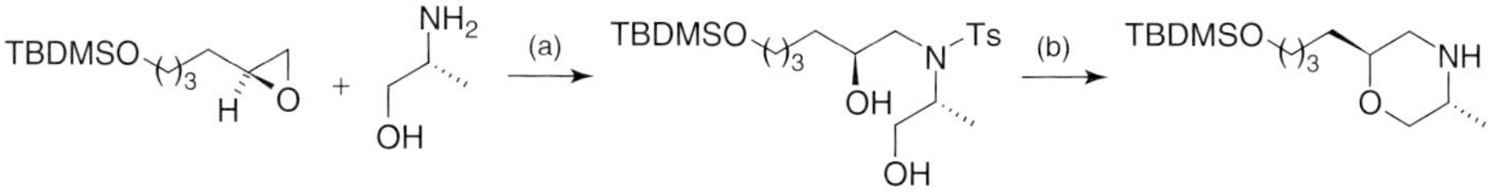

Reagents and conditions: (a) 1.nPrOH/rfx (99%), 2. TsCl/Et$_3$N/CH$_2$Cl$_2$ (77%),
(b) 1. NaH/THF/TsIm (99%), 2. Na/NH$_3$/EtOH (100%).

Scheme 7.103

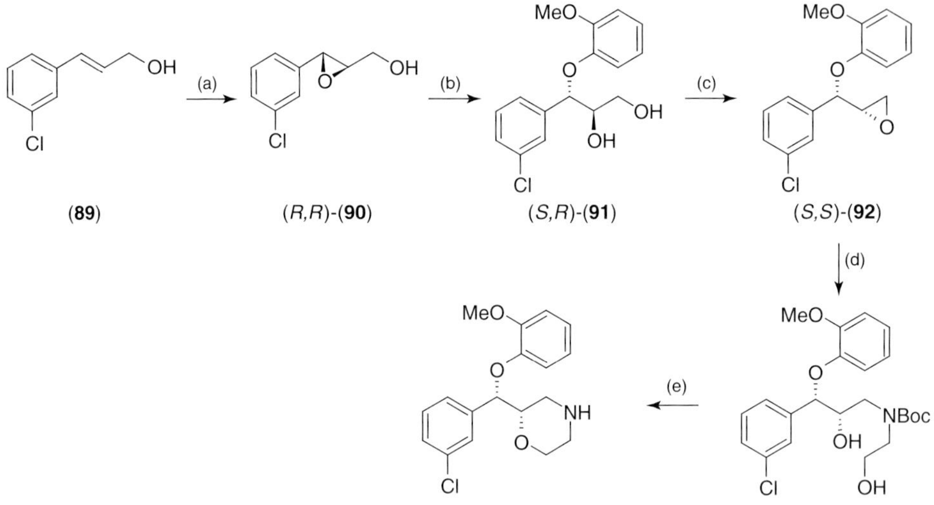

Scheme 7.104

Reagents and conditions: (a) Ti(OiPr)$_4$/cumene hydroperoxide/D-DET/CH$_2$Cl$_2$/mol. sieves (72%, 85% ee), (b) guaiacol/NaOH/CH$_2$Cl$_2$, (c) 1. TMSCl/Et$_3$N/EtOAc, 2. MsCl/Et$_3$N/EtOAc, 3. 1N HCl, 4. NaOH/toluene/MeBu$_3$NCl (60%), (d) 1. etanolamine/iPrOH, 2. Boc$_2$O/CH$_2$Cl$_2$ (83%), (e) 1. NaH/TsIm/THF (61%), 2. TFA/CH$_2$Cl$_2$ (93%).

could be obtained using L-diethyl tartrate (DET) instead of D-DET as the additive. Ring opening of (R,R)-**90** with guaiacol gave diol (S,R)-**91**, which was then transformed to epoxide (S,S)-**92** in four steps. Reaction of (S,S)-**92** with ethanolamine, and the subsequent N-protection and cyclization of (S,S)-**93** with tosylimidazol gave (S,S)-**94** [208].

Starting from olefin **95**, Sharpless dihydroxylation was performed affording diol **96** in 98% ee. Selective tosylation on the primary hydroxy group, followed by nucleophilic substitution with ethanolamine and N-protection gave compound **97**, which was cyclized under Mitsunobu conditions, and then deprotected to morpholine derivative (R)-**98** [170, 209].

Ring Closure via Nucleophilic Substitution with C–N Bond Formation Another pathway for the synthesis of (S)-2-hydroxymethylmorpholine (**72**) from enantiomerically pure starting material involved the alkylation of the protected amino alcohol **99** derived from L-serine with tert-butyl bromoacetate to ester **100**, which was then reduced in two stages to the corresponding alcohol. Subsequent mesylation to **101** and base-mediated cyclization followed by deprotection gave (S)-**72** [210].

Using (R)-1-benzylglycerol ((R)-**102**), S- and R-hydroxymethylmorpholine derivatives (S)-**105** and (R)-**108**, respectively, could be prepared selectively via the key intermediate **103** by varying the removal of the applied protecting groups. Thus, the

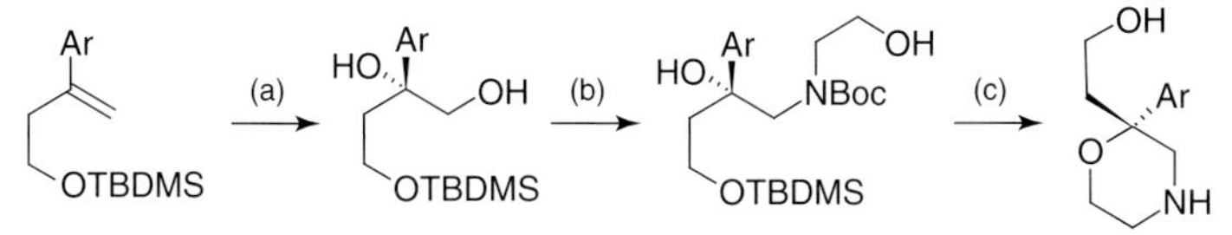

7.1 Six-Membered Rings with Another Heteroatom in the Same Ring

(95) → **(96)** → **(97)** → **(R)-(98)**

Ar = 3,4-dichlorophenyl

Reagents and conditions: (a) AD-mix-β/tBuOH/H$_2$O (94%, 98% ee),
(b) 1. TsCl/pyridine, 2. etanolamine/LiClO$_4$/MeCN,
3. Boc$_2$O/Et$_3$N/CH$_2$Cl$_2$ (86%), (c) 1. DEAD/PPh$_3$/toluene,
2. 4 N HCl/dioxane, 3. 5% NaOH.

Scheme 7.105

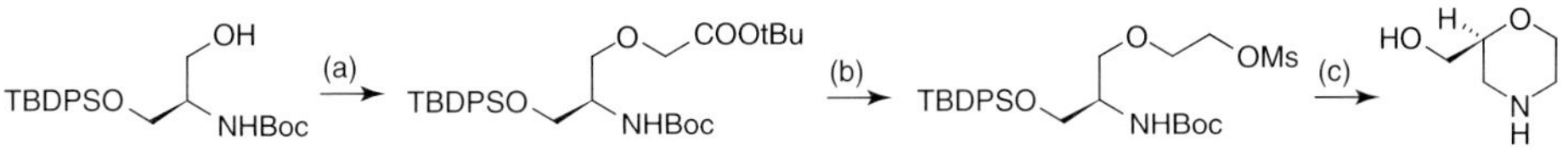

(99) → **(100)** → **(101)** → **(S)-(72)**

Reagents and conditions: (a) BrCH$_2$COOt-Bu/aq NaOH/toluene/BuN$_4$I (87%), (b) 1. DIBAL-H/CH$_2$Cl$_2$,
2. LiBH$_4$/Et$_2$O (85%), 3. MsCl/Et$_3$N/CH$_2$Cl$_2$ (93%),
(c) 1. TFA/CH$_2$Cl$_2$, 2. DIEA/MeOH, 3. nBu$_4$NF/THF (87%).

Scheme 7.106

first debenzylation of **103** to **104** followed by ditosylation and subsequent treatment with benzylamine gave (S)-N-benzyl-2-tert-butyldiphenylsilyloxymethylmorpholine ((S)-**105**). On the contrary, when the tert-butyldiphenylsilyl (TBDPS) group of **103** was removed followed by ditosylation of **106**, then ring closure of **107** in the presence of benzylamine was performed, and (R)-N-benzyl-2-benzyloxymethylmorpholine ((R)-**108**) could be obtained [211].

Synthesis of (2R,6R)-dimethylmorpholine ((2R,6R)-**113**) was carried out by a reaction sequence starting with S$_N$2 reaction of (R)-ethyl lactate ((R)-**109**) on the triflate of (S)-ethyl lactate ((S)-**110**). The conditions, particularly the used solvent, had great influence on the stereochemistry of the reaction. Complete inversion could be achieved by using a 4:1 mixture of n-pentane and 1,2-dichloroethane as the solvent and K$_2$CO$_3$ as base. Reduction of diester (R,R)-**111** to diol and subsequent mesylation gave dimesylate (R,R)-**112**, which could be cyclized with ninefold excess of benzylamine and then debenzylated to (2R,6R)-**113** [212].

Iodination of **114** using N-iodosuccinimide, and subsequent reaction with N-Boc-ethanolamine gave compound **115**. Ring closure with base, deprotection and subsequent recrystallization with D-(−)-tartaric acid provided the stereochemically pure product (R)-**116** [170, 213].

(2S,2′S)-2, 2′-Bimorpholine ((2S,2′S)-**119**) could be synthesized from diazido diol **117** via NaH-induced double cyclization of intermediate **118** by intramolecular displacements of the mesyloxy groups with the Boc-protected nitrogens. In an analogous way, the isomeric (3S,3′S)-3, 3′-bimorpholine ((3S,3′S)-**122**) was prepared

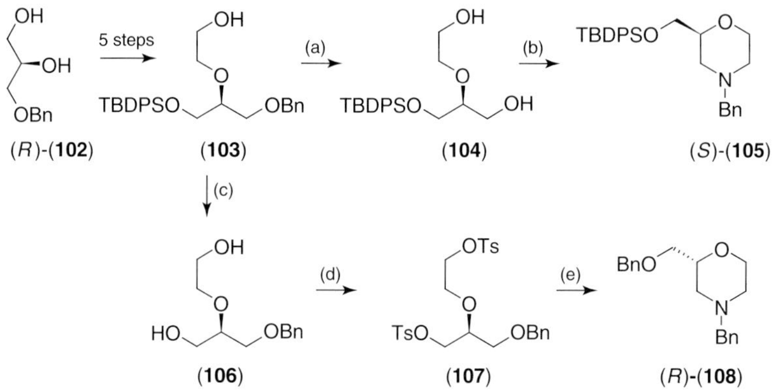

Scheme 7.107

Reagents and conditions: (a) H₂/Pd–C (98%), (b) 1. TsCl/pyridine (57%), 2. BnNH₂ (84%),
(c) HF/MeCN (87%), (d) TsCl/pyridine (56%),
(e) BnNH₂/Na₂CO₃/MeCN (85%).

Scheme 7.108

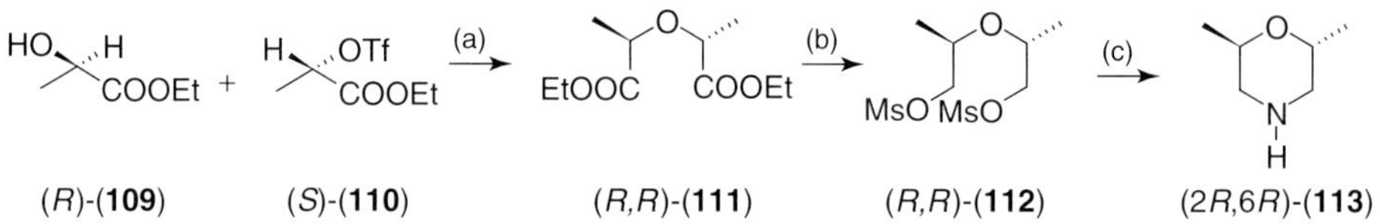

Reagents and conditions: (a) K₂CO₃/npentane-1,2-dichloroethane (4:1) (89%),
(b) 1. LiAlH₄/Et₂O, 2. MsCl/pyridine (94%),
(c) 1. BnNH₂/dioxane (85%), 2. H₂/Pd–C/AcOH/MeOH.

Scheme 7.109

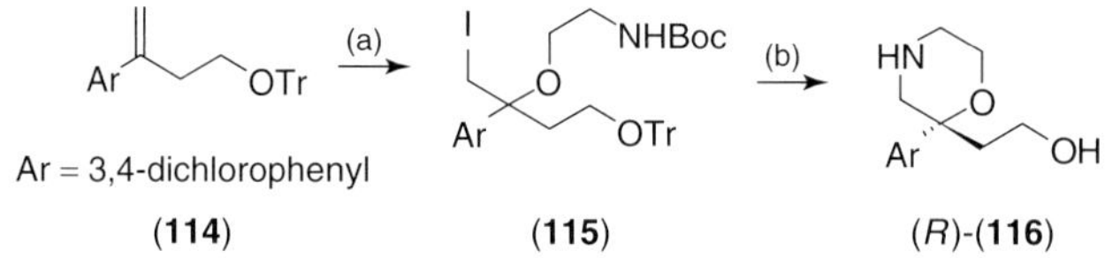

Reagents and conditions: (a) N-Boc-etanolamine/NIS/MeCN (72%), (b) 1. NaH/DMF (77%),
2. 4 N HCl/dioxane/EtOH, 3. recryst. with D-(−)-tartaric acid,
4. 1 N NaOH (82%, 99% ee).

from **120** via the intermediate **121**. In both cases, an enantiomeric excess of over 98% was achieved [214].

Ring Closure by Intramolecular Michael Addition   Intramolecular conjugate addition of either nitrogen or oxygen to α,β-unsaturated ester was employed to morpholine ring formation. Thus, N-alkylation of **123** with **124** and subsequent intramolecular

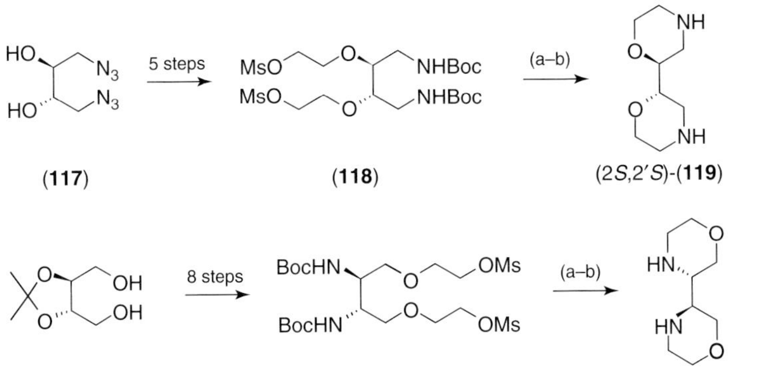

(117) (118) (2S,2'S)-(119)

(120) (121) (3S,3'S)-(122)

Reagents and conditions: (a) NaH/THF (91–99%),
(b) 1. TFA/CH$_2$Cl$_2$, 2. 3 M NaOH/Et$_2$O (29–47%, >98% ee).

Scheme 7.110

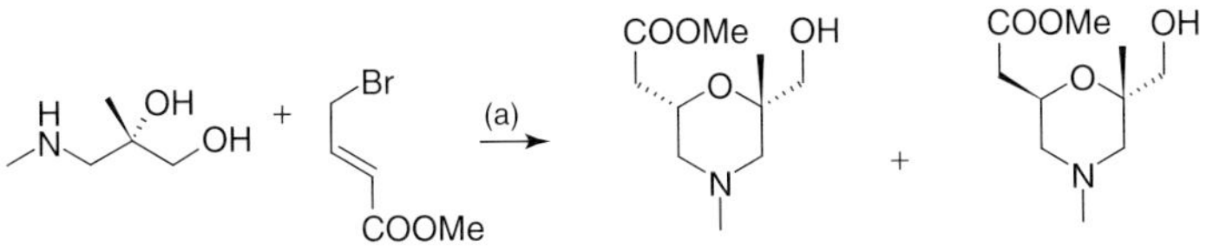

(123) (124) (2S,6S)-(125) (2R,6S)-(125)

Reagents and conditions: (a) K$_2$CO$_3$/Et$_2$O/MeOH (100%, dr 6:1).

Scheme 7.111

Michael addition afforded a 6:1 mixture of (2S,6S)-**125** and (2R,6S)-**125**, which could be separated by flash chromatography [170, 215].

Selenium-promoted addition of (R)-phenylglycinol to alkene **126** bearing an electron-withdrawing group gave selenides **128** and **132**, which after separation were oxidized to unsaturated nitrile derivatives **129** or **133**, respectively. Intramolecular Michael addition of the *trans* isomer **129** gave a 1:1 diastereomeric mixture of **130** and **131**, while the *cis* Michael acceptor **133** resulted in a 7:3 mixture of **134** and **135** [216].

Miscellaneous Methods for Ring Closure Morpholines **138** and **139** could be synthesized in excellent regio- and stereoselectivities (dr: 22.7:1 to 1:14.4) by opening the epoxide **136** in the presence of enantiomerically pure palladium complexes using ligands **140**, and subsequent ring closure with KCN [170, 217].

Three-component boro-Mannich reaction between N-protected ethanolamines **141**, glyoxal derivatives **142**, and boronic acids **143** via a tetracoordinated intermediate (**144**) provided 2-hydroxymorpholines **145** in various stereoselectivities (10–75% de) depending on the nature of the substituents [170, 218]. The stereoisomer lactols are interconvertible through the ring-opened form, and in some cases, the facile equilibration of the isomers in solution in combination with the crystallization

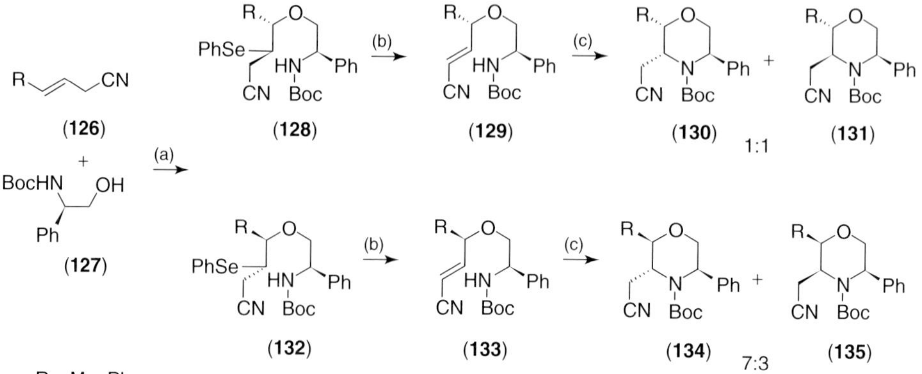

Scheme 7.112

Reagents and conditions: (a) N-PSP/BF$_3$·OEt$_2$ (20–20%), (b) H$_2$O$_2$/CH$_2$Cl$_2$ (86–93%), (c) NaH/THF (92–95%).

R = Me, Ph
N-PSP = N-(phenylseleno)phthalimide.

Scheme 7.113

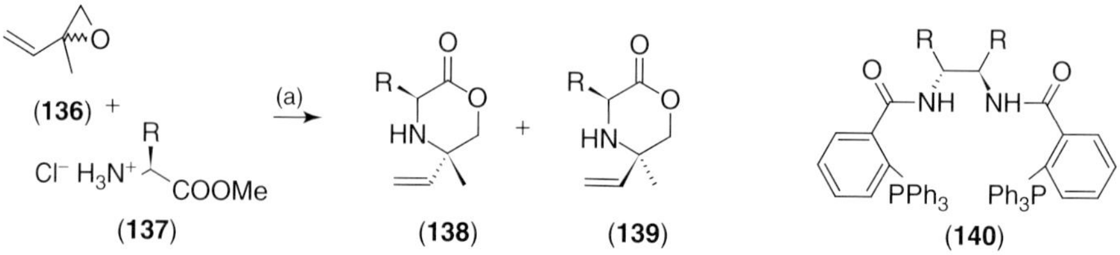

Reagents and conditions: (a) 1. [η^3-CH$_3$H$_5$PdCl]$_2$/Et$_3$N/CH$_2$Cl$_2$/140, 2. KCN/THF/MeCN (39–91%, 22.7:1 to 1:14.4).

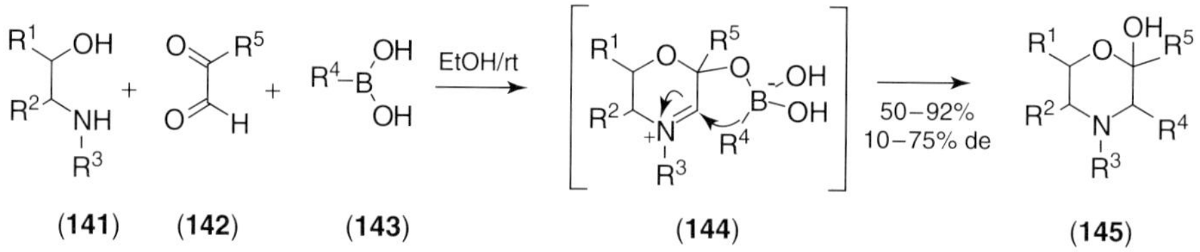

Scheme 7.114

of the desired diastereomer could lead to a crystallization-induced transformation resulting in a single isomer from a complex mixture [219].

Employing a similar method, alkenyl moiety deriving from boronic acid **146** could be introduced to the morpholine ring by reacting with glyoxylic acid and amino alcohol **147**. From the obtained diastereomeric mixture, compound (3S,5S)-**148** could be transformed to (3R,5S)-**148** by treatment with triethyl amine, increasing the diastereoselectivity to 100% in this way [170, 220].

Olefins **149** and N-protected ethanolamines **150** reacted via palladium-catalyzed allylic substitution in the presence of a chiral palladium(0) complex to give

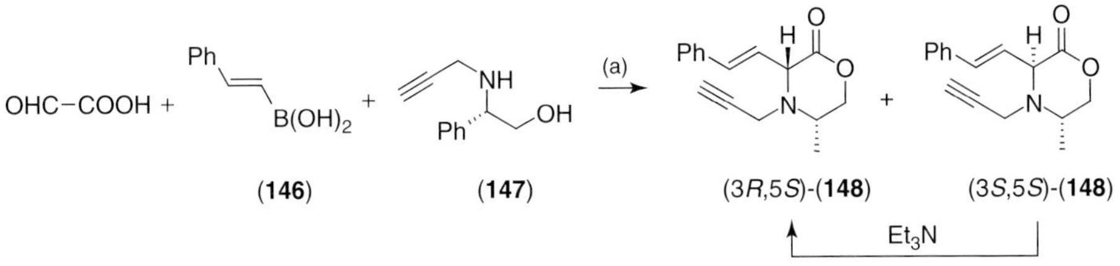

(146) (147) (3R,5S)-(148) (3S,5S)-(148)

Reagents and conditions: (a) CH_2Cl_2 (87%, 1.5:1 or 1:0).

Scheme 7.115

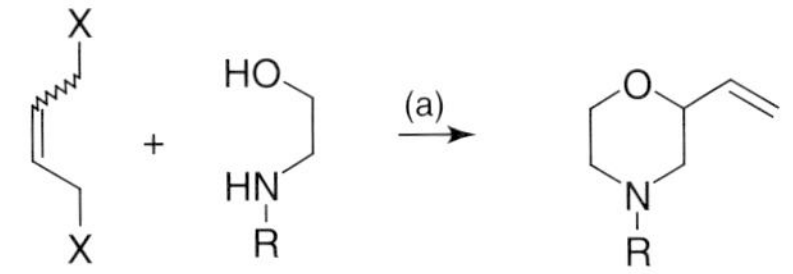

(149) (150) (151)

Reagents and conditions: (a) Pd(0)/L/THF(20–80%, 16–90% ee).
X = AcO, MeOOCO, *t*BuOOCO.

Scheme 7.116

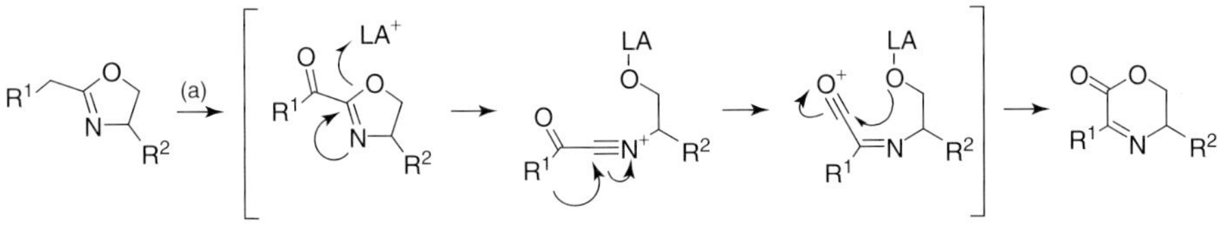

(152) (153) (154) (155) (156)

R^1= H, Me, *i*Pr, *t*Bu, Ph, napth
R^2 =(*S*)-*i*Pr, (*S*)-*i*Pr, (*R*)-Ph, (*R*)-$BnOCH_2$

Reagents and conditions: (a) SeO_2/dioxane/D (22–93%).

Scheme 7.117

2-vinylmorpholines **151** with moderate to good (16–90%) enantioselectivities [92–94,170, 221, 222].

A SeO_2-promoted oxidative rearrangement of 2-alkyl and 2-(arylmethyl)oxazolines **152** to unsaturated morpholinones **156** was reported. The proposed mechanism as follows: α-keto carbonyl derivative **153** formed by SeO_2 oxidation of **152** undergoes Lewis acid-catalyzed ring opening to nitrilium ion **154**, which rearranges to acylium ion **155** followed by a ring closure to **156** [223].

Morpholine *N*-oxides could be stereoselectively prepared from (1*S*,2*S*)-pseudoephedrine ((*S*,*S*)-**157**). Reaction of (*S*,*S*)-**157** with acrylonitrile gave the *N*-cyanoethyl derivative **158**, which was then O-alkylated with allyl bromide to **159**. Treatment of **159** with *m*-chloroperbenzoic acid yielded the *N*-oxide **160**, which underwent *in situ* Cope elimination to hydroxylamine **161** and subsequent reverse Cope elimination by heating in dichloromethane, providing morpholine *N*-oxide (4*R*)-**162** as the

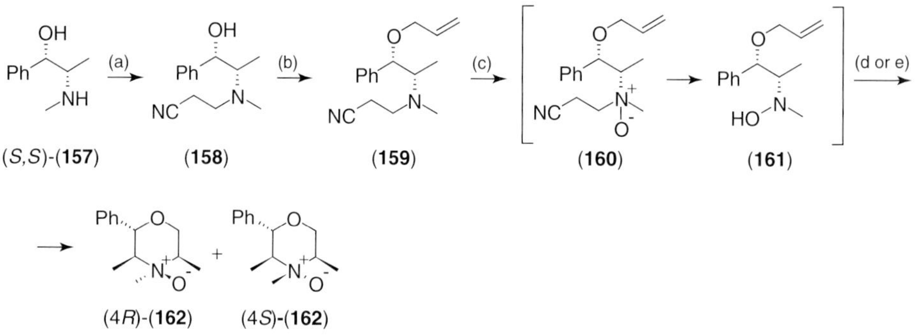

Reagents and conditions: (a) CH$_2$CHCN/MeOH (100%), (b) 1. NaH/THF, 2. CH$_2$CHCH$_2$Br,
(c) m-CPBA/CH$_2$Cl$_2$/K$_2$CO$_3$, (d) CH$_2$Cl$_2$/D (32%, dr 1:0),
(e) MeOH/D (82%, dr 3:2).

Scheme 7.118

single diastereomer. However, heating **161** in methanol gave a 3:2 mixture of the diastereomers (4R)-**162** and (4S)-**162** [224].

Aminodiazoacetoacetates **165** prepared from amino alcohols **163** by treatment with ketene **164** followed by diazotransfer with MsN$_3$ underwent cyclization to morpholinones **166** via copper-catalyzed carbene-transfer reaction with limited diastereoselectivity (dr: 1:1 to 3:3) [225].

7.1.5.2 Asymmetric Transformations with the Involvement of the Morpholine Ring

C-Alkylations of Morpholinones Asymmetric C-alkylation of a stereochemically pure substituted morpholine derivative and subsequent hydrolysis is a widely used strategy for the synthesis of enantiomerically pure amino acids. Thus, enantiomerically pure 5,6-diphenylmorpholin-2-ones **167** (called Williams chiral auxiliaries) can be stereoselectively alkylated on the ring carbon α-position to the oxo group by reacting the corresponding enolates **168** with alkylhalogenides, thereby providing the products **169** with the newly entered alkyl group *trans* to the phenyl groups. In most cases, high diastereoselectivity was obtained. Alkylated

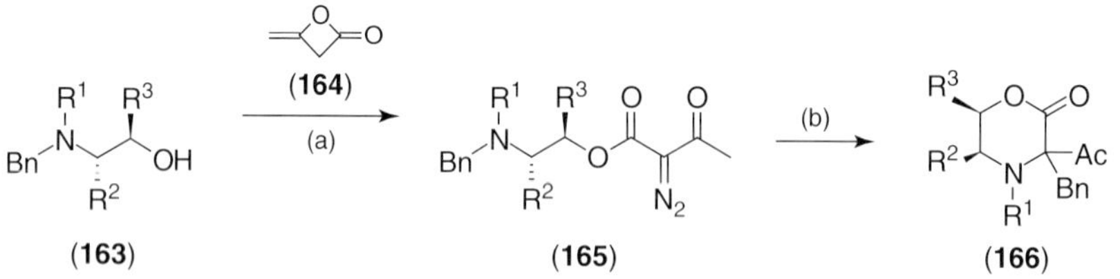

R^1 = Me
R^2 = Me, Ph, Bn
R^3 = H, Ph,

Reagents and conditions: (a) 1. **164**/Et$_3$N/CH$_2$Cl$_2$, 2. MeSO$_2$N$_3$/Et$_3$N/aq MeCN (64–83%),
(b) Cu-powder/toluene/rfx (70–71%, dr 1:1 to 3:1).

Scheme 7.119

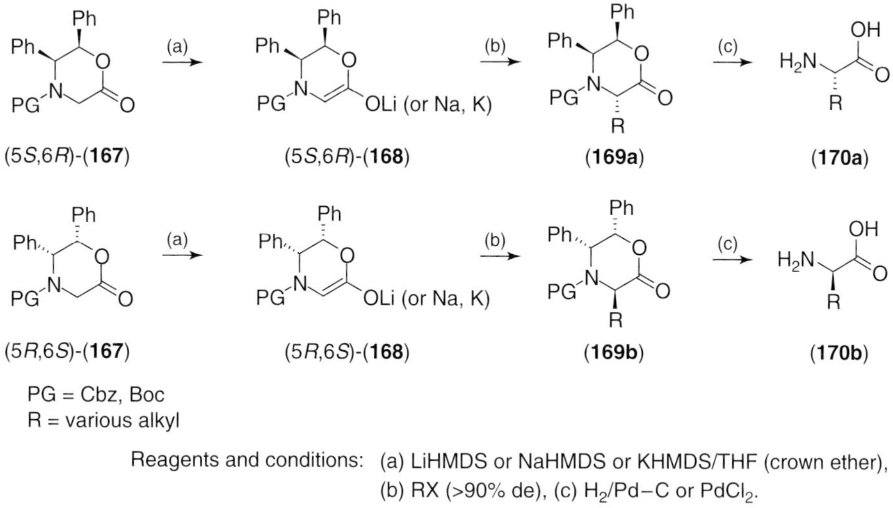

Scheme 7.120

Reagents and conditions: (a) LiHMDS or NaHMDS or KHMDS/THF (crown ether), (b) RX (>90% de), (c) H$_2$/Pd–C or PdCl$_2$.

PG = Cbz, Boc
R = various alkyl

morpholinones, after hydrogenolysis, could be transformed to the corresponding amino acids **170** [175, 226–237].

Other differently substituted, optically active morpholinones could also be alkylated at the α-position with various stereoselectivities [196, 238, 239].

Dialkylation at 2-position could be diastereoselectively achieved by deprotonation of monoalkylated derivative **169** and subsequent reaction with various nucleophiles affording **171** [240, 241].

Najera et al. used mild PTC conditions in the presence of a weak base for diastereoselective alkylation of morpholinones **172** with activated halides. Pd-catalyzed allylation of the same substrate was also performed by reacting **172** with allyl carbonates in the presence of Pd(PPh$_3$)$_4$ and 1,2-bis(diphenylphosphino)ethane (dppe) to give the product **173** [242].

However, in the alkylation reactions of **172** with inactivated alkyl halide, the PTC conditions were not suitable; instead, 2-tert-butylimino-2-diethylamino-1,3-dimethylperhydro-1,3,2-diazaphosphorine (BEMP) or 1,8-diazabicyclo(5.4.0)undec-7ene (DBU) was used as the base. In this case, a competitive O-alkylation was observed, which could be suppressed by adding LiI [243].

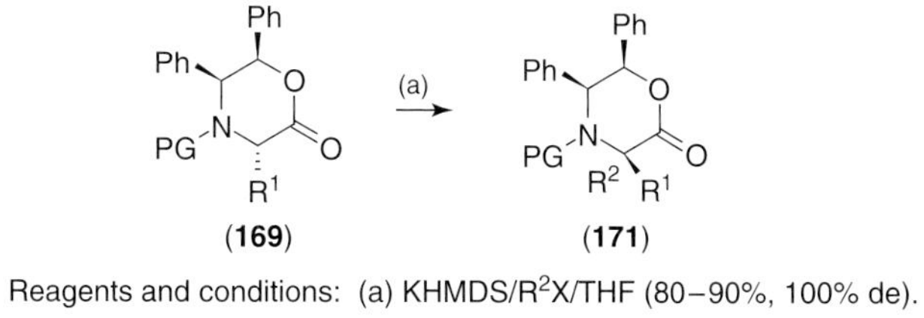

Reagents and conditions: (a) KHMDS/R^{2}X/THF (80–90%, 100% de).

Scheme 7.121

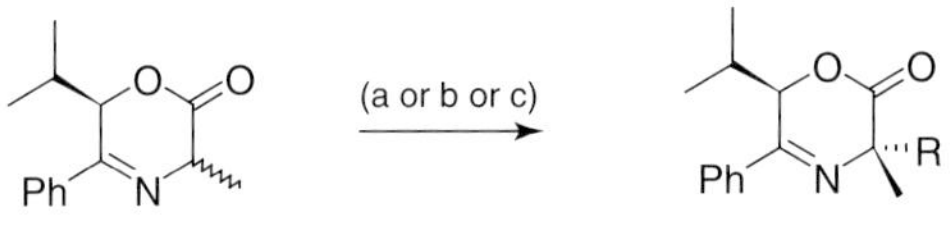

(172) → **(173)**

(a or b or c)

R = for (a) or (b): (subst)allyl, propargyl, Bn, EtCOOCCH$_2$ for (c): Et, iPr, nBu, iBu.

Reagents and conditions: (a) RX/K$_2$CO$_3$/nBu$_4$NBr/MeCN or CH$_2$Cl$_2$/rt (60–75%, >92% de),
(b) allyl carbonates/Pd(PPh$_3$)$_4$/dppe/THF/rt (53–65%, >91% de),
(c) RX/BEMP or DBU/ (LiI)/NMP (28–65%, >96% de).

dppe: 1,2-bis(diphenylphosphino)ethane.
BEMP: 2-tert-butylimino-2-diethylamino-1,3-dimethylperhydro-1,3,2-diazaphosphorine.
DBU: 1,8-diazabicyclo(5.4.0)undec-7-ene.

Scheme 7.122

Besides halogenides, allyl silanes are also suitable reagents for allylation of the morpholine ring. Reaction of acetoxy hemiacetals **174** or acetals **176** with allyltrimethylsilane in the presence of BF$_3$•OEt$_2$ or TiCl$_4$, gave allylated products **175** or **177** as single diastereomers, respectively [244, 245]. The same methods were applied to chiral morpholines substituted in a different way [246, 247].

Titanium, aluminum, or boron enolates of diphenylmorpholinone **167** could be reacted with orthoester, aldimine, or aldehyde, affording stereoselectively the corresponding alkylated product **178, 179**, or **180**, respectively, with the entering group *trans* to the phenyl groups. In case of compounds **179** and **180**, a 3.1:1 or 8:1 diastereomeric mixture was obtained, respectively, with regard to the newly formed chiral center in the side chain [248–250].

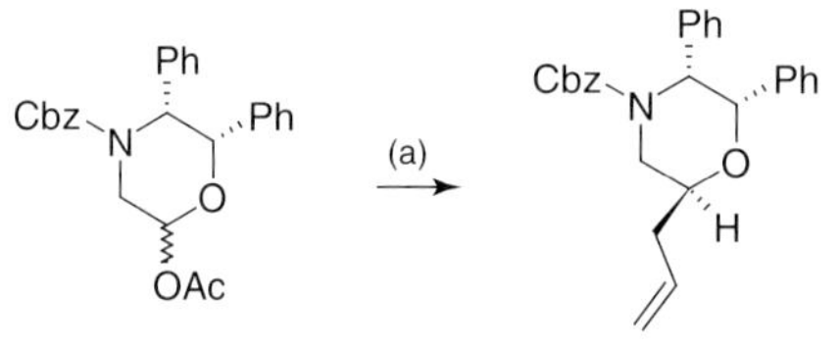

(174) → **(175)**

Reagents and conditions: (a) allyltrimethylsilane/BF$_3$·Et$_2$O/MeCN (98%).

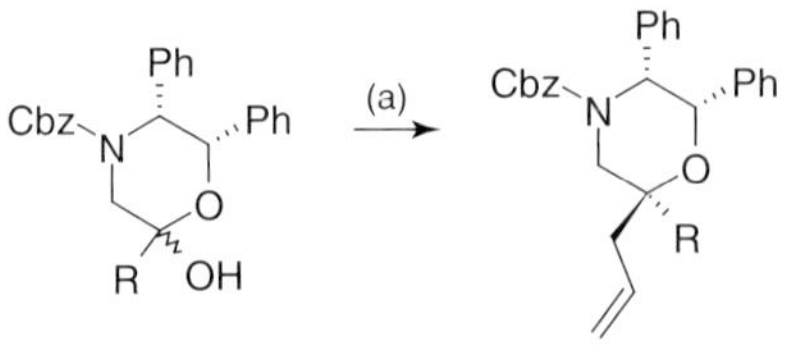

(176) → **(177)**

R = alkyl

Reagents and conditions: (a) allyltrimethylsilane/TiCl$_4$/CH$_2$Cl$_2$ (65–67%).

Scheme 7.123

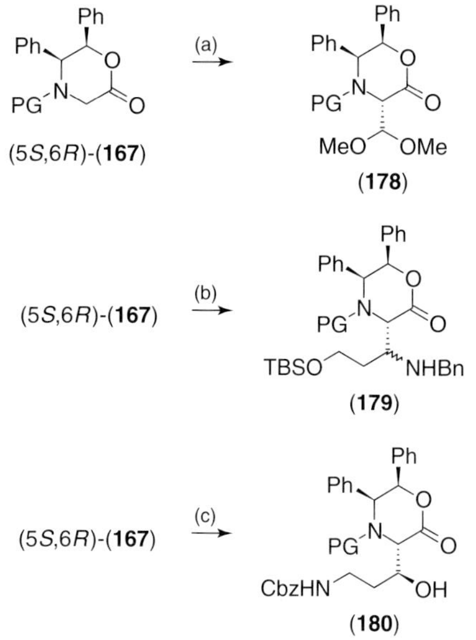

7.1 Six-Membered Rings with Another Heteroatom in the Same Ring | 355

Scheme 7.124

Reagents and conditions: (a) 1. TiCl$_4$/Et$_3$N, 2. (MeO)$_3$CH/CH$_2$Cl$_2$ (94%),
(b) 1. LiHMDS/THF, 2. Me$_2$AlCl, 3. TBSO(CH$_2$)$_2$CH=NBn (60%, dr 3.1:1),
(c) 1. Bu$_2$BOTf/Et$_3$N/CH$_2$Cl$_2$, 2. Cbz(CH$_2$)$_2$CHO (69%, dr 8:1).

Alkyl radicals generated from alkyl iodides were added to the imine bond in unsaturated morpholinones (S)-**156a** and (5R,6S)-**182** in the presence of Bu$_3$SnH/AIBN. In the reaction of disubstituted morpholinone (5R,6S)-**182** higher diastereoselectivity than that of (S)-**156a** was observed, affording the products **181** and **183**, respectively, by preferential attack *anti* with respect to the substituents. Complete stereoselectivity, but lower yield, was obtained by alkylating compound (5R,6S)-**182** at low temperature with Et$_3$B as complexing agent and a radical initiator [251].

Radical addition to morpholines **184** was performed in the presence of BF$_3$•OEt as Lewis acid using triethyl borane as the radical initiator providing the ethylated product **185** (R = Et) and ethylated imine **186** (R = iPr). Isopropyl radical addition proceeded smoothly in the presence of isopropyl iodide [252]. Under the same conditions, except using Lewis acid, radical addition to unsaturated 5,6-diphenylmorpholinone N-oxides were carried out in high (95%) diastereoselectivity [253].

Arylations Stereoselective introduction of aryl groups to position 3 was performed on substrate **187** in two alternative ways. Reaction of **187** with various aromatic compounds possessing bromine, methoxy or hydroxy groups as the substituents in the presence of TFA or reaction of boronic acid gave the corresponding substituted products **188** in 60–90% de [170, 254].

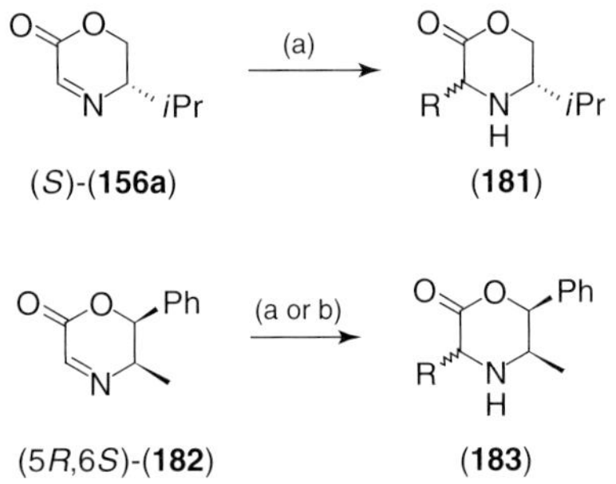

356 | 7 Asymmetric Synthesis of Ring Nitrogen Heterocycles

(S)-(156a) → (181)

(5R,6S)-(182) → (183)

R = nBu, tBu, cHex

Reagents and conditions: (a) RI/nBu$_3$SnH/AIBN/benzene/80 °C
(41–62%, dr 55:45 to 60:40 for 181, 75:25 to 87:13 for 183),
(b) RI/Et$_3$B/CH$_2$Cl$_2$/–40 °C (25–27%, dr 90:10 to 100:0).

Scheme 7.125

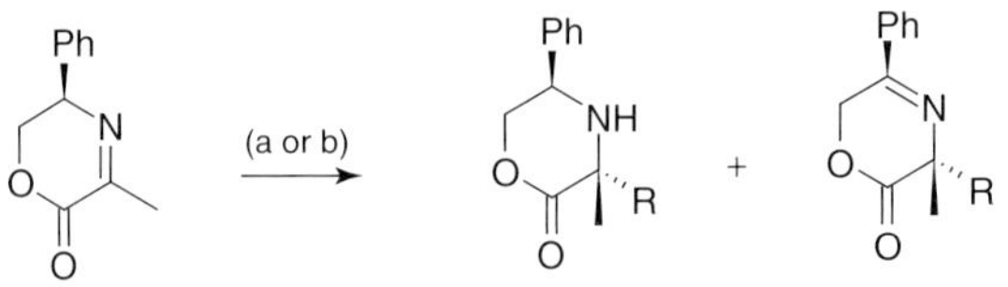

(184) → (185) + (186)

Reagents and conditions: (a) BF$_3$·OEt$_2$/Et$_3$B/hexane/CH$_2$Cl$_2$/0 °C-rfx
(42–74%, 81–87% de, 185:186=10:6 to 10:1),
(b) BF$_3$·OEt$_2$/iPrI/Et$_3$B/hexane/CH$_2$Cl$_2$/rfx
(185: 61%, 82% de, 186: 15%).

Scheme 7.126

Arylation of oxazinones **189** was achieved by treatment of the corresponding enolates with (arene)Mn(CO)$_3$ complexes, followed by oxidative demetalation of the intermediate substituted cyclohexadienyl-Mn(CO)$_3$ complex **190** to afford arylmorpholinones **191** in 75–99% de [255].

Stereoselective Reductions Reduction of alkylidene substituents in compounds **192** or **194** to the alkyl ones could be carried out either by hydrogen gas in the

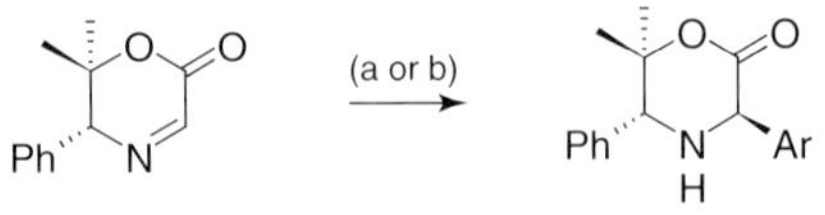

(187) → (188)

Reagents and conditions: (a) TFA/CH$_2$Cl$_2$/ArH (83–98%, 60–100% de),
(b) 4-MeOC$_6$H$_4$B(OH)$_2$/Cl(CH$_2$)$_2$Cl/TFA (85%, 90% de).

Scheme 7.127

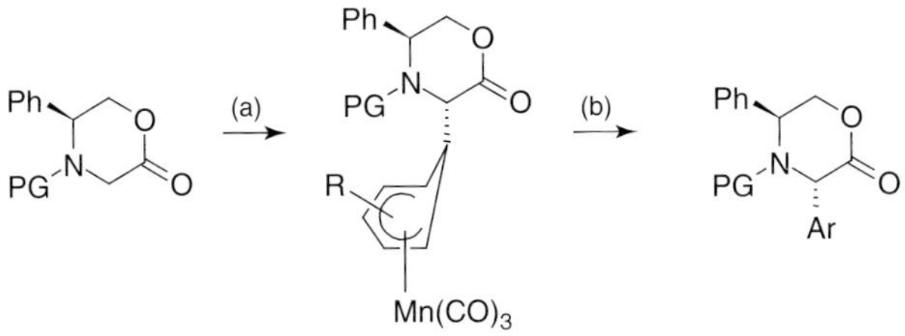

7.1 Six-Membered Rings with Another Heteroatom in the Same Ring

(189) → (a) → **(190)** → (b) → **(191)**

PG = Boc, Cbz
Reagents and conditions: (a) 1. NaHMDS/THF, 2. (arene)Mn(CO)$_3^+$PF$_6^-$,
(b) NBS/Et$_2$O (44–65%, 75–99% de).

Scheme 7.128

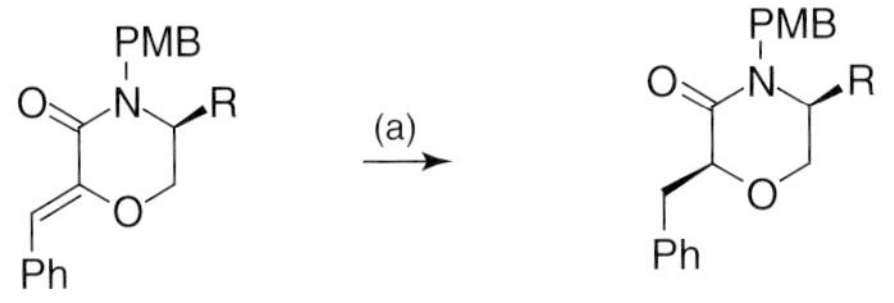

(192) → (a) → **(193)**

R = iBu, sBu
Reagents and conditions: (a) H$_2$/Pd–C (97–98%, dr 8:1 to 10:1).

Scheme 7.129

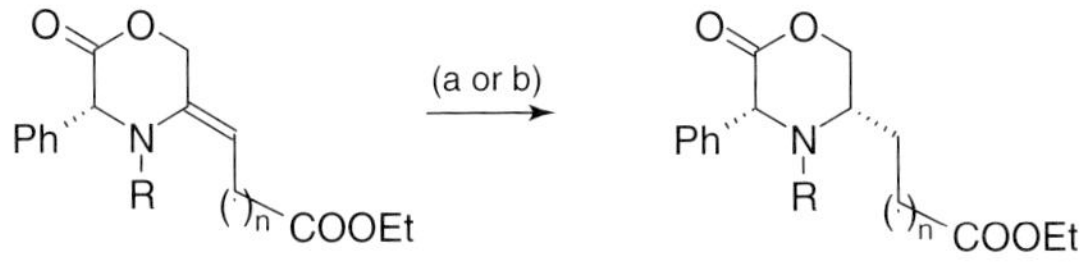

(194) → (a or b) → **(195)**

R = H, Bn
n = 0, 1, 2

Reagents and conditions: (a) BH$_3$·THF/MeCN (76–80%, 90–92% de),
(b) NaBH$_4$/AcOH/MeCN/TFA (71%, 94% de).

Scheme 7.130

presence of a Pd–C catalyst or with NaBH$_4$, or using the borane–THF complex as the reducing agent, providing the *cis* products **193** or **195**, respectively, in good stereoselectivity [186, 256].

Iridium complex-catalyzed asymmetric hydrogenation of achiral exocyclic double bond in **196** allowed the introduction of a chiral center on an achiral structure. The reaction was performed in the presence of chiral phosphine ligands and afforded the corresponding saturated derivatives (*S*)-**197** in good (79–85.5%) enantiomeric excesses [257].

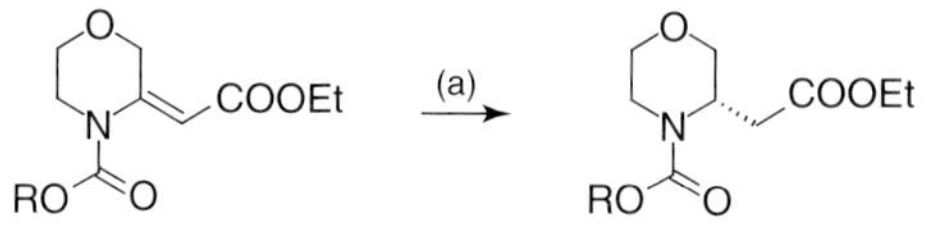

Scheme 7.131

Reagents and conditions: (a) 40–50 bar H$_2$/((COD)Ir(Cl))$_2$/(S)-BINAP or (S)-TolBINAP (100% conv, 79–85.5% ee).

Both endo- and exocyclic double bond in **198** was saturated using PtO$_2$ as the catalyst to **199** in excellent stereoselectivities (dr: 96:4 to 97:3). In the presence of formaldehyde, N-methylated derivative **200** was obtained with diastereomeric ratios of 91:9 to 98:2 [192, 193].

Diastereoselective reductions of chiral unsaturated morpholinones **201** to **202** were carried out by three different methods. PtO$_2$/H$_2$, NaBH(OAc)$_3$, or BH$_3$ proved to be the suitable reducing agent [256, 258].

For saturation of **203** to **204** or **205** to **206**, respectively, either PdCl$_2$ or Pd–C was applied successfully as hydrogenation catalyst [183, 259, 260].

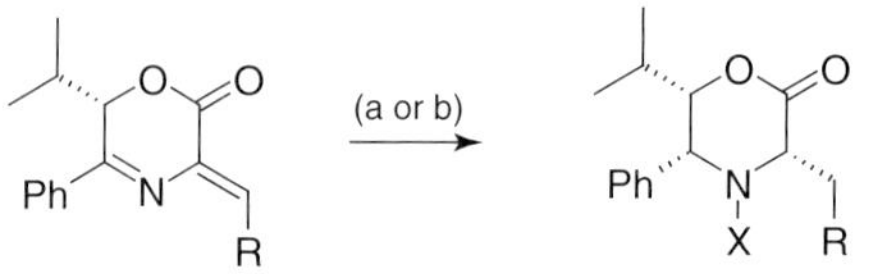

(198)
R = Me, Ph, iPr, iBu
X = H, Me

(199) X = H
(200) X = Me

Reagents and conditions: (a) PtO$_2$/H$_2$/MeOH (62–95%, dr 96:4 to 97:3),
(b) PtO$_2$/H$_2$/CH$_2$O/MeOH (63–75%, dr 91:9 to 98:2).

Scheme 7.132

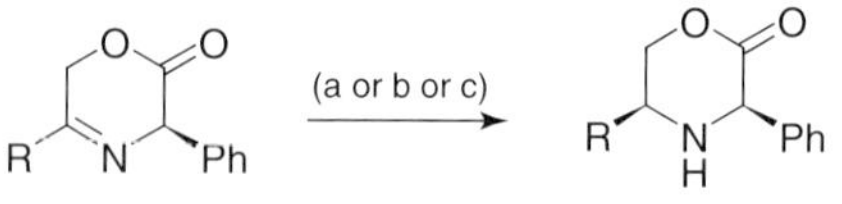

(201)   (202)

R = Me, Et, iPr, nBu, tBu

Reagents and conditions: (a) H$_2$/PtO$_2$/MeOH (73–89%, cis:trans = 83:17 to 93:7),
(b) NaBH(OAc)$_3$/TMSCl (73–83%, 60–86% de),
(c) BH$_3$·THF/MeCN (71–93%, 84–98% de).

Scheme 7.133

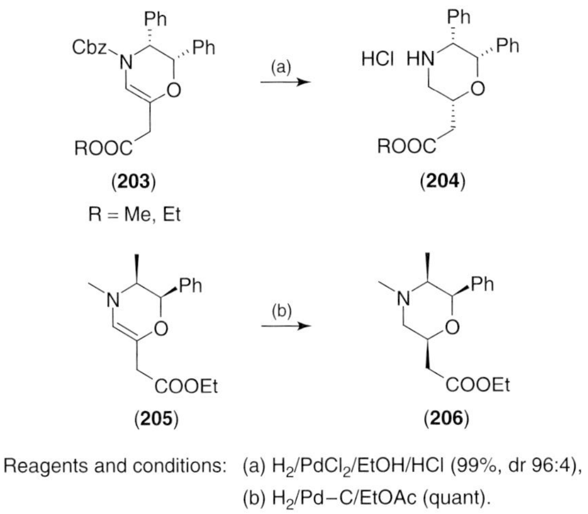

Reagents and conditions: (a) H$_2$/PdCl$_2$/EtOH/HCl (99%, dr 96:4),
(b) H$_2$/Pd–C/EtOAc (quant).

Scheme 7.134

References

1. Hale, K. J., Cai, J., Delisser, V., Manaviazar, S., Peak, S. A., Bhatia, G. S., Collins, T. C. and Jogiya, N. (**1996**) *Tetrahedron*, **52**, 1047–68.
2. Hale, K. J., Delisser, V. M. and Manaviazar, S. (**1992**) *Tetrahedron Lett.*, **33**, 7613–14.
3. Banteli, R., Brun, I., Hall, P. and Metternich, R. (**1999**) *Tetrahedron Lett.*, **40**, 2109–12.
4. Coats, R. A., Lee, S. L., Davis, K. A., Patel, K. M., Rhoads, E. K. and Howard, M. H. (**2004**) *J. Org. Chem.*, **69**, 1734–37.
5. Schmidt, U., Braun, C. and Sutoris, H. (**1996**) *Synthesis*, **2**, 223–29.
6. Ciufolini, M. A. and Xi, N. (**1997**) *J. Org. Chem.*, **62**, 2320–21.
7. Aoyagi, Y., Saitoh, Y., Ueno, T., Horiguchi, M., Takeya, K. and Williams, R. M. (**2003**) *J. Org. Chem.*, **68**, 6899–904.
8. Makino, K., Jiang, H., Suzuki, T. and Hamada, Y. (**2006**) *Tetrahedron: Asymmetry*, **17**, 1644–49.
9. Owings, F. F., Fox, M., Kowalski, C. J. and Baine, N. H. (**1991**) *J. Org. Chem.*, **56**, 1963–66.
10. Yoshida, N., Awano, K., Kobayashi, T. and Fujimori, K. (**2004**) *Synlett*, **10**, 1554–56.
11. Gardiner, J. and Abell, A. D. (**2003**) *Tetrahedron Lett.*, **44**, 4227–30.
12. Alvarez-Ibarra, C., Csaky, A. G., Gomez de la Oliva, C. and Rodriguez, E. (**2001**) *Tetrahedron Lett.*, **42**, 2129–31.
13. Hitchcock, P. B., Papadopoulos, K. and Young, D. W. (**2003**) *Org. Biomol. Chem.*, **1**, 2670–81.
14. Ahmed, O., Hitchcock, P. B. and Young, D. W. (**2006**) *Org. Biomol. Chem.*, **4**, 1596–603.
15. Oelke, A. J., Kumarn, S., Longbottom, D. A. and Ley, S. V. (**2006**) *Synlett*, **16**, 2548–52.
16. Aspinall, I. H., Cowley, P. M., Mitchell, G., Raynor, C. M. and Stoodley, R. J. (**1999**) *J. Chem. Soc. Perkin Trans. 1*, **18**, 2591–99.
17. Robiette, R., Cheboub-Benchaba, K., Peeters, D. and Marchand-Brynaert, J. (**2003**) *J. Org. Chem.*, **68**, 9809–12.
18. Arroyo, Y., Rodriguez, J. F., Santos, M., Sanz Tejedor, M. A., Vaca, I. and Garcia Ruano, J. L. (**2004**) *Tetrahedron: Asymmetry*, **15**, 1059–63.
19. Makino, K., Henmi, Y., Terasawa, M., Hara, O. and Hamada, Y. (**2005**) *Tetrahedron Lett.*, **46**, 555–58.

20. Kaname, M., Arakawa, Y. and Yoshifuji, S. (**2001**) *Tetrahedron Lett.*, **42**, 2713–16.
21. Kawasaki, M. and Yamamoto, H. (**2006**) *J. Am. Chem. Soc.*, **128**, 16482–83.
22. Chu, K. S., Negrete, G. R., Konopelski, J. P., Lakner, F. J., Woo, N. T. and Olmstead, M. M. (**1992**) *J. Am. Chem. Soc.*, **114**, 1800–12.
23. Lakner, F. J., Chu, K. S., Negrete, G. R. and Konopelski, J. P. (**1996**) *Org. Synth.*, **73**, 201–14.
24. Iglesias-Arteaga, M. A., Castellanos, E. and Juaristi, E. (**2003**) *Tetrahedron: Asymmetry*, **14**, 577–80.
25. Mahindaratne, M. P. D., Quinones, B. A., Recio, A., Rodriguez, E. A., Lakner, F. J. and Negrete, G. R. III (**2005**) *ARKIVOC*, **11**, 321–28.
26. Mahindaratne, M. P. D., Quinones, B. A., Recio, A., Rodriguez, E. A., Lakner, F. J. and Negrete, G. R. (**2005**) *Tetrahedron*, **61**, 9495–501.
27. Avila-Ortiz, C. G., Reyes-Rangel, G. and Juaristi, E. (**2005**) *Tetrahedron*, **61**, 8372–81.
28. Diaz-Sanchez, B. R., Iglesias-Arteaga, M. A., Melgar-Fernandez, R. and Juaristi, E. (**2007**) *J. Org. Chem.*, **72**, 4822–25.
29. Juaristi, E. and Escalante, J. (**1993**) *J. Org. Chem.*, **58**, 2282–85.
30. Escalante, J. and Juaristi, E. (**1995**) *Tetrahedron Lett.*, **36**, 4397–400.
31. Capriati, V., Degennaro, L., Florio, S., Luisi, R. and Cuocci, C. (**2007**) *Tetrahedron Lett.*, **48**, 8655–58.
32. Patino-Molina, R., Cubero-Lajo, I., Perez de Vega, M. J., Garcia-Lopez, M. T. and Gonzalez-Muniz, R. (**2007**) *Tetrahedron Lett.*, **48**, 3613–16.
33. Garratt, P. J., Thorn, S. N. and Wrigglesworth, R. (**1991**) *Tetrahedron Lett.*, **32**, 691–94.
34. Belov, V. N., Brands, M., Raddatz, S., Kruger, J., Nikolskaya, S., Sokolov, V. and de Meijere, A. (**2004**) *Tetrahedron*, **60**, 7579–89.
35. DeMong, D. E. and Williams, R. M. (**2001**) *Tetrahedron Lett.*, **42**, 3529–32.
36. Barluenga, J., Olano, B., Fustero, S., Foces-Foces, M. C. and Hernandez Cano, F. (**1988**) *J. Chem. Soc., Chem. Commun.*, **6**, 410–12.
37. Barluenga, J., Viado, A. L., Aguilar, E., Fustero, S. and Olano, B. (**1988**) *J. Org. Chem.*, **58**, 5972–75.
38. Dondoni, A., Massi, A., Sabbatini, S. and Bertolasi, V. (**2002**) *J. Org. Chem.*, **67**, 6979–94.
39. Dondoni, A., Massi, A. Minghini, E., Sabbatini, S. and Bertolasi, V. (**2003**) *J. Org. Chem.*, **68**, 6172–83.
40. Munoz-Muniz, O. and Juaristi, E. (**2003**) *ARKIVOC*, **11**, 16–26.
41. Munoz-Muniz, O. and Juaristi, E. (**2003**) *Tetrahedron*, **59**, 4223–29.
42. Juaristi, E., Balderas, M. and Ramirez-Quiros, Y. (**1998**) *Tetrahedron: Asymmetry*, **9**, 3881–88.
43. Enders, D., Wortmann, L., Ducker, B. and Raabe, G. (**1999**) *Helv. Chim. Acta*, **82**, 1195–201.
44. Beaulieu, F., Arora, J., Veith, U., Taylor, N. J., Chapell, B. J. and Snieckus, V. (**1996**) *J. Am. Chem. Soc.*, **118**, 8727–28.
45. Trost, B. M. and Schroeder, G. M. (**2000**) *J. Org. Chem.*, **65**, 1569–73.
46. Brunner, H., Ittner, K.-P., Lunz, D., Schmatloch, S., Schmidt, T. and Zabel, M. (**2003**) *Eur. J. Org. Chem.*, **5**, 855–62.
47. Cardillo, G., Gentilucci, L., Tolomelli, A. and Tomasini, C. (**1998**) *J. Org. Chem.*, **63**, 3458–62.
48. Weigl, M. and Wünsch, B. (**2002**) *Tetrahedron*, **58**, 1173–83.
49. Falorni, M., Satta, M., Conti, S. and Giacomelli, G. (**1993**) *Tetrahedron: Asymmetry*, **4**, 2389–98.
50. Viso, A., Fernandez de la Pradilla, R., Flores, A. and Garcia, A. (**2007**) *Tetrahedron*, **63**, 8017–26.
51. Schöllkopf, U. and Neubauer, H. J. (**1982**) *Synthesis*, **10**, 861–64.
52. Ma, C., Liu, X., Li, X., Flippen-Anderson, J., Yu, S. and Cook, J. M. (**2001**) *J. Org. Chem.*, **66**, 4525–42.
53. Li, X., Yin, W., Sarma, P. V. V. S., Zhou, H., Ma, J. and Cook, J. M. (**2004**) *Tetrahedron Lett.*, **45**, 8569–73.
54. Rose, J. E., Leeson, P. D. and Gani, D. (**1992**) *J. Chem. Soc. Perkin Transactions 1: Org. Bio-Org. Chem.*, **13**, 1563–65.

55 Rose, J. E., Leeson, P. D. and Gani, D. (**1995**) *J. Chem. Soc. Perkin Transactions 1: Org. Bio-Org. Chem.*, **2**, 157–65.
56 Couladouros, E. A. and Magos, A. D. (**2005**) *Mol. Diver.*, **9**(1–3), 99–109.
57 Kim, Y., Ha, H. J., Han, K., Ko, S. W., Yun, H., Yoon, H. J., Kim, M. S. and Lee, W. K. (**2005**) *Tetrahedron Lett.*, **46**, 4407–9.
58 Paradisi, F., Porzi, G. and Sandri, S. (**2001**) *Tetrahedron: Asymmetry*, **12**, 3319–24.
59 Jiang, X. H., Song, Y. L., Feng, D. Z. and Long, Y. Q. (**2005**) *Tetrahedron*, **61**, 1281–88.
60 Vaz, E., Fernandez-Suarez, M. and Munoz, L. (**2003**) *Tetrahedron: Asymmetry*, **14**, 1935–42.
61 Sewald, N., Seymour, L. C., Burger, K., Osipov, S. N., Kolomiets, A. F. and Fokin, A. V. (**1994**) *Tetrahedron: Asymmetry*, **5**, 1051–60.
62 Govek, S. P. and Overman, L. E. (**2007**) *Tetrahedron*, **63**, 8499–513.
63 Mickelson, J. W., Belonga, K. L. and Jacobsen, E. J. (**1995**) *J. Org. Chem.*, **60**, 4177–83.
64 Mickelson, J. W. and Jacobsen, E. J. (**1995**) *Tetrahedron: Asymmetry*, **6**, 19–22.
65 Liu, Q., Marchington, A. P., Boden, N. and Rayner, C. M. (**1995**) *Synlett*, **10**, 1037–39.
66 Mizutani, H., Takayama, J. and Honda, T. (**2005**) *Synlett*, **2**, 328–30.
67 Kogan, T. P. and Rawson, T. E. (**1992**) *Tetrahedron Lett.*, **33**, 7089–92.
68 Viso, A., Fernandez de la Pradilla, R., Lopez-Rodriguez, M. L., Garcia, A. and Tortosa, M. (**2002**) *Synlett*, **5**, 755–58.
69 Viso, A., Fernandez de la Pradilla, R., Flores, A., Garcia, A., Tortosa, M. and Lopez-Rodriguez, M. L. (**2006**) *J. Org. Chem.*, **71**, 1442–48.
70 Oishi, T. and Hirama, M. (**1992**) *Tetrahedron Lett.*, **33**, 639–42.
71 Cho, S., Keum, G., Kang, S. B., Han, S. Y. and Kim, Y. (**2003**) *Mol. Diver.*, **6**(3–4), 283–86.
72 Sollis, S. L. (**2005**) *J. Org. Chem.*, **70**, 4735–40.
73 Argouarch, G., Gibson, C. L., Stones, G. and Sherrington, D. C. (**2002**) *Tetrahedron Lett.*, **43**, 3795–98.
74 Abellan, T., Najera, C. and Sansano, J. M. (**1998**) *Tetrahedron: Asymmetry*, **9**, 2211–14.
75 Najera, C., Abellan, T. and Sansano, J. M. (**2000**) *Eur. J. Org. Chem.*, **15**, 2809–20.
76 Abellan, T., Mancheno, B., Najera, C. and Sansano, J. M. (**2001**) *Tetrahedron*, **57**, 6627–40.
77 DiMaio, J. and Belleau, B. (**1989**) *J. Chem. Soc. Perkin Trans. 1: Org. Bio-Org. Chem.*, **9**, 1687–89.
78 Lee, B. K., Kim, M. S., Hahm, H. S., Kim, D. S., Lee, W. K. and Ha, H. J. (**2006**) *Tetrahedron*, **62**, 8393–97.
79 Patino-Molina, R., Herranz, R., Gracía-López, M. T. and González-Muniz, R. (**1999**) *Tetrahedron*, **55**, 15001–10.
80 Jiang, B., Yang, C. G. and Wang, J. (**2001**) *J. Org. Chem.*, **66**, 4865–69.
81 Kouko, T., Matsumura, K. and Kawasaki, T. (**2005**) *Tetrahedron*, **61**, 2309–18.
82 Jiang, B., Yang, C. G. and Wang, J. (**2002**) *J. Org. Chem.*, **67**, 1396–98.
83 Higuchi, K., Takei, R., Kouko, T. and Kawasaki, T. (**2007**) *Synthesis*, **5**, 669–74.
84 Yang, C. G., Wang, J., Tang, X. X. and Jiang, B. (**2002**) *Tetrahedron: Asymmetry*, **13**, 383–94.
85 Askin, D., Eng, K. K., Rossen, K., Purick, R. M., Well, R. P., Volante, R. P. and Reider, P. J. (**1994**) *Tetrahedron Lett.*, **35**, 673–76.
86 Bigge, C. F., Johnson, G., Ortwine, D. F., Drummond, J. T., Retz, D. M., Brahce, L. J., Marcoux, F. W. and Probert, A. W. Jr (**1992**) *J. Med. Chem.*, **35**, 1371–84.
87 Bruce, M. A., St Laurent, D. R., Poindexter, G. S., Monkovic, I., Huang, S. and Balasubramanian, N. (**1995**) *Synth. Commun.*, **25**, 2673–84.
88 Warshawsky, A. M., Patel, M. V. and Chen, T.-M. (**1997**) *J. Org. Chem.*, **62**, 6439–40.
89 Gurjar, M. K., Karmakar, S., Mohapatra, D. K. and Phalgune,

U. D. (2002) *Tetrahedron Lett.*, **43**, 1897–900.

90 Lukina, T. V., Sviridov, S. I., Shorshnev, S. V., Alexandrov, G. G. and Stepanov, A. E. (2005) *Tetrahedron Lett.*, **46**, 1205–7.

91 Lukina, T. V., Sviridov, S. I., Shorshnev, S. V., Stepanov, A. E. and Alexandrov, G. G. (2006) *Tetrahedron Lett.*, **47**, 51–54.

92 Yamazaki, A. and Achiwa, K. (1995) *Tetrahedron: Asymmetry*, **6**, 1021–24.

93 Uozumi, Y., Tanahashi, A. and Hayashi, T. (1993) *J. Org. Chem.*, **58**, 6826–32.

94 Nakano, H., Yokoyama, J., Fujita, R. and Hongo, H. (2002) *Tetrahedron Lett.*, **43**, 7761–64.

95 Kukula, P. and Prins, R. (2002) *J. Catal.*, **208**, 404–11.

96 Santes, V., Gomez, E., Zarate, V., Santillan, R., Farfan, N. and Rojas-Lima, S. (2001) *Tetrahedron: Asymmetry*, **12**, 241–47.

97 Schöllkopf, U., Hartwig, W. and Groth, U. (1979) *Angew. Chem.*, **91**, 922–23.

98 Schöllkopf, U. and Nozulak, J. (1982) *Synthesis*, **10**, 866–68.

99 Groth, U., Schmeck, R. C. and Schöllkopf, U. (1993) *Liebigs Ann. Chem.*, **3**, 321–23.

100 Hammer, K. and Undheim, K. (1997) *Tetrahedron*, **53**, 2309–22.

101 Otsubo, K., Morita, S., Uchida, M., Yamasaki, K., Kanbe, T. and Shimizu, T. (1991) *Chem. Pharm. Bull.*, **39**, 2906–09.

102 Hartwig, W. and Mittendorf, J. (1991) *Synthesis*, **11**, 939–41.

103 Cushman, M. and Lee, E. S. (1992) *Tetrahedron Lett.*, **33**, 1193–96.

104 Shapiro, G., Buechler, D., Ojea, V., Pombo-Villar, E., Ruiz, M. and Weber, H. P. (1993) *Tetrahedron Lett.*, **34**, 6255–28.

105 Baldwin, J. E., Adlington, R. M., Bebbington, D. and Russell, A. T. (1994) *Tetrahedron*, **50**, 12015–28.

106 Zhang, P., Liu, R. and Cook, J. M. (1995) *Tetrahedron Lett.*, **36**, 9133–36.

107 Amici, R., Pevarello, P., Colombo, M. and Varasi, M. (1996) *Synthesis*, **10**, 1177–79.

108 Moeller, B. S., Benneche, T. and Undheim, K. (1996) *Tetrahedron*, **52**, 8807–12.

109 Ohaba, M., Imasho, M. and Fujii, T. (1996) *Heterocycles*, **42**, 219–28.

110 Bull, S. D., Davies, S. G., Garner, A. C. and Mujtaba, N. (2001) *Synlett*, **6**, 781–84.

111 Wild, N. and Groth, U. (2003) *Eur. J. Org. Chem.*, **22**, 4445–49.

112 Kim, S., Kim, E. Y., Ko, H. and Jung, Y. H. (2003) *Synthesis*, **14**, 2194–98.

113 Jam, F., Tullberg, M., Luthman, K. and Grotli, M. (2007) *Tetrahedron*, **63**, 9881–89.

114 Bull, S. D., Davies, S. G., Epstein, S. W. and Ouzman, J. V. A. (1998) *Chem. Commun.*, **6**, 659–60.

115 Bull, S. D., Davies, S. G., Garner, A. C. and O'Shea, M. D. (2001) *J. Chem. Soc., Perkin Trans. 1*, **24**, 3281–87.

116 Bull, S. D., Davies, S. G., Garner, A. C., O'Shea, M. D., Savory, E. D. and Snow, E. J. (2002) *J. Chem. Soc., Perkin Trans. 1*, **22**, 2442–48.

117 Bull, S. D., Davies, S. G., Epstein, S. W., Garner, A. C., Mujtaba, N., Roberts, P. M., Savory, E. D., Smith, A. D., Tamayo, J. A. and Watkin, D. J. (2006) *Tetrahedron*, **62**, 7911–25.

118 Galeazzi, R., Garavelli, M., Grandi, A., Monari, M., Porzi, G. and Sandri, S. (2003) *Tetrahedron: Asymmetry*, **14**, 2639–49.

119 Balducci, D., Porzi, G. and Sandri, S. (2004) *Tetrahedron: Asymmetry*, **15**, 1085–93.

120 Balducci, D., Crupi, S., Galeazzi, R., Piccinelli, F., Porzi, G. and Sandri, S. (2005) *Tetrahedron: Asymmetry*, **16**, 1103–12.

121 Balducci, D., Grandi, A., Porzi, G. and Sandri, S. (2005) *Tetrahedron: Asymmetry*, **16**, 1453–62.

122 Balducci, D., Grandi, A., Porzi, G. and Sandri, S. (2006) *Tetrahedron: Asymmetry*, **17**, 1521–28.

123 Balducci, D., Bottoni, A., Calvaresi, M., Porzi, G. and Sandri, S. (2007) *Tetrahedron: Asymmetry*, **18**, 1448–56.

124 Schanen, V., Riche, C., Chiaroni, A., Quirion, J. C. and Husson, H. P. (1994) *Tetrahedron Lett.*, **35**, 2533–36.

125 Schanen, V., Cherrier, M. P., De Melo, S. J., Quirion, J. C. and Husson, H. P. (**1996**) *Synthesis*, **7**, 833–37.
126 Franceschini, N., Sonnet, P. and Guillaume, D. (**2005**) *Org. Biomol. Chem.*, **3**, 787–93.
127 Gull, R. and Schöllkopf, U. (**1985**) *Synthesis*, **11**, 1052–55.
128 Hammer, K., Romming, C. and Undheim, K. (**1998**) *Tetrahedron*, **54**, 10837–50.
129 Andrei, M. and Undheim, K. (**2004**) *Tetrahedron: Asymmetry*, **15**, 53–63.
130 Schöllkopf, U., Westphalen, K. O., Schroeder, J. and Horn, K. (**1988**) *Liebigs Ann. Chem.*, **8**, 781–86.
131 Pettig, D. and Schöllkopf, U. (**1988**) *Synthesis*, **3**, 173–75.
132 Ojea, V., Ruiz, M., Shapiro, G. and Pombo-Villar, E. (**1994**) *Tetrahedron Lett.*, **35**, 3273–76.
133 Ruiz, M., Ojea, V., Shapiro, G., Weber, H. P. and Pombo-Villar, E. (**1994**) *Tetrahedron Lett.*, **35**, 4551–54.
134 Schöllkopf, U., Pettig, D., Busse, U., Egert, E. and Dyrbusch, M. (**1986**) *Synthesis*, **9**, 737–40.
135 Wild H. and Born, L. (**1991**) *Angew. Chem., Int. Ed. Engl.*, **30**, 1685–87.
136 Busch, K., Groth, U. M., Kühnle, W. and Schöllkopf, U. (**1992**) *Tetrahedron*, **48**, 5607–18.
137 Schöllkopf, U., Kühnle, W., Egert, E. and Dyrbusch, M. (**1987**) *Angew. Chem., Int. Ed. Engl.*, **26**, 480–82.
138 Pearson, A. J., Bruhn, P. R., Gouzoules, F. and Lee, S. H. (**1989**) *J. Chem. Soc. Chem. Comm.*, **10**, 659–61.
139 Pearson, A. J., Lee, S. H. and Gouzoules, F. (**1990**) *J. Chem. Soc., Perkin Trans. 1: Org. Bio-Org. Chem.*, **8**, 2251–54.
140 Lee, S. H. and Lee, E. K. (**2001**) *Bull. Korean Chem. Soc.*, **22**, 551–52.
141 Groth, U. and Schöllkopf, U. (**1982**) *Synthesis*, **10**, 864–66.
142 Neubauer, H. J., Baeza, J., Freer, J. and Schöllkopf, U. (**1985**) *Liebigs Ann Chem.*, **7**, 1508–11.
143 Kotha, S. and Kuki, A. (**1992**) *J. Chem. Soc., Chem. Commun.*, **5**, 404–6.
144 Cremonesi, G., Dalla Croce, P., Fontana, F. and La Rosa, C. (**2006**) *Tetrahedron: Asymmetry*, **17**, 2637–41.
145 Cremonesi, G., lla Croce, P., Fontana, F., Forni, A. and La Rosa, C. (**2007**) *Tetrahedron: Asymmetry*, **18**, 1667–75.
146 Schöllkopf, U., Nozulak, J. and Grauert, M. (**1985**) *Synthesis*, **1**, 55–56.
147 Grauert, M. and Schöllkopf, U. (**1985**) *Liebigs Ann. Chem.*, **9**, 1817–24.
148 Schöllkopf, U. and Bardenhagen, J. (**1987**) *Liebigs Ann. Chem.*, **5**, 393–97.
149 Beulshausen, T., Groth, U. and Schöllkopf, U. (**1991**) *Liebigs Ann. Chem.*, **11**, 1207–9.
150 Schöllkopf, U., Neubauer, H. J. and Hauptreif, M. (**1985**) *Angew. Chem., Int. Ed. Engl.*, **24**, 1066–67.
151 Leeming, P., Fronczek, F. R., Ager, D. J. and Laneman, S. A. (**2000**) *Top. Catal.*, **13**, 175–77.
152 Davies, S. G., Rodriguez-Solla, H., Tamayo, J. A., Garner, A. C. and Smith, A. D. (**2004**) *Chem. Commun.*, **21**, 2502–3.
153 Davies, S. G., Rodriguez-Solla, H., Tamayo, J. A., Cowley, A. R., Concellon, C., Garner, A. C., Parkes, A. L. and Smith, A. D. (**2005**) *Org. Biomol. Chem.*, **3**, 1435–47.
154 Kuwano, R., Uemura, T., Saitoh, M. and Ito, Y. (**2004**) *Tetrahedron: Asymmetry*, **15**, 2263–71.
155 Barbaric, M., Kraljevic, S., Grce, M. and Zorc, B. (**2003**) *Acta Pharm.*, **53**, 175–86.
156 Guanti, G., Banfi, L. and Narisano, E. (**1989**) *Tetrahedron Lett.*, **30**, 5511–14.
157 Guanti, G., Banfi, L., Narisano, E. and Thea, S. (**1992**) *Synlett*, **4**, 311–12.
158 Hitchcock, S. R., Nora, G. P., Casper, D. M., Squire, M. D., Maroules, C. D., Ferrence, G. M., Szczeoura, L. F. and Standard, J. M. (**2001**) *Tetrahedron*, **57**, 9789–98.
159 Hitchcock, S. R., Casper, D. M., Vaughn, J. F., Finefield, J. M.,

and Katsuki, T. (**2004**) *Tetrahedron Lett.*, **45**, 7277–81.
223 Shafer, C. M. and Molinski, T. F. (**1996**) *J. Org. Chem.*, **61**, 2044–50.
224 Henry, N. and O'Neil, I. A. (**2007**) *Tetrahedron Lett.*, **48**, 1691–94.
225 Glaeske, K. W., Naidu, B. N. and West, F. G. (**2003**) *Tetrahedron: Asymmetry*, **14**, 917–20.
226 Williams, R. M. and Yuan, C. (**1992**) *J. Org. Chem.*, **57**, 6519–27.
227 Baldwin, J. E., Lee, V. and Schofield, C. J. (**1992**) *Synlett*, **3**, 249–51.
228 Williams, R. M., Fegley, G. J., Gallegos, R., Schaefer, F. and Pruess, D. L. (**1996**) *Tetrahedron*, **52**, 1149–64.
229 Bender, D. M. and Williams, R. M. (**1997**) *J. Org. Chem.*, **62**, 6690–91.
230 Van den Nieuwendijk, A. M. C. H., Kriek, N. M. A. J., Brussee, J., Van Boom, J. H. and Van der Gen, A. (**2000**) *Eur. J. Org. Chem.*, **22**, 3683–91.
231 Sui, G., Kele, P., Orbulescu, J., Huo, Q. and Leblanc, R. M. (**2002**) *Lett. Pept. Sci.*, **8**, 47–51.
232 Jin, W. and Williams, R. M. (**2003**) *Tetrahedron Lett.*, **44**, 4635–39.
233 Williams, R. M., Sinclair, P. J. and DeMong, D. E. (**2003**) *Org. Synth.*, **80**, 31–37.
234 Gustafsson, T., Schou, M., Almqvist, F. and Kihlberg, J. (**2004**) *J. Org. Chem.*, **69**, 8694–701.
235 Looper, R. E., Runnegar, M. T. C. and Williams, R. M. (**2006**) *Tetrahedron*, **62**, 4549–62.
236 Aoyagi, Y. and Williams, R. M. (**1998**) *Synlett*, **10**, 1099–101.
237 Lee, S. H., Lee, E. K. and Jeun, S. M. (**2002**) *Bull. Kor. Chem. Soc.*, **23**, 931–32.
238 Porzi, G. and Sandri, S. (**1996**) *Tetrahedron: Asymmetry*, **7**, 189–96.
239 Arcelli, A., Balducci, D., Estevao Neto, S. F., Porzi, G. and Sandri, M. (**2007**) *Tetrahedron: Asymmetry*, **18**, 562–68.
240 Williams, R. M. and Im, M. N. (**1991**) *J. Am. Chem. Soc.*, **113**, 9276–86.
241 Oishi, S., Kang, S. U., Liu, H., Zhang, M., Yang, D., Deschamps, J. R. and Burke, T. R. (**2004**) *Tetrahedron*, **60**, 2971–77.
242 Chinchilla, R., Falvello, L. R., Galindo, N. and Najera, C. (**1997**) *Angew. Chem., Int. Ed. Engl.*, **36**, 995–97.
243 Chinchilla, R., Galindo, N. and Najera, C. (**1998**) *Tetrahedron: Asymmetry*, **9**, 2769–72.
244 Jain, R. P. and Williams, R. M. (**2001**) *Tetrahedron*, **57**, 6505–9.
245 Aoyagi, Y. and Williams, R. M. (**1998**) *Tetrahedron*, **54**, 10419–433.
246 Pansare, S. V., Jain, R. P. and Ravi, R. G. (**1999**) *Tetrahedron: Asymmetry*, **10**, 3103–6.
247 Pansare, S. V. and Adsool, V. A. (**2007**) *Tetrahedron Lett.*, **48**, 7099–101.
248 DeMong, D. E. and Williams, R. M. (**2002**) *Tetrahedron Lett.*, **43**, 2355–57.
249 DeMong, D. E. and Williams, R. M. (**2003**) *J. Am. Chem. Soc.*, **125**, 8561–65.
250 DeMong, D. E. and Williams, R. M. (**2001**) *Tetrahedron Lett.*, **42**, 183–85.
251 Bertrand, M. P., Feray, L., Nouguier, R. and Stella, L. (**1998**) *Synlett*, **7**, 780–82.
252 Miyabe, H., Yamaoka, Y., Takemoto, Y., (**2005**) *J. Org. Chem.*, **70**, 3324–3327.
253 Ueda, M., Miyabe, H., Teramachi, M., Miyata, O. and Naito, T. (**2005**) *J. Org. Chem.*, **70**, 6653–60.
254 Toma, S., Endo, A., Kan, T. and Fukuyama, T. (**2001**) *Synlett*, **7**, 1179.
255 Lee, S. H. and Ahn, D. J. (**1999**) *Bull. Kor. Chem. Soc.*, **20**, 264–66.
256 Segat-Dioury, F., Lingibé, O., Graffe, B., Sachet, M. C. and Lhommet, G. (**2000**) *Tetrahedron*, **56**, 233–48.
257 Pousset, C., Callens, R., Marinetti, A. and Larcheveque, M. (**2004**) *Synlett*, **15**, 2766–70.
258 Lingibe, O., Graffe, B., Sacquet, M. C. and Lhommet, G. (**1994**) *Heterocycles*, **37**, 1469–72.
259 Jain, R. P. and Williams, R. M. (**2001**) *Tetrahedron Lett.*, **42**, 4437–40.
260 Jain, R. P. and Williams, R. M. (**2002**) *J. Org. Chem.*, **67**, 6361–65.

8
Asymmetric Synthesis of Seven-Membered Rings with More Than One Heteroatom

Jacques Royer

8.1
Diazepines

The diazepines represent a large class of compounds for which many syntheses were described in the literature. They are natural and synthetic products with interesting biological activity. Among diazepines, the 1,2-diazepine are rare while 1,4-diazepines are the most often reported isomers. The 1,3-diazepines are urea or amine derivatives and their preparation was on the basis of the access to these functional groups. Concerning the 1,4-diazepines, many methods of preparation were reported but most of them were derived from amino acids, which furthermore allowed the introduction of chirality.

8.1.1
1,2-Diazepines

Asymmetric synthesis of chiral 1,2-diazepine is rarely reported in the literature though some of them are known as *biologically active compounds*. For example, talampanel was found to possess potential antiepilectic, neuroprotectant and skeletal muscle relaxant activities [1] and cilazapril (Figure 8.1) is effective in the treatment of hypertension and other cardiovascular disorders [2].

A very few methods were reported and were classical ones on the basis of a unique strategy. They are mainly prepared through the bis-condensation of a hydrazine to a 1,5-dicarbonyl derivative. The latter is thus the chiral part of the molecule and no new chiral center is formed during the process. The bis-condensation may be attained during a one-pot reaction; for example, the pyrrolodiazepine **3**, a potential acetylcholine esterase (ACE) inhibitor, was prepared through the reaction of a monosubstituted unprotected hydrazine with the cetoacide derivative **1** [3] to give the required heterocycle **2** in modest yield (Scheme 8.1).

Asymmetric Synthesis of Nitrogen Heterocycles. Edited by Jacques Royer
Copyright © 2009 WILEY-VCH Verlag GmbH & Co. KGaA, Weinheim
ISBN: 978-3-527-32036-3

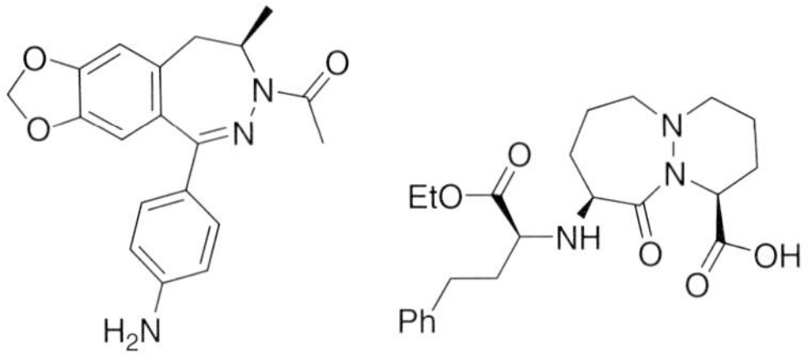

368 | *8 Asymmetric Synthesis of Seven-Membered Rings with More Than One Heteroatom*

Fig. 8.1 Biologically active 1,2-diazepines.

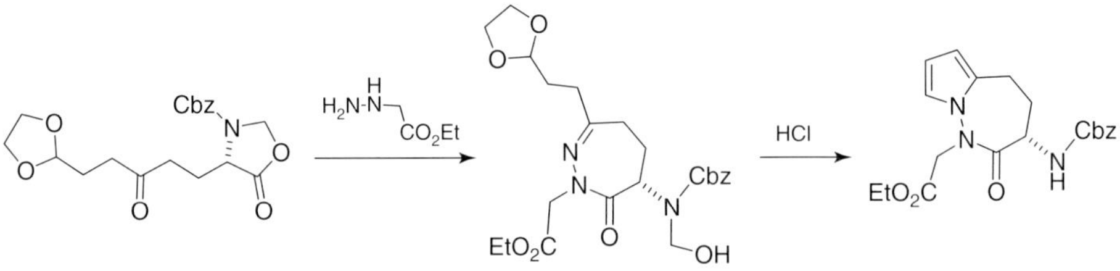

Scheme 8.1

Indeed, a two-step process seemed more efficient and is currently used. The 1,5-dicarbonyl compound may be a suitably protected glutamic acid or a synthetic product. The hydrazine part may also be chiral as in the following two examples. In these cases, devoted to the preparation of interleukin-1β converting enzyme (ICE) inhibitors the condensation is sequential and ends up by lactamization [4, 5] or by reductive alkylation [6] (Scheme 8.2).

In the preparation of **9**, a sugar (D-xylose) was the starting material, which was treated with a monoprotected hydrazine (Scheme 8.3). The basic treatment of mesitylsulfonate **7** gave in good yield and without loss of optical purity, the seven-membered ring **8** probably via the intermediate formation of an epoxide [7].

8.1.2
1,3-Diazepines

These heterocycles are widely represented by compounds found as potent HIV protease inhibitors. This is particularly the case for cyclic ureas such as DMP 323

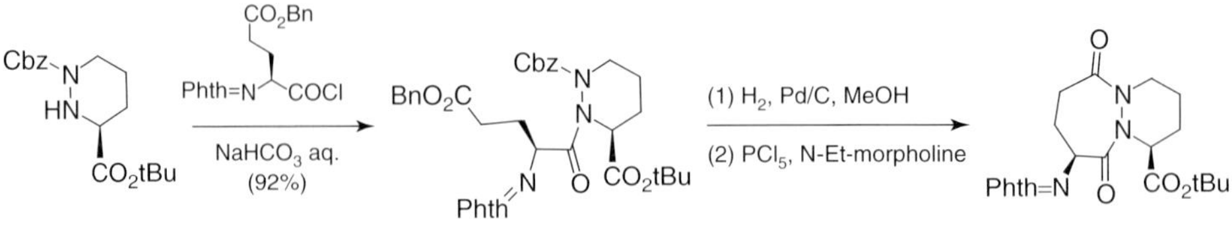

Scheme 8.2

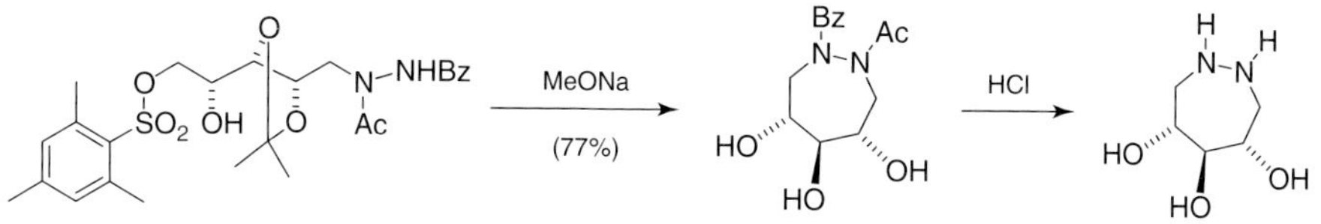

8.1 Diazepines | 369

Scheme 8.3

and DMP 450. The main characteristics of their structures are the seven-membered ring, the cyclic urea functionality, and a C2 symmetry. The synthesis of these compounds and their analogs is, in most cases, the same and consists of the preparation of a chiral 1,4-diamine, which is transformed into a cyclic urea with a suitable reagent, such as carbonyldiimidazole, [8] phosgene [9], trialkylorthocarbonate [10], or dimethylcyanodithioiminocarbonate via the cyanoguanidine [11]. The following sequence [8c] (Scheme 8.4) is representative of the published syntheses. In this example, the chiral diamine **11** was obtained through the low valent vanadium pinacol coupling with the aldehyde **10** derived from phenylalanine. Several examples are also reported on the preparation of analogous diamines from tartaric acid or sugars.

The preparation of structurally related C2-symmetric guanidines was also reported. These compounds are potent glycosidase inhibitors. Diamine **13** (Scheme 8.5), derived from mannitol, when treated with carbon disulfide in pyridine gave thiourea **14**, which could be transformed to guanidine **15** [12].

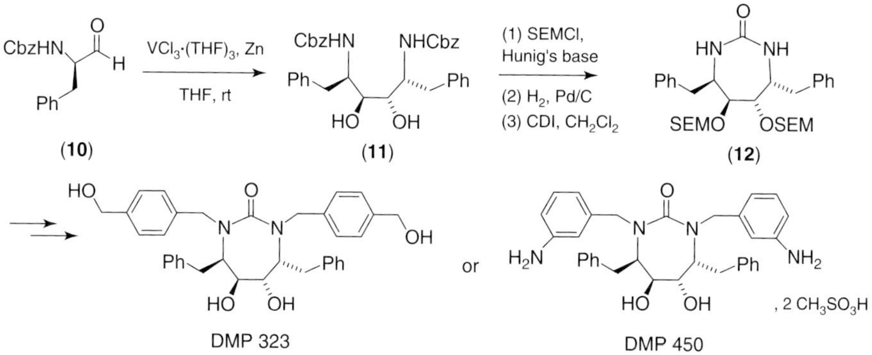

Scheme 8.4

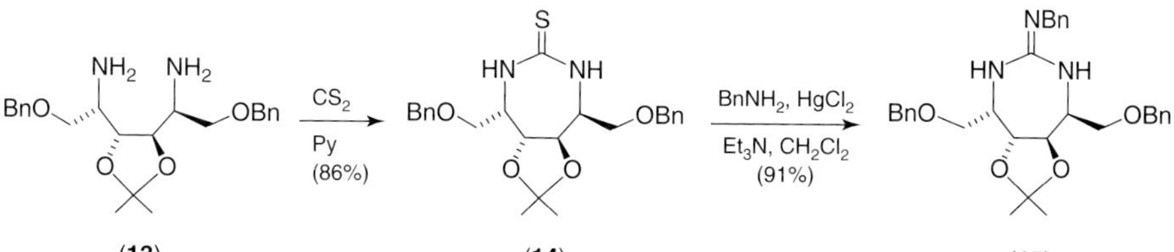

Scheme 8.5

Fused diazepine–pyrrolidine compounds have been obtained by Wasserman by the use of the vinyl tricarbonyl reagent he introduced some years ago (Scheme 8.6). Hence, the amide derived from glutamic acid was condensed with this tricarbonyl reagent **17** to give a bicyclic compound in a fair 33% yield [13]. A single stereomer was obtained but its configuration not reported.

A Pd-catalyzed cyclization process of 2-vinylpyrrolidine **19** and aryl isocyanate or carbodiimide was recently proposed [14] and gave a vinyl 1,3-diazepine **20** (Scheme 8.7). This interesting reaction, which formally implied the insertion of the carbodiimide in the pyrrolidine ring, was also conducted in the presence of chiral phophine ligands such as (*R*,*R*)-NORPHOS; unfortunately, only a modest chiral induction was observed.

A much more classical reaction was described in 2005 to achieve the asymmetric synthesis of a 1,3-diazepine system. The synthesis was conducted in solid phase and the cyclization to form the azepine ring was the result of the trapping of an *in situ* formed acyliminium (Scheme 8.8). Acidic deprotection of oxazidine A gave an aldehyde that could condense with the amide to form an acyliminium, which slowly cyclized in TFA to form the bicyclic heterocycle in high yield and as a unique stereomer [15].

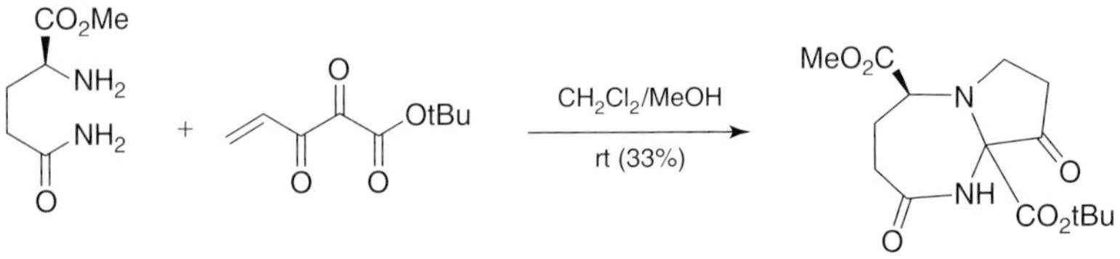

(**16**) (**17**) (**18**)

Scheme 8.6

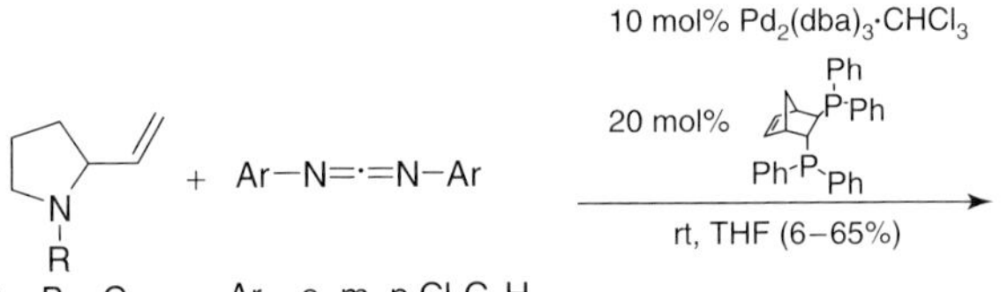

R = Bu, Cy
(**19**)

Ar = o, m, p Cl-C_6H_4

(**20**) (6–41% ee)

Scheme 8.7

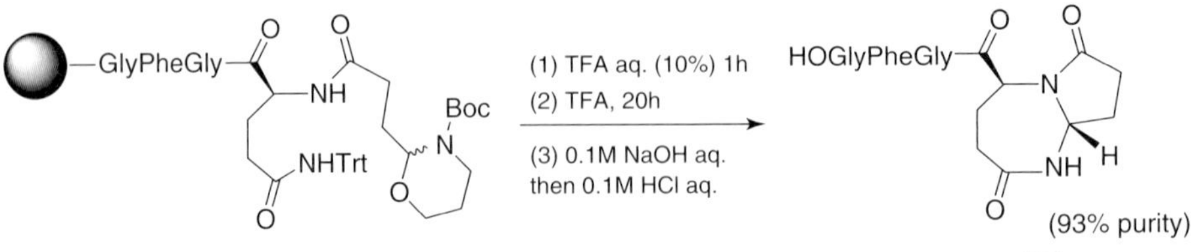

(**21**) (**22**)

Scheme 8.8

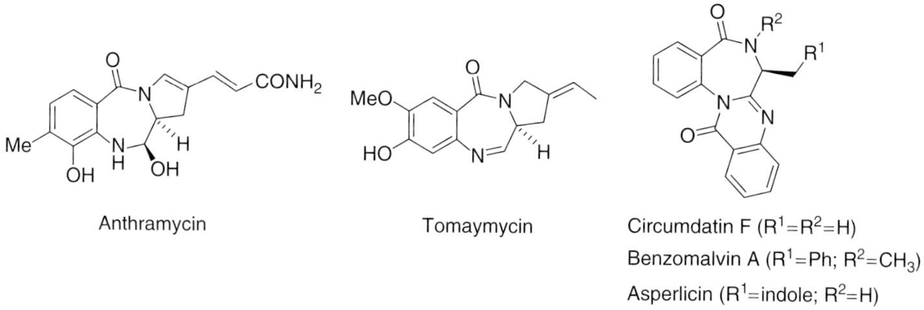

8.1 Diazepines

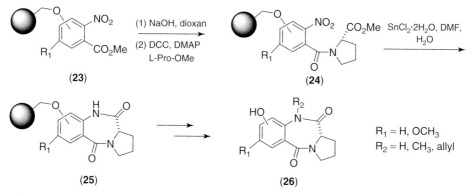

Fig. 8.2 Natural benzodiazepines.

8.1.3
1,4-Diazepines

These heterocycles are of great importance as a very impressive number of 1,4-diazepines are biologically active products. Among them, the benzodiazepine series is a very well known series and has been proved to exhibit important activities ranging from those of central nervous system (CNS) to those of anticancer agents, while antibiotics, anti-HIV, or cardiovascular agents were also reported. Several natural products were also known and some of them are depicted in Figure 8.2.

Because of the huge importance of this series of products, a very rich literature exists describing numerous asymmetric syntheses. Indeed, compared to the number of papers dealing with 1,4-diazepines, the number of available methods of preparation is not so large; the more general and recent ones are gathered below. Because of their paramount importance, the benzodiazepine products are described separately.

8.1.3.1 1,4-Benzodiazepines
Via 2-Nitrobenzoic Acid Derivatives or Isatoic Anhydride Formally, a benzodiazepine is the result of the condensation between anthranilic acid and an α-amino acid. This

Scheme 8.9

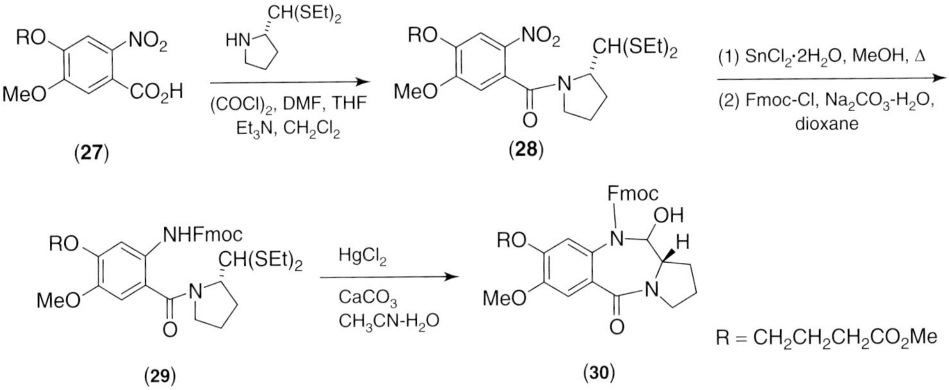

Scheme 8.10

methodology is not directly used but derived methods were more often described. A 2-nitrobenzoic acid (or ester or acyl chloride) [16] is usually condensed with an amino acid through classical coupling methods as in the recent example depicted in Scheme 8.9 showing the possible use of a solid phase synthesis. The reduction of the nitro group then allowed cyclization by lactamization [16b]. The same type of reaction was also described starting from an o-azido benzoic acid [17].

The use of isatoic anhydride (substituted or not) as a more activated starting material has been reported [18]. The condensation with an α-amino acid may be conducted in a single step in refluxing pyridine [19].

The preparation of anthramycin, tomaymycin, and analogs and more generally of antitumor- and gene-targeted drugs possessing an imine or N, O-acetal (or hemiacetal) function [20] requires the use of a chiral aminoaldehyde instead of the parent amino acid. The strategy remained the same and is illustrated in Scheme 8.10 [20d]. The nitrobenzoic acid **27** was first coupled with a protected aldehyde derived from proline and the nitro group was reduced to a primary amino group. The Fmoc protection, followed by the selective aldehyde regeneration allowed the cyclization to occur with formation of the benzodiazepine ring.

Via an S_NAr Reaction Some authors described the synthesis of benzodiazepines through an S_NAr reaction. The attachment of the appropriate amino acid to a benzoic acid derivative was followed by the S_NAr reaction. This reaction usually necessitated the presence of a fluorine atom and a nitro activating group. This may

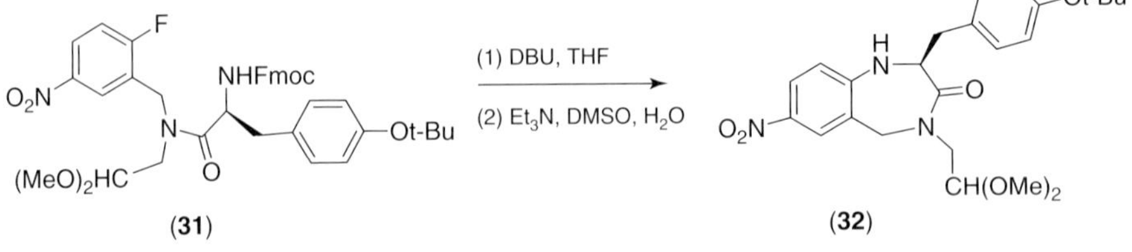

Scheme 8.11

allow the specific substituents on the aromatic ring to be suitably positioned. The S$_N$Ar reaction could occur for the cyclization step or for the introduction of the amino acid moiety onto the aromatic ring. The sequence described in Scheme 8.11 used the S$_N$Ar reaction as the final cyclization step and furnished a benzodiazepine as a mimetic of type-II β-turn [21].

To achieve the cyclization, an intramolecular CuI-catalyzed arylamination according to the Ullmann reaction has also been described [22]. For instance, 1,5-benzodiazepines **34** (Scheme 8.12) has been prepared as a peptidomimetic with caspase-1 inhibitor properties. Along with this synthesis the S$_N$Ar reaction of 1-fluoro-2-nitrobenzene with Boc-(L)-2,3-diamino propionic acid afforded amino acid **33** [23]. The latter was reduced and then cyclized to give **34**.

Via a Strecker Reaction R. Herranz proposed an original strategy of preparation of 1,4-benzodiazepines using an aminonitrile route. Methyl anthranilate was condensed with a chiral N-Boc-amino aldehyde in the presence of TMSCN to give the aminonitrile **35** in good yield but as a 2:1 mixture of stereomers (Scheme 8.13). Reduction (Raney Ni) and cyclization allowed the isolation of the desired benzodiazepines **36**, which were chromatographically resolved to both epimers [24].

Ugi Reaction In a series of papers devoted to the enantioselective preparation of 1,4-benzodiazepine-2,5-diones as Hdm2 antagonist, Lu and coworkers [25] reported the use of a Ugi reaction according to the process established by Amstrong [26], which used 1-cyclohexyl isocyanide. The Ugi reaction was followed by *in situ* cyclization to the diaza heterocycle. Chiral amine **37**, benzaldehyde **38**, and acid **39** were condensed with isocyanocyclohexene **40** for two days at room temperature and followed by cyclization with acetyl chloride to give the expected diazepine **41** in good yield but as a 1:1 mixture of epimers, which could be separated by chromatography (Scheme 8.14).

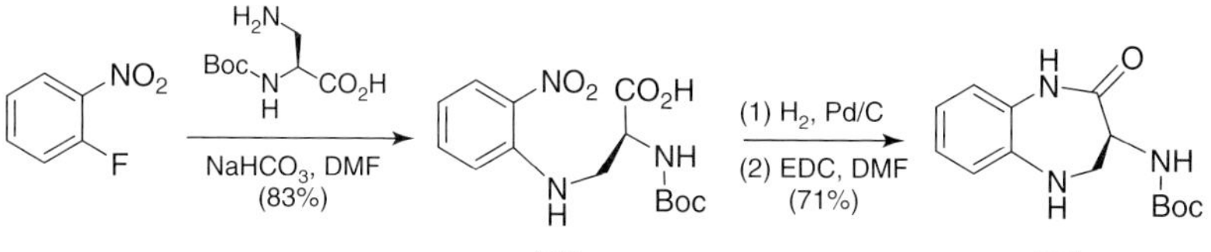

Scheme 8.12

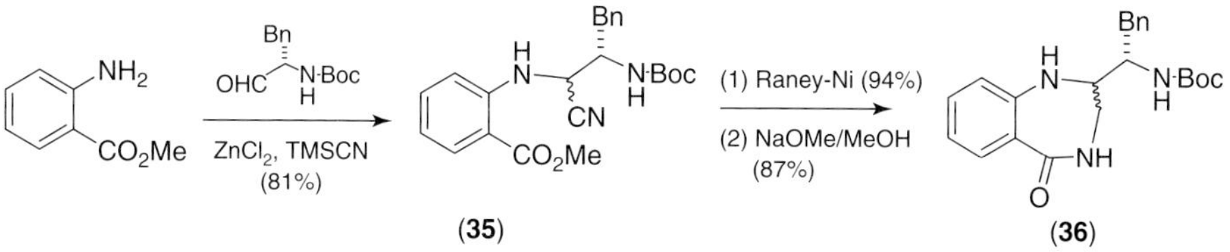

Scheme 8.13

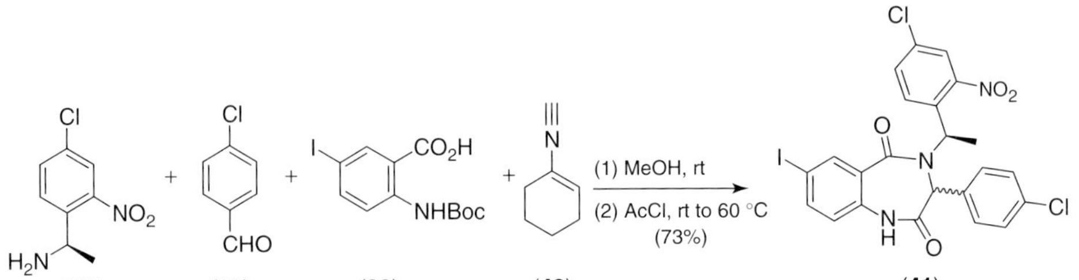

(37) (38) (39) (40) (41)

Scheme 8.14

Diastereoselective Alkylation Carlier reported the enantioselective synthesis of diversely substituted quaternary 1,4-benzodiazepine -2-one **43** by alkylation of chiral diazepine **42** following the concept of memory of chirality defined by K. Fuji. The use of the di-(*p*-anisyl)methyl (DAM) group on the lactam nitrogen was proved to provide quaternary benzodiazepines in higher enantioselectivities and yields upon the deprotonation/alkylation sequence as ee > 99% was obtained as in the following example (Scheme 8.15) [27].

Photocyclization The same concept of memory of chirality was invoked in a photochemical process that may involve a 1,7-triplet biradical. The proline derived potassium salt **44** gave the pyrrolobenzodiazepine **45** upon a decarboxylative photocyclization (Scheme 8.16). A high degree of chirality was retained as compound **45** was obtained with 86% ee [28].

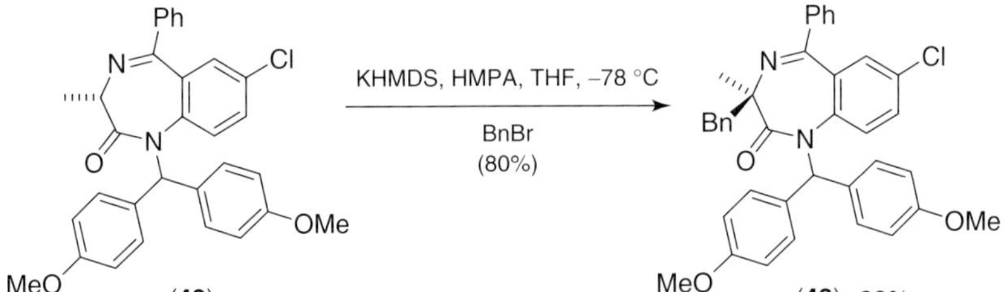

(42) (43) 99% ee

Scheme 8.15

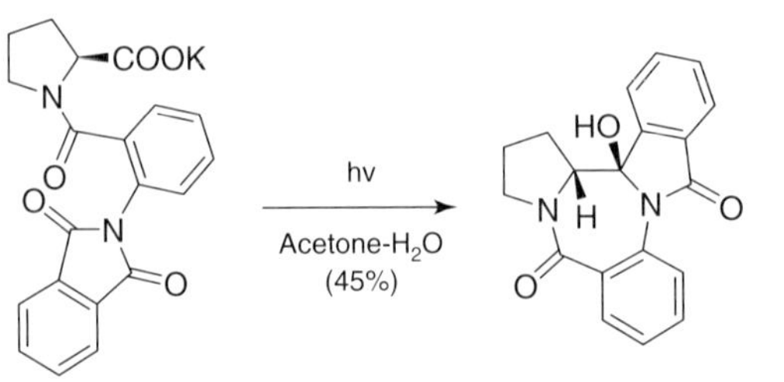

(44) (45) 86% ee

Scheme 8.16

1,3-Dipolar Cycloaddition Homochiral N-alkenoyl azide 46 (Scheme 8.17), easily obtained from the corresponding amines via diazotation, underwent an intramolecular 3+2 cycloaddition to furnish original triazol-benzodiazepine-4-ones 47a,b. Under very mild conditions, and thanks to the presence of a chiral appendage on nitrogen, diastereomers were obtained and could be separated. Though the method was somewhat hampered by a modest diastereoselectivity, it offered an original and straightforward approach to this type of compounds [29].

Meyers' Bicyclic Lactam Strategy In efforts toward the asymmetric synthesis of natural products of the circumdatin series, the Meyers' bicyclic lactam strategy was used to attain the benzodiazepine skeleton in very good diastereoselectivity albeit low yield. The methyl-quinazoline 48 was easily prepared by the condensation of two molecules of anthranilic acid and then oxidized to the aldehyde 49 with selenium oxide (Scheme 8.18). Then aldehyde 49 was condensed with phenylglycinol to afford the quinazolone–benzodiazepine 50 in a modest 25% yield but an excellent 95% de [30]. Importantly, the conditions allowed the formation of the natural product biaryl aS configuration.

Nitrilium Insertion An original new approach to pyrrolobenzodiazepines (PDB) known as *antitumor compounds* and their heterocycle-fused pyrrolodiazepine analogs appeared in 2005 [31]. In this approach, the cyclization process consisted in intramolecularly trapping of a nitrilium species by an electron-enriched aromatic group. Proline derivative 51 was readily obtained by classical amino acid coupling and transformed to the methoxyamide 52. Hypervalent iodine (phenyliodine bis(trifluoroacetate) (PIFA) (Scheme 8.19) oxidation was found to be effective

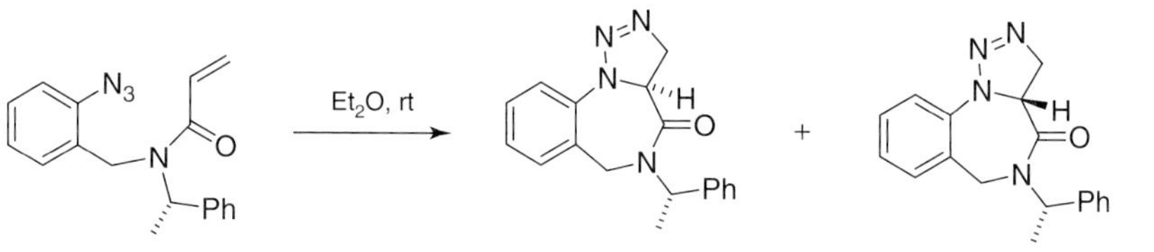

Scheme 8.17

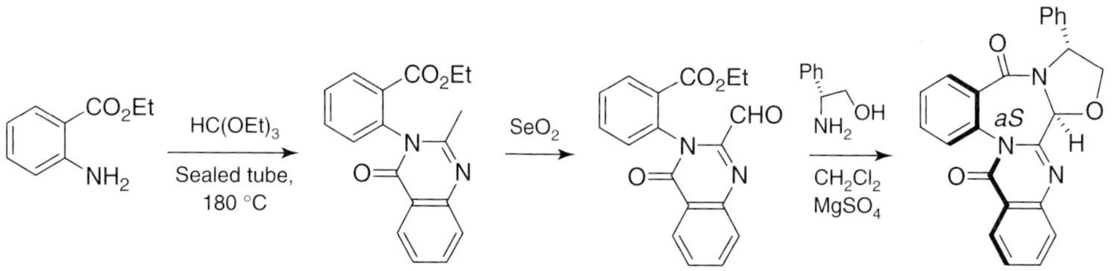

Scheme 8.18

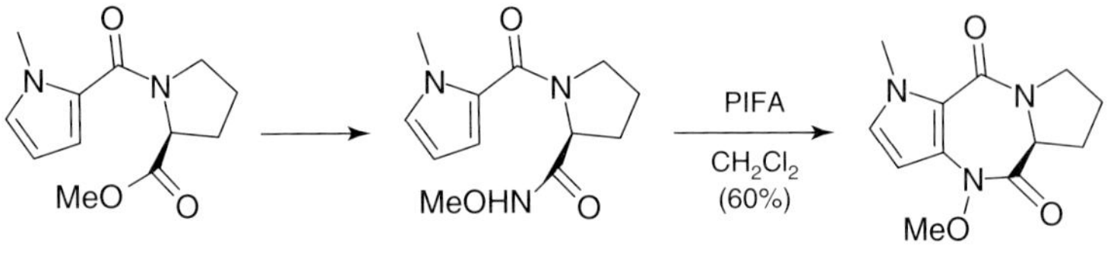

Scheme 8.19

to form the positively charged N-acetyl nitrenium species that was trapped *in situ* to form the pyrrolo benzodiazepine (PBD) analog **53**.

8.1.3.2 Other 1,4-Diazepines

Amino Acid Strategies The methods already described in the benzodiazepine series could be used in the preparation of diazepines in general. The more general and obvious method is the condensation of a β-amino acid with an α-amino acid; both of them could be asymmetric [32] but more frequently only the α-amino acid is the chiral partner with no formation of a new chiral center [33]. In the following example (Scheme 8.20), the preparation of a novel μ-opiod receptor antagonist was on the basis of the asymmetric access to the highly substituted pipecolic acid **54**, coupled with 3-amino propionic acid and cyclized to form the diazepine skeleton [34].

The amino acid could be condensed with a 3-haloamine or 3-acrylonitrile [35] giving rise after function manipulations to intermediary derivatives ready to cyclize into the diazepine ring. The final cyclization is generally a lactamization, as described above, but a Mitsunobu reaction with a serine hydroxyl was reported [36].

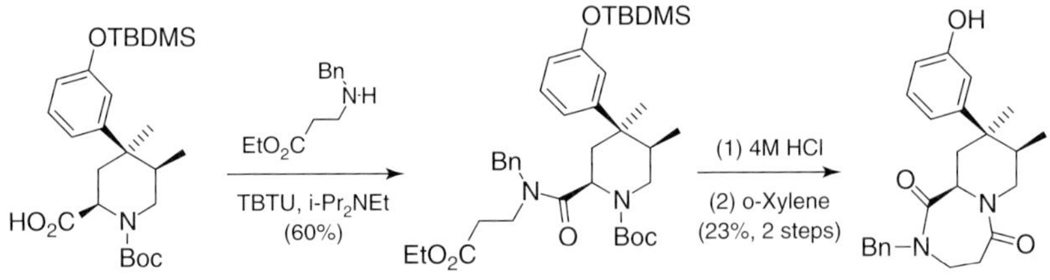

Scheme 8.20

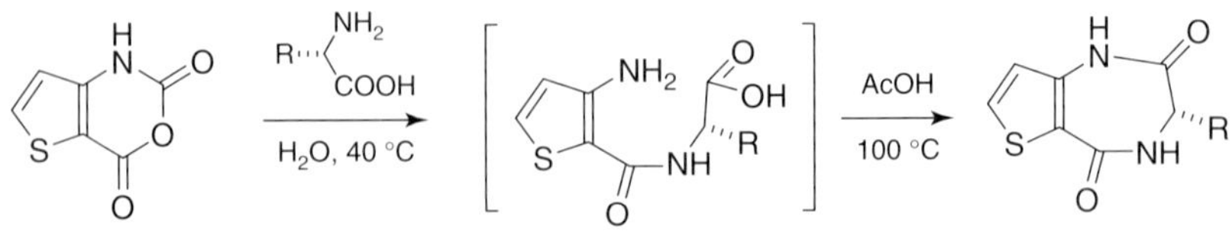

Scheme 8.21

In a recent paper, Martinez [37] described the reaction of various amino acids with thiaisatoic anhydride (Scheme 8.21). This two-step, one-pot procedure allowed an expeditive and efficient access to thienodiazepines.

Cyclization through a Nitrogen Nucleophilic Displacement Several diazepines were prepared through a non–amino acid route. In most of these cases, the formation of a carbon–nitrogen bond was the ultimate cyclization step giving rise to the diazepine skeleton [38]. In the following example (Scheme 8.22), 1,4-diazepane annulated β-lactam was obtained by construction of the diazepine ring on the β-lactam. The (R)-glyceraldehyde acetonide (**59**) was the starting material for the β-lactam ring formation through the Staudinger reaction. The oxidative cleavage of the diol and alkylation of the nitrogen of the lactam allowed the formation of a bromo aldehyde and then the bromo imine **61**. The latter was reduced with NaBH$_4$ to allow the cyclization and furnish the optically active bicyclic β-lactam **62** [39].

Cyclization through a Strecker type reaction was used in the total synthesis of manzacidin A and C reported in 2000 by Ohfune [40]. In the synthesis of liposidomycins, a new class of complex nucleoside-type antibiotics, the construction of the diazepinone ring was attained by the double opening of bis-epoxide **63** by methylamine as shown in Scheme 8.23 [41].

Paal–Knorr Cyclization Lubell has described the synthesis of pyrrolodiazepinone via an intramolecular Paal–Knorr condensation [42]. The synthesis started by the condensation of an α- and a β-amino acid to **65** (Scheme 8.24), which was transformed to the homoallylic ketone **66** precursor of 4-keto aldehyde or 1,4

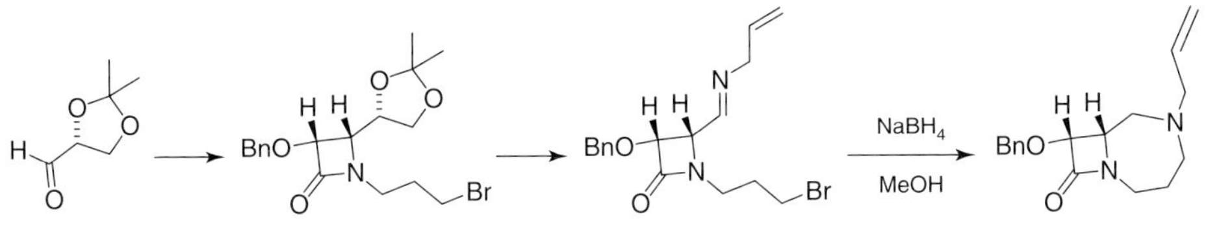

Scheme 8.22

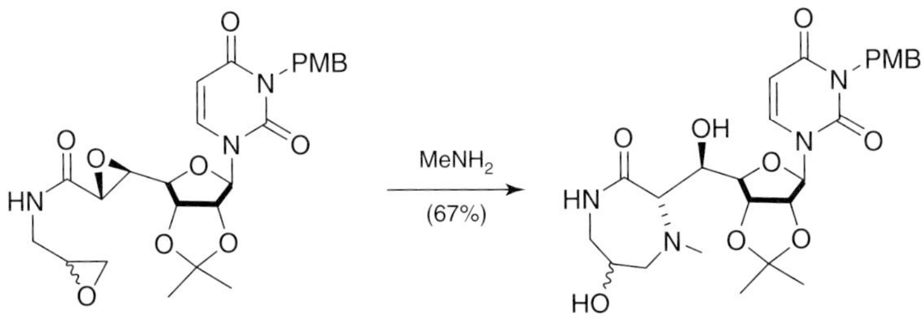

Scheme 8.23

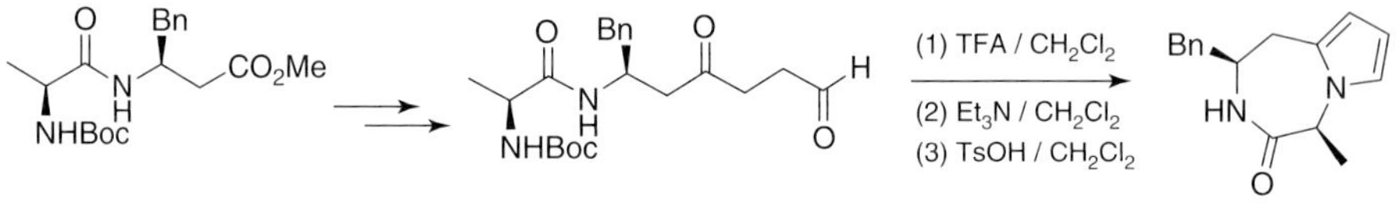

(65) (66) (67)

Scheme 8.24

diketone through oxidative processes. Deprotection of the nitrogen allowed the Paal–Knorr condensation to occur in good yield.

8.2
Oxazepines

Oxazepines constitute a large family of compounds while not so large as the diazepine ones. Their preparations are interesting and basically different from those of diazepine derivatives.

8.2.1
1,2-Oxazepine

This type of compounds has mainly been prepared by cycloaddition: 4+2, 3+2 and 3+4 cycloadditions were found to be used to give bicyclic derivatives with good diastereoselectivities.

8.2.1.1 Diels–Alder Cycloaddition

Several workers reported that Diels–Alder cycloaddition of cycloheptatriene with chiral acyl- or chloronitroso derivatives afforded oxaza-bicyclo[3.2.2] nonanes in good yields and stereoselectivities. Lallemand [43] investigated various acylnitroso compounds (derived from alanine, mandelic acid, and ribose). The hydroxamic acid **69**, derived from alanine (Scheme 8.25), was reported to be oxidized under Swern conditions and *in situ* cyclized with cycloheptatriene to give the bicycloadduct **70** as a 3 : 1 mixture of stereomers [44]. Recrystallization of the mixture allowed the diastereomeric ratio to be raised to 50 : 1, and was followed by deprotection and oxidative cleavage of the double bond to furnish the diester **71**.

During the preparation of the same type of oxaza-bicyclo compounds, Wang [45] reported that both the acylnitroso and the chloronitroso compounds derived from

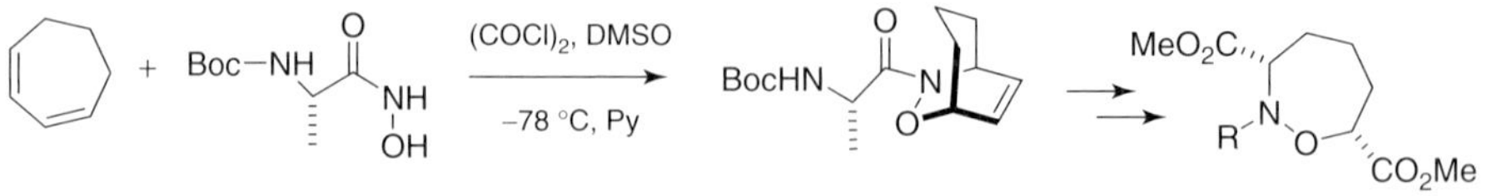

(68) (69) (70) (71)

Scheme 8.25

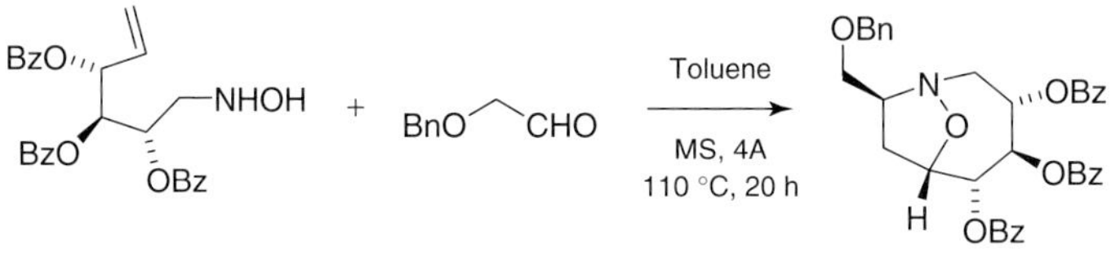

Scheme 8.26

the same ketopinic acid added to cycloheptadiene to give the desired compounds in good yield and excellent but inverted diastereoselectivity.

8.2.1.2 Intramolecular 3+2 Cycloaddition

Several papers reported the intramolecular 3+2 cycloaddition between a nitrone and an olefine [46]. The condensation of hydroxylamine 72, derived from methyl α-D-glucopyranose, with 2-(benzyloxy)-acetaldehyde in dry toluene gave cycloadduct 73 as the only isolable product (Scheme 8.26).

8.2.1.3 Pd-Catalyzed 4+3 Cycloaddition

Inspired by the cycloaddition of the Palladium trimethylenemethane complexes reported by B. Trost, R. Shintani, and T. Hayashi [47] described the cycloaddition of nitrones with γ-methylene-δ-valerolactone 74. The use of chiral ligands such as 77 furnished the oxazepine 76 in excellent yield, as well as dia- and enantioselectivity (Scheme 8.27).

8.2.1.4 Rearrangements

Some rearrangements, in natural product series, have been found to provide the 1,2-oxazepines in good preparative yields and deserve to be noticed herein [48]. Thus, (+)-nupharidine (78) (Scheme 8.28) furnished 1,2-oxazepine 79 in 65% yield

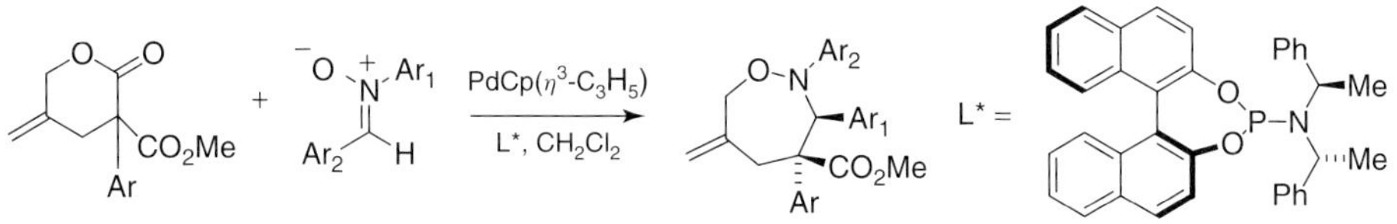

Scheme 8.27

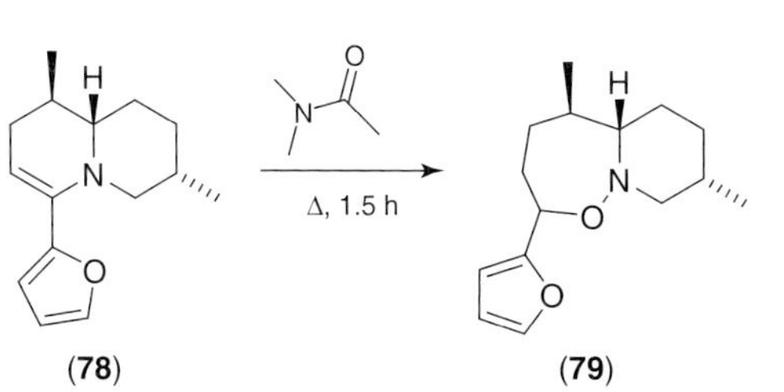

Scheme 8.28

upon simple heating in dimethyl acetamide (DMA) for 1.5 h. A stereoselective reaction was observed but the configuration of the new formed chiral center was not determined [49].

8.2.2
1,3-Oxazepines

The two heteroatoms in this type of compounds can form a *N,O*-acetal or a carbamate function. This would dictate the mode of formation of the heterocycle.

8.2.2.1 *N,O*-Acetals

Iminium Cyclization The formation of a *N,O*-acetal was classically obtained through an iminium cyclization. The iminium ion can be classically formed by oxidative or reductive methods or by *in situ* condensation of a secondary amine with an aldehyde. This is illustrated herein in the preparation of similar peptidomimetic derivatives.

In the synthesis of peptidomimetic **80** described by Moeller [50], the iminium was formed by the anodic oxidation of the proline-homoserine dipeptide **81** and trapped *in situ* to give regio- and stereoselectively the pyrrolooxazepine (Scheme 8.29). The same product was obtained by Zhang [51] in 2001, but in this case the partial and selective reduction of the lactam function of a pyroglutamic-homoserine derivative **82** furnished the iminium.

A similar compound was recently prepared through the Rh catalyzed hydroformylation of the dipeptide **83** (Scheme 8.30). The so-formed intermediate aldehyde **84** allowed the formation of the *N,O*-acetal of the piperidinooxazepine **85** derivative with complete stereoselectivity [52].

In the synthesis of alkaloid (+)-tacamonine, an interesting insertion of two formaldehyde units were reported to occur in high yield to give the pentacycle **87**,

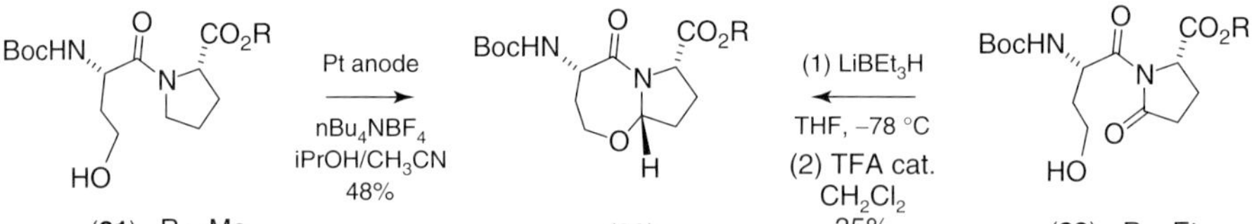

(**81**) R = Me (**80**) (**82**) R = Et

Scheme 8.29

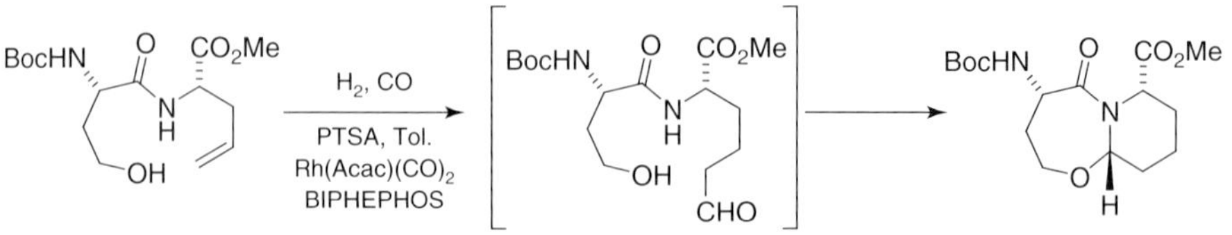

(**83**) (**84**) (**85**)

Scheme 8.30

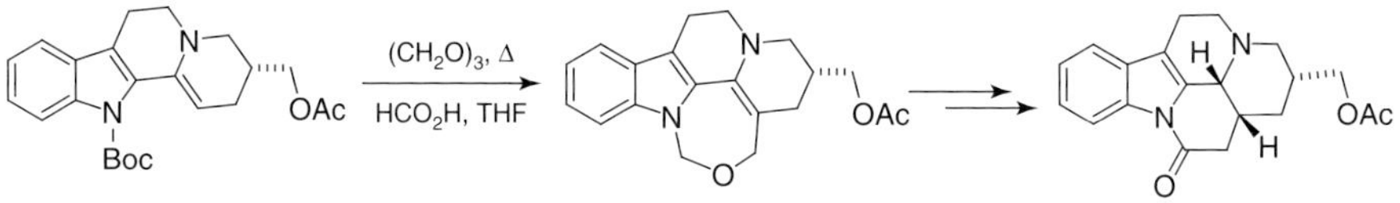

8.2 Oxazepines

Scheme 8.31

which was stereoselectively reduced to the indoloquinolizidinooxazepine allowing the introduction of two of the stereogenic centers of natural tacamonine (Scheme 8.31) [53].

Formaldehyde was also used to attain the oxazepine skeleton in the synthesis of analogs of quinocarcin [54]. The treatment of the amino alcohol, resulting from the Cbz deprotection of **88**, with formol furnished the tricyclic compound **89** in 52% overall yield and >95% ee (Scheme 8.32).

An interesting while undesired cyclization to an oxazepine was observed in the course of the synthesis of pinnaic acid. This was obtained during the trimethylsilyliodide (TMSI) cleavage of the O-benzyl group of intermediate **90** to give **91** in 83% yield [55] (Scheme 8.33). This reaction may constitute an interesting access to oxazepine derivatives and, moreover, shows the high electrophilicity of the trifluoroacetyl group.

Nucleophilic Displacement The 1,3-oxazepine skeleton has been formed in very good yield by an intramolecular nitrogen alkylation of β-lactam **92** (Scheme 8.34) [56]. Only very minor amount (3%) of the expected six-membered ring **94** was found in this process while the seven-membered ring **93** was isolated in 80% yield.

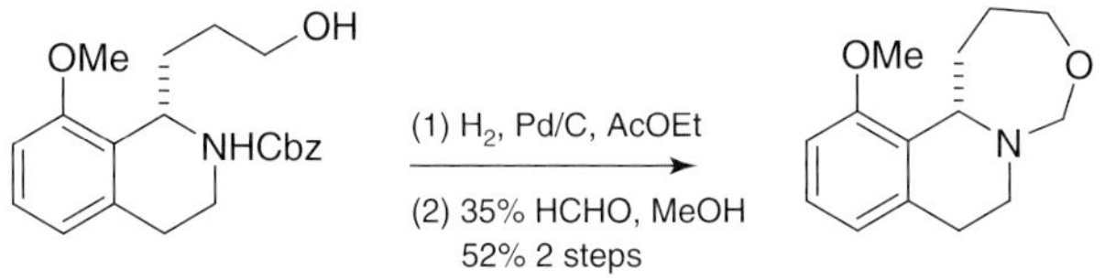

Scheme 8.32

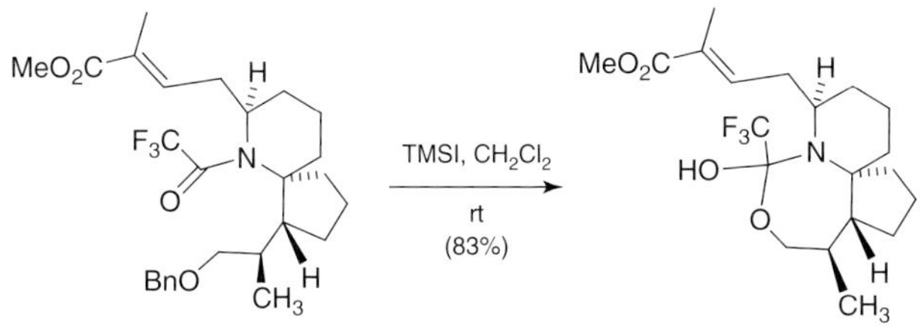

Scheme 8.33

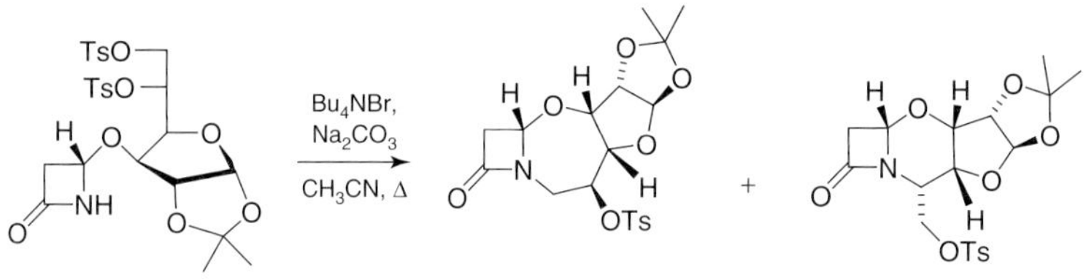

8 Asymmetric Synthesis of Seven-Membered Rings with More Than One Heteroatom

Scheme 8.34

Radical Cyclization A tandem radical cyclization was observed when the phenyselenium derivative **95** was treated with Bu$_3$SnH and AIBN in benzene. The second cyclization gave the seven-membered ring compound **96**, which upon Tamao conditions furnished the cyclonucleoside **97** in 35% overall yield (Scheme 8.35) [57].

Carbamates The nitrogen and the oxygen of the 1,3-oxazepine may be linked in a carbamate function. This function can be classically obtained from the corresponding amino alcohol. Indeed, in the examples found in the literature, the oxazepine was an undesired by-product obtained by the nucleophilic attack of an alcohol on the amide or carbamate protecting group [58]. Similarly, the photolysis of azido derivative **98** gave rise to the olefine aziridination in 85% yield, instead of the expected nitrene insertion at the allylic position (Scheme 8.36) [59].

Others In sugar series, the thermal treatment of diazides such as **100** led to the formation of a tetrazole ring with enlargement of the sugar ring to a 1,3-oxazepine skeleton in a very good yield (Scheme 8.37) [60]. The reaction was also reported through photolysis giving 1,3- and 1,2-oxazepines [61].

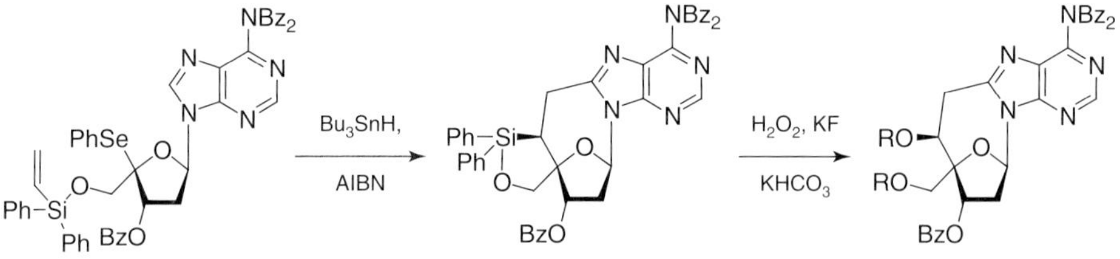

Scheme 8.35

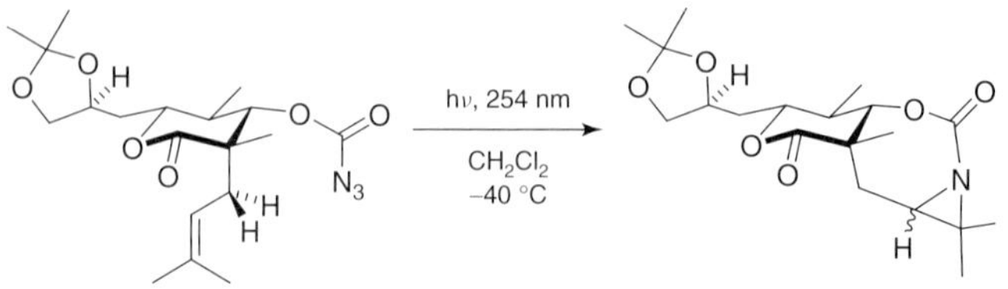

Scheme 8.36

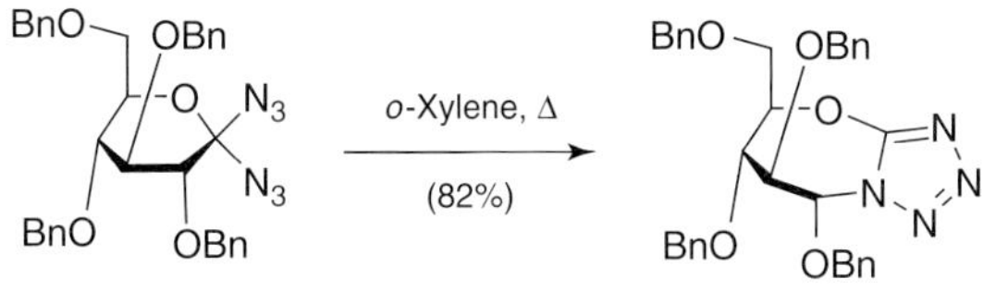

8.2 Oxazepines

Scheme 8.37

8.2.3
1,4-Oxazepines

8.2.3.1 Amino Alcohol Double Condensation

Formally, the condensation of a β-amino-alcohol with a 1,3-bis-electrophilic derivative should lead to the 1,4-oxazepine skeleton. This strategy was classically used with different electrophilic functions. In the following example (Scheme 8.38), ephedrine (**102**) was condensed with dimethyl malonate to construct the oxazepine skeleton in a sequential three-step reaction. The first two steps were the amide formation and the saponification to the acid **103**. The latter was eventually lactonized through the use of 2-chloro-1-methylpyridinium tosylate to give the optically pure oxazepinedione **104** in 55% overall yield [62].

The double condensation of a β-amino alcohol with a α-haloacetate was also reported. The reaction could be conducted in one step and necessitated the use of a base reagent such as NaH [63]. A more sophisticated and interesting synthesis was also reported by Bartlett [64] starting from alaninol. The condensation of the D-amino alcohol with aldehydo-α,β-unsaturated amide **105** was catalyzed by tin triflate and furnished the bicyclic pyrrolidino-oxazepine **106** in very good yield and diastereoselectivity (Scheme 8.39). In this synthesis, a thermodynamic control allowed the reaction to occur stereoselectively with formation two new chiral centers. Interestingly, condensation of the same compound with L-alaninol led to the enantiomer **109** of the target compound as the result of a difficult cyclization of **107**, which equilibrated to **108** after ring opening to the iminium intermediate.

8.2.3.2 Other Cyclization Methods

Lactamization, lactonization, and transacetalization are classical reactions allowing cyclization to 1,4-oxazepines.

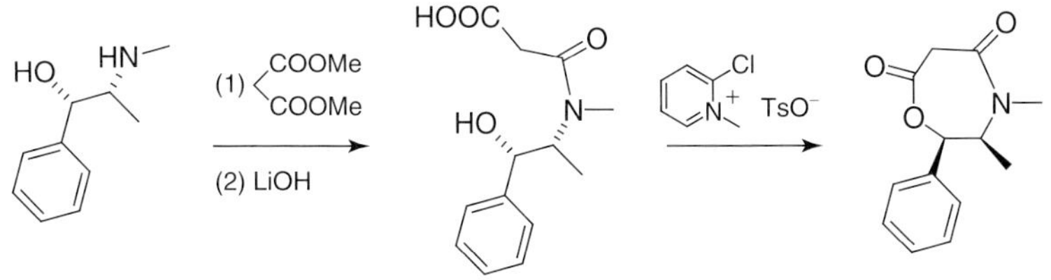

Scheme 8.38

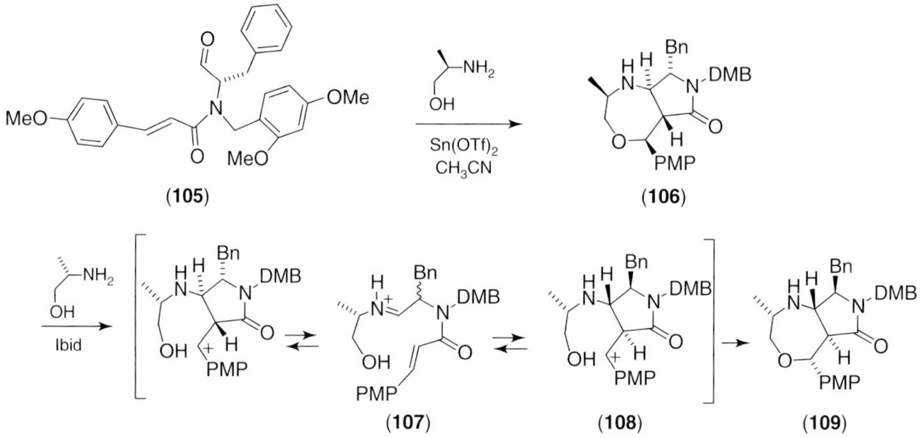

Scheme 8.39

Lactamization of chiral aminoesters was often used to attain the oxazepinone skeleton. The chirality was already installed on the framework of the amino esters [65]. In the following scheme (Scheme 8.40) [66], upon reduction of the nitro group of the chiral nitroketal **110** derived from dimethyl tartrate, spontaneous lactamization occurred to give almost quantitatively the optically active bicyclic oxazepinone **111**.

Similarly, the final cyclization may be the result of lactonization [67] or transacetalization [68, 69]. In all cases, the chirality was brought from chiral amino alcohol which was or not protected. Intramolecular opening of a chiral epoxide by an amino group was also reported to give the oxazepine in good yield [70].

8.2.3.3 Pd-Catalyzed Allene Cyclization

Tanaka et al. [71] have investigated the reaction of a bromoallene, which can act as a dication equivalent in the presence of Pd catalyst. When the bromoallene also bore an amino alcohol group, a cyclization could occur to give a ring containing a nitrogen and an oxygen. Following this methodology, the preparation of oxazepine was reported. Thus, (S,aS)-bromoallene **112**, in the presence of sodium methylate in methanol and of 5 mol% of Pd(PPh$_3$)$_4$, was transformed into oxazepine **113** in 73% yield, which represented the major regioisomer (9% of isomer **114** was also isolated) (Scheme 8.41).

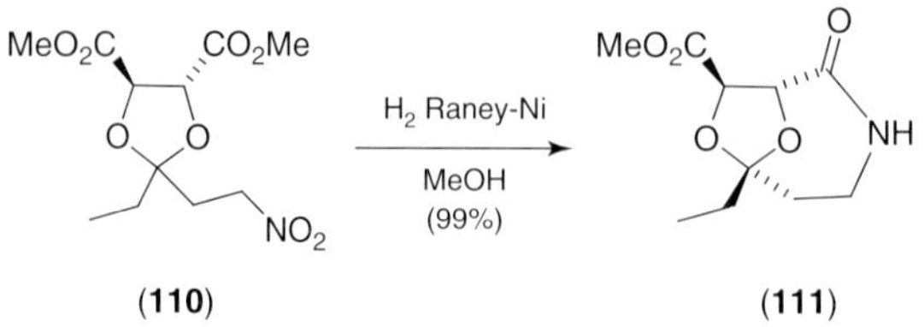

Scheme 8.40

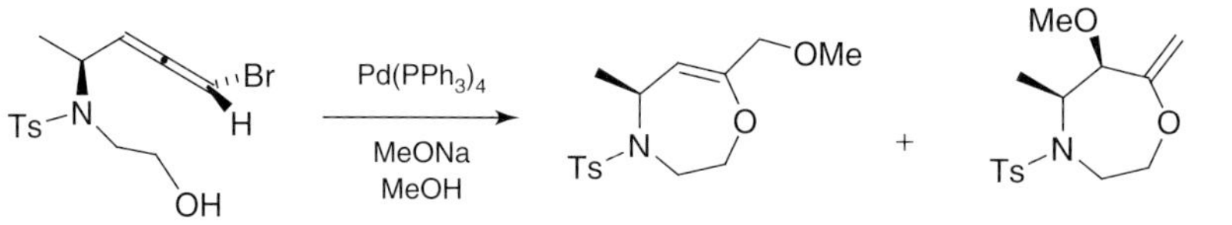

8.2 Oxazepines

Scheme 8.41

8.2.3.4 Radical Cyclization

The intramolecular nucleophilic carbonyl trapping of an α-ketenyl radical by an amino group allowed Ryu [72] to propose an original and efficient preparation method of nitrogen heterocycles. This method is general enough to be used for the access of oxazepines. Alkynylamine **115**, prepared from prolinol, was treated in benzene at 90 °C for 3 h with tributyltin hydride under CO pressure of 75 atm using AIBN as initiator, to give the α-methylene-oxazepinone **116** in 52% yield after protodestannylation (TMSCl, MeOH) (Scheme 8.42).

8.2.3.5 Ring Enlargement

Several papers reported on the preparation of asymmetric 1,4-oxazepine via the ring enlargement of a six-membered ring. One way to this ring transformation was the Baeyer–Villiger oxidation of a piperidinone to the oxazepine. Thus, the m-chloroperbenzoic acid (MCPBA) oxidation of the piperidine-4-one **117**, in the presence of sulfuric acid (to protect the basic nitrogen), gave a 64% yield of oxazepine **118** as a unique regio- and stereomer [73] (Scheme 8.43).

The [1,2]-Stevens rearrangement was also reported to perform a six- to seven-membered ring enlargement. In the following example (Scheme 8.44), the rearrangement of a cyclic ammonium ylide offered a simple and efficient route to chiral nonracemic cyclic amines. The key intermediate **121** was obtained in 80% yield by

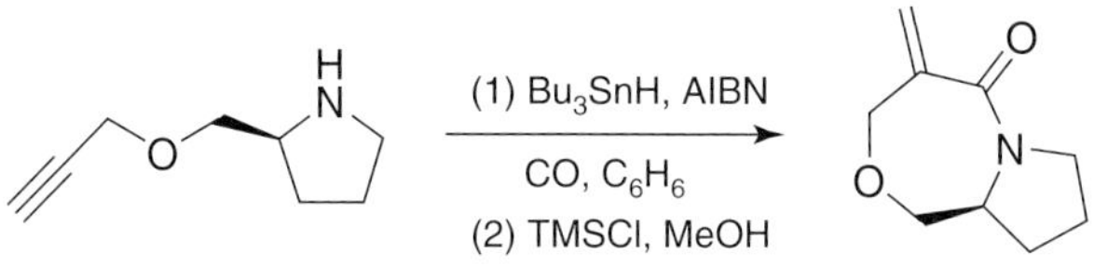

Scheme 8.42

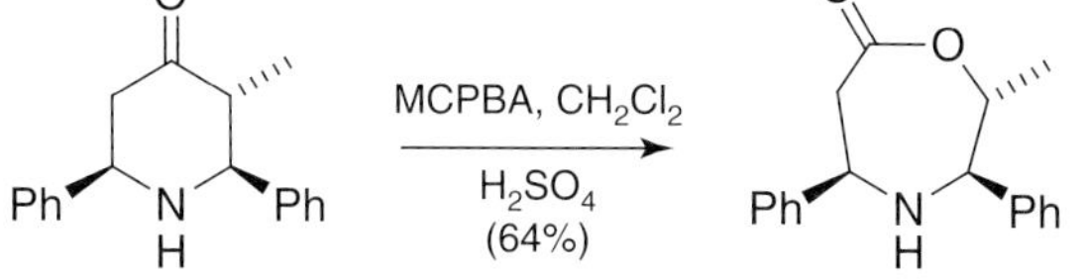

Scheme 8.43

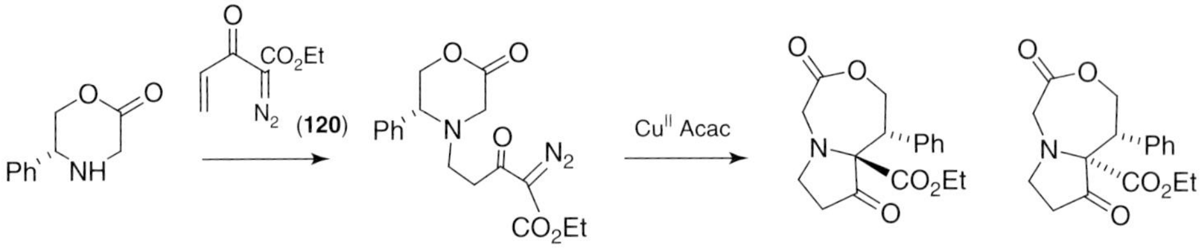

Scheme 8.44

the conjugate addition of (R)-5-phenylmorpholin-2-one **119** to the diazo compound **120**. The copper(II) acetyl acetonate (ACAC) catalyzed reaction of the diazo intermediate **121** in boiling toluene provided quantitatively the easily separable bicyclic rearranged diastereomers **122** and **123** [74].

8.2.3.6 Cycloaddition

The cycloaddition of azides with an unsaturated bond (Huysgens condensation) to give rise to triazol derivatives also offers an efficient method to access various heterocycles when it proceeds in an intramolecular fashion. In particular, performing the 1,3-dipolar cycloaddition reaction on carbohydrate-derived azido-alkynes (or alkene [75]) should offer good opportunity to the construction of bicyclic chiral triazolo-oxazepines. Starting from xylofuranosyl diol **124**, the primary alcohol function was transformed to the azido derivative via a classical sequence and the secondary alcohol function then converted to the propargyl ether **125** (Scheme 8.45). The cycloaddition was obtained by simple heating of **125** in toluene at 100 °C for 2 h resulting in the formation of **126** in 95% yield [76].

8.3
Thiazepines

Few works have been reported on the asymmetric preparation of chiral thiazepines probably because sulfur-containing starting materials are rare. Most of the thiazepines described so far are 1,4-thiazepines. Despite the similarity with oxazepines, most of the preparations are different.

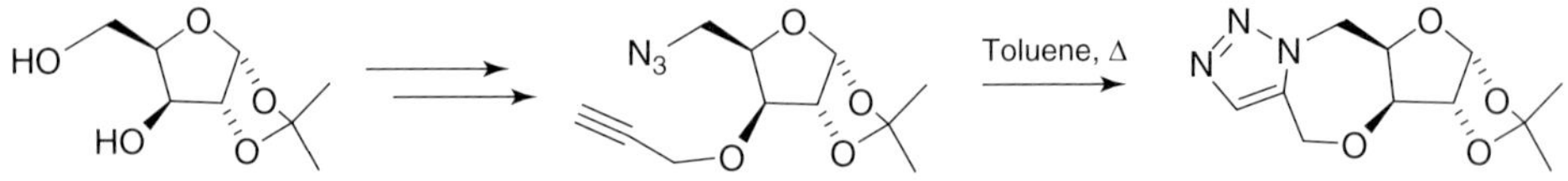

Scheme 8.45

8.3.1
1,2-Thiazepines

The 1,2-thiazepines are only represented by cyclic sulfonamides and their preparation methods are classical ones, the chiral pool being used in order to introduce the chirality.

In the synthesis of bicyclic sulfonamide **129** reported by S. Hannessian [77] as a constrained proline analog, 2-propenyl derivative **127** was prepared from proline and treated with but-3-ene-1-sulfonyl chloride to give **128** in 55% yield. An ring closure metathesis (RCM) with first generation Grubbs catalyst furnished the bicyclic sulfonamide **129** in 81% yield.

A cyclic sulfonamide was described in the synthesis of enantiopure 4-phenyl pyrrolidine-2-yl-methanol. In this synthesis (Scheme 8.47), the sulfonamide **130** obtained in two steps from hydroxyproline was transformed to the 2,5-dihydropyrrole **131**, which was subjected to a Heck reaction furnishing a 1 : 1 mixture of benzosulfonamides **132**. Interestingly, protection of the primary alcohol as a bulky pivaloyl ester gave a stereoselective Heck reaction.

8.3.2
1,3-Thiazepines

Bicyclic thiazepinone was prepared as dipeptide mimetics with metalloprotease inhibitor properties [78]. The synthesis, in this case, parallels the synthesis of oxazepine via the cyclization into iminium ions (cf Section on Iminium Cyclization). Homocysteine derivative **135** was resolved from racemic material and condensed with hydroxylnorleucine **134** (enzymatically resolved) and the dipeptide was oxidized to aldehyde **136** (Scheme 8.48). The deprotection of the thiol was followed by the acid promoted cyclization and Cbz removal to **137**.

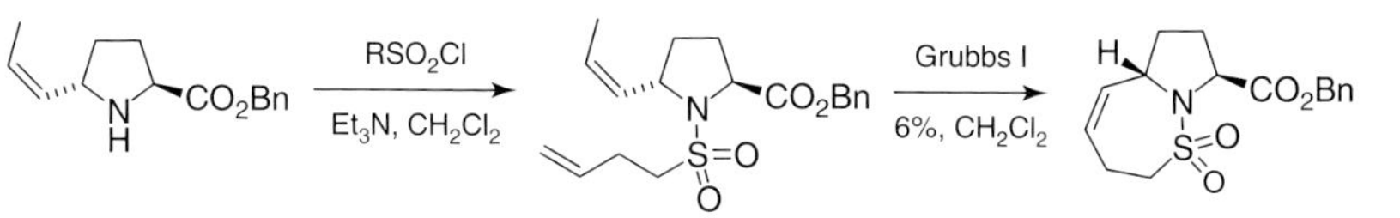

Scheme 8.46

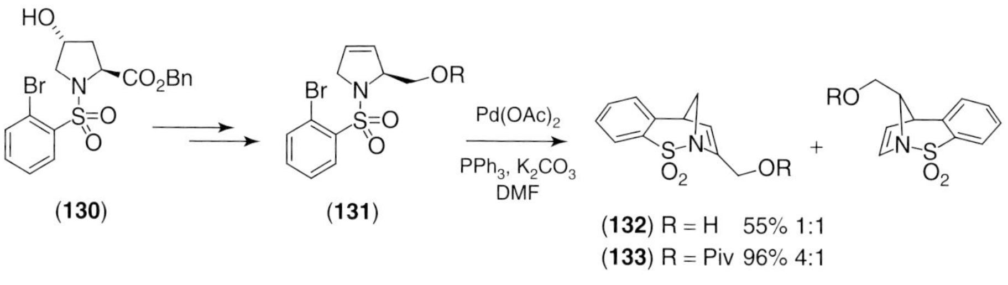

Scheme 8.47

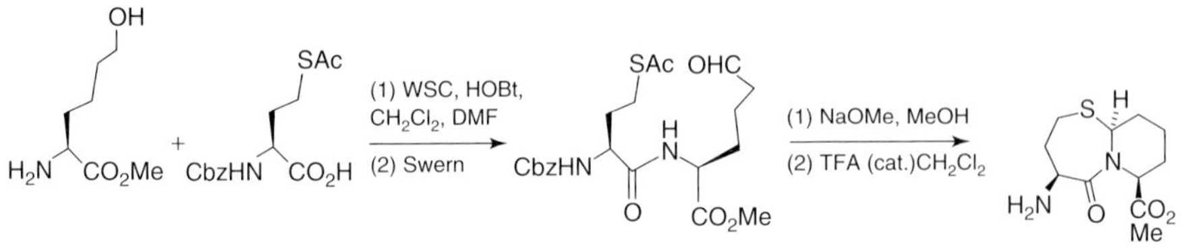

Scheme 8.48

8.3.3
1,4-Thiazepines

The 1,4-thiazepines are the most frequently described thiazepines. Several asymmetric syntheses are available. In most of the cases, the chirality was introduced as a fragment arising from the chiral pool; cysteine and penicillamine are the most common. Curiously, the syntheses are specific ones and very few follow the strategy encountered with 1,4-diazepine or oxazepine counterparts.

8.3.3.1 From Mercaptopropionic Acid Derivatives

We will find in this category the achiral mercapto acids as well as cysteine or penicillamine. The two last compounds react at the thiol and acidic functions of the mercapto acids; thus the acids react with a 3-atom fragment possessing a nitrogen and an electrophilic function.

Condensation with Haloethylamine, β-Amino Alcohol, and Derivatives The condensation of cysteine with haloethylamine has been used recently in a one-pot process to prepare thiazepine with ICE (interleukin converting enzyme) inhibitor properties. A simple reflux in methanol, in the presence of NEt$_3$, of L-cysteine methyl ester hydrochloride and 2-chloroethylamine hydrochloride allowed the condensation to occur furnishing 1,4-thiazepine 138 [79] (Scheme 8.49). This reaction illustrates a type of condensation that was used with various reagents and generally in a multistep process.

The condensation of cysteine with a chiral β-haloamine, for example, allowed the introduction of new chiral centers on the seven-membered ring. The first step of the sequence was the sulfur alkylation, which was followed by a peptide coupling [80]. Reaction of cysteine with a chiral amino alcohol was also reported; in this case the final cyclization was obtained after activation of the OH function as a mesylate [81]. Zhu [82] recently reported on the N-carbamate-assisted stereoselective substitution

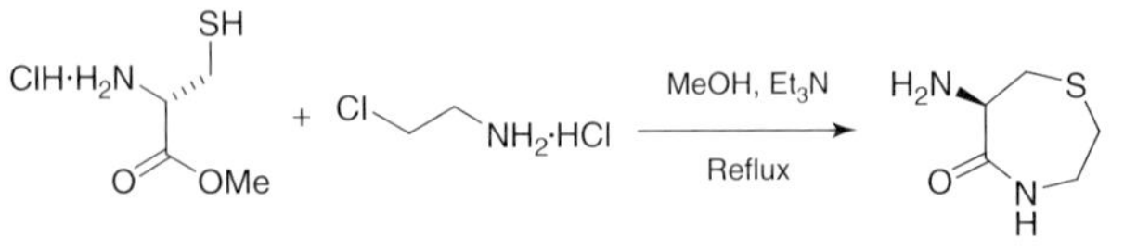

Scheme 8.49

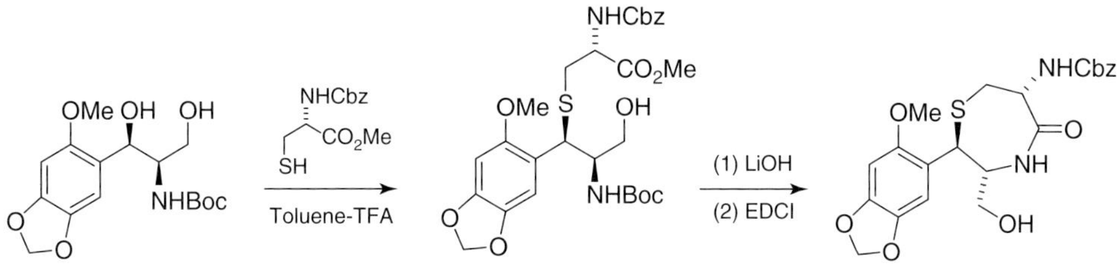

(139) (140) (141)

Scheme 8.50

of benzylic hydroxyl group by thiol. Simple treatment of a benzylic alcohol having a neighboring carbamate function with a thiol in the presence of TFA led to the introduction of the thiol with retention of configuration. This was applied to compound **139** and cysteine, which gave **140** with complete stereocontrol and the latter was cyclized with 1-ethyl-3-(3-dimethylaminopropyl)carbodiimide (EDC) to furnish the thiazepine **141**.

Reactions with achiral mercaptopropionic acid were also reported; chiral aziridine [83] or amino acid [84] was used as the 3-atom fragment. In the preparation of thiazepanes as inhibitors of nitric oxide synthase, the amino alcohol **142** was the precursor of the aziridine **143** via a Mitsunobu reaction (Scheme 8.51). Addition of β-mercaptopropionic acid furnished the amino acid **144**, which was cyclized to **145** via conventional methods.

Condensation with Nitro Olefine The research of new active inhibitors of ACE led Japanese researchers to propose an interesting access to thiazepine through the Michael addition of cysteine onto a nitro olefine [85]. In this sequence a new chiral center was formed. Unfortunately, there is no report of the diastereoselectivity of the Michael addition. The reduction of the nitro group into a primary amine allowed the final cyclization through lactamization. The final product was obtained in enantiomerically pure form after recrystallization.

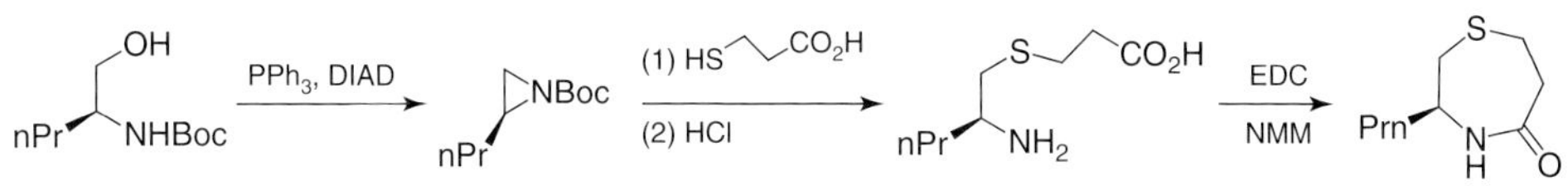

(142) (143) (144) (145)

Scheme 8.51

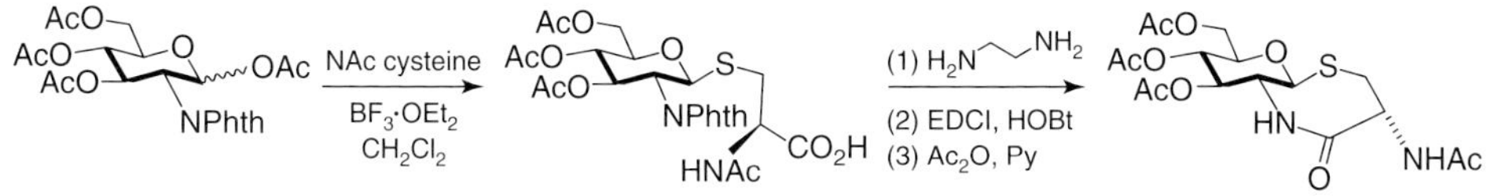

(146) (147) (148)

Scheme 8.52

Condensation with Amino Sugar Fused bicyclic glycosides–thiazepines have been prepared by the reaction of the phthalimido-protected sugar **146**, used as donor, with N-acetylated L-cysteine [86] (Scheme 8.52). This ensured the introduction of cysteine at the anomeric center as the desired β-thioglycoside. Selective removal of the N-phthalimido group was obtained by treatment with ethylenediamine in methanol and the free amino group was reacted with EDCI/hydroxybenzotriazole (HOBt) to give the bicyclic product in 50% yield.

Cysteine in S_NAr Reaction Benzothiazepine compounds have been prepared in the search of new effective inhibitors of ACE. One of these syntheses paralleled the synthesis of benzodiazepine (cf Section on Via an S_NAr Reaction) starting with a S_NAr reaction of N-Ac cysteine on 2-fluoronitrobenzene. The reduction of the nitro group into primary amine was followed by lactamization to give the required benzothiazepine [87].

8.3.3.2 From Amino Thiols

Aminothiols were also broadly used as sulfur reagents in the preparation of thiazepines. Since the aminothiol is a bis-nucleophile derivative, the counterpart in this condensation should possess two electrophilic functions.

Aminothiols and Acrylic Acid Derivatives The most frequently used reactants were acrylic acid and its analogs [88]. Thus, as illustrated in Scheme 8.53, D-penicillamine was reacted with α-phenylacetamidoacrylate (**149**) in the presence of Et₃N to give the expected thiazepine **150**, unfortunately as a mixture of stereomers without indication of the ratio [89].

Aminothiols and Other Dielectrophilic Compounds Diltiazem, a calcium channel blocker, is probably the most representative compound of thiazepines. Several syntheses were published but very few were asymmetric ones. Researchers from Hoffmann-La Roche published an efficient synthesis [90] in which the condensation of 2-aminobenzenethiol to an enantiomerically pure epoxy ester **152** allowed the preparation of diltiazem in asymmetric form (Scheme 8.54). The synthesis was based on the lipase-catalyzed kinetic resolution of trans-2-phenylcyclohexanol to give the pure (−)-1R,2S enantiomer **151**. This compound was the chiral source to attain pure glycidic ester **152** via a Darzens reaction. 2-Aminobenzenethiol reacted with **152** in refluxing toluene to give **153** with the expected configuration. Deprotection of **153** was followed by acidic treatment, which allowed cyclization to occur.

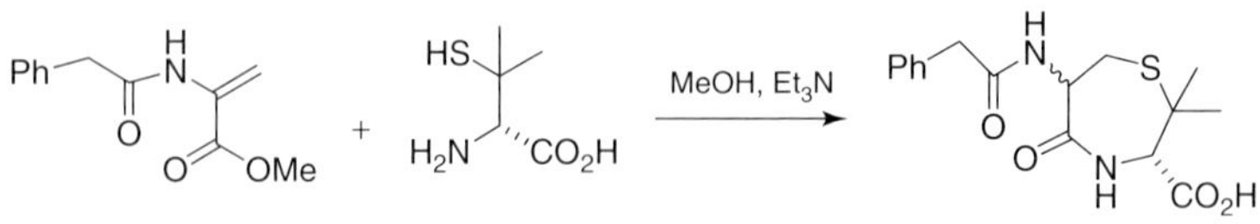

Scheme 8.53

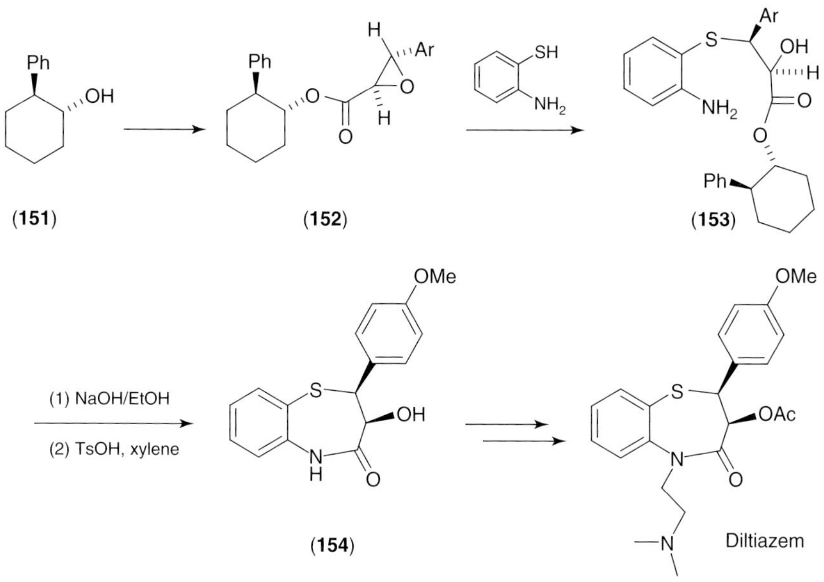

Scheme 8.54

Dihalopropan and halopropanol were also used in the condensation of aminothiols. A one-step reaction was claimed to occur by condensation of 3-chlorobromopropane with penicillamine [91]. In the following example (Scheme 8.55) [92], the first step was the sulfur alkylation of cysteine with bromopropanol, which was followed by N-sulfonylation to give **155**. The final cyclization was achieved by a Mitsunobu reaction to give the thiazepine **156** as depicted in Scheme 8.55.

8.3.3.3 Others

Recently, Gleason [93] reported a study devoted to the asymmetric construction of quaternary centers on the basis of the use of the 5,7-bicyclic thioglycolate **159**. An interesting, short and straightforward preparation of **159** was proposed, which was like the Meyer's lactam synthesis. In the present preparation, achiral methyl thioglycolate was alkylated with 2-bromoethyl dioxolane and the obtained ester **157** used to N-acylate S-valinol (Scheme 8.56). The resulting amide **158** underwent a

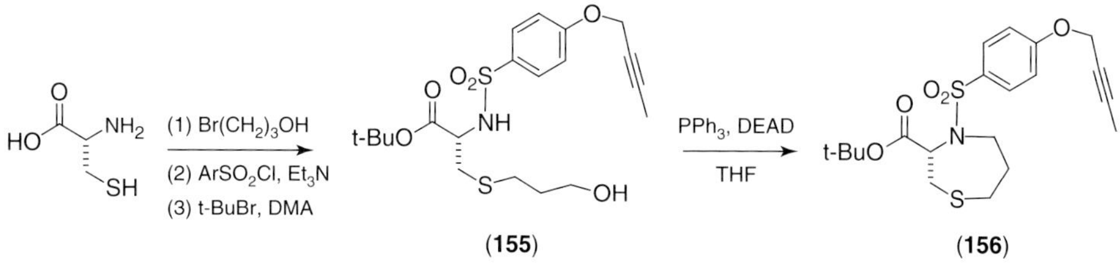

Scheme 8.55

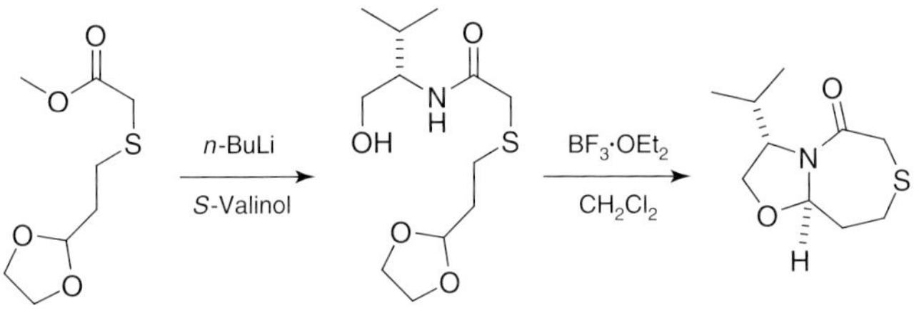

Scheme 8.56

transacetalization to give the expected bicyclic compound **159**, in an excellent 71% overall yield and as a single stereoisomer.

References

1 Srinivasan, S. and Argade, N. P. Z. (2003) *Tetrahedron: Asymmetry*, **14**, 333–37.
2 Attwood, M. R., Francis, R. J., Hassall, C. H., KrGhn, A., Lawton, G., Natoff, I. L., Nixon, J. S., Redshaw, S. and Thomas, W. A. (1984) *FEBS Lett.*, **165**, 201–6.
3 Bolòs, J., Perez, A., Gubert, S., Anglada, L., Sacristàn, A. and Ortiz, J. A. (1992) *J. Org. Chem.*, **57**, 3535–39.
4 Attwood, M. R., Hassall, C. H., Krohn, A., Lawton, G. and Redshaw, S. (1986) *J. Chem. Soc. Perkin Trans. 1*, 1011–19.
5 Chen, M. H., Goel, O. P., Hyun, J.-W., Magano, J. and Rubin, J. R. (1999) *Bioorg. Med. Chem. Lett.*, **9**, 1587–92.
6 Liu, B., Brandt, J. D. and Moeller, K. D. (2003) *Tetrahedron*, **59**, 8515–23.
7 Ernholt, B. V., Thomsen, B., Lohse, A., Plesner, I. W., Jensen, K. B., Hazell, R. G., Liang, X., Jakobsen, A. and Bols, M. (2000) *Chem. Eur. J.*, **6**, 278–87.
8 (a) Nugiel, D. A., Jacobs, K., Worley, T., Patel, M., Kaltenbach, III, R. F., Meyer, D. T., Jadhav, P. K., De Lucca, G. V., Smyser, T. E., Klabe, R. M., Bacheler, L. T., Rayner, M. M. and Seitz, S. P. (1996) *J. Med. Chem.*, **39**, 2156–69; (b) Hultén, J., Bonham, N. M., Nillroth, U., Hansson, T., Zuccarello, G., Bouzide, A., Aqvist, J., Classon, B., Danielson, U. H., Karlén, A., Kvarnström, I., Samuelsson, B. and Hallberg, A. (1997) *J. Med. Chem.*, **40**, 885–97; (c) Lam, P. Y. S., Ru, Y., Jadhav, P. K., Aldrich, P. E., DeLucca, G. V., Eyermann, C. J., Chang, C.-H., Emmett, G., Holler, E. R., Daneker, W. F., Li, L., Confalone, P. N., McHugh, R. J., Han, Q., Li, R., Markwalder, J. A., Seitz, S. P., Sharpe, T. R., Bacheler, L. T., Rayner, M. M., Klabe, R. M., Shum, L., Winslow, D. L., Kornhauser, D. M., Jackson, D. A., Erickson-Viitanaen, S. and Hodge, C. N. (1996) *J. Med. Chem.*, **39**, 3514–25; (d) De Lucca, G. V., Liang, J., Aldrich, P. E., Calabrese, J., Cordova, B., Klabe, R. M., Rayner, M. M. and Chang, C.-H. (1997) *J. Med. Chem.*, **40**, 1707–19; (e) Kalmtenbach R. F. III, Patel, M., Waltermire, R. E., Harris, G. D., Stone, B. R. P., Klabe, R. M., Garber, S., Bacheler, L. T., Cordova, B. C., Logue, K., Wright, M. R., Erickson-Viitanen, S. and Trainor, G. L. (2003) *Bioorg. Med. Chem. Lett.*, **13**, 605–8; (f) Han, W., Pelletier, J. C. and Hodge, C. N. (1998) *Bioorg. Med. Chem. Lett.*, **8**, 3615–20; (g) Patel, M., Bacheler, T., Rayner, M. M., Cordova, B. C., Klabe, R. M., Erickson-Viitanen, S. and Seitz,

S. P. (**1998**) *Bioorg. Med. Chem. Lett.*, **8**, 823–28; (h) Patel, M., Kalmtenbach, R. F. III, Nugiel, D. A., McHugh R. J. Jr, Jadhav, P. K., Bacheler, L. T., Cordova, B. C., Klabe, R. M., Erickson-Viitanen, S., Garber, S., Reid, C. and Seitz, S. P. (**1998**) *Bioorg. Med. Chem. Lett.*, **8**, 1077–82; (i) Pierce, M. E., Harris, G. D., Islam, Q., Radesca, L. A., Storace, L., Waltermire, R. E., Wat, E., Jadhav, P. K. and Emmett, G. C. (**1996**) *J. Org. Chem.*, **61**, 444–50; (j) Kaltenbach, R. F., Nugiel, D. A., Lam, P. Y. S., Klabe, R. M. and Seitz, S. P. (**1998**) *J. Med. Chem.*, **41**, 5113–17; (k) Rossano, L. T., Lo, Y. S., Anzalone, L., Lee, Y.-C., Meloni, D. J., Moore, J. R., Gale, T. M. and Arnett, J. F. (**1995**) *Tetrahedron Lett.*, **36**, 4967–70.

9 (a) Schreiner, E. P. and Pruckner, A. (**1997**) *J. Org. Chem.*, **62**, 5380–84; (b) Dondoni, A., Perrone, D. and Rinaldi, M. (**1998**) *J. Org. Chem.*, **63**, 9252–64.

10 Stone, B. R. P., Harris, G. D., Cann, R. O., Smyser, T. E. and Confalone, P. N. (**1998**) *Tetrahedron Lett.*, **39**, 6127–30.

11 Jadhav, P. K., Woerner, F. J., Lam, P. Y. S., Hodge, C. N., Eyermann, C. J., Man, H.-W., Daneker, W. F., Bacheler, L. T., Rayner, M. M., Meek, J. L., Erickson-Viitanen, S., Jackson, D. A., Calabrese, J. C., Schadt, M. and Chang, C.-H. (**1998**) *J. Med. Chem.*, **41**, 1446–55.

12 Le Merrer, Y., Gauzy, L., Gravier-Pelletier, C. and Depezay, J.-C. (**2000**) *Bioorg. Med. Chem.*, **8**, 307–20.

13 Wasserman, H. H., Henke, S. L., Nakanishi, E. and Schultze, G. (**1992**) *J. Org. Chem.*, **57**, 2641–45.

14 Zhou, H.-B. and Alper, H. (**2004**) *Tetrahedron*, **60**, 73–79.

15 Nielsen, T. E., Le Quement, S. and Meldal, M. (**2005**) *Org. Lett.*, **7**, 3601–4.

16 (a) Kamal, A., Ramesh, G., Laxman, N., Ramulu, P., Srinivas, O., Neelima, K., Kondapi, A. K., Sreenu, V. B. and Nagarajaram, H. A. (**2002**) *J. Med. Chem.*, **45**, 4679–88; (b) Kamal, A., Reddy, G. S. K. and Raghavan, S. (**2001**) *Bioorg. Med. Chem. Lett.*, **11**, 387–89; (c) Mishra, J. K., Garg, P., Dohare, P., Kumar, A., Siddiqi, M. I., Ray, M. and Panda, G. (**2007**) *Bioorg. Med. Chem. Lett*, **17**, 1326–31; (d) Dyatkin, A. B., Hoekstra, W. J., Hlasta, D. J., Andrade-Gordon, P., de Garavilla, L., Demarest, K. T., Gunnet, J. W., Hageman, W., Look, R. and Maryanoff, B. E. (**2002**) *Bioorg. Med. Chem. Lett*, **12**, 3081–84; (e) Madani, H., Thompson, A. S. and Threadgill, M. D. (**2002**) *Tetrahedron*, **58**, 8107–11; (f) Kitamura, T., Sato, Y. and Mori, M. (**2004**) *Tetrahedron*, **60**, 9649–57.

17 (a) Kamal, A., Babu, A. H., Ramana, A. V., Ramana, K. V., Bharathi, E. V. and Kumar, M. S. (**2005**) *Bioorg. Med. Chem. Lett.*, **15**, 2621–23; (b) Kamal, A., Ramana, A. V., Reddy, K. S., Ramana, K. V., Babu, A. H. and Prasad, B. R. (**2004**) *Tetrahedron Lett.*, **45**, 8187–90; (c) Kamal, A., Shankaraiah, N., Reddy, K. L. and Devaiah, V. (**2006**) *Tetrahedron Lett.*, **47**, 4253–57.

18 Sabb, A. L., Vogel, R. L., Welmaker, G. S., Sabalski, J. E., Coupet, J., Dunlop, J., Rosenzweig-Lipsonb, S. and Harrison, B. (**2004**) *Bioorg. Med. Chem. Lett.*, **14**, 2603–7.

19 Hunt, J. T., Ding, C. Z., Batorsky, R., Bednarz, M., Bhide, R., Cho, Y., Chong, S., Chao, S., Gullo-Brown, J., Guo, P., Kim, S. H., Lee, F. Y. F., Leftheris, K., Miller, A., Mitt, T., Patel, M., Penhallow, B. A., Ricca, C., Rose, W. C., Schmidt, R., Slusarchyk, W. A., Vite, G. and Manne, V. (**2000**) *J. Med. Chem.*, **43**, 3587–95.

20 (a) Wells, G., Martin, C. R. H., Howard, P. W., Sands, Z. A., Laughton, C. A., Tiberghien, A., Woo, C. K., Masterson, L. A., Stephenson, M. J., Hartley, J. A., Jenkins, T. C., Shnyder, S. D., Loadman, P. M., Waring, M. J. and Thurston, D. E. (**2006**) *J. Med. Chem.*, **49**, 5442–61; (b) Masterson, L. A., Croker, S. J., Jenkins, T. C.,

Howard, P. W. and Thurston, D. E. (2004) *Bioorg. Med. Chem. Lett.*, **14**, 901–4; (c) Chen, Z., Gregson, S. J., Howard, P. W. and Thurston, D. E. (2004) *Bioorg. Med. Chem. Lett.*, **14**, 1547–49; (d) Zhou, Q., Duan, W., Simmons, D., Shayo, Y., Raymond, M. A., Dorr, R. T. and Hurley, L. H. (2001) *J. Am. Chem. Soc.*, **123**, 4865–66.

21 (a) Rosenström, U., Sköld, C., Lindeberg, G., Botros, M., Nyberg, F., Karlén, A. and Hallberg, A. (2004). *J. Med. Chem.*, **47**, 859–70; (b) ibid **2006**, **49**, 6133–37; (c) Abrous, L., Jokiel, P. A., Friedrich, S. R., Hynes, J. Jr., Smith, A. B., III and Hirschmann, R. (2004). *J. Org. Chem.*, **69**, 280–302.

22 Ma, D. and Xia, C. (2001) *Org. Lett.*, **3**, 2583–86.

23 Lauffer, D. A. and Mullican, M. D. (2002) *Bioorg. Med. Chem. Lett.*, **12**, 1225–27.

24 (a) Herrero, S., García-López, M. T., Cenarruzabeitia, E., Del Río, J. and Herranz, R. (2003) *Tetrahedron*, **59**, 4491–99; (b) Herrero, S., Garcia-Lopez, M. T. and Herranz, R. (2003) *J. Org. Chem.*, **68**, 4582–85.

25 Leonard, K., Marugan, J. J., Raboisson, P., Calvo, R., Gushue, J. M., Koblish, H. K., Lattanze, J., Zhao, S., Cummings, M. D., Player, M. R., Maroney, A. C. and Lu, T. (2006) *Bioorg. Med. Chem. Lett.*, **16**, 3463–68.

26 Keating, T. A. and Armstrong, R. W. (1996) *J. Am. Chem. Soc.*, **118**, 2574–83.

27 Carlier, P. R., Zhao, H., MacQuarrie-Hunter, S. L., DeGuzman, J. C. and Hsu, D. C. (2006) *J. Am. Chem. Soc.*, **128**, 15215–20.

28 Griesbeck, A. G., Kramer, W., Bartoschek, A. and Schmickler, H. (2001) *Org. Lett.*, **3**, 537–39.

29 (a) Broggini, G., Garanti, L., Molteni, G. and Pilati, T. (2001) *Tetrahedron : Asymmetry*, **12**, 1201–6; (b) Broggini, G., Casalone, G., Garanti, L., Molteni, G., Pilati, T. and Zecchi, G. (1999) *Tetrahedron : Asymmetry*, **10**, 4447–54; (c) Molteni, G., Broggini, G. and Pilati, T. (2002) *Tetrahedron: Asymmetry*, **13**, 2491–95.

30 Penhoat, M., Bohn, P., Dupas, G., Papamicaël, C., Marsais, F. and Levacher, V. (2006) *Tetrahedron: Asymmetry*, **17**, 281–86.

31 Correa, A., Tellitu, I., Domínguez, E., Moreno, I. and SanMartin, R. (2005) *J. Org. Chem.*, **70**, 2256–64.

32 (a) Taillefumier, C., Thielges, S. and Chapleur, Y. (2004) *Tetrahedron*, **60**, 2213–24; (b) Drouillat, B., Bourdreux, Y., Perdon, D. and Greck, C. (2007) *Tetrahedron: Asymmetry*, **18**, 1955–63.

33 Wattanasin, S., Kallen, J., Myers, S., Guo, Q., Sabio, M., Ehrhardt, C., Albert, R., Hommel, U., Weckbecker, G., Welzenbach, K. and Weitz-Schmidt, G. (2005) *Bioorg. Med. Chem. Lett.*, **15**, 1217–20.

34 Le Bourdonnec, B., Goodman, A. J., Graczyk, T. M., Belanger, S., Seida, P. R., DeHaven, R. N. and Dolle, R. E. (2006) *J. Med. Chem.*, **49**, 7290–306.

35 Biftu, T., Feng, D., Qian, X., Liang, G.-B., Kieczykowski, G., Eiermann, G., He, H., Leiting, B., Lyons, K., Petrov, A., Sinha-Roy, R., Zhang, B., Scapin, G., Patel, S., Gao, Y.-D., Singh, S., Wu, J., Zhang, X., Thornberry, N. A. and Weber, A. E. (2007) *Bioorg. Med. Chem. Lett.*, **17**, 49–52.

36 Lampariello, L. R., Piras, D., Rodriquez, M. and Taddei, M. (2003) *J. Org. Chem.*, **68**, 7893–95.

37 Brouillette, Y., Lisowski, V., Fulcrand, P. and Martinez, J. (2007) *J. Org. Chem.*, **72**, 2662–65.

38 Natsugari, H., Ikeura, Y., Kamo, I., Ishimaru, T., Ishichi, Y., Fujishima, A., Tanaka, T., Kasahara, F., Kawada, M. and Doi, T. (1999) *J. Med. Chem.*, **42**, 3982–93.

39 Van Brabandt, W., Vanwalleghem, M., D'hooghe, M. and De Kimpe, N. (2006) *J. Org. Chem.*, **71**, 7083–86.

40. Namba, K., Shinada, T., Teramoto, T. and Ohfune, Y. (**2000**) *J. Am. Chem. Soc.*, **122**, 10708–09.
41. Sarabia, F., Martín-Ortiz, L. and López-Herrera, F. J. (**2003**) *Org. Lett.*, **5**, 3927–30.
42. Iden, H. S. and Lubell, W. D. (**2006**) *Org. Lett.*, **8**, 3425–28.
43. Faitg, T., Soulié, J., Lallemand, J.-Y. and Ricard, L. (**1999**) *Tetrahedron: Asymmetry*, **10**, 2165–74.
44. Shireman, B. T. and Miller, M. J. (**2001**) *J. Org. Chem.*, **66**, 4809–13.
45. Wang, Y.-C., Lu, T.-M., Elango, S., Lin, C.-K., Tsai, C.-T. and Yan, T.-H. (**2002**) *Tetrahedron: Asymmetry*, **13**, 691–95.
46. (a) Pádár, P., Hornyák, M., Forgó, P., Kele, Z., Paragi, G., Howarth, N. M. and Kovács, L. (**2005**) *Tetrahedron*, **61**, 6816–23; (b) Pádár, P., Bokros, A., Paragi, G., Forgo, P., Kele, Z., Howarth, N. M. and Kovács, L. (**2006**) *J. Org. Chem.*, **71**, 8669–72.
47. Shintani, R., Murakami, M. and Hayashi, T. (**2007**) *J. Am. Chem. Soc.*, **129**, 12356–57.
48. Takano, I., Yasuda, I., Nishijima, M., Hitotsuyanagi, Y., Takeya, K. and Itokawa, H. (**1997**) *J. Org. Chem.*, **62**, 8251–54.
49. Lalonde, R. T., Woolever, J. T., Auer, E. and Wong, C. F. (**1972**) *Tetrahedron Lett.*, **13**, 1503–6.
50. (a) Sun, H., Martin, C., Kesselring, D., Keller, R. and Moeller, K. D. (**2006**) *J. Am. Chem. Soc.*, **128**, 13761–71; (b) Cornille, F., Slomczynska, U., Smythe, M. L., Beusen, D. D., Moeller, K. D. and Marshall, G. R. (**1995**) *J. Am. Chem. Soc.*, **117**, 909–17; (c) Cornille, F., Fobian, Y. M., Slomczynska, U., Beusen, D. D., Marshall, G. R. and Moeller, K. D. (**1994**) *Tetrahedron Lett.*, **35**, 6989–92.
51. Zhang, X., Jiang, W. and Schmitt, A. C. (**2001**) *Tetrahedron Lett.*, **42**, 4943–45.
52. Chiou, W.-H., Mizutani, N. and Ojima, I. (**2007**) *J. Org. Chem.*, **72**, 1871–82.
53. Danieli, B., Lesma, G., Passarella, D., Sacchetti, A. and Silvani, A. (**2001**) *Tetrahedron Lett.*, **42**, 7237–40.
54. Saito, S., Tamura, O., Kobayashi, Y., Matsuda, F., Katoh, T. and Terashima, S. (**1994**) *Tetrahedron*, **50**, 6193–208.
55. Zhang, H.-L., Zhao, G., Ding, Y. and Wu, B. (**2005**) *J. Org. Chem.*, **70**, 4954–61.
56. Furman, B., Molotov, S., Thürmer, R., Kaluza, Z., Voelter, W. and Chmielewski, M. (**1997**) *Tetrahedron*, **53**, 5883–90.
57. Shuto, S., Kanazaki, M., Ichikawa, S., Minakawa, N. and Matsuda, A. (**1998**) *J. Org. Chem.*, **63**, 746–54.
58. (a) Kusano, G., Aimi, N. and Sato, Y. (**1970**) *J. Org. Chem.*, **35**, 2624–26; (b) Sato, Y. and Nagai, M. (**1972**) *J. Org. Chem*, **37**, 2629–31; (c) Middleton, M. D., Peppers, B. P. and Diver, S. T. (**2006**) *Tetrahedron*, **62**, 10528–40.
59. Williams, D. R., Rojas, C. M. and Bogen, S. L. (**1999**) *J. Org. Chem.*, **64**, 736–46.
60. Yokoyama, M., Hirano, S., Matsushita, M., Hachiya, T., Kobayashi, N., Kubo, M., Togo, H. and Seki, H. (**1995**) *J. Chem. Soc. Perkin Trans. 1*, 1747.
61. (a) Yokoyama, M., Matsushita, M., Hirano, S. and Togo, H. (**1993**) *Tetrahedron Lett.*, **34**, 5097–100; (b) Praly, J.-P., Stéfano, C., Descotes, G. and Faure, R. (**1994**) *Tetrahedron Lett.*, **35**, 89–92.
62. Brown, R. T. and Ford, M. J. (**1988**) *Synth. Commun.*, **18**, 1801–6.
63. (a) Stajer, G., Virag, M., Szabo, A. E., Bernath, A. E., Sohar, P. and Sillanpaeaer, R. (**1996**) *Acta Chem. Scand.*, **50**, 922–30; (b) Alcaide, B., Garcia-Gravalos, M. D., Lopez, B., Plumet, J. and Del Valle, A. (**1992**) *Heterocycles*, **33**, 56–58.
64. Spaller, M. R., Thielemann, W. T., Brennam, P. E. and Bartlett, P. A. (**2002**) *J. Comb. Chem.*, **4**, 516–22.
65. (a) Räcker, R., Döring, K. and Reiser, O. (**2000**) *J. Org. Chem.*, **65**, 6932–39; (b) Burkholder, T. P.,

Huber, E. W. and Flynn, G. A. (**1993**) *Bioorg. Med. Chem. Lett.*, **3**, 231–34.

66 Scarpi, D., Stranges, D., Cecchi, L. and Guarna, A. (**2004**) *Tetrahedron*, **60**, 2583–91.

67 Alcaide, B., Palanco, C. and Sierra, M. A. (**1998**) *Eur. J. Org. Chem.*, 2913–21.

68 Gandon, L. A., Russell, A. G. and Snaith, J. S. (**2004**) *Org. Biomol. Chem.*, 2270–71.

69 Schultz, A. G. and Suderaraman, P. (**1984**) *Tetrahedron Lett.*, **25**, 4591–94.

70 Trtek, T., Cerny, M., Trnka, T. and Budesinsky, M. (**2004**) *Collect. Czech. Chem. Commun.*, **69**, 1818–28.

71 Ohno, H., Hamaguchi, H., Ohata, M. and Tanaka, T. (**2003**) *Angew. Chem. Int. Ed.*, **42**, 1749–53.

72 Tojino, M., Uenoyama, Y., Fukuyama, T. and Ryu, I. (**2004**) *Chem. Commun.*, 2482–83.

73 Reddy, D. B., Reddy, A. S. and Padmavathi, V. (**1999**) *Indian J. Chem., Sect. B*, **38**, 141–47.

74 Chelucci, G., Saba, A., Valanti, R. and Bacchi, A. (**2000**) *Tetrahedron: Asymmetry*, **11**, 3449–53.

75 Tripathi, S., Singha, K., Achari, B. and Mandal, S. B. (**2004**) *Tetrahedron*, **60**, 4959–65.

76 Hotha, S., Anegundi, R. I. and Natu, A. A. (**2005**) *Tetrahedron Lett.*, **46**, 4585–88.

77 Hanessian, S., Sailes, H. and Therrien, E. (**2003**) *Tetrahedron*, **59**, 7047–56.

78 Robl, J. A., Sun, C.-Q., Stevenson, J., Ryono, D. E., Simpkins, L. M., Cimarusti, M. P., Dejneka, T., Slusarchyk, W. A., Chao, S., Stratton, L., Misra, R. N., Bednarz, M. S., Assad, M. M., Cheung, H. S., Abboa-Offei, B. E., Smith, P. L., Mathers, P. D., Fox, M., Schaeffer, T. R., Seymour, A. A. and Trippodo, N. C. (**1997**) *J. Med. Chem.*, **40**, 1570–77.

79 (a) Ellis, C. D., Oppong, K. A., Laufersweiler, M. C., O'Neil, S. V., Soper, D. L., Wang, Y., Wos, J. A., Fancher, A. N., Iu, W., Suchanek, M. K., Wang, R. L., De, B. and Demuth, T. P. Jr. (**2006**) *Bioorg. Med. Chem. Lett.*, **16**, 4728–32; (b) Ahmed, S. A., Esaki, N., Tanaka, H. and Soda, K. (**1984**) *FEBS Lett.*, **174**, 76–79.

80 Dugave, C. and Ménez, A. (**1997**) *Tetrahedron: Asymmetry*, **8**, 1453–65.

81 Corelli, F., Crescenza, A., Dei, D., Taddei, M. and Botta, M. (**1994**) *Tetrahedron: Asymmetry*, **5**, 1469–72.

82 De Paolis, M., Blankenstein, J., Bois-Choussy, M. and Zhu, J. (**2002**) *Org. Lett.*, **4**, 1235–38.

83 Shankaran, K., Donnelly, K. L., Shah, S. K., Caldwell, C. G., Chen, P., Hagmann, W. K., MacCoss, M., Humes, J. L., Padholok, S. G., Kelly, T. M., Grant, S. K. and Wong, K. K. (**2004**) *Bioorg. Med. Chem. Lett.*, **14**, 5907–11.

84 Neamati, N., Turpin, J. A., Winslow, H. E., Christensen, J. L., Williamson, K., Orr, A., Rice, W. G., Pommier, Y., Garofalo, A., Brizzi, A., Campiani, G., Fiorini, I. and Nacci, V. (**1999**) *J. Med. Chem.*, **42**, 3334–41.

85 Yanagisawa, H., Ishihara, S., Ando, A., Kanazaki, T., Miyamoto, S., Koike, H., Iijima, Y., Oisumi, K., Matsushita, Y. and Hata, T. (**1987**) *J. Med. Chem.*, **30**, 1984–91.

86 Slättegård, R., Gammon, D. W. and Oscarson, S. (**2007**) *Carbohydr. Res.*, **342**, 1943–46.

87 Slade, J., Stanton, J. L., Ben-David, D. and Mazzenga, G. C. (**1985**) *J. Med. Chem.*, **28**, 1517–21.

88 (a) Klar, B. and Imming, P. (**1997**) *Liegigs Ann. Recl.*, 1711–18; (b) Bohrisch, J., Faltz, H., Pätzel, M. and Liebscher, J. (**1994**) *Tetrahedron*, **50**, 10701–8; (c) Leonard, N. J. and Wilson, G. E. (**1964**) *Tetrahedron Lett.*, 1465–68; (d) Starckenmann, C. (**2003**) *J. Agric. Food Chem.*, **51**, 7146–55.

89 Leonard, N. J. and Ning, R. Y. (**1966**) *J. Org. Chem.*, **31**, 3928–35.

90 Schwartz, A., Madan, P. B., Mohacsi, E., O'Brien, J. P., Todaro, L. J. and Coffen, D. L. (**1992**) *J. Org. Chem.*, **57**, 851–56.

91 Almstead, N. G., Bradley, R. S., Pikul, S., De, B., Natchus, M. G., Taiwo, Y. O., Gu, F., Williams, L. E., Hynd, B. A., Janusz, M. J., Dunaway, M. and Mieling, G. E. (**1999**) *J. Med. Chem.*, **42**, 4547–62.

92 Zask, A., Kaplan, J., Du, X. M., MacEwan, G., Sandanayaka, V., Eudy, N., Levin, J., Jin, G., Xu, J., Cummons, T., Barone, D., Ayral-Kaloustian, S. and Skotnicki, J. (**2005**) *Bioorg. Med. Chem. Lett.*, **15**, 1641–45.

93 Arpin, A., Manthorpe, J. M. and Gleason, J. L. (**2006**) *Org. Lett.*, **8**, 1359–62.

Index

a
acylation 342f.
N-acylaziridine 5
acyliminium 131
– addition reaction 128
– N-acyliminium allylation 129
aldimine 19
aldol reactions 277
alkaloids
– marine 118, 122, 133
– poison-frog 96, 103f., 114f., 119, 126, 133
– polycyclic complex 140f.
– Stemona 141, 147, 157, 163
alkene 12, 59
alkylation 69f., 80, 242, 337f., 348, 352ff.
– allylic 309
– asymmetric 277, 298, 309
– amido- 73
– deprotonation/alkylation sequence 374
– dialkylation 353
– diastereoselective 328, 374
– phase-transfer catalyzed 70
– piperazine ring 327f.
– protected amino alcohol 346f.
– reaction 200
– reductive 53, 71f., 74, 83, 108f., 368
– stereoselective 307f.
alkylidene malonates 11
alkynes 96
allylic
– alcohol 28, 110, 115
– alkylation 309
– strain 115
– substitution 110, 231, 350
Alzheimer's desease 51, 141
amination
– catalytic asymmetric hydro- 58f., 112
– diastereoselective 332
– Pd-catalyzed 57f.
– reductive 53, 110, 315, 322, 343
amino alcohol 38, 105
– β-amino alcohol 7, 40
– camphor-derived 231
– enantiopure 7
– N-substituted 224
– polyfunctionalized 39
aminohalo-derivatives 38
aminothiols 390f.
ammonolysis 197
angiotensin-converting enzyme (ACE) 140, 389f.
anisomycin 51f., 58
annulation approach 66f., 108
– imines 106
– pyrrolines 79f.
antibacterial activity 4, 223, 241, 264
antibiotics 3, 51f., 171
anticytotoxic activity 51
antidepressant 55
antifungal 37, 57
anti-HIV activity 270, 279
– protease inhibitors 278, 368
anti-inflammatory 223
antineoplastic activity 4
antitumor activity 3, 142, 170, 270
antiworming 51
asymmetric allylation reaction 55, 108, 225
asymmetric aziridination 6ff.
– benzylidene derivatives 14
– catalytic 22f.
– cycloaddition methods 12ff.
– protocol 9
– stereospezific 13
– styrene derivatives 14
asymmetric desymmetrization 241
asymmetric induction 20, 23f., 123, 308
asymmetric reduction 39
asymmetric synthesis of azirines 33ff.

Asymmetric Synthesis of Nitrogen Heterocycles. Edited by Jacques Royer
Copyright © 2009 WILEY-VCH Verlag GmbH & Co. KGaA, Weinheim
ISBN: 978-3-527-32036-3

azabicycles   148
aza-Claisen rearrangement   179f.
azacycloheptatriene, see azepine
aza-Darzens-type reaction   17, 19ff.
– diazo compounds   21ff.
– α-haloenolate   17ff.
aza-Payne
– displacement   29
– /hydroamination   58f.
– reaction   5
– rearrangement   66
azasugar   54, 110
– stereoselective synthesis   131
azepanes   130f.
– 3-hydroxy-   150, 153
– 4-hydroxy-   153
– substituted   150, 166ff.
azepines   139ff.
– dihydro-   139
– fragment   143, 146, 158
– hexahydro-   139, 148f., 157
– perhydro-   157
– ring closure metathesis (RCM)   155ff.
– spiro-   159
– substituted   139ff.
– tetrahydro-   139, 160, 171
– trans-hexahydro-   146
azepino-indoline   147
azetidines
– 2-Cyano   41
– cyclization methods   38ff.
– hydroxy-   43
– monocyclic   36
– precursor   39
– unsaturated derivatives   36f.
azinomycin   3f.
aziridine
– N-benzyl-aziridine-2-esters   7
– N-Boc-   7
– -2-carboxylates   6ff.
– cis-   21f.
– cis-vinyl-   26
– elemination   34f.
– -2-esters   29
– -2,2-dicarboxylate   5, 11
– N-halo-   34
– -2-imides   9
– N-substituted   5
– oxidation   36
– -2-phosphonates   17
– polyfunctionalized   18f.
– precursors   12
– N-protected   13, 16
– ring systems   3f.

– -2-tert-butylate   8
– N-tosyl-   13, 17
– trans-   29
– transformations   4f.
– 3'-unsubstituted-N-Boc-   10
azirines   3f., 33ff.
– -carboxylates   34
– 2H-   33f.
– motif   4
– 2-phosphinyl-2H-   33
azirinomycin   4
azocines   171ff.

b
bacterial strains   32
Baeyer-Villiger oxidation   385
Baldwin's adaptation   42
barbiturate   309f.
Baylis-Hillman-aldol reaction   62, 277
Beckmann rearrangement   178
– asymmetric   165f.
– thermal regioselective   165
benzazocine   173
benzodiazepine, see diazepine
Biginelli reaction   306
bioactive   4, 325
– alkaloids   39
– derivatives   28
– peptides   8
– peptidomimetics   8
biocatalysts   31
biological activity
– alkaloids   51
– diazepine   367ff.
– peptides   264
biotransformation   32
bisoxazoline   13
Boc
– deprotection   151, 213
– oxidation   105
– protection   105, 314f.
bond formation
– C-C bond formation   40f., 107, 113ff.
– C_1/C_5 bond formation   52ff.
– C_2-C_3 bond formation   61, 119
– C_3-C_4 bond formation   62, 102
– C-N bond formation   39, 142
α-bromo-alkene   8
Brønsted
– acid-catalyzed annulation   21
– acid-catalyzed reduction   101
building blocks   3, 28, 81
– β-hydroxylamino-carbonyl   11
– piperacic acid   294

Index | 401

– prochiral 60
– versatile 40f., 78, 108, 113, 123, 129
Burgess' reagent 271f.

c
camphor-derived chiral ligands 15
camphor sultam 9, 20, 276
– α-bromoenolates 20
– N-crotonyl- 9
– Oppolzer's L- 65
Candida Cylindracea Lipase (CCL) 31
carbamates 382
– pyrrolidine 77f.
carbanion-mediated sulphonamide
    intermolecular coupling (CSIC) 279
carbenes 16, 21
– chiral 262f.
– heterocyclic 262
– insertion 63
carbocyclic-fused systems 36, 107
carbonylation 6, 11
carboxylation 236
catalyst
– chiral 13, 16, 43, 244
– Grubbs 2nd 129f., 133, 156f., 159f.,
  175, 387
– Grubbs-Hoveyda catalyst 98, 114
– organocatalyst 244
– Pearlman's 316
cation-π complex 96
cellular growth control 139
chain
– carbon side 18
– functionalized 6
– nitrogen imine 24
chelation 124, 146
– -controlled model 100
chemical resolution 42
chemoselective 12
chiral
– auxiliary 9ff.
– bisoxazoline catalyst 13
– center 123, 367, 376
– copper complexes 14
– epoxides 29
– heterosubstituted aziridines 18
– imides 9
– imines 16f., 115
– metal catalyst 13
– metal complexes 11, 14, 16
– morpholinone 306
– nonracemic alicyclic precursor 52
– nonracemic azetidinones reduction 44
– nucleophiles 16

– organosilanes 115
– piperidines 112, 117
– polymer-bound amines 33
– pool-derived N,O-acetals 72f.
– pyridinium salt 96f.
– pyrrolidine derivatives 12
– sulfinamide 98
– sulfinimines 17, 114
chirality 7, 24, 28
– ylides 25
chiron 52, 69
chiroptical properties 193
chromatographic separation 43
chromatography
– chiral 190
– column 258
– flash 9, 144, 168, 250
– high performance liquid (HPLC) 151
– ion-exchange 176
Cieplak hypothesis 119, 126f.
circular dichroism (CD) 193
CN(R,S) method 123, 226
condensation
– β-amino alcohol double 383, 388f.
– amino sugar 390
– bis- 367f.
– haloethylamine 388
– Huysgens 386
– in situ 380
– intramollecular 340
– nitro olefine 389
– Paal-Knorr 377f.
configuration 70, 72
– retention 72
conformation 124
conjugate addition 9ff.
– O-benzylhydroxylamine 12
– hydroxylamines 11
conversion 29, 43, 333
Cope elimination 245, 351
Corey-Chaykovsky reaction 25
Cromwell general method 42
cross-metathesis reaction 108, 114
crystallization 190, 276, 339
– -induced dynamic resolution (CIDR) 337
– -induced transformation 349
Curtius rearrangement 237
cyclization
– alkylative 105
– amino alcohols 7
– amino halides 7
– azetidines 38ff.
– Barbier-type 62
– catalytic-oxidative 240

cyclization (contd.)
– electro- 105
– endo 147f., 170
– 7-endo-dig 149
– exo/anti 246
– Heck 151f., 387
– hydrozirconation- 62
– imine/enamine formation 321ff.
– iminium 380f.
– in situ 144
– intramolecular 103, 144, 149, 172, 227, 232, 239, 251
– iodine-induced 151
– Lewis-acid 155
– meta-catalyzed 110
– nitrogen nucleophilic displacement 8, 33, 304f., 319, 377, 381
– nucleophilic 84
– Paal-Knorr 377
– partial 316, 319
– Pd-catalyzed 83, 370, 384
– Pictet-Spengler 154
– pinacol-type 146
– piperidines 107ff.
– pyrrolidine 52ff.
– pyrrolines 79
– radical 115f., 145f., 382f., 384
– reductive amino- 151f.
– reverse-Cope 60
– Richman-Atkins 321
– transannular 84f.
– two-step 104
cycloaddition 5 12f., 107, 117ff.
– [3 + 2] 119, 375, 378f.
– [4 + 2] 163, 378
– [4 + 3] 162, 378f.
– [5 + 2] 162
– acylnitroso 117f.
– azadiene 117
– azides 386
– azocines 174f.
– 1,3-dipolar 5, 128, 134f., 161, 235, 248, 279, 375
– intermolecular 244, 248, 325
– intramolecular 244, 259, 278, 325
– Lewis-acid catalyzed 259
– nitrile oxide 247
– nitrone 161ff.
– photo- 162
– piperazinone 325f.
– pyrrolidines 64ff.
– regioelective 245, 262
– stereoselective 245
cycloadduct 118

cyclocondensation reaction 271, 294
cyclopropanation 277
cycloreversion-cycloaddition process 257
cytotoxic compound 4, 35

d

DAG methodology 24
Darzens reaction 390
decarboxylative 60
– oxidation 74
dehydroxylation reaction 73
deprotection/protection 8, 54, 110, 130, 317, 336, 340f.
deprotonation 25, 70, 242, 345, 353
– achiral imidazolidine 249
– cycloalkylation 63
– deprotonation/alkylation sequence 374
– selective 169
– -substitution 77
– sulfonamide nitrogen 216
deracemization processes 101f.
Dess-Martin periodinane 99, 296
desymmetrization 108
dialkylation 38
diastereofacial selectivity 14, 245
diastereoisomer 20, 26, 43, 263, 279
– cis 20, 265
– oxaziridine 205
– separation 205
– single 21, 56
– thiazolidines 263f.
– trans 265
diastereomer 69f., 129, 319
– adduct 10
– cis 173, 322
– (S,R) 315
– single 257
– trans 69
diastereomeric
– aziridines 15
– esters 146
– excess 20
– mixture 20, 69
– ratio 19, 309, 319, 358, 378
diastereoselective
– addition 109, 216
– allylation 99
– amination 332
– aziridination 15
– conjugate reduction 331
– coupling 335
– epoxidation-nucleophilic addition 96
– 3-exo-tet ring closure 10
– hydrogenation 105

– intramolecular Michael cyclation 102
– nucleophilic 1,2-addition 280
– reduction 358
– synthesis 9f., 19
– trans- 54, 57
diastereoselectivity 10, 16, 21, 24, 56, 66, 69, 77, 97, 115, 378
diazepines 367ff.
– 1,4 benzodiazepines 371ff.
– 1,2-diazepines 367f.
– 1,3-diazepines 368ff.
– 1,4-diazepines 367, 371ff.
– fused diazepine-pyrrolidine 370
– pyrrolobenzodiazepines (PDB) 375f.
– pyrrolodiazepine 375
diazeridination 191ff.
diazeridines 189ff.
– chiral nonracemic 191
– monocyclic 190
diazetidines 208f.
diazirines 193f.
diazocompounds 21f., 26f.
diazoketene insertion reaction 45
diazotation 375
Dieckmann reaction 62
Diels-Adler reaction 117f., 234f., 251, 254, 277, 293, 378f.
– asymmetric 170
– homo- 231
– intramolecular (IMDA) 161, 163f., 175, 279
– stereoselective hetero- 231, 301, 299
– pyridazine 299ff.
diethoxyphenylphosphorane (DTTP) 8
diketopiperazines (DKPs) 311ff, 319
dipolarophile 244, 246, 259, 278

e

electrochemical oxidation 131
electrocyclization 33, 105
electron
– -donating 17
– -enriched aromatic group 375
– -withdrawing group (EWG) 17, 25f., 194f., 228, 238, 244
electrophilic 6, 383, 388
– amination 191, 195
– center 102
– dielectrophilic 390
– imine carbon 17
– nitrogen 10
electrophiles 6, 98, 150, 298, 327
electrophilic
– imine carbon 23
– reactants 311

electrophilicity 17, 381
enamine 102ff.
– functionalized cyclic 103
– intramolecular 102
– reaction 102f.
enantiocontrol 16
enantiomer 28, 108
– natural 133
– (R,R) 317
– single 69
enantiomeric 11
– excess 11ff.
– purity 26, 121, 214, 344
– ratio 307
enantiopure
– N-acylaziridine 19
– amino alcohol 7
– azetidines 36, 39ff.
– azides 28
– aziridine-2-carboxylates 21
– aziridine-2-phosphonates 17
– benzyl aziridine-2-carboxylates 9
– N-benzyl-aziridine-2-esters 7
– -2-carboxylates 8
– diamines 252
– ethynylazetidines 39
– imidazolidinones 253
– imines 18
– oxiranes 28, 261
– solfoxide 279
– sultam 276, 278f.
– ylides 25
enantioselective
– addition 264
– allyltitanation 108
– aziridination 12ff.
– catalytic reduction 43
– hydrogenation 142
– hydrolysis 32
– nucleophilic substitution 131
– reactions 77
– reduction 69, 213
– synthesis 110
– transformations 280
enantioselectivity 112, 239f., 249, 253, 256
enantiospecific synthesis 159
endocyclic enecarbamates 76
endo-selective 78
ene-reaction 277
enolate 69, 329
– aza-enolate anion 327
– aliphatic 20
– amination 204
– intermediate 11

enolate (contd.)
– lithium   18, 330
enolization   10
enones
– acyclic   129
– cyclic   129
enzymatic resolution   31, 42, 44, 195
enzymatic synthesis   54
– azasugar   54
– chemo-   54, 167f.
– morpholine   344
enzyme inhibitors   4
environmental impact   13
epimerization   41, 104, 124, 142, 242, 333, 339
epoxidation   277
epoxide   27f., 54, 345f.
– racemic   30
– reductive opening   100
– ring opening   345
Eschenmoser sulfide contraction reaction   71
esterification   303
Evans aldol reaction   160
exo-trig reaction pathway   146

f

Food and Drug Administration (FDA)   242
formaldehyde   380f.
Friedel-Crafts acylation   296

g

Gabriel-Cromwell reaction   8f.
Ganem's approach   53, 63
Garner's stereocomplementary asymmetric approach   65
Geissman-Waiss lactone synthesis   73, 77
Gennari-Evans-Vederas reaction   295
Gly-Gly-derived macrocycles   165
Grignard
– addition/cyclization reaction   59
– addition to nitrones   75
– addition-transamidation-reduction sequence   336
– reagent   19, 58f., 73, 75, 96f., 119, 125, 227, 247
guanidine
– cyclic derivative   29
– C2-symmetric   369

h

haloacetates   20
halogenation   310f.
Hantzsch diethyl ester   101, 253
Heck reaction   151f., 387

Helmchen's auxiliaries   9
homochiral nucleophilic templates   327
Horner-Wadsworth-Emmons reaction   45, 160, 313
hydrazine   367f.
hydrazone formation   294
hydroformylation   380
hydrogenation   39, 73, 104f., 142, 150
– asymmetric   169, 276f., 294
– asymmetric transfer   277
– catalyzed Pd/C   144, 301, 314, 322f., 358
hydrogenolysis   4, 110, 160, 303f.
hydrolysis   41
– alkaline   100
– enzymatic   31
– ester   314
hydrolytic cleavage   110
hydropyridines
– chiral 1,4-   96f.
– dihydropyridines   95ff.
– tetrahydropyridines   98ff.
– substituted   96
hydroxylation
– amino   143
– pyrimidinone   310f.

i

imidazolidines   249ff.
imidazolidinones   249, 252ff.
imine   16f., 19, 21ff.
– annulation   106
– chiral   16, 115
– imine/enamine formation   321ff.
– formation   311
iminium   251, 380f., 383
– addition   113
– reduction   133
– strategies   128
– sulfinyliminium salt   131f.
imino
– eight-membered iminoalditols   176
– α-imino esters   22
– iminothiazolidines   267
– sugar   168
inhibitor
– glycosidase   151, 369
– neutral endopeptidase (NEP)   140
– selective   140
– vasopeptidase   140, 142
inhibitory acitvity   37, 139
interconversion   12
intramolecular
– alkylation   40
– amidation   52, 319

– amination   52
– condensation   74
– conjugate addition reaction   133, 272, 348f.
– displacement   38, 53, 236, 342
– hydroboration-cycloalkylation   59
– nucleophilic substitution   172
– Pd-catalyzed coupling   177f.
– substitution   232
inversion barriers   190, 194, 217
iodocarbamation   108
iodocyclation   227
isatoic anhydride   371f.
isomer   128, 131
– chiral   250
– constitutional   212, 217
– diazetidine   208
– DKPs (diketopiperazines)   319
– meso   250
– oxazolidine   224f.
– six-membered   257
– thiazetidines   212f.
isomerization   5, 114, 149, 191
– photo-   190
isoxazolidines   243ff.
– chiral   244
isoxazolines   243, 246ff.
– N-oxides   248

j

Jacobsen's asymmetric epoxidation of olefin   120
Jacobsen's (salen)-Co^{III} chiral comples   238
Julia
– coupling reaction   133
– olefin synthesis   71, 99

k

keto function   18
ketones
– moiety   43, 104, 114
– prochiral   42
kinase C   37, 139
kinetic
– aminolytic kinetic resolution (AKR)   238
– asymmetric transformation   80
– control   24
– racemization   190
– resolution   29, 31f., 35, 43, 70, 247, 296, 390
Kulinkovich reaction   62

l

lactamization   142f., 368, 372, 376, 383f., 389

lactam
– chiral bicyclic   134, 227
– chiral caprolactams   164ff.
– eight-membered   173f., 176
– enantiopure   123
– formation   311f., 317, 338f.
– β-lactams   266
– δ-lactams   227
– γ-lactam   122
– Meyers'-bicyclic lactam strategy   124, 375
– ring enlargement   122
– ring formation   335, 338f.
– seven-membered   140, 142ff.
– ten-membered   179
lactone
– chiral   167
– formation   337
– ring transformation   121, 335, 338
lactonization   368, 383f.
Lee's chiral synthesis   100
Lewis acid   11, 22
– activation   37
– catalyzed annulation   67
– -Lewis base bifunctional asymmetric catalyst   96
Ley's sulfone chemistry   71
ligands   11f.
– BINAP-type   268
– Box   233f., 248, 256, 260, 262
– chiral diamine   249
– VANOL   23
– VAPOL   23
lowest unfilled molecular orbital (LUMO)   253

m

Ma's synthesis   104
Mannich-type
– boro-   349
– cyclization   113f.f.
– reaction   82, 256
mercaptopropionic acid derivatives   388f.
mesylation   319
metal-carbene intermediate complex   21
methylvinyl ketone (MVK)   62
Meyer's chiral bicyclic lactams   78
Michael
– addition   11., 55, 101ff.
– Aldol Retro-Dieckmann (MARDi) sequence   169
– aza-   12, 56ff.
– enantioselective Michael addition   10
– intramolecular   42, 102
– retro-Michael elimination reaction   102

microwave irradiation   42, 67, 319
– hydrogenation   122
– rearrangement   67
Mitsunobu
– condition   7, 39
– intramolecular reaction   101, 236, 271, 296, 305, 317, 321, 339, 376, 391
Mizuno's dinuclear peroxotungstate catalyst   54
Moffat oxidation   121
monoperoxycamphoric acid (MPCA)   195, 199
Morita-Baylis-Hillman type reaction   129
morpholines   335ff.
– asymmetric transformations   352ff.
– ring formation   335ff.
morpholinone   335, 337, 340ff.
– C-alkylations   352
Mukaiyama
– reagent   339
– -type aldol addition metholodogy   81

n

Neber reaction   33
Negishi-coupling reaction   101
nitrene
– formation   12f., 16
– insertion   382
– singlet state   12
– triplet state   12
nitrilium insertion   375f.
nitrogen
– α-carbanion   113
– insertion process   164
– inversion barrier   3
nitrones   74ff.
– cycloaddition   127f., 161
– photocyclation   207f.
– regioisomeric   128
nitrosation rearrangement   197
NMR spectroscopy   203
nosylation   8
Nozaki-Hiyama-Kishi allylation   234
nucleophile   5
– addition   17
– chiral   16
– nitrogen   108
– piperazine ring   330f.
– prochiral   204
nucleophilic
– addition   30, 131, 237
– addition/cycloaddition   74
– attack   22, 124, 382

– substitution   54, 108, 311
– nucleophilic reaction   71f.

o

olefin   120
– aryl-substituted olefin substrate   13
– endocyclic   152
– Grubb's olefin metathesis   157
– Horner-Wadsworth-Emmons   160
– Jacobsen's asymmetric epoxidation   120
– Julia olefin synthesis   71, 99
– ring-closing metathesis reaction   79
olefination   42
one-pot
– construction   133
– procedure   102, 104, 106, 165, 209, 301, 322
optically active
– 2-alkyl diamines   250
– aminoalcohols   39
– azepine   151
– aziridine   7, 10, 66
– azirines   33, 35
– imines   24
– morpholinones   353
– oxazepinone   384
– oxazilines   232
– oxaziridine   207
– oxaziridinium salt   200ff.
– oxazoline   304
– pyrimidine   304
– trans-aziridine   9, 11
– N-sulfinyl-aziridine   19
organocatalytic processes   7, 12
organolithium reagent   247
organometallic reagent   18f., 73
organosilanes   115
oxadiazines   332ff.
oxadiazinones   334
oxathianes   25
oxazepines   378ff.
– 1,2-oxazepines   378f.
– 1,3-oxazepines   380f.
– 1,4-oxazepines   383
oxazetidines   210ff.
oxazidine   370
oxaziridines   194ff.
– chiral nonracemic spirocyclic   198, 200
– N- acyl   203f.
– N-alkoxycarbonyl   203f.
– N-alkyl   197ff.
– N-phosphinoyl   207
– N-quaternarized   200
– N-sulfonyl   204f.
– N-unsubstituted   196f.

- photorearrangement 200
- preparation 195f.
oxazolidine 123, 165, 223ff.
- *N*-alkyl- 224
- *N*-Boc 228
- *N*-Tosyl 228
- oxazolidin-2-ones 235ff.
- oxazolidin-4-ones 242f.
- oxazolidin-5-ones 242f.
- photolysis 165
oxazolidinone 7, 235ff.
- bicyclic 242
- *trans*- 237
oxazolines 5, 7, 230ff
oxazolizinone ring 100
oxidative methods 131
- anodic 132
- electrochemical 131f.
- phenol coupling reaction 155f.
oxyfunctionalization 200
ozonolysis 297

p

Paternò-Büchi reaction 76
penicillin derivates 263
pharmacological activity 263
pharmacophores 139
phase transfer
- -catalyzed conditions 298
- solid-liquid 25
phenylglycinol 123
photochemical rearrangement 194
photocyclation 207, 374
- chiral nitrones 207f.
photolysis 165, 382
Pictet-Spengler reaction 152, 154
piperazine 311ff.
- asymmetric hydrogenation 331ff.
- DKPs (diketopiperazines) 311ff.
- ring formation 311ff.
- stereoselective ring transformations 327ff.
- substituted 322f.
piperazinone ring system 325
piperidine 95, 104, 107ff.
- alkaloids 124
- enantiopure 125
- functionalized 95, 110, 120
- monocyclic 107
- ring system 102, 123
- saturated 95
- substituted 95f., 109, 115, 123f.
piperidones 121f.
polyoxin C 37
pyrazolidines 255ff.

- chiral 255
- conjugate addition 260
pyrazolidinones 260ff.
pyrazolines 255, 257ff.
pyridazines 293ff.
- dihydropyridazine derivates 298
- tetrahydropyridazine derivates 295ff.
pyridine 264
pyridinium salt 96f.
pyrimidines 302ff.
- ring formation 302, 304ff.
pyrimidinones 306f.
- halogenation 310f.
- hydroxylation 310f.
pyrrolidine
- amino- 55
- bioactive 52
- carbamates 77f.
- cyclization methods 52ff.
- dihydroxylated 68
- functionalized 51, 121
- nitrones 74
- -2-ones 52f.
- polyhydroxylated 54, 57
- polysubstituted 57ff.
- pyrrolidine-2-ones 66ff.
- pyrrolidine-2-ones/pyrrolidine 52f., 62f., 68
- pyrrolidine/1-pyrrolidine derivatives 52
- vinyl 179f.
pyrrolidinones 51
pyrrolines synthesis 79f.
pyrrolizidines 82ff.
- polyhydroxylated 85f.

r

racemate resolution 31, 42
racemic
- alcohols 146
- azetidine mixtures 38, 43f.
- aziridine 22
- azirines 35f.
- biotransformation 32
- chiral non- 123
- epoxides 238
- isoxazoline 247
- mixture 31
radical
- addition 277, 355
- alkyl 355
radicamine alkaloids 75f.
radicamization free-methods 121
Raney nickel 123, 373

receptor
- antagonist   120
- oxytocin (OT)   140
- selective antagonist   96
- selective 5HT$_{2C}$ 140
recrystallization   43, 378, 389
reduction
- concomitant   160
- stereoselective   52, 160, 380
regiocontrol   4, 279
regioselective
- nitrone formation   76
- oxidative transformation   123
- reduction   76, 129
regioselectivity   5, 35f., 39, 62, 74f., 380
Reissert reaction   96
resolution
- chemical optical   146f.
- enzymatic optical   146f.
rhazinilam synthesis   177f.
ring
- aromatic   20
- backbone   44
- cleavage   37
- -closing metathesis (RCM)   98ff.
- contraction   68, 85, 177
- enlargement   120f., 385
- 4-exo-tet   40
- expansion   5, 44, 67, 166ff.
- fused-azetidine   39
- morpholine   306
- piperazine   311ff.
- α-position   40, 77, 128
- protonated   37
- pyrimidine   306ff.
- rearrangement metathesis   101
- transformation   27f. 44f., 67ff.
ring heterocycles
- bicyclic   117
- *cis*   25
- eight-membered azacycles   139
- five-membered   40, 113, 223f., 249f., 263ff.
- four-membered   37f., 208ff.
- functionalized   41f.
- nine-membered azacycles   139
- seven-membered   139f., 367ff.
- six-membered   40, 95f., 293ff.
- substitution   68f., 80f., 123f.
- C$_2$-symmetric disubstituted   42
- three-membered   10f., 17, 19, 37, 39, 117, 189ff.
ring closure   6f., 10, 133
- aminoalcohols   39
- aminoallenes   40

- asymmetric ring-closing metathesis (ARCM)   23, 175
- base-mediated selective   238
- imine/enamine bond formation   340ff.
- intramolecular Michael addition   348f.
- intramolecular nucleophilic displacement 304ff.
- intramolecular nucleophilic substitution 319ff.
- lactam formation   335, 338
- lactone formation   335, 338
- methods   306, 326f., 349f.
- nucleophilic substitution with C-N bond 346ff.
- nucleophilic substitution with C-O bond 342ff.
- olefin ring-closing metathesis (RCM) reaction   79
ring opening
- carbon nucleophile   4f.
- epoxide   38, 171
- hetero-nucleophile   5
- intermolecular   108
- intramolecular   54, 108
- Lewis acid-catalyzed   351
- nucleophilic   66
- reductive   53
- regioselective   27
- Vederas' serine lactone   325

S
SAMP
- -hydrazone   214, 308
- -hydrazonosulfonates   280
- method   110
saponification   142
Schmidt rearrangement   165f.
Schöllkopf chiral auxiliaries   312, 327f., 330
Seebach's SRS (self-regeneration of stereochemistry) methodology   69
selectivity
- *cis/trans*   24
- endo   279
- facial   279
- *syn*-   58
- trans   146
Sharpless asymmetric
- amino hydroxylation   57, 238
- dihydroxylation   54, 323
- epoxidation   28, 54, 60, 79, 157, 345
S$_N$2 reactions   7, 108, 236
S$_N$Ar reactions   372f., 390
Staudinger reaction   28f., 67, 266
Staudinger-aza-Wittig formation   151, 324

stereochemistry 4, 12, 18, 148
– aminalcohol 39
– C_2/C_5 cis 53
– trans 150
stereocontrol 4, 17f., 24, 26, 279, 341
– partial 319
– total synthesis 133
stereodivergent construction 126
stereoelectronic 127
stereogenic centers 3, 5, 24, 31, 36, 55, 247, 342, 381
– aziridune 5
– benzylic 199
– carboxylic derivates 230
– nitrogen 179, 194
– β-sultams 213
stereoisomer 42, 309, 342
– N-acetyl derivate 315
– single 246
stereoselective
– alkylation 307f.
– arylation 329, 355f.
– conjugate addition 337
– hydrogenation 73, 109, 341
– Grinard addition to nitrones 75
– reduction 52, 160, 356f.
– synthesis 3, 11, 131, 215
– transformation 189, 302, 307f., 327f.
stereoselectivity 5, 11, 13, 20, 26f., 378, 380
stereospecific 58
– oxidation 197
Stevens rearrangement 169, 385
Stork-type approach 55
Strecker reaction 55, 296, 336, 341, 373
sulfonamides 387
sultams 276ff.
– bicyclic 278f.
– chiral γ-sultams 276f.
– cis 216
– Oppolzer's camphor sultam auxiliary 276
– β-sultams 213ff.
Suzuki cross-coupling reaction 143

t
Tamao oxidation 66
Tamura's Beckmann reagent 67
tautomer 139
– enamine 200
– imine 224
thiazepines 386ff.
– 1,2 thiazepines 387
– 1,3 thiazepines 387
– 1,4 thiazepines 386, 388
thiazetidines 212ff.

thiazolidines 263ff.
thiazolidinethiones 268
thiazolidinones 269f.
thiazolines 270ff.
– cyclic oligo- 274
– mono- 274
– 2-thiazolines 270
– 3-thiazolines 275f
Thorpe reaction 62
Thorpe-Ingold effect 224
N-tosyl amino acids 7
Toyooka endgame sequence 133
trans isomer 20
transacetalization 383f.
transacylation 44
transamidation 122
transesterification 43
transmetalation 306

u
Ugi-type multicomponent reaction 209, 319, 373
Ullmann reaction 373
UV irradiation 16

v
vinyl
– azide 34
– aziridine 24
– 1,3-diazipine 370
– tricarbonyl 370

w
Weinreb amide 62
Wittig reaction 45, 73
– -aza Michael 56
– domino addition- 83
– olefination 42
Wolff rearrangement reaction 60
Woodward-Prevost reaction 58

x
X-ray diffraction analysis 190, 224, 314

y
ylides
– chirality 25
– guanidine 29
– heteroatom 23
– methylene sulfur 24
– semistabilized 26
– sulfonium 25f.

z
zwitterionic 179